Superfractals

Superfractals is the long awaited successor to *Fractals Everywhere*, in which the power and beauty of iterated function systems (IFSs) were introduced and applied to the production of startling and original images that reflect complex structures found for example in nature. This provoked the question whether there is a deeper connection between topology, geometry, IFSs and codes on the one hand and biology, DNA and protein development on the other. Now, 20 years later, Professor Barnsley brings the story up to date by explaining how IFSs have developed in order to address this issue. New ideas such as fractal tops and superIFSs are introduced, and the classical deterministic approach is combined with probabilistic ideas to produce new mathematics and algorithms that reveal a theory which could have applications in computer graphics, bioinformatics, economics, signal processing and beyond. For the first time these ideas are explained in book form and illustrated with breathtaking pictures. The text is accessible to all mathematical scientists with some knowledge of calculus and will open up new ways in which the world can be seen.

MICHAEL FIELDING BARNSLEY
Professor of Mathematics, Australian National University, Canberra

Superfractals

CAMBRIDGE
UNIVERSITY PRESS

CAMBRIDGE UNIVERSITY PRESS
Cambridge, New York, Melbourne, Madrid, Cape Town, Singapore, São Paulo

Cambridge University Press
The Edinburgh Building, Cambridge CB2 2RU, UK

Published in the United States of America by Cambridge University Press, New York

www.cambridge.org
Information on this title: www.cambridge.org/9780521844932

First published 2006

Printed in the United Kingdom at the University Press, Cambridge

A catalogue record for this publication is available from the British Library

ISBN-13 978-0-521-84493-2 hardback
ISBN-10 0-521-84493-2 hardback

For my daughters, Diana and Rose

CONTENTS

ACKNOWLEDGEMENTS

Firstly I thank Louisa Anson Barnsley, my wife, for illustrating this book. She helped to select the underlying computational images and turned them into informative and beautiful illustrations. She also produced nearly all the diagrams.

There are many others whose contribution to the book was vital. I thank especially John Hutchinson and Örjan Stenflo for ideas that have influenced the content of this book. I also thank particularly John Hutchinson, and also Tim Brown, Alan Carey, Peter Hall and Kelly Wicks, for facilitating the excellent research and teaching environment at the Australian National University. I am grateful to the Australian National University and the Australian Research Council for their support of this work. Tony Guttman and Derek Chan of the University of Melbourne provided warm and practical support. Maria Navascués and Herb Kunze generously and enthusiastically read the typescript; any remaining errors are mine alone. I am grateful to Mochi Signori and Ruifeng Xie for discussions and Dong Sheng Cai for encouragement. The Institute for Mathematics and Its Applications kindly supported a workshop in 2001 which convinced me that fractal geometry has a broad future in mathematics, science and engineering. Nigel Lesmoir-Gordon stimulated my thinking with many discussions about fractal geometry. I would like to thank David Tranah of Cambridge University Press for his support, skill and vision, in producing this book. I note particularly the many gifted contributions of Susan Parkinson, who has improved the images, style and precision of the book.

I also thank and acknowledge all those on whose work this book relies, or whose work has influenced me, although I may not have explicitly referenced it.

I thank my wife again for many critiques, correcting errors and for her patience and continual support while this book was written.

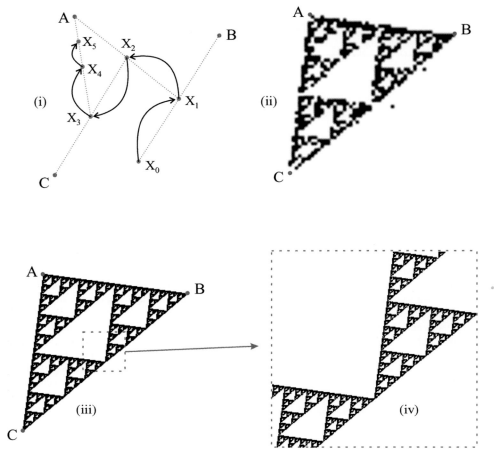

Figure 0.1 In the chaos game, one of a few simple rules is selected at random and applied to a point, to yield a new point. This random step is repeated over and over again, to produce an 'attractor'. Here the attractor is a Sierpinski triangle. The figures illustrate (i) the first few points, (ii) the result after 1000 iterations, (iii) the same result at a higher resolution with outliers discarded, (iv) a magnification of (iii). What happens if you change the rules?

INTRODUCTION

0.1 The chaos game

The following process, which I call the 'chaos game', provides a simple introduction to the idea of an **iterated function system** (IFS) and its **attractor**.

Mark four points on a sheet of paper. Label three of them **A**, **B** and **C** and label the remaining point $\mathbf{X}_0$, as in Figure 0.1(i). Label two faces of a six-sided die A, two other faces B and the remaining two faces C, or devise your own way of producing a random sequence of the symbols A, B and C.

Roll the die, to choose randomly a symbol A, B or C. On the paper, mark the midpoint between $\mathbf{X}_0$ and the point labelled by the selected symbol. Call this midpoint $\mathbf{X}_1$. For example, if the result of rolling the die is B then $\mathbf{X}_1$ is the midpoint between $\mathbf{X}_0$ and **B**.

Roll the die again. Plot the midpoint between $\mathbf{X}_1$ and the point whose label shows on the die. Call this new point $\mathbf{X}_2$. You get the idea. Roll the die again, and again, ..., and plot a new midpoint on the paper each time. The result, on the sheet of paper, is very likely to look something like Figure 0.1(ii). It is an approximate picture of a **Sierpinski triangle**, with some extra 'outlier' points.

Suppose that you carry out a similar experiment using a computer. Then you can compute accurately a sequence of millions of points

$$\mathbf{X}_0, \quad \mathbf{X}_1, \quad \mathbf{X}_2, \ldots, \quad \mathbf{X}_{10\,000\,000}, \quad \mathbf{X}_{10\,000\,001}, \quad \ldots$$

and print them out as a high-resolution picture. If the points **A**, **B** and **C** are fixed then each time you run the experiment you are likely to obtain a different picture of the Sierpinski triangle, but only slightly different. In fact, if you work at a resolution of 256×256, compute ten million points and discard the first sixteen points, then it is probable that the resulting picture will look the same each time you run the experiment. An illustration of such a result is shown in Figure 0.1(iii), (iv).

Almost always, regardless of the choice of starting point $\mathbf{X}_0$ and regardless of the particular sequence of random choices, the sequence of points $\mathbf{X}_0, \mathbf{X}_1, \mathbf{X}_2, \ldots$ seems to be drawn towards, or 'attracted to', the Sierpinski triangle; after sufficiently many random iterations, the successive points appear, at viewing resolution, to lie exactly on the Sierpinski triangle, and to dance around on it forever.

1

Table 0.1   *Coefficients of the IFS that created Figures 0.2 and 0.3*

n	a_n	b_n	c_n	d_n	e_n	k_n	g_n	h_n	j_n	p_n
1	19.05	0.72	1.86	-0.15	16.9	-0.28	5.63	2.01	20.0	$\frac{60}{100}$
2	0.2	4.4	7.5	-0.3	-4.4	-10.4	0.2	8.8	15.4	$\frac{1}{100}$
3	96.5	35.2	5.8	-131.4	-6.5	19.1	134.8	30.7	7.5	$\frac{20}{100}$
4	-32.5	5.81	-2.9	122.9	-0.1	-19.9	-128.1	-24.3	-5.8	$\frac{19}{100}$

0.2   Attractors of iterated function systems

In the above example the IFS consists of three simple rules, each of which moves the current point to a new location.

>   *Rule 1: Move to the point midway between the current location and* **A***.*
>   *Rule 2: Move to the point midway between the current location and* **B***.*
>   *Rule 3: Move to the point midway between the current location and* **C***.*

We can write these rules in terms of three functions f_1, f_2, f_3 that map the euclidean plane into itself. For example, using coordinate notation, suppose that $\mathbf{A} = (2, 1)$, $\mathbf{B} = (3, 0)$ and $\mathbf{C} = (4, 0)$. Then we define

$$f_1(x, y) = \left(\frac{x + 2}{2}, \frac{y + 1}{2} \right), \quad f_2(x, y) = \left(\frac{x + 3}{2}, \frac{y}{2} \right),$$

$$f_3(x, y) = \left(\frac{x + 4}{2}, \frac{y}{2} \right).$$

Using this notation the repeated step in the chaos game can be expressed as

$$(x_{i+1}, y_{i+1}) = f_{\sigma_i}(x_i, y_i) \quad \text{for } i = 0, 1, 2, \ldots$$

where σ_i is a number randomly chosen from the set $\{1, 2, 3\}$ and $\mathbf{X}_i = (x_i, y_i)$. The collection of functions f_1, f_2 and f_3 is called an **iterated function system** (**IFS**). It is denoted by $\{\mathbb{R}^2; f_1, f_2, f_3\}$, where $\mathbb{R}^2$ is the euclidean plane, the space on which the functions act. The Sierpinski triangle is an **attractor** of this IFS.

A different example of an IFS is $\{\square; f_1, f_2, f_3, f_4\}$, where $\square$ is the unit square, defined in Section 1.2, and the functions f_n are given by

$$f_n(x, y) = \left(\frac{a_n x + b_n y + c_n}{g_n x + h_n y + j_n}, \frac{d_n x + e_n y + k_n}{g_n x + h_n y + j_n} \right) \quad \text{for } n = 1, 2, 3, 4.$$

The coefficients are given in Table 0.1. In this case, to implement the chaos game we apply one of the functions f_1, f_2, f_3, f_4 to the current point $\mathbf{X}_i$, to obtain the the next point $\mathbf{X}_{i+1}$ for $i = 1, 2, 3, \ldots$ We apply f_1 with probability p_1, f_2 with probability p_2, f_3 with probability p_3 and f_4 with probability p_4 . For each step,

(i)   (ii)   (iii)

ZOOM to
Figure 0.3

Figure 0.2   Pictures of attractors of an IFS: (i) the set attractor, (ii) the measure attractor and (iii) the fractal top.

More holes
with fractal
boundaries
are revealed

Figure 0.3   Zoom in on the fractal top in Figure 0.2.

the choice of function is made independently of the choices made at all other steps. The probabilities p_n are given in Table 0.1. This time, almost certainly, the sequence of points $\mathbf{X}_0, \mathbf{X}_1, \mathbf{X}_2, \ldots$ will be attracted to a set that looks like the left-hand picture in Figure 0.2. This is a picture of an attractor of the IFS represented by Table 0.1.

Amazingly, this picture is unlikely to change significantly if the probabilities are adjusted, provided sufficiently many points are plotted. The colours in Figures 0.2(iii) and 0.3 were 'stolen' from Figure 0.4. In Chapter 4 you will discover what this means.

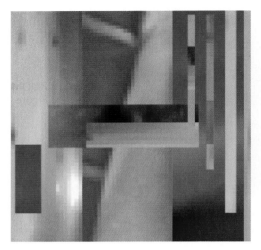

Figure 0.4 Colours were 'stolen' from this picture to produce Figure 0.3 and the image in Figure 0.2(iii).

In this book you will discover different kinds of attractor associated with an IFS. For example, Figure 0.2 illustrates the set attractor, the measure attractor and the fractal top for the IFS in Table 0.1. These beautiful objects may be computed by variants of the chaos game and by other means too. Quite generally, although the IFSs themselves are simple to write down, their attractors are geometrically and topologically complicated. Typically, computer pictures of them can be magnified up endlessly to reveal more and more intricate detail. For example, Figure 0.3 illustrates a tiny hole in the fractal top in Figure 0.2, greatly magnified. Often, simultaneously, such pictures are reminiscent of biological structures and convey the feeling of real-world images, with repetition and disorder combined and the property that one may look ever closer, revealing more and more mysteries. They are suggestive of diverse applications in biology and imaging.

The mathematics in this book is separate from the pictures that illustrate it and the biology that inspired it. Indeed, we will treat all pictures as though they actually are mathematical objects. The attractor of an IFS may be topologically conjugate to a fractal fern without ever leaving the abstract world in which it lives, trapped in mathematical amber, so to speak. All the theorems are independent of the pictures. The mathematics describes something much more general, something bigger, than the pictures.

In this book I try to capture in a precise way a fascinating combination of geometry, topology, probability and pictures. I think that just over the horizon, in the direction in which this book points, there is an unambiguous, new, branch of geometry that combines colour and space. In trying to move towards this goal, I present much new material including the theory of fractal tops, fractal home-omorphisms, orbital pictures and superfractals. At the time of writing only one major paper about superfractals has appeared in print, although a number are in the pipeline.

It is important to read the book from the beginning. Read enough on each page to be sure that you do not miss the themes that build steadily towards two 'peaks' and then the superfractal 'summit'. In Chapter 1 we introduce and explore code space and topology and develop familiarity with metric spaces whose elements are collections of objects. Code space is a major theme of the book. The second major theme, developed in Chapter 2, is elementary transformations and how, specifically and precisely, they act on sets, pictures and measures. Then in Chapter 3 we bring code spaces and transformations together in the framework of IFS semigroups of transformations acting on sets, pictures and measures. It is in the combination of code space and transformations that beautiful new mathematical structures such as orbital pictures, the first 'peak', are discovered.

In Chapter 4 we reach the second 'peak': fractal tops, colour-stealing and fractal homeomorphisms. We discover that we can handle algebraically the topology of some IFS attractors with the same ease that Descartes handled geometrical objects in his Cartesian plane. One application is to computer graphics, via the production of diverse families of beautiful homeomorphisms between images. This chapter combines the chaos game, transformations, identification topologies on code space and basic IFS theory. In effect we study certain limit sets belonging to the objects introduced earlier.

In Chapter 5 we reach the 'summit', which is superfractals. We combine the themes already developed with the concept of V-variability. This enables us to describe and synthesize vast collections of related mathematical objects, be they galleries full of random variations of seascapes or families of related ferns, as illustrated in Figure 0.5. With the aid of our knowledge of transformations, IFS semigroups, code space structure and V-variability we discover that we can produce vast families of homeomorphic objects with random, but not too random, variations. Superfractals provide a bridge, made of IFSs, from deterministic fractals to the world of random fractals. Previously I did not know how to get there.

0.3 Another chaos game

Here is a simple variant of the chaos game. Mark four points on a sheet of paper. Label three of them **A**, **B** and **C** and label the remaining point $\mathbf{X}_0$. We add two more rules to the three in Section 0.2 above:

Rule 4: Shift by $2(\mathbf{B} - \mathbf{C})$.

Rule 5: Rotate by $180°$ *degrees about the point* $\dfrac{\mathbf{A} + 5\mathbf{B} - 4\mathbf{C}}{2}$.

This time, when you play the game, *remember what the dice showed the last time you rolled it*. Begin by rolling the dice once, to give you a starting value.

Figure 0.5   The chaos game produces a sequence of mathematical objects, successively closer and closer to random fractal ferns lying on a 'superfractal'.

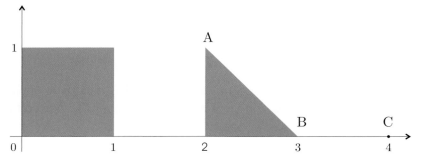

Figure 0.6   The chaos game is played with slightly more complicated rules. The random point now dances on *both* of two classical euclidean objects, a square and a triangle.

Now, each time you roll the dice, if last time it showed *A* or *B* and this time it shows *A* then apply *Rule 1*. If it shows *B* then apply *Rule 2* but if it shows *C* then apply *Rule 4*. If last time it showed *C*, however, and this time it produces *A* or *B* then apply *Rule 5* but if it produces *C* then apply *Rule 3*.

In this new game, if you discard the initial few points then you will obtain a result quite as astonishing as the Sierpinski triangle; you will obtain *simultaneously* two classical geometrical objects, a filled parallelogram and a triangle. See Figure 0.6, which shows the resulting picture when $A = (2, 1)$, $B = (3, 0)$ and $C = (4, 0)$. By following the rules above, the current point will continually dance on the square and the triangle, sometimes moving from one to the other, sometimes moving back again to the first – forever.

So you see, Diana, Rose and gentle reader, this book is about much more than basic fractal geometry. It is about extraordinary transformations of pictures, homeomorphisms between complicated objects and the magic of code space. It is about superfractals.

CHAPTER 1

Codes, metrics and topologies

1.1 Introduction

Any picture may be conceived as a mathematical object, lying on part of the euclidean plane, each point having its own colour. Then it is a strange and wonderful entity. It is mysterious, for you probably cannot see it. And worse, you cannot even describe it in the type of language with which you normally talk about objects you can see; at least, not without making a lot of assumptions. But we want to be able to see, to describe and to make pictures on paper of fractals and other mathematical objects that we feel ought to be capable of representation as pictures. We want to make mathematical models for real-world images, biological entities such as leaves and many other types of data. To be able to do this we need certain parts of the language of mathematics, related to set theory, metric spaces and topology.

Code space   There is a remarkable set, called a code space, which consists of an uncountable infinity of points and which can be embedded in the tiniest real interval. A code space can be reorganized in an endless variety of amazing geometrical, topological, ways, to form sets that look like leaves, ferns, cells, flowers and so on. For this reason we think of a code space as being somehow protoplasmic, plastic, impressionable and capable of diverse re-expressions, like the meristem of a plant; see Figure 1.1. This idea is a theme of this chapter and of the whole book.

Structure and contents of this chapter

In this chapter we introduce and discuss spaces, with the focus on those that we will be using later. They include the euclidean plane, code spaces and spaces whose points are certain subsets of other spaces. In particular, we discuss spaces that consist of infinitely many points, such as the real interval [0, 1] and the euclidean plane $\mathbb{R}^2$. We also introduce notation that we shall use throughout the book.

A space may have one or more of the following three properties: (i) its points are organized by means of a system of **addresses** or coordinates; (ii) the relationship between the points of the space is described by means of a **metric** or distance function; (iii) the relationship between the points of the space is described by

Figure 1.1 'Meristem, a specialized section of plant tissue characterized by cell division and growth . . . In one type of lateral meristem, called cambium, or vascular cambium, the cells divide and differentiate to form the conducting tissues of the plant, i.e. the wood or xylem, and the phloem.' (*Columbia Encyclopedia*, sixth edition, 2004)

means of a **topology**, with certain subsets labelled 'open'. Typically properties (i), (ii) and (iii) are not independent. Moreover different systems of addresses, diverse metrics and various topologies may be possible on the same space.

We discuss addressing schemes and spaces of addresses, namely code spaces, in Section 1.4. In particular, we explain how addresses for the points in a line segment in $\mathbb{R}^2$ may be defined geometrically via successive bisections. In Section 1.6 we show how diverse metrics may be defined on a code space by embedding it in a space such as $\mathbb{R}^2$. We treat code spaces as very important because of their central role in describing fractals, fractal tops and superfractals in later chapters.

We introduce metric spaces in Section 1.5 and topological spaces in Section 1.8. In Section 1.9 we describe a number of basic, readily constructed, topologies, including identification topologies. An identification topology on a space may be obtained by treating some pairs of points as single points, that is, by 'gluing them together'. In this manner a code space may be given the topology of a line segment, a Möbius strip or a fractal tree. Identification topologies play a very important role in Chapter 4, where we discuss fractal homeomorphisms.

The possible organizational schemes (i), (ii) and (iii) are brought to life by transformations, introduced in Section 1.3. Some of the fundamental properties of a space are those that are preserved by rich collections of transformations such as addressing functions, metric transformations, continuous transformations and homeomorphisms. From this point of view we discuss properties of metric spaces in Section 1.7 and those of topological spaces in Section 1.10. The properties of completeness, defined in Section 1.7, and compactness, defined in Section 1.11, are needed to establish the existence of fractal objects. We describe the conditions under which these properties occur.

Over and above the themes of code spaces, properties of spaces that are preserved under transformations and the nature of euclidean space, a central focus of this chapter, which will carry on throughout the book, is the idea that *the points in a space may themselves be mathematical objects*. For example, they may be mathematical pictures, or measures, defined in Chapter 2. Or they may simply be the nonempty compact subsets of another underlying space.

Thus, the points of a space H_X may be constructed using sets of the points of an underlying space X. Organizational principles such as addresses, metrics and topologies may be inherited from X and provide structure to H_X. Properties of the underlying space X such as compactness and completeness may also be inherited by the space H_X. Moreover, transformations acting on X may be used to define transformations on H_X. These inheritances are important because they enable us to establish the existence of diverse types of fractal in later chapters.

For example, in Section 1.13 we show that the property of being a compact metric space may be inherited from X by a certain space $\mathbb{H}(X)$. The inherited metric, the Hausdorff metric, is discussed earlier, in Section 1.12, with a view to developing our intuition about how it works. This remarkable inheritance continues from generation to generation, from X to $\mathbb{H}(X)$ to $\mathbb{H}(\mathbb{H}(X))$ and so on. It enables us to establish the existence of superfractal sets in Chapter 5.

1.2   Points and spaces

In this section we introduce the notation and nomenclature for points, sets and spaces that we shall use throughout the book.

A **space** is a set. The elements of the set are called the **points** of the space. We use the notation $\mathbb{X}$ to denote a space. The expression $x \in \mathbb{X}$ means that x belongs to the set $\mathbb{X}$ or equivalently that x is a point of the space $\mathbb{X}$. Similarly the expression $x, y \in \mathbb{X}$ means that both x and y are points of $\mathbb{X}$. We say that two points $x, y \in \mathbb{X}$ are **distinct** if $x \neq y$, that is, x is not equal to y. When we consider several spaces at once, we may denote them by $\mathbb{X}, \mathbb{Y}, \ldots$ A space may be empty, that is, it may contain no points.

For illustration, some spaces are shown in Figure 1.2. An important example of a space is $\mathbb{R}$, the set of all finite real numbers. A point $x \in \mathbb{R}$ is simply any number, positive or negative, that can be expressed by a decimal expansion, either finite as in $x = 1.5$ or unending as in $x = -7.93121059912791101 \cdots$. We can write

$$\mathbb{R} = \{x : -\infty < x < +\infty\}.$$

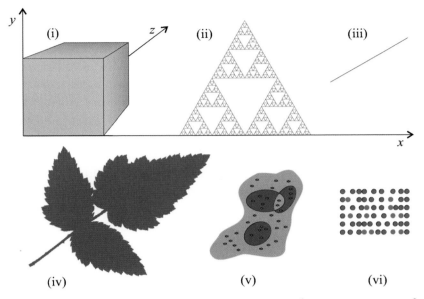

Figure 1.2 Shown here are illustrations of spaces: (i) a cube in $\mathbb{R}^3$; (ii) a fractal subset of $\mathbb{R}^2$; (iii) a line segment; (iv) a subset of $\mathbb{R}^2$ that looks like a leaf; (v) the space of subsets of a set; (vi) a code space.

We use the notation $\{elements : conditions\}$ to mean a set of *elements*, or objects, on the left of the colon, that obey the *conditions* on the right of the colon. We may think of the points of $\mathbb{R}$ as being organized to lie on a straight line, the x-axis in coordinate geometry; see Section 1.4.

We denote the four intervals defined by $a, b \in \mathbb{R}$, with $a < b$, by $[a, b] = \{x \in \mathbb{R} : a \leq x \leq b\}, [a, b) = \{x \in \mathbb{R} : a \leq x < b\}, (a, b] = \{x \in \mathbb{R} : a < x \leq b\}$ and $(a, b) = \{x \in \mathbb{R} : a < x < b\}$. Each interval is an example of a space.

An important space is the euclidean plane, which we denote by $\mathbb{R}^2$. It should be familiar to you from calculus and geometry. It is sometimes called the xy-plane. It is the place where straight lines and circles exist and where one imagines graphs of functions like $y = x^2 + 1$. Each point in the euclidean plane can be represented by a pair of coordinates (x, y), x and y being finite real numbers. We can write

$$\mathbb{R}^2 = \{(x, y) : -\infty < x < +\infty, \ -\infty < y < +\infty\}.$$

If $\mathbb{X}$ and $\mathbb{Y}$ are spaces then $\mathbb{X} \times \mathbb{Y}$ denotes the space of ordered pairs of points, which are denoted by either $x \times y$ or (x, y), where $x \in \mathbb{X}$ and $y \in \mathbb{Y}$. We write $\mathbb{X}^1 = \mathbb{X}, \mathbb{X}^2 = \mathbb{X} \times \mathbb{X}$ and $\mathbb{X}^{n+1} = \mathbb{X}^n \times \mathbb{X}$, for $n \in \{1, 2, \ldots\}$. So for example we have the space $\mathbb{R}^2 = \mathbb{R} \times \mathbb{R}$. Note that we can write

$$\mathbb{R}^2 = \{(x_1, x_2) : x_1 \in \mathbb{R}, \ x_2 \in \mathbb{R}\}. \tag{1.2.1}$$

We use any of the notations x, (x_1, x_2), (x, y) to denote a point in $\mathbb{R}^2$ and the obvious extension of this notation in $\mathbb{R}^3, \mathbb{R}^4, \ldots$

EXERCISE 1.2.1　*Express $\mathbb{R}^3$ in a similar way to $\mathbb{R}^2$ in Equation (1.2.1).*

The spaces $\mathbb{R}, \mathbb{R}^2, \mathbb{R}^3, \ldots$ occur throughout the mathematical sciences and serve numerous purposes, many related ultimately to models of reality. With $\mathbb{R}$ we model distance, time, mass, temperature and other scalar physical quantities. Using $\mathbb{R}^2$ we model observations of flat things, patterns for making clothes, pictures, maps, photographs and so on. And $\mathbb{R}^3$ is the oldest model for the physical space about us, in which we live, design buildings and fly space missions. Also, the spaces $\mathbb{R}, \mathbb{R}^2, \mathbb{R}^3, \ldots$ are the underlying mathematical fabric from which are constructed prime examples in topology, geometry, measure theory and many other areas; in them we formulate the basic equations of physics. They are incredibly rich in structure and properties.

Most fractals that we study in this book are either subsets of $\mathbb{R}, \mathbb{R}^2, \mathbb{R}^3, \ldots$ or else built upon them, and many properties of fractals are inherited from these spaces. We learn something new about these spaces, the fabric of which they are made, by studying fractals.

The spaces that interest us most are those that are in some way self-similar. In this book we describe the euclidean plane as $\mathbb{R}^2$. But we may consider this space unadorned by coordinates, so that we have a blank space, like an endless, perfectly flat, homogeneous sheet of paper. Then one part of the space is like any other and we have no way of knowing whether, for example, a circle inscribed on this plane is big or small, or even where it is. The space is just like itself everywhere and at all scales of observation.

An example of a space with a finite boundary is the **unit square**

$$\square := \{(x, y) : 0 \leq x \leq 1, 0 \leq y \leq 1\}.$$

The symbol ':=' means 'is defined to be'. Imagine an empty picture that represents $\square$. Its homogeneous quality represents the uniformity of the euclidean plane before it is invaded by theories and marks, like a new beach after the tide has gone down on which no one has yet walked. One mathematician looking at it might imagine open sets, topology and connected paths; another, lines, triangles and intersections; and yet another, myriads of points of some algebraic variety. But let us, just for a moment, imagine nothing.

Let $\mathbb{S}(\mathbb{X})$ denote the set of all subsets of the space $\mathbb{X}$; then $\mathbb{S}(\mathbb{X})$ is also a space. In $\mathbb{S}(\mathbb{X})$ both the empty set $\varnothing$, the subset of $\mathbb{X}$ that contains no points of $\mathbb{X}$, and $\mathbb{X}$ itself are single points!

Some spaces that we shall consider, such as sets of points or sets of circles in the euclidean plane, have an explicit geometrical character while others, such as $\mathbb{S}(\mathbb{X})$, are more abstract. But we will try to think geometrically about spaces, for

Figure 1.3 A green image in a white surround may be thought of as an approximate description of a set of points in the space □. All points that are not white belong to the set. A set of points in □ represents a single point in $\mathbb{S}(\square)$, the space whose points are the subsets of □.

example by assigning distance functions or imagining pictorial representations; *we could think of* the space $\mathbb{S}(\square)$ as the set of all green drawings on □. In this way of thinking, green of varying strength (the strongest green is the lightest in appearance) replaces white and each green dot in such a drawing represents a point in □, that is, an element of a set in $\mathbb{S}(\square)$. A blank white image, where no drawing has occurred, represents the empty set, and an entirely green image represents the point $\square \in \mathbb{S}(\square)$. A green line from the lower left corner to the upper right corner of □ represents the point $\{(x, y) \in \square : x = y\} \in \mathbb{S}(\square)$; and an image such as Figure 1.3 serves as an approximate description of a single point in $\mathbb{S}(\square)$.

1.3 Functions, mappings and transformations

In this section we introduce notation and definitions related to functions. We use the notation

$$f : \mathbb{X} \to \mathbb{Y}$$

to denote a function f that acts on the space $\mathbb{X}$ to produce values in the space $\mathbb{Y}$; f assigns to each point $x \in \mathbb{X}$ a unique point $f(x)$ in $\mathbb{Y}$.

The **graph** of f is defined by

$$G_f := \{(x, f(x)) \in \mathbb{X} \times \mathbb{Y} : x \in \mathbb{X}\}.$$

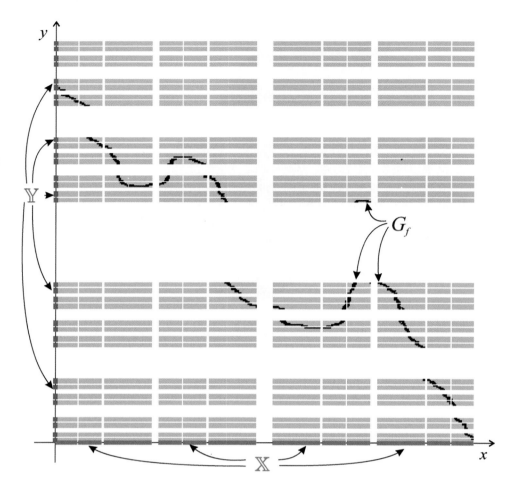

Figure 1.4 The space $\mathbb{X} \times \mathbb{Y}$, where both $\mathbb{X} \subset \mathbb{R}$ and $\mathbb{Y} \subset \mathbb{R}$ are unions of intervals. Also shown is the graph G_f of a function $f : \mathbb{X} \to \mathbb{Y}$. This picture is a reminder that $\mathbb{X} \times \mathbb{Y}$ may be much 'bigger' than $\mathbb{X}$ or $\mathbb{Y}$ and that complicated domains and ranges may occur and yield fragmented graphs.

To know f is equivalent to knowing G_f. That is, to specify a function $f : \mathbb{X} \to \mathbb{Y}$ is equivalent to specifying a certain type of subset of $\mathbb{X} \times \mathbb{Y}$, one whose 'shadow' or 'projection' on $\mathbb{X}$ is all $\mathbb{X}$ and such that for each $x \in \mathbb{X}$ there is a unique 'height' value in $\mathbb{Y}$. In Figure 1.4 we illustrate the graph of such a function in the case where both the domain and the range of the function are disconnected subsets of $\mathbb{R}$.

We also call $f : \mathbb{X} \to \mathbb{Y}$ a **transformation** from the space $\mathbb{X}$ to the space $\mathbb{Y}$ or a **mapping** from the space $\mathbb{X}$ to the space $\mathbb{Y}$. We define the **domain** D_f of the function f to be the set of points upon which it acts. If $f : \mathbb{X} \to \mathbb{Y}$ then $D_f = \mathbb{X}$. The **range** of f is defined by

$$R_f := \{y \in \mathbb{Y} : f(x) = y \text{ for some } x \in \mathbb{X}\} =: f(\mathbb{X}).$$

Let $S \subset \mathbb{X}$ denote a subset of $\mathbb{X}$. Note that S might be the empty set $\varnothing$ or it might be $\mathbb{X}$ itself. Let $f : S \to \mathbb{Y}$. Then $D_f = S$. In such cases we might use the notation $f : S \subset \mathbb{X} \to \mathbb{Y}$ and when we also wish to refer explicitly to the range of f we will sometimes write $f : S \subset \mathbb{X} \to R_f \subset \mathbb{Y}$.

Generally, we will extend the definition of a function $f : \mathbb{X} \to \mathbb{Y}$ to encompass a function $f : \mathbb{S}(\mathbb{X}) \to \mathbb{S}(\mathbb{Y})$ defined by

$$f(S) = \{f(x) : x \in S\},$$

for $S \in \mathbb{S}(\mathbb{X})$, where $\mathbb{S}(\mathbb{X})$ is the space of all subsets of $\mathbb{X}$. We intend that it should be clear from the context whether we mean a point-valued or set-valued function.

We say that a function f is **one-to-one** if and only if (iff) for each $y \in R_f$ there is a unique point $x \in D_f$ such that $f(x) = y$. In this case the inverse function $f^{-1} : R_f \subset \mathbb{Y} \to D_f \subset \mathbb{X}$ is defined by $f^{-1}(y) = x$.

When $f : \mathbb{X} \to R_f \subset \mathbb{Y}$ is one-to-one we will sometimes call f an **embedding** function. Then we may use the points of $f(\mathbb{X}) = R_f$ to **represent** the points of $\mathbb{X}$. We think of $\mathbb{X}$ as being embedded in the space $\mathbb{Y}$, where it is represented by the set $f(\mathbb{X})$.

We say that $f : D_f \subset \mathbb{X} \to \mathbb{Y}$ is **onto** when $R_f = \mathbb{Y}$. Even when $f : \mathbb{X} \to \mathbb{Y}$ is neither one-to-one nor onto, we define the set-valued inverse function

$$f^{-1} : \mathbb{S}(\mathbb{Y}) \to \mathbb{S}(\mathbb{X})$$

by

$$f^{-1}(S) = \{x \in \mathbb{X} : f(x) \in S\},$$

for all $S \in \mathbb{S}(\mathbb{X})$. We will sometimes write $f^{-1}(x)$ in place of $f^{-1}(\{x\})$ when $\{x\}$ is a singleton set, that is, the set consisting of the single point $x \in \mathbb{X}$. For us, such an inverse function always exists but its values may consist of a set of more than one point or the empty set.

EXERCISE 1.3.1   *Let $f : \mathbb{R} \to \mathbb{R}$ be defined by $f(x) = 1 + x^2$. Show that f is not one-to-one and not onto. Show also that $f^{-1}(x) = \{\sqrt{x-1}, -\sqrt{x-1}\}$ when $x \geq 1$ and $f^{-1}(x) = \varnothing$ when $x < 1$. Also, can you explain the point of this exercise? Define $\xi : \mathbb{R} \to \mathbb{R}^2$ by $\xi(x) = (x, 0)$. Show that ξ is an embedding function.*

Now we introduce the union symbol $\cup$ and the intersection symbol $\cap$. The expression $\mathbb{X} \cup \mathbb{Y}$ means the set that consists of all the points in $\mathbb{X}$ and all the points in $\mathbb{Y}$:

$$\mathbb{X} \cup \mathbb{Y} = \{x : x \in \mathbb{X} \text{ or } x \in \mathbb{Y}\}.$$

Note that a point that belongs to both $\mathbb{X}$ and $\mathbb{Y}$ also belongs to $\mathbb{X} \cup \mathbb{Y}$. Let $\mathcal{I}$ denote an **index** set, that is, a set of objects that we call **indices**. Let S_i be a set for each

$i \in \mathcal{I}$. Then we will use the notation

$$\bigcup_{i \in \mathcal{I}} S_i := \{x : x \in S_i \text{ for at least one } i \in \mathcal{I}\}.$$

Similarly we define the intersection of two sets by

$$\mathbb{X} \cap \mathbb{Y} = \{x : x \in \mathbb{X} \text{ and } x \in \mathbb{Y}\},$$

and we write

$$\bigcap_{i \in \mathcal{I}} S_i := \{x : x \in S_i \text{ for all } i \in \mathcal{I}\}.$$

When $A, B \subset \mathbb{X}$, the notation $A \backslash B$ means the set of points of $\mathbb{X}$ that are in A and not in B.

EXERCISE 1.3.2 *Let* $f : \mathbb{X} \to \mathbb{Y}$, *let* $S, T \subset \mathbb{X}$ *and let* $V, W \subset \mathbb{Y}$. *Prove, and learn forever, that:*
  (i) $f(S \cup T) = f(S) \cup f(T)$;
 (ii) $f(S \cap T) \subset f(S) \cap f(T)$;
(iii) $f^{-1}(V \cup W) = f^{-1}(V) \cup f^{-1}(W)$;
 (iv) $f^{-1}(V \cap W) = f^{-1}(V) \cap f^{-1}(W)$;
  (v) $f^{-1}(\mathbb{Y} \backslash V) = \mathbb{X} \backslash f^{-1}(V)$.

Let $f : \mathbb{X} \to \mathbb{Y}$ and let $S \subset \mathbb{X}$. Then we can define a function $f|_S : S \to \mathbb{Y}$ by $f|_S(x) = f(x)$ for all $x \in S$. $f|_S$ is called the **restriction** of f to S. We will often denote $f|_S$ simply by f.

1.4   Addresses and code spaces

In this section we describe how the points of a space may be organized by means of **addresses**. Addresses are themselves members of certain types of spaces that we call **code spaces**.

When a space consists of many points, as in the cases of $\mathbb{R}$ and $\mathbb{R}^2$, it is often convenient to have addresses for the points in the space. An address of a point is a means to identify the point, just as a postal address identifies a mailbox. It is in effect an algorithm or formula for locating the point precisely. It may be a string of numbers or symbols, either finite or infinite, that uniquely specifies the point, via some procedure that is implicitly understood and unstated. For example, the address of a point $x \in \mathbb{R}$ may be its decimal or binary expansion. Points in $\mathbb{R}^2$ may be addressed by ordered pairs of decimal expansions.

A single point may have more than one address; for example the same point in $\mathbb{R}$ has the two binary addresses $1.\overline{0} = 1.0000\cdots$ and $0.\overline{1} = 0.1111\cdots$. Here an overbar means that the symbol or finite string of symbols is repeated endlessly, so that

$$\bar{s} = sssssssssssss \cdots.$$

We shall introduce some useful spaces of addresses, namely code spaces. These spaces will be needed later to represent sets of points on fractals. An address is made from an **alphabet** of symbols. An alphabet $\mathcal{A}$ consists of a nonempty finite set of symbols such as $\{1, 2, \ldots, N\}$, $\{0, 1, \ldots, N\}$ or $\{A, B, \ldots, Z\}$ where each symbol is distinct. The number of symbols in the alphabet is $|\mathcal{A}|$. For example, $|\{0, 1, 2, \ldots, N\}| = N + 1$.

Let $\Omega'_{\mathcal{A}}$ denote the set of all finite strings made of symbols from the alphabet $\mathcal{A}$. The set $\Omega'_{\mathcal{A}}$ includes the empty string $\varnothing$. That is, $\Omega'_{\mathcal{A}}$ consists of all expressions of the form

$$\sigma = \sigma_1 \sigma_2 \cdots \sigma_K,$$

where $\sigma_n \in \mathcal{A}$ for all $n \in \{1, 2, \ldots, K\}$ with K a positive integer, as well as $\varnothing$. We will write $|\sigma|$ to denote the length of the string $\sigma \in \Omega'_{\mathcal{A}}$.

Examples of points in $\Omega'_{\{A,B,\ldots,Z\}}$ are A, $DOOR$, $AAAAAA$, $\varnothing$ and YOU. Examples of points in $\Omega'_{\{1,2,3\}}$ are 1111111, 123, 1231111 and 3. A convenient address for a point $\sigma \in \Omega'_{\mathcal{A}}$ is σ itself. An example of a point in $\Omega'_{\{0,1\}}$ is $\sigma = 1011010111$, which we refer to as a **finite binary string**. Notice that $\Omega'_{\{0,1\}}$ is a convenient space for addressing the space of all computer files.

Throughout the book we will often refer to the spaces $\Omega_{\{1,2,\ldots,N\}}$ and $\Omega'_{\{1,2,\ldots,N\}}$. Make sure now that you really do know what these symbols signify.

Given two strings $\sigma, \omega \in \Omega'_{\mathcal{A}}$ we will write $\sigma\omega$ to denote the **concatenated string**

$$\sigma\omega := \sigma_1 \sigma_2 \cdots \sigma_{|\sigma|} \omega_1 \omega_2 \cdots \omega_{|\omega|}.$$

So for example if $\sigma, \omega \in \Omega'_{\{0,1\}}$ with $\sigma = 000$ and $\omega = 11$ then $\sigma\omega = 00011$ and $\omega\sigma = 11000$. And if $\sigma = 000$ and $\omega = \varnothing$ then $\omega\sigma = \sigma\omega = 000$.

An important space, which we denote by $\Omega_{\mathcal{A}}$, consists of all **infinite** strings of symbols from the alphabet $\mathcal{A}$. That is, $\sigma \in \Omega_{\mathcal{A}}$ if and only if it can be written

$$\sigma = \sigma_1 \sigma_2 \cdots \sigma_n \cdots$$

where $\sigma_n \in \mathcal{A}$ for all $n \in \{1, 2, \ldots\}$. An example of a point in $\Omega_{\{0,1\}}$ is $\sigma = 1011010111 \cdots$. A point in $\Omega_{\{A,B,C\}}$ is $\bar{A}$.

Ω' is countable but, when $|\mathcal{A}| > 1$, $\Omega_{\mathcal{A}}$ is uncountable.

DEFINITION 1.4.1 Let $\varphi : \Omega \to \mathbb{X}$ be a mapping from $\Omega \subset \Omega'_{\mathcal{A}} \cup \Omega_{\mathcal{A}}$ onto a space $\mathbb{X}$. Then φ is called an **address function** for $\mathbb{X}$, and points in Ω are called addresses. Ω is called a **code space**. Any point $\sigma \in \Omega$ such that $\varphi(\sigma) = x$ is called an **address** of $x \in \mathbb{X}$. The set of all addresses of $x \in \mathbb{X}$ is $\varphi^{-1}(\{x\})$.

$$\Omega \qquad\qquad\qquad\qquad\qquad\qquad \mathbb{X}$$

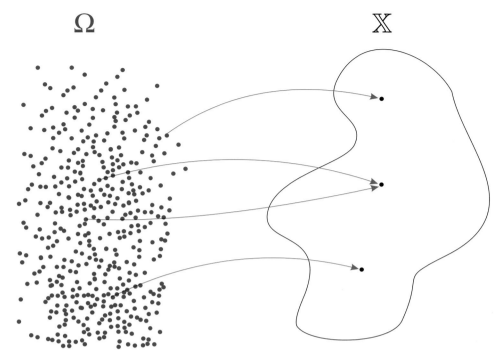

Figure 1.5 Points in the space $\mathbb{X}$ may be assigned addresses. The function $f : \Omega \to \mathbb{X}$ that maps from code space Ω, the space of addresses, to points in $\mathbb{X}$ is called an addressing function. Each point in $\mathbb{X}$ has at least one address.

Figure 1.5 illustrates the concepts in this definition.

EXERCISE 1.4.2 *Define a code space and address function for each of the following spaces:*

(i) $\mathbb{X} = [0, 1] = \{x : 0 \leq x \leq 1\}$;

(ii) $\mathbb{X} = \{(x, y) \in \mathbb{R}^2 : x^2 + y^2 = 1\} \cap \{(x, y) \in \mathbb{R}^2 : y > 0\}$;

(iii) $\mathbb{X} = \{(x, y) \in \mathbb{R}^2 : x^2 + y^2 \leq 1\}$;

(iv) $\mathbb{X} = \mathbb{Z}^+ = \{1, 2, 3, \dots\}$, *the set of positive integers;*

(v) *the set of real numbers that can be written in the form $x = m/2^n$ for some $m \in \{0, 1, \dots, 2^n - 1\}$ and $n \in \{0, 1, 2, \dots\}$. How many addresses does the point $x = 0.25$ have, according to your addressing scheme?*

Addresses of points on a line

In this subsection we illustrate an addressing scheme for the points on a line segment in the euclidean plane. One goal is to demonstrate how coordinates may depend on geometrical properties of the space. But also we illustrate how real space may be broken up into smaller and smaller similar parts.

Let A and B denote a pair of distinct points in the euclidean plane. Let $L[A, B]$ denote the set of points in the line segment that joins A and B. Then $L[A, B]$ is a space.

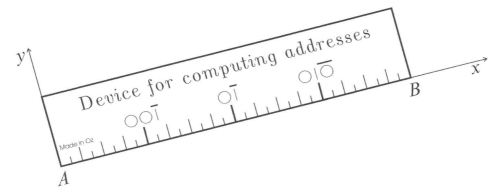

Figure 1.6 A binary ruler makes finding an address of a point in the line segment $L[A, B]$ easy! Or does it?

Each point $x \in L[A, B]$ can be represented by an address in $\Omega_{\{0,1,\ldots,9\}}$, and each address in $\Omega_{\{0,1,\ldots,9\}}$ defines a unique point in $L[A, B]$. A simple way to see this is to identify $L[A, B]$ with the unit interval $[0, 1] := \{x \in \mathbb{R} : 0 \leq x \leq 1\}$. Simply take A to be the origin of coordinates, let the x-axis pass through B and define B to be the point $x = 1$. Then an address of x for $0 \leq x < 1$ is the sequence of digits after the decimal point in a decimal expansion of x, and the point $x = 1$ is assigned the address $\overline{9} \in \Omega_{\{0,1,\ldots,9\}}$. Alternatively, we may use an address function $\varphi : \Omega_{\{0,1\}} \to L[A, B]$ defined by using the binary expansion of x.

These addressing schemes and others like them will be used often later on. So here we describe a bit more deeply the construction of the address function $\varphi : \Omega_{\{0,1\}} \to L[A, B]$. The description in the previous paragraph assumed that we already have a ruler or measuring stick, namely the unit interval addressed by real numbers; see Figure 1.6. But this ruler can be constructed using a straight-edge and compass, which reveals the geometrical origin of such addresses.

$L_\varnothing := L[A, B]$ may be bisected, as illustrated in Figure 1.7, by constructing two circles, one centred at A and passing through B and the other centred at B and passing through A. Denote the two points of intersection of these circles by C and D. Then construct the line segment $L[C, D]$, and let this meet $L[A, B]$ at the point E. Then E is the bisection point of $L[A, B]$. The result is two intervals, which we denote by L_0 and L_1, with, say, L_0 to the left of L_1.

This latter assertion is of a geometrical kind – it derives from axiomatic properties of line segments. See for example [26], p. 22, the end of the first paragraph. We have

$$L_\varnothing = L_0 \cup L_1 \quad \text{and} \quad E = L_0 \cap L_1;$$

see Figure 1.8. Both L_0 and L_1 contain the midpoint of $L_\varnothing$. We next similarly bisect L_0 to produce two intervals L_{00} and L_{01}, where L_{00} lies to the left of

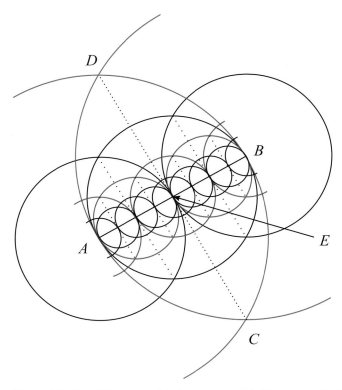

Figure 1.7 Nested bisections of the line segment $L[A, B]$ constructed with straight-edge and compass. Identify here some of the shields shown in Figure 1.9.

L_{01}, and we bisect L_1 to produce two intervals, L_{10} lying to the left of L_{11}. This successive bisection process is done in such a way that the geometrical ordering of the intervals, from say left to right, starting at A and going to B, as $L_{00}, L_{01}, L_{10}, L_{11}$ corresponds to the lexicographic ordering of the strings $00, 01, 10, 11 \in \Omega'_{\{0,1\}}$. We now have

$$L_0 = L_{00} \cup L_{01} \quad \text{and} \quad L_1 = L_{10} \cup L_{11}$$

as well as

$$L_\varnothing = L_{00} \cup L_{01} \cup L_{10} \cup L_{11}.$$

We can repeat this bisection process inductively. At the nth generation we obtain 2^n intervals, denoted by $\{L_\sigma : \sigma \in \Omega'_{\{0,1\}}, |\sigma| = n\}$. These intervals form a partition of $L[A, B]$, that is,

$$
\begin{aligned}
L[A, B] &= L_{00\cdots0} \cup L_{00\cdots1} \cup \cdots \cup L_{11\cdots0} \cup L_{11\cdots1} \\
&= \cup \{L_\sigma : \sigma \in \Omega'_{\{0,1\}}, |\sigma| = n\};
\end{aligned}
\tag{1.4.1}
$$

L_σ is of length $1/2^{|\sigma|}$ times the length of $L[A, B]$.

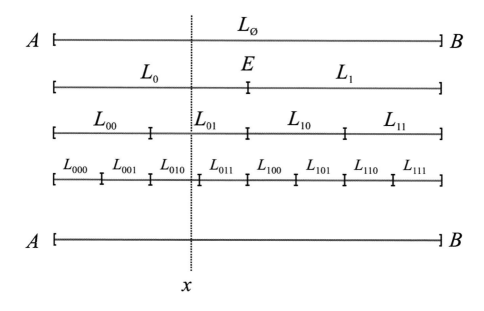

$$L_0 \supset L_{01} \supset L_{010} \supset \cdots \supset x$$

Figure 1.8 $L_\varnothing$ is partitioned into smaller and smaller subintervals L_σ, where σ belongs to the code space $\Omega'_{\{0,1\}}$. Any point $x \in L_\varnothing$ belongs to an infinite sequence of such intervals, and the sequence of addresses of these intervals determines an address in $\Omega_{\{0,1\}}$ for the point x. (In the case illustrated the address begins $010 \cdots$.) Conversely, given any address in $\Omega_{\{0,1\}}$ we can define uniquely a corresponding sequence of nested intervals and a point $x \in L_\varnothing$.

We call the elements of the space $\mathcal{C} := \{L_\sigma : \sigma \in \Omega'_{\{0,1\}}\}$ the **cylinder sets** of $L[A, B]$.

By construction, we have

$$L_\sigma \subset L_{\tilde{\sigma}} \iff \sigma = \tilde{\sigma}\omega \quad \text{for some } \omega \in \Omega'_{\{0,1\}}. \tag{1.4.2}$$

The symbol $\iff$ means 'if and only if'; it says that the expressions on either side of it are equivalent.

Now let $\sigma \in \Omega_{\{0,1\}}$ and suppose that $\sigma = \sigma_1\sigma_2\sigma_3 \cdots \sigma_n \cdots$. For each $n = 1, 2, \ldots$ let $\omega_n \in \Omega'_{\{0,1\}}$ be defined by

$$\omega_n := \sigma_1\sigma_2\sigma_3 \cdots \sigma_n.$$

Then by Equation (1.4.2) we have

$$L_{\omega_1} \supset L_{\omega_2} \supset \cdots \supset L_{\omega_n} \supset \cdots. \tag{1.4.3}$$

That is, L_{ω_1} contains L_{ω_2} and so on. We say that the sequence of subsegments $\{L_{\omega_n} : n = 1, 2, \ldots\}$ forms a **decreasing sequence** of sets.

As we will explain in Section 1.11 each set in this sequence is **compact** and, since the length of L_{ω_n} shrinks towards zero as n increases towards infinity, this

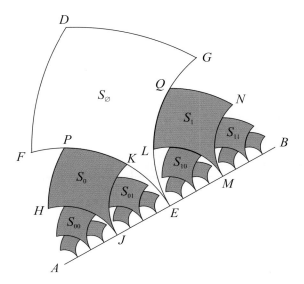

Figure 1.9 Space made of a 'tree' of shield-shaped tiles. The shields are defined by regions produced in the repeated bisection construction illustrated in Figure 1.6. The **limit set** of the tiles – reached after the construction is repeated infinitely many times – is the set of points in the line segment $L[A, B]$. Finite binary strings provide addresses for the tiles, while infinite binary strings address the points in the line segment.

sequence defines a single point $x \in L[A, B]$. It is this procedure that defines the mapping $\varphi : \Omega_{\{0,1\}} \to L[A, B]$.

To show that $\varphi : \Omega_{\{0,1\}} \to L[A, B]$ is indeed an address function, we need to show that it is an onto mapping. But, given any point $x \in L[A, B]$ and any $n = 1, 2, 3, \ldots$, Equation (1.4.1) tells us that we can find at least one string $\omega_n \in \Omega'_{\{0,1\}}$ of length n such that $x \in L_{\omega_n}$, and clearly we can do this in such a way that $L_{\omega_1} \supset L_{\omega_2} \supset \cdots \supset L_{\omega_n} \supset \cdots$. Equation (1.4.2) implies that

$$\omega_{n+1} = \omega_n \sigma_{n+1},$$

where $\sigma_{n+1} \in \{0, 1\}$. As above, this sequence of subsegments defines a unique point, and that point must be x. So $\varphi : \Omega_{\{0,1\}} \to L[A, B]$ is onto, and hence provides an address function for $L[A, B]$. This completes our excursion into how an addressing scheme may depend upon geometrical properties of the space.

Here is another example of a space and an addressing function. In Figure 1.9 we show a branching tree of *shields,* tiles defined by four circular arcs. The circular arcs are produced during the iterative bisection construction described above. DFEG is the single zeroth-generation shield, the 'base' of the tree, to which we assign the address $\varnothing$. We denote this zeroth-generation shield $S_{\varnothing}$. It is formed by arcs from circles used to construct the zeroth and first generations of bisection points. The two first-generation shields, PHJK and QLMN, are denoted S_0 and S_1

respectively and are formed by arcs of the circles used to construct the first and second generations of bisection points. The four second-generation shields, of one quarter the linear dimensions of DFEG, are denoted $S_{00}, S_{01}, S_{10}, S_{11}$. Similarly there are eight third-generation shields, $S_{000}, S_{001}, S_{010}, \ldots, S_{111}$, sixteen fourth-generation shields and so on.

In this way an unique shield S_σ is defined corresponding to each $\sigma \in \Omega'_{\{0,1\}}$. Then

$$S := \left\{ S_\sigma : \sigma \in \Omega'_{\{0,1\}} \right\}$$

is a space. We might call it *shield space*. A convenient addressing function is $\varphi : \Omega'_{\{0,1\}} \to S$, defined by $\varphi(\sigma) = S_\sigma$. In this example the elements of the space are sets and the addressing function maps codes onto sets.

EXERCISE 1.4.3   *Let $\varphi : \Omega_{\{0,1\}} \to L[A, B]$ be the address function defined above. Which points in $L[A, B]$ have more than one address? Show that the point $C \in L[A, B]$, which is one third of the way from A to B, has only one address. What is the address of C?*

EXERCISE 1.4.4   *Show that the two circles used above to bisect the line segment $L[A, B]$ in Figure 1.7 intersect at $120°$.*

EXERCISE 1.4.5   *Show that $\varphi(\sigma\overline{0}) \in S_\sigma \cap L[A, B]$.*

EXERCISE 1.4.6   *Let $\mathbb{X}$ denote the set of all functions $f : S \to \{0, 1\}$. Then $\mathbb{X}$ may be used to model the set of pictures of S in which some shields are coloured red and the others green. Devise an address function and code space for $\mathbb{X}$. Using this address function, give a possible address of the point $x \in \mathbb{X}$ represented by the picture in Figure 1.9 with $S_\varnothing$ coloured green.*

1.5   Metric spaces

In this section we introduce a second property which a space may possess and through which we may consider its points to be organized. It is the property of possessing a metric.

DEFINITION 1.5.1   A **metric space** $(\mathbb{X}, d)$ consists of a space $\mathbb{X}$ together with a **metric** or **distance function** $d : \mathbb{X} \times \mathbb{X} \to \mathbb{R}$ that measures the distance $d(x, y)$ between pairs of points $x, y \in \mathbb{X}$ and has the following properties:
   (i) $d(x, y) = d(y, x)$ for all $x, y \in \mathbb{X}$ (i.e. the distance from x to y is the same as the distance from y to x);
  (ii) $0 < d(x, y) < +\infty$ whenever x and y are distinct points of $\mathbb{X}$ (i.e. distance is always greater than zero when $x \neq y$);
 (iii) $d(x, x) = 0$ for all $x \in \mathbb{X}$;

Figure 1.10 A metric space $(\mathbb{X}, d)$ consists of a space $\mathbb{X}$ together with a function $d : \mathbb{X} \times \mathbb{X} \to \mathbb{R}$ having certain properties that make it behave like 'distance'. Here $\mathbb{X}$ is a leafy-looking subspace of $\mathbb{R}^2$ and distance is measured using a conveniently positioned binary ruler of length unity. Determine the approximate binary distance between the tips of the second and fourth fronds, counting up from the bottom. (The figure depicts two fractal sets with colours given by IFS colouring; see Section 4.6.)

(iv) $d(x, y)$ obeys the **triangle inequality**, namely $d(x, y) \le d(x, z) + d(z, y)$ for all $x, y, z \in \mathbb{X}$.

When it is clear from the context what the metric is, or the particular metric does not matter, we may write $\mathbb{X}$ in place of $(\mathbb{X}, d)$.

Metric spaces of diverse types play a fundamental role in fractal geometry. They include familiar spaces like $\mathbb{R}$ and $\mathbb{R}^2$, code spaces and many other examples; see Figure 1.10. One example of a metric space is $(\mathbb{R}, \ d(x, y) = |x - y|)$, where $|x - y|$ denotes the absolute value or norm of the real number $x - y$. Suppose that $x, y \in [0, 1]$ are both represented in base N, that is,

$$x = 0.x_1 x_2 x_3 \cdots (\text{base } N) := \sum_{n=1}^{\infty} \frac{x_n}{N^n} \quad \text{where } x_n \in \{0, 1, 2, \ldots, N - 1\} \text{ for all } n,$$

with a similar expression for y. Then

$$|x - y| = \left| \sum_{n=1}^{\infty} \frac{x_n - y_n}{N^n} \right|. \qquad (1.5.1)$$

EXERCISE 1.5.2 *Compute the distance between the base-2 numbers $x =$ $0.101\overline{0}$ and $y = 0.10\overline{1}$ using Equation (1.5.1). What does this distance become if you interpret these expansions of x and y as being in base 3? Explain what is going on here.*

Another example of a metric space is $(\mathbb{R}^2, d_{euclidean})$, where

$$d_{euclidean}(x, y) := \sqrt{(x_1 - y_1)^2 + (x_2 - y_2)^2} \quad \text{for all } x, y \in \mathbb{R}^2.$$

Quite generally, $(\mathbb{R}^n, d_p)$ is a metric space for all $n = 1, 2, 3, \ldots$ and $p > 0$, where

$$d_p(x, y) := \sqrt[p]{\sum_{m=1}^{n} |x_m - y_m|^p} \quad \text{for all } x, y \in \mathbb{R}^n.$$

In $\mathbb{R}^n$ we define $|x - y| := d_{euclidean}(x, y) = d_2(x, y)$.

EXERCISE 1.5.3 *Show that the following are both metrics in $\mathbb{R}^2$:*
 (i) $d_{max}(x, y) := \max\{|x_1 - y_1|, |x_2 - y_2|\}$ for all $x, y \in \mathbb{R}^2$,
 (ii) $d_{manhattan}(x, y) := |x_1 - y_1| + |x_2 - y_2|$ for all $x, y \in \mathbb{R}^2$.

EXERCISE 1.5.4 *Check whether you agree that if $(\mathbb{X}_0, d)$ is a metric space and $\mathbb{X} \subset \mathbb{X}_0$ then $(\mathbb{X}, d|_{\mathbb{X} \times \mathbb{X}})$ is a metric space. We say that $(\mathbb{X}, d|_{\mathbb{X} \times \mathbb{X}})$ is a **subspace** of $(\mathbb{X}_0, d)$.*

We now draw attention to the following wonderful method for constructing metrics. We will use it to make 'geometrical' metrics on code spaces in Section 1.6.

THEOREM 1.5.5 *Suppose that $\mathbb{X}$ is a space, that $(\mathbb{Y}, d_{\mathbb{Y}})$ is a metric space and that $\xi : \mathbb{X} \to (\mathbb{Y}, d_{\mathbb{Y}})$ is an embedding function. Then $(\mathbb{X}, d_{\mathbb{X}})$ is a metric space, where*

$$d_{\mathbb{X}}(x, y) := d_{\mathbb{Y}}(\xi(x), \xi(y)) \quad \text{for all } x, y \in \mathbb{X}.$$

PROOF   This is straightforward. (i) $d_{\mathbb{X}}(x, y) = d_{\mathbb{Y}}(\xi(x), \xi(y)) = d_{\mathbb{Y}}(\xi(y), \xi(x)) = d_{\mathbb{X}}(y, x)$ for all $x, y \in \mathbb{X}$. (ii) Suppose that x and y are distinct points of $\mathbb{X}$. Then $\xi(x)$ and $\xi(y)$ are distinct points of $\mathbb{Y}$ because ξ, being an embedding function, is one-to-one. Hence $0 < d_{\mathbb{Y}}(\xi(x), \xi(y)) < \infty$, and so $0 < d_{\mathbb{Y}}(x, y) < \infty$. (iii) $d_{\mathbb{X}}(x, x) = d_{\mathbb{Y}}(\xi(x), \xi(x)) = 0$ for all $x \in \mathbb{X}$. (iv) $d_{\mathbb{X}}(x, y) = d_{\mathbb{Y}}(\xi(x), \xi(y)) \leq d_{\mathbb{Y}}(\xi(x), \xi(z)) + d_{\mathbb{Y}}(\xi(z), \xi(y)) = d_{\mathbb{X}}(x, z) + d_{\mathbb{X}}(z, y)$ for all $x, y, z \in \mathbb{X}$. $\square$

Figure 1.11 An inchworm tries to work out the shortest distance to a delicious morsel that she has spotted.

EXERCISE 1.5.6 *Suppose that $\mathbb{X}$ is a subset of $\mathbb{R}^2$ that 'looks like' a ragged leaf; see Figure 1.11. Argue that the following is a metric:*

$$d_{caterpillar}(x, y) = length\ of\ shortest\ path,\ on\ the\ leaf,\ from\ x\ to\ y.$$

EXERCISE 1.5.7 *Let $(\mathbb{X}, d)$ be a metric space. Define $d' : \mathbb{X} \times \mathbb{X} \to \mathbb{R}$ by*

$$d'(x, y) = \frac{d(x, y)}{(1 + d(x, y))} \quad for\ all\ x, y \in \mathbb{X}.$$

Show that $(\mathbb{X}, d')$ is a metric space and that $d'(x, y) \in [0, 1)$ for all $x, y \in \mathbb{X}$.

EXERCISE 1.5.8 *Let $d'(x, y) = 1$ when $d_{euclidean}(x, y) > 1$ and $d'(x, y) = d_{euclidean}(x, y)$ when $d_{euclidean}(x, y) \leq 1$. Show that $(\mathbb{R}^3, d'(x, y))$ is a metric space.*

We will sometimes write

$$f : (\mathbb{X}, d_{\mathbb{X}}) \to (\mathbb{Y}, d_{\mathbb{Y}})$$

to denote a transformation between two metric spaces $(\mathbb{X}, d_{\mathbb{X}})$ and $(\mathbb{Y}, d_{\mathbb{Y}})$.

DEFINITION 1.5.9   Two metrics d and $\widetilde{d}$ are said to be **equivalent** iff there exists a finite positive constant C such that

$$\frac{1}{C}d(x, y) \le \widetilde{d}(x, y) \le Cd(x, y) \quad \text{for all } x, y \in \mathbb{X}. \tag{1.5.2}$$

A function $f : (\mathbb{X}, d_{\mathbb{X}}) \to (\mathbb{Y}, d_{\mathbb{Y}})$ is called a **metric transformation**, and $(\mathbb{X}, d_{\mathbb{X}})$ and $(\mathbb{Y}, d_{\mathbb{Y}})$ are called **equivalent metric spaces**, iff f is one-to-one and onto and the metric $d_{\mathbb{X}}$ is equivalent to the metric $\widetilde{d}$ given by

$$\widetilde{d}(x, y) = d_{\mathbb{Y}}(f(x), f(y)) \quad \text{for all } x, y \in \mathbb{X}. \tag{1.5.3}$$

In Section 1.14 we will discover that each bounded subset of $\mathbb{R}^n$ has associated with it a number, called its **fractal dimension**, whose value depends upon the underlying metric. This number is unchanged when the metric is altered to another equivalent metric, and hence *fractal dimension is invariant under any metric transformation.*

A metric transformation for which $C = 1$ in Equation (1.5.2) is called an **isometry** or **isometric** transformation. Distance is invariant under an isometric transformation.

Throughout this book we will be mentioning properties of mathematical objects – points in appropriate spaces – that are invariant under transformations of one type or another. This is a recurring theme. Quite generally, geometry studies the properties of sets that are invariant under **groups** of transformations; see Chapter 3. Geometrical properties are properties that are invariant under a group. Here we are getting our first taste of this idea: the set of metric transformations forms a group and so does the set of isometries. Fractal dimension is a geometrical property of metric transformations just as distance is a geometrical property of isometries.

EXERCISE 1.5.10   *Prove that if Equation (1.5.2) is true then it is also true when d and $\widetilde{d}$ are swapped.*

EXERCISE 1.5.11   *Prove that Equation (1.5.3) indeed defines a metric.*

EXERCISE 1.5.12   *Let $(\mathbb{X}, d)$ be a metric space. Show that $(\mathbb{X}, \widetilde{d})$ is an equivalent metric space, where $\widetilde{d}(x, y) = 2d(x, y)$.*

EXERCISE 1.5.13   *Let d and $\widetilde{d}$ be equivalent metrics on $\mathbb{X}$. Let $e : (\mathbb{X}, d) \to (\mathbb{X}, \widetilde{d})$ be defined by $e(x) = x$ for all $x \in \mathbb{X}$. Show that e is a metric transformation.*

EXERCISE 1.5.14   *Let $f : \mathbb{R}^2 \to \mathbb{R}^2$ be defined by $f(x, y) = (2x, 2y + 1)$. Show that $f : (\mathbb{R}^2, d_{euclidean}) \to (\mathbb{R}^2, d_{euclidean})$ is a metric transformation.*

EXERCISE 1.5.15   *Let $S \subset \mathbb{R}^2$ denote a circle with radius 1. Let $d_{shortest}(x, y)$ denote the shortest path in S from x to y. Show that $(S, d_{shortest})$ and $(S, d_{euclidean})$ are not equivalent.*

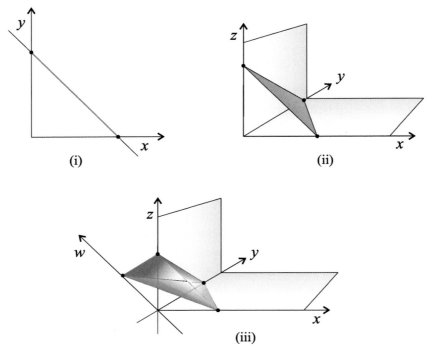

Figure 1.12 In $\mathbb{R}^n$, it is possible to find $n + 1$ points $\{x_1, x_2, \ldots, x_n\}$ such that $d_{euclidean}(x_i, x_j) = 1$ for all $i \neq j$, for $n = 1, 2, 3, \ldots$ In (i) and (ii) we illustrate a way to do this when $n = 1$ and $n = 2$ respectively. Use the hint provided by (iii) to find the coordinates of such a set of points in the case $n = 3$.

EXERCISE 1.5.16 *Let $\mathbb{X}$ be a space. Define $d(x, y) = 1$ for all $x, y \in \mathbb{X}$, with $x \neq y$ and $d(x, x) = 0$. Prove that $(\mathbb{X}, d)$ is a metric space.*

EXERCISE 1.5.17 *Prove that in $\mathbb{R}^n$ there does not exist a set of $n + 2$ points $\{x_1, x_2, \ldots, x_{n+2}\}$ such that $d_{euclidean}(x_i, x_j) = 1$ for all $i \neq j$, where $i, j \in \{1, 2, \ldots, n + 2\}$ for all $n = 1, 2, 3, \ldots$ See also Figure 1.12.*

1.6 Metrics on code space

In this section we show how any code space $\Omega \subset \Omega_A \cup \Omega'_A$ can be embedded in $\mathbb{R}^2$ in diverse ways and consequently can be endowed with numerous different metrics. A simple metric on Ω_A is defined by $d_\Omega(\sigma, \sigma) = 0$ for all $\sigma \in \Omega_A$, and

$$d_\Omega(\sigma, \omega) := \frac{1}{2^m} \quad \text{if } \sigma \neq \omega, \tag{1.6.1}$$

for $\sigma = \sigma_1 \sigma_2 \sigma_3 \cdots$ and $\omega = \omega_1 \omega_2 \omega_3 \cdots \in \Omega_A$, where m is the smallest positive integer such that $\sigma_m \neq \omega_m$.

EXERCISE 1.6.1 *Show that (Ω_A, d_Ω) is indeed a metric space.*

EXERCISE 1.6.2 *Evaluate $d_\Omega(101\overline{0}, 10\overline{1})$ when $\mathcal{A} = \{0, 1\}$ and when $\mathcal{A} = \{0, 1, 2\}$.*

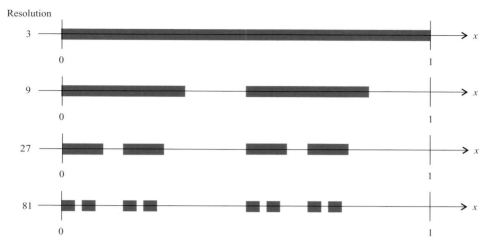

Figure 1.13 This shows the code space $\Omega_{\{0,1\}}$ represented as a subset of the real interval $0 \le x \le 1$. To obtain this figure we represented the points of $[0, 1]$ in base 3 and then plotted all those points whose representation does not include the symbol 2.

We readily extend d_Ω to $\Omega'_\mathcal{A} \cup \Omega_\mathcal{A}$ by adding a symbol, which we will call Z, to the alphabet $\mathcal{A}$ to make a new alphabet $\widetilde{\mathcal{A}} = \mathcal{A} \cup \{Z\}$. Then we embed $\Omega'_\mathcal{A} \cup \Omega_\mathcal{A}$ in $\Omega_{\widetilde{\mathcal{A}}}$ via the mapping $\varepsilon : \Omega'_\mathcal{A} \cup \Omega_\mathcal{A} \to \Omega_{\widetilde{\mathcal{A}}}$ defined by

$$\varepsilon(\sigma) = \sigma\, ZZZZZZ \cdots = \sigma\overline{Z} \quad \text{if } \sigma \in \Omega'_\mathcal{A}$$

$$\varepsilon(\sigma) = \sigma \quad \text{if } \sigma \in \Omega_\mathcal{A}, \tag{1.6.2}$$

and we define

$$d_\Omega(\sigma, \omega) = d_\Omega(\varepsilon(\sigma), \varepsilon(\omega)) \quad \text{for all } \sigma, \omega \in \Omega'_\mathcal{A} \cup \Omega_\mathcal{A}. \tag{1.6.3}$$

It is readily verified that ε is one-to-one and hence that d_Ω does indeed furnish a metric on $\Omega'_\mathcal{A} \cup \Omega_\mathcal{A}$. This metric is a very simple one to work with.

But there is another metric, of a different type and with a more geometrical character, that we can define on $\Omega'_\mathcal{A} \cup \Omega_\mathcal{A}$. It is constructed with the aid of the embedding technique of Theorem 1.5.5. It depends explicitly on the number of elements $|\mathcal{A}|$ in the alphabet $\mathcal{A}$, so we denote it by $d_{|\mathcal{A}|}$.

Assume, without loss of generality, that $\mathcal{A} = \{0, 1, \ldots, N - 1\}$. Then we treat the addresses in $\Omega_\mathcal{A}$ as representing points in the real interval $[0, 1] = \{x : 0 \le x \le 1\}$ in base $N + 1$; and we take the distance between two addresses to be the euclidean distance between their representations. Note that the base number is one more than $|\mathcal{A}|$, the number of elements in the alphabet. See Figure 1.13.

Thus, we **embed** the code space $\Omega_{\{0,1,\ldots,N-1\}}$ in the real interval using the map $\xi : \Omega_{\{0,1,\ldots,N-1\}} \to [0, 1]$ defined by

$$\xi(\sigma) = \sum_{n=1}^{\infty} \frac{\sigma_n}{(N+1)^n}. \tag{1.6.4}$$

This map is one-to-one, as we now demonstrate. Let $\sigma, \omega \in \Omega_{\{0,1,\ldots,N-1\}}$, with $\sigma \neq \omega$, and let $\sigma = \sigma_1\sigma_2\sigma_3 \cdots$ and $\omega = \omega_1\omega_2\omega_3 \cdots$, where $\sigma_n, \omega_n \in \{0, 1, 2, \ldots, N-1\}$ for all n. Then

$$|\xi(\sigma) - \xi(\omega)| = \left| \sum_{n=1}^{\infty} \frac{\sigma_n - \omega_n}{(N+1)^n} \right|$$

$$\geq \left| \frac{\sigma_m - \omega_m}{(N+1)^m} + \sum_{n=m+1}^{\infty} \frac{\sigma_n - \omega_n}{(N+1)^n} \right|,$$

where m is the lowest positive integer such that $\sigma_m \neq \omega_m$. We now use the inequality $|a + b| \geq |a| - |b|$, which is valid for all $a, b \in \mathbb{R}$, to yield

$$|\xi(\sigma) - \xi(\omega)| \geq \frac{|\sigma_m - \omega_m|}{(N+1)^m} - \sum_{n=m+1}^{\infty} \frac{|\sigma_n - \omega_n|}{(N+1)^n}$$

$$\geq \frac{1}{(N+1)^m} - \frac{N-1}{(N+1)^{m+1}} \sum_{n=0}^{\infty} \frac{1}{(N+1)^n}$$

$$= \frac{1}{(N+1)^m} \left(1 - \frac{N-1}{N+1} \frac{N+1}{N} \right) = \frac{1}{N(N+1)^m} > 0. \tag{1.6.5}$$

Correspondingly, we have that $(\Omega_{\mathcal{A}}, d_{|\mathcal{A}|})$ is a metric space, where

$$d_{|\mathcal{A}|}(\sigma, \omega) = \left| \sum_{n=1}^{\infty} \frac{\sigma_n - \omega_n}{(|\mathcal{A}|+1)^n} \right| \quad \text{for all } \sigma, \omega \in \Omega_{\mathcal{A}}. \tag{1.6.6}$$

This expression should be compared with Equation (1.5.1). Notice now how any two distinct addresses are a positive distance apart because ξ is one-to-one. If we were to change $|\mathcal{A}| + 1 = N + 1$ in the denominator in Equation (1.6.6) to N then this would no longer be true.

EXERCISE 1.6.3 *Evaluate the metric* $d_{|\mathcal{A}|}(101\overline{0}, 10\overline{1})$ *when* $\mathcal{A} = \{0, 1\}$ *and when* $\mathcal{A} = \{0, 1, 2\}$.

Finally we extend $d_{|\mathcal{A}|}$ to the space $\Omega'_{\mathcal{A}} \cup \Omega_{\mathcal{A}}$ by defining $\xi : \Omega'_{\mathcal{A}} \to [0, 1]$ such that

$$\xi(\sigma_1\sigma_2\sigma_3 \cdots \sigma_m) = 0.\sigma_1\sigma_2\sigma_3 \cdots \sigma_m\overline{N},$$

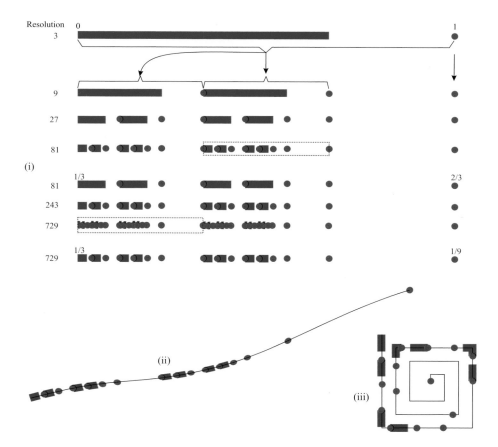

Figure 1.14   Different embeddings $\xi : \Omega \to \mathbb{R}^2$ of code space $\Omega = \Omega_{\{0,1\}} \cup \Omega'_{\{0,1\}}$ in the euclidean plane lead to different metrics on Ω. Panel (i) shows Ω embedded in $[0, 1] \subset \mathbb{R}$ at various resolutions, indicated by the numbers at the left-hand side. Zooms are shown of parts of the lines, as indicated, at resolutions 81 and 729. The red dots indicate points of $\xi(\Omega'_{\{0,1\}})$, while the green intervals represent approximations to sets of points in $\xi(\Omega_{\{0,1\}})$. The two lower panels show cartoons that represent Ω embedded in (ii) a curve and (iii) a squared spiral. One could embed Ω in a double helix in $\mathbb{R}^3$ to produce an interesting metric.

that is,

$$\xi(\sigma) = \sum_{n=1}^{m} \frac{\sigma_n}{(N+1)^n} + \frac{1}{(N+1)^m}$$

for all $\sigma = \sigma_1\sigma_2\sigma_3 \cdots \sigma_m \in \Omega'_{\mathcal{A}}$. See Figure 1.14. It is readily verified that $\xi : \Omega'_{\mathcal{A}} \cup \Omega_{\mathcal{A}} \to [0, 1]$ is one-to-one and consequently that $(\Omega'_{\mathcal{A}} \cup \Omega_{\mathcal{A}}, d_{|\mathcal{A}|})$ is a metric space, where

$$d_{|\mathcal{A}|}(\sigma, \omega) = |\xi(\sigma) - \xi(\omega)| = d_{euclidean}(\xi(\sigma), \xi(\omega)) \quad \text{for all } \sigma, \omega \in \Omega'_{\mathcal{A}} \cup \Omega_{\mathcal{A}}.$$

We now summarize what we have proved:

THEOREM 1.6.4   *Both* $(\Omega_{\mathcal{A}} \cup \Omega'_{\mathcal{A}}, d_\Omega)$ *and* $(\Omega_{\mathcal{A}} \cup \Omega'_{\mathcal{A}}, d_{|\mathcal{A}|})$ *are metric spaces.*

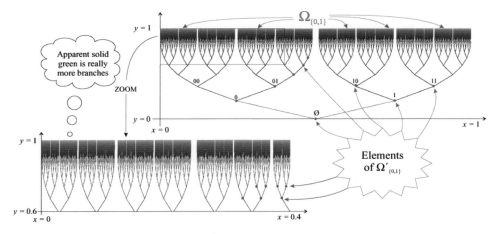

Figure 1.15   The code space $\Omega_{\{0,1\}} \cup \Omega'_{\{0,1\}}$ has here been embedded in a tree-like structure in $\mathbb{R}^2$.

EXERCISE 1.6.5   *Compute $d_\Omega(\sigma, \omega)$ and $d_{|\mathcal{A}|}(\sigma, \omega)$, where $\sigma = 010$ and*

$$\omega = 0010101010 \cdots = 00\overline{10}.$$

EXERCISE 1.6.6   *Show that $(\Omega_\mathcal{A}, d')$ is a metric space, where*

$$d'(\sigma, \omega) := \sum_{n=1}^{\infty} \frac{|\sigma_n - \omega_n|}{(|\mathcal{A}| + 1)^n} \quad \text{for all } \sigma, \omega \in \Omega_\mathcal{A}.$$

Show that $d_{|\mathcal{A}|}$ and d' are equivalent metrics.

EXERCISE 1.6.7   *Show that $(\Omega_\mathcal{A}, d_{|\mathcal{A}|})$ and $(\Omega_\mathcal{A}, d_\Omega)$ are not equivalent. Explain 'geometrically' why this is so. Hint: Think how you might try to embed $\Omega_\mathcal{A}$ in say $\mathbb{R}^2$ in such a way that the euclidean metric induces the metric d_Ω on $\Omega_\mathcal{A}$.*

EXERCISE 1.6.8   *Here we describe an embedding of $\Omega_{\{0,1\}} \cup \Omega'_{\{0,1\}}$ in $\mathbb{R}^2$ that looks like all the nodes, $\Omega'_{\{0,1\}}$, of a 'tree' together with the tips of all of the 'twigs', $\Omega_{\{0,1\}}$, of the tree, as illustrated in Figure 1.15. This figure shows the relationship between $\Omega'_{\{0,1\}}$ and $\Omega_{\{0,1\}}$. We define $\xi : \Omega_{\{0,1\}} \cup \Omega'_\mathcal{A} \to \mathbb{R}^2$ simply by*

$$\xi(\sigma_1 \sigma_2 \cdots \sigma_m) = \left(\frac{1}{2} \prod_{k=1}^{m} \left(\frac{\sigma_k}{2} + 0.499 \right), 1 - \left(\frac{5}{8} \right)^m \right),$$

$$\xi(\varnothing) = \left(\frac{1}{2}, 0 \right) \quad \text{and} \quad \xi(\sigma_1 \sigma_2 \sigma_3 \cdots) = \left(\frac{1}{2} \prod_{k=1}^{\infty} \left(\frac{\sigma_k}{2} + 0.499 \right), 1 \right).$$

Verify that ξ is one-to-one and write down the corresponding metric on $\Omega_{\{0,1\}} \cup \Omega'_{\{0,1\}}$. Is this new metric equivalent to d_2?

Later on we will introduce other metric spaces. Given a mathematical set-up it is often worth looking for associated metric spaces, since then one has not only a more geometrical way of looking at the set-up but also the possibility of using contraction mapping methods, as will be discussed in Chapter 2. Contraction mapping methods are used in a number of different settings in this book to prove the existence of various types of fractal.

1.7 Cauchy sequences, limits and continuity

In this section we define Cauchy sequences, limits, completeness and continuity. These important concepts are related in particular to the construction and existence of various types of fractal object. We also note some ways in which these concepts relate to code spaces.

DEFINITION 1.7.1 Let $(\mathbb{X}, d)$ be a metric space. Then a sequence of points $\{x_n\}_{n=1}^{\infty} \subset \mathbb{X}$ is said to be a **Cauchy sequence** iff given any $\epsilon > 0$ there is a positive integer $N > 0$ such that

$$d(x_n, x_m) < \epsilon \quad \text{whenever } n, m > N.$$

The sequence $\{x_n\}_{n=1}^{\infty} \subset \mathbb{X}$ is said to **converge** (in the metric d) to a point $x \in \mathbb{X}$ iff given any $\epsilon > 0$ there is a positive integer $N > 0$ such that

$$d(x_n, x) < \epsilon \quad \text{whenever } n > N.$$

In this case x is called the **limit** of $\{x_n\}_{n=1}^{\infty}$, and we write

$$\lim_{n \to \infty} x_n = x.$$

EXERCISE 1.7.2 *Show that the sequence of points* $\{x_n = 1/n : n = 1, 2, \dots\} \subset \mathbb{R}$ *converges to the point* $x = 0$ *in the euclidean metric.*

EXERCISE 1.7.3 *Show that the sequence of points*

$$\left\{ \sigma_n = \underbrace{ABAB \cdots AB}_{n \text{ times}} A\overline{B} \right\}_{n=1}^{\infty} \subset \Omega_{\{A,B\}}$$

is a Cauchy sequence in each of the metric spaces d_Ω *and* $d_{|\mathcal{A}|}$.

It is easy to see by using the triangle inequality, $d(x_n, x_m) \leq d(x_n, x) + d(x_m, x)$, that if $\{x_n\}_{n=1}^{\infty}$ converges to x then $\{x_n\}_{n=1}^{\infty}$ is a Cauchy sequence. But the converse is not true. For example, $\{x_n = 1/n : n = 1, 2, \dots\}$ is a Cauchy sequence in the metric space $((0, 1), d_{euclidean})$ but it has no limit in the space. So we make the following definition:

DEFINITION 1.7.4 A metric space $(\mathbb{X}, d)$ is said to be **complete** iff whenever $\{x_n\}_{n=1}^{\infty} \subset \mathbb{X}$ is a Cauchy sequence it converges to a point $x \in \mathbb{X}$. We say that a subset $S \subset \mathbb{X}$ is complete if the space (S, d) is complete.

Figure 1.16 A subspace of $\mathbb{R}^2$ is represented in green. It consists of a centrally symmetrical pattern of 'snowflakes' that converge, along radial paths, towards the central point. But imagine that the point at the centre is missing, that it is not part of the subspace. Then the subspace is incomplete. Each snowflake is made of smaller snowflakes. Imagine that the centres of *all* the snowflakes are missing, regardless of the sizes, however small. Then, does *any* Cauchy sequence of points in the subspace have a limit point in the subspace?

The spaces $(\mathbb{R}^n, d_{euclidean})$ for $n = 1, 2, 3, \ldots$ are complete. So are $([0, 1], d_{euclidean})$ and $(\Box, d_{euclidean})$. But the spaces $((0, 1), d_{euclidean})$ and $(B := \{(x, y) \in \mathbb{R}^2 : x^2 + y^2 < 1\}, d_{euclidean})$ are not complete. Figure 1.16 illustrates an incomplete metric space.

A useful example of a complete metric space is $(C[a, b], d_{\max})$, where $C[a, b]$ denotes the set of all continuous functions $f : [a, b] \to \mathbb{R}, -\infty < a < b < +\infty$, and

$$d_{\max}(f, g) = \max\{|f(x) - g(x)| : x \in [a, b]\}.$$

This maximum is a finite real number, as you will remember from elementary calculus. The fact that $(C[a, b], d_{\max})$ is complete provides a simple demonstration of the existence of certain fractal interpolation functions.

THEOREM 1.7.5 *Let d be either d_Ω or $d_{|\mathcal{A}|}$. Then the metric spaces $(\Omega_{\mathcal{A}} \cup \Omega'_{\mathcal{A}}, d)$ and $(\Omega_{\mathcal{A}}, d)$ are complete. But the metric space $(\Omega'_{\mathcal{A}}, d)$ is not complete.*

The closure, defined in Section 1.8, of $\Omega'_{\mathcal{A}}$ in $(\Omega_{\mathcal{A}} \cup \Omega'_{\mathcal{A}}, d)$ is $\Omega_{\mathcal{A}} \cup \Omega'_{\mathcal{A}}$, that is,

$$\overline{\Omega'_{\mathcal{A}}} = \Omega_{\mathcal{A}} \cup \Omega'_{\mathcal{A}}.$$

PROOF   We prove that $(\Omega_{\mathcal{A}}, d)$ is complete where d is either d_Ω or $d_{|\mathcal{A}|}$. Let N be given. Then, in both cases, we can choose $\delta > 0$ so small that $\sigma, \omega \in \Omega_{\mathcal{A}}$ must agree through the first N terms of their expansions whenever $d(\sigma, \omega) < \delta$.

Now let $\{\sigma_n\}_{n=1}^{\infty} \subset \Omega_{\mathcal{A}}$. Then we can find an integer $M(N)$ such that $d(\sigma_n, \sigma_m) < \delta$ whenever $n, m \geq M(N)$, and consequently

$$\sigma_{n,k} = \sigma_{m,k} \quad \text{for } k = 1, 2, \ldots, N \quad \text{whenever } n, m \geq M(N),$$

where we write

$$\sigma_n = \sigma_{n,1}\sigma_{n,2}\sigma_{n,3} \cdots.$$

Now let

$$\sigma = \sigma_{M(1),1}\sigma_{M(2),2}\sigma_{M(3),3} \cdots.$$

Then σ_n agrees with σ through the first N terms whenever $n \geq M(N)$.

Now let $\epsilon > 0$ be given. Then we can choose N such that $d(\sigma, \omega) < \epsilon$ whenever σ and ω agree through the first N terms. It follows that $d(\sigma, \sigma_n) < \epsilon$ whenever $n \geq M(N)$, from which it follows that $\lim_{n\to\infty} \sigma_n = \sigma$.

To establish that $(\Omega_{\mathcal{A}} \cup \Omega'_{\mathcal{A}}, d)$ is complete we simply note that the above argument applies equally well in the more general setting if we adopt the following conventions. (i) We say that the expansions of $\sigma \in \Omega'_{\mathcal{A}}$ and $\omega \in \Omega'_{\mathcal{A}}$ agree through K terms iff either (a) K is less than or equal to both $|\sigma|$ and $|\omega|$ and $\sigma_n = \omega_n$ for $n = 1, 2, \ldots, K$ or (b) $\sigma = \omega$. (ii) We say that the expansions of $\sigma \in \Omega'_{\mathcal{A}}$ and $\omega \in \Omega_{\mathcal{A}}$ agree through K terms iff $\sigma_n = \omega_n$ for $n = 1, 2, \ldots, \min\{|\sigma|, K\}$.

Finally, $\Omega'_{\mathcal{A}}$ is not complete since the alphabet $\mathcal{A}$ contains at least one symbol A, and $\Omega'_{\mathcal{A}}$ does not contain the limit of the Cauchy sequence

$$\left\{ \sigma_n = \underbrace{AAA \cdots A}_{n \text{ times}} \right\}_{n=1}^{\infty} \subset \Omega'_{\mathcal{A}}. \qquad\qquad \square$$

We omit the proof of the last assertion in the theorem.

DEFINITION 1.7.6   Let $(\mathbb{X}, d_{\mathbb{X}})$ and $(\mathbb{Y}, d_{\mathbb{Y}})$ be metric spaces. Then a function

$$f : (\mathbb{X}, d_{\mathbb{X}}) \to (\mathbb{Y}, d_{\mathbb{Y}})$$

is said to be **continuous at a point** x iff, given any $\epsilon > 0$, there is a $\delta > 0$ (which may vary depending on x and ϵ) such that

$$d_{\mathbb{Y}}(f(x), f(y)) < \epsilon \quad \text{whenever } d_{\mathbb{X}}(x, y) < \delta \quad \text{with } x, y \in \mathbb{X};$$

$f : \mathbb{X} \to \mathbb{Y}$ is said to be **continuous** iff it is continuous at every point $x \in \mathbb{X}$, and it is said to be **uniformly continuous** iff moreover it is possible to choose δ so that it does not depend on x.

A transformation from a metric space into itself can be thought of as picking up a duplicate copy of the space, deforming it, breaking it up, perhaps, and putting it back into the original space, so that each point of the space may be moved to a new point. A continuous transformation is one that does not tear or rip the space, in the sense that nearby points are carried to nearby points. But it can stretch and squeeze it hugely. For example the transformation $f : (0, \infty) \to (0, \infty)$ defined by $f(x) = 1/x$ is continuous.

It is easy to see that if $f : \mathbb{X} \to \mathbb{Y}$ is continuous then it is continuous on any subset $S \subset \mathbb{X}$, that is, $f|_S : S \subset \mathbb{X} \to \mathbb{Y}$ is continuous.

THEOREM 1.7.7 *The embedding mapping* $\xi : (\Omega_\mathcal{A} \cup \Omega'_\mathcal{A}, d) \to ([0, 1], d_{euclidean})$ *is continuous where d is either d_Ω or $d_{|\mathcal{A}|}$.*

PROOF First consider the case of the mapping $\xi : (\Omega_\mathcal{A}, d_\Omega) \to [0, 1]$. We have

$$d_{euclidean}(\xi(\sigma), \xi(\omega)) = |\xi(\sigma) - \xi(\omega)|$$
$$= \left| \sum_{n=1}^{\infty} \frac{\sigma_n - \omega_n}{(N + 1)^n} \right| \leq \sum_{n=m+1}^{\infty} \frac{N}{(N + 1)^n} = \frac{1}{(N + 1)^m},$$

where m is the number of initial successive agreements between σ and ω. The right-hand side here is smaller than $\epsilon > 0$ for all $m > M$, when M is chosen to be sufficiently large. But by choosing $d_\Omega(\sigma, \omega)$ smaller than $\delta = 1/2^M$ we ensure that m is larger than M.

Similarly, consider the mapping $\xi : (\Omega'_\mathcal{A}, d_\Omega) \to [0, 1]$ and let $\sigma, \omega \in \Omega'_\mathcal{A}$. Without loss of generality we assume that $|\sigma| \leq |\omega|$. Then, much as above, we find that

$$d_{euclidean}(\xi(\sigma), \xi(\omega)) = |\xi(\sigma) - \xi(\omega)| \leq \sum_{n=m+1}^{\max\{|\sigma|, |\omega|\}} \frac{N}{(N + 1)^n} < \frac{1}{(N + 1)^m},$$

where m is the number of initial successive agreements between σ and ω, and the value of $\delta > 0$ is the same for a given $\epsilon > 0$.

Finally we consider the case $\xi : (\Omega_\mathcal{A} \cup \Omega'_\mathcal{A}, d_{|\mathcal{A}|}) \to [0, 1]$. But now

$$d_{euclidean}(\xi(\sigma), \xi(\omega)) = d_{|\mathcal{A}|}(\sigma, \omega),$$

so whenever the right-hand side is smaller than $\delta = \epsilon$ the left-hand side is too!

$\square$

EXERCISE 1.7.8 *Imagine a transformation $f : \square \subset \mathbb{R}^2 \to \mathbb{R}^2$ of the following type: all the points in $\square$ are transferred to a magical (infinitely thin) sheet of paper,*

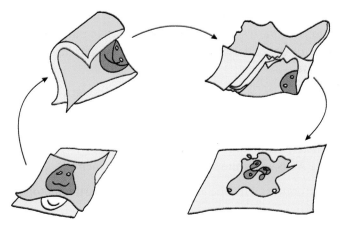

Figure 1.17   Picture of a 'clingfilm' transformation.

which is then picked up, folded, crumpled and squashed perfectly flat back onto $\mathbb{R}^2$, *thus defining the transformed locations of the original points. Argue that* f *is continuous but that if the paper was ripped in the process then* f *would not be continuous. What happens if, in addition to being crumpled, the paper is stretched, this way and that, with no ripping? See Figure 1.17.*

EXERCISE 1.7.9   *Let* $\mathbb{X}$ *be the space of continuous functions* $f : [0, 1] \to \mathbb{R}$ *and let* $d(f, g) := \max\{|f(x) - g(x)| : x \in [0, 1]\}$ *for all* $f, g \in \mathbb{X}$. *Show that* $(\mathbb{X}, d)$ *is a metric space.*

EXERCISE 1.7.10   *Let* $\Omega = \Omega_A \cup \Omega'_A$, *and let* $\alpha \in \mathcal{A}$. *Define* $w_\alpha : \Omega \to \Omega$ *by* $w_\alpha(\sigma) = \alpha\sigma$ *for all* $\sigma \in \Omega$. *Show that* w_α *is continuous with respect to the metric* d_Ω.

1.8   Topological spaces

In this section we introduce a third type of property that a space may possess, namely, a topology.

A topology provides a wonderful method and language for organizing the points of a space by studying and describing properties of the space that are somehow geometrical but to which most of the 'standard' geometrical concepts do not apply. The space is considered to be more like a jelly, even protoplasmic, rather than rigid. There is no sense of length, angle, fractal dimension, area and so on. What is under consideration is the concept of how points are related to other points by virtue of the kinds of subset of the space to which they belong. In particular, topology is the study of properties of mathematical objects that are preserved by a general class of transformations called homeomorphisms. Later we will become more and more geometrical, studying properties of sets that are preserved by more restrictive classes of transformations – many of which are homeomorphisms – such as euclidean transformations, affine transformations or projective transformations.

Topological concepts are absolutely essential as part of our language for describing fractals.

DEFINITION 1.8.1   A **topological space** $(\mathbb{X}, \mathbb{T})$ is a space $\mathbb{X}$ together with a topology $\mathbb{T} = \mathbb{T}(\mathbb{X})$. A **topology** $\mathbb{T}$ for $\mathbb{X}$ is a set of subsets of $\mathbb{X}$ such that
  (i) $\varnothing, \mathbb{X} \in \mathbb{T}$,
 (ii) if $\{O_i : i \in \mathcal{I}\} \subset \mathbb{T}$ is any collection of members of $\mathbb{T}$ then $\bigcup_{i \in \mathcal{I}} O_i \in \mathbb{T}$,
(iii) if $\{O_n : n = 1, 2, \ldots, N\}$ is any finite set of members of $\mathbb{T}$ then $\bigcap_{n=1}^{N} O_n \in \mathbb{T}$.

When it is clear from the context what the topology is, or the particular topology does not matter, we sometimes write $\mathbb{X}$ in place of $(\mathbb{X}, \mathbb{T})$.

The sets of $\mathbb{T}$ are called **open** sets.

Any set $C \subset \mathbb{X}$ that can be written, for some $O \in \mathbb{T}$, in the form

$$C = \mathbb{X} \backslash O := \{x \in \mathbb{X} : x \notin O\}$$

(which reads: C equals the set of elements of $\mathbb{X}$ that are not in O) is called a **closed** set.

Let $O \in \mathbb{T}$ and let $x \in O$; then O is called a **neighbourhood** of x. The **closure** of a set $S \subset \mathbb{X}$ is defined to be the 'smallest' closed set that contains S and is denoted by $\overline{S}$, not to be confused with $SSSSS \cdots$. That is,

$$\overline{S} := \bigcap_{\{C \supset S : \, C \text{ is closed}\}} C.$$

A point $x \in \mathbb{X}$ is said to be an **accumulation point** of a set $S \subset \mathbb{X}$ if every neighbourhood of x contains infinitely many points of S. Notice that an accumulation point of S may not belong to S.

EXERCISE 1.8.2   *Show that $\overline{S} = S \cup \{$accumulation points of $S\}$.*

A metric space $(\mathbb{X}, d)$ has associated with it a **natural topology** $\mathbb{T}_d(\mathbb{X})$ in which a set $O \subset \mathbb{X}$ is called open iff, for every $x \in O$, there is a real number $r > 0$ such that

$$B(x, r) := \{y \in \mathbb{X} : d(y, x) < r\} \subset O.$$

$B(x, r)$ is called the (open) ball of radius r centred at x. Then one readily proves that, for $S \subset \mathbb{X}$,

$$\overline{S} := \{x \in \mathbb{X} : B(x, r) \cap S \neq \varnothing \quad \text{for all } r > 0\}.$$

In general, when we are dealing with a metric space and we refer to topological concepts, it will be the natural topology to which we refer. When we wish to specify the underlying metric we may write $\mathbb{T} = \mathbb{T}_d(\mathbb{X})$. So for example the metric space $(\mathbb{X}, d)$ is associated with the topological space $(\mathbb{X}, \mathbb{T}_d(\mathbb{X}))$.

The natural topology on any subset of $\mathbb{R}^n$ will always be taken to be the topology associated with the euclidean metric.

A topological space $\mathbb{X}$ is called a **Hausdorff** space when for each pair of distinct points x, $y \in \mathbb{X}$ there is always a neighbourhood of x and one of y that have no point in common.

EXERCISE 1.8.3 *Show that if* $(\mathbb{X}, d)$ *is a metric space then* $(\mathbb{X}, \mathbb{T}_d(\mathbb{X}))$ *is a Hausdorff space.*

EXERCISE 1.8.4 *Let* $(\mathbb{X}, d)$ *be a metric space and let* $\{x_n\}_{n=1}^{\infty} \subset \mathbb{X}$ *converge to* x, *with* $x_n \neq x_m$ *for all* $m, n = 1, 2, \ldots$ *with* $m \neq n$. *Show that* x *is an accumulation point of* $\{x_n\}_{n=1}^{\infty} \subset \mathbb{X}$.

Topological language allows us to generalize concepts from metric spaces to more general settings. For example:

DEFINITION 1.8.5 Let $(\mathbb{X}, \mathbb{T})$ and $(\mathbb{Y}, \mathbb{T}')$ be (topological) spaces. Then a function

$$f : (\mathbb{X}, \mathbb{T}) \to (\mathbb{Y}, \mathbb{T}')$$

is said to be **continuous** if $f^{-1}(O) \in \mathbb{T}$ whenever $O \in \mathbb{T}'$ (i.e. if the inverse image of every open set is an open set).

One readily proves that if $f : (\mathbb{X}, d) \to (\mathbb{Y}, d')$ is continuous according to Definition 1.7.6 then it is continuous according to Definition 1.8.5 (i.e. $f : (\mathbb{X}, \mathbb{T}_d(\mathbb{X}))$ $\to (\mathbb{Y}, \mathbb{T}_{d'}(\mathbb{Y}))$ is continuous.) One reason for using topological language, even in the case of metric spaces, is that it is more efficient for describing properties because it allows us to avoid 'epsilon and delta' language.

DEFINITION 1.8.6 A mapping $f : \mathbb{X} \to \mathbb{Y}$ is called **open** iff it carries open sets to open sets.

An example of an open mapping is any metric transformation. But the continuous function $f : (\mathbb{R}, d_{euclidean}) \to (\mathbb{R}, d_{euclidean})$ defined by $f(x) = 4x(1 - x)$ is not open because $f((0, 1)) = (0, 1]$. In Chapter 4 we will encounter very interesting transformations on code spaces that are continuous but not open. They have applications to painting fractals in very beautiful ways.

DEFINITION 1.8.7 A mapping $f : \mathbb{X} \to \mathbb{Y}$ is called a **homeomorphism** iff it is one-to-one, onto, continuous and open.

Let $f : \mathbb{X} \to \mathbb{Y}$ be a homeomorphism between two spaces $\mathbb{X}$ and $\mathbb{Y}$. Then a set $O \subset \mathbb{X}$ is open iff $f(O)$ is open, i.e. $O \in \mathbb{T}(\mathbb{X}) \iff f(O) \in \mathbb{T}(\mathbb{Y})$. That is, f is a homeomorphism iff it induces a one-to-one invertible transformation between $\mathbb{T}(\mathbb{X})$ and $\mathbb{T}(\mathbb{Y})$.

Figure 1.18   The right-hand picture shows the result of applying a certain homeomorphism $f : \square \subset \mathbb{R}^2 \to \square$ to the picture on the left. Transformations of pictures are defined in Chapter 2. The transformation here is not a metric transformation and allows extreme stretching on very small scales. It is actually an example of a fractal transformation, as discussed in Chapter 4.

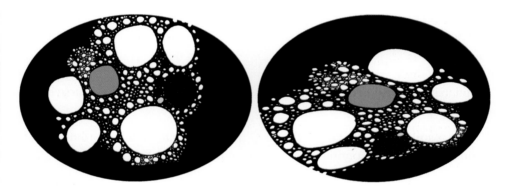

Figure 1.19   This represents an elliptical-shaped subspace $\mathbb{X} \subset \mathbb{R}^2$ before and after a homeomorphism is applied. Some open sets belonging to the natural topology are represented by the white and coloured regions. Of course there are vastly many more open sets, of endless diversity. The homeomorphism is of the form $A \circ M \circ A^{-1}$, where A is projective and M is a Möbius transformation (see Section 2.6).

We say that a property of a set $S \subset \mathbb{X}$ is **invariant** under or is **preserved** by a transformation $f : \mathbb{X} \to \mathbb{Y}$ iff {the property is true of S iff it is true for $f(S)$}. For example, we have already mentioned that fractal dimension is invariant under any metric transformation. Homeomorphisms preserve topological properties, i.e. properties that can be defined in terms of being open and closed. Figures 1.18 and 1.19 show the result of applying certain homeomorphisms $f : \square \subset \mathbb{R}^2 \to \square$, where the points in a space have been assigned colours.

THEOREM 1.8.8   *The metric spaces $(\Omega_{\mathcal{A}}, d_{\Omega})$ and $(\Omega_{\mathcal{A}}, d_{|\mathcal{A}|})$ have the same natural topology; that is, $\mathbb{T}_{d_{\Omega}}(\Omega_{\mathcal{A}}) = \mathbb{T}_{d_{|\mathcal{A}|}}(\Omega_{\mathcal{A}})$.*

PROOF   Let $e : (\Omega_{\mathcal{A}}, d_{\Omega}) \to (\Omega_{\mathcal{A}}, d_{|\mathcal{A}|})$ denote the identity map, so that $e(x) = x$ for all $x \in \Omega_{\mathcal{A}}$. Then we show that e is a homeomorphism. The continuity of e follows at once from the statement, see Equation (1.6.1),

$$d_\Omega(x, y) < \frac{1}{2^m} \quad \Rightarrow \quad d_{|\mathcal{A}|}(x, y) < \frac{1}{(N+1)^m},$$

where m is a positive integer and $x, y \in \Omega_\mathcal{A}$.

To prove that e^{-1} is continuous we use the 'reverse' inference,

$$d_{|\mathcal{A}|}(x, y) < \frac{1}{N(N+1)^m} \quad \Rightarrow \quad d_\Omega(x, y) < \frac{1}{2^m},$$

which follows from Equation (1.6.5).   □

Similar arguments show that $(\Omega'_\mathcal{A}, d_\Omega)$ and $(\Omega'_\mathcal{A}, d_{|\mathcal{A}|})$ have the same natural topology, as do $(\Omega_\mathcal{A} \cup \Omega'_\mathcal{A}, d_\Omega)$ and $(\Omega_\mathcal{A} \cup \Omega'_\mathcal{A}, d_{|\mathcal{A}|})$. We will refer to this topology as *the* natural topology $\mathbb{T}_\Omega$ on any subset of $\Omega = \Omega_\mathcal{A} \cup \Omega'_\mathcal{A}$. Now that you know this, you should not fuss much about which metric we use, d_Ω or $d_{|\mathcal{A}|}$.

A map may be one-to-one, onto and continuous, but not open. For example, the map $f : ([0, 1), d_{euclidean}) \rightarrow (S^1, d_{shortest})$, where S^1 is the circle of radius 1 centred at the origin in $\mathbb{R}^2$, defined by $f(x) = (\cos 2\pi x, \sin 2\pi x)$ for all $x \in [0, 1)$ is one-to-one, onto and continuous but not open. To see this, note that $[0, 0.5)$ is an open subset of $[0, 1)$ in the relative topology (see below) in say $\mathbb{R}^1$ or $\mathbb{R}^2$ but $f([0, 0.5))$ is not an open subset of the circle.

Notice that if two metric spaces are equivalent then they are homeomorphic, but the converse is not true. For example $f : ((0, 1], d_{euclidean}) \rightarrow ([1, \infty), d_{euclidean})$ defined by $f(x) = 1/x$ is a homeomorphism but not a metric transformation.

1.9   Important basic topologies

We have already introduced the natural topology associated with a metric space. But there are five other key topologies that are easy to build and that we will need for our discussion of fractals.

Discrete topology

Let $\mathbb{X}$ be a space. Then the **discrete topology** $\mathbb{T}_{discrete}$ on $\mathbb{X}$ is obtained by defining all the subsets of $\mathbb{X}$ to be open. It follows that all subsets of $\mathbb{X}$ are also closed.

The discrete topology is the natural topology associated with $(\mathbb{X}, d)$ for which $d(x, y) = 1$ whenever $x, y \in \mathbb{X}$, with $x \neq y$, and of course $d(x, x) = 0$. This is the ultimate Hausdorff space! Every point $x \in \mathbb{X}$ lives in its own private open-and-closed set $\{x\}$, nicely separated from every other point. Such a space may seem very artificial, but we will soon use the discrete topology on an alphabet $\mathcal{A}$ to build a natural topology on $\Omega_\mathcal{A}$.

Relative topology

Let $(\mathbb{X}, \mathbb{T})$ be a topological space and let $S \subset \mathbb{X}$. Then we can convert S into a topological space, which we may denote by $(S, \mathbb{T}|_S)$ and which is a **subspace** of the original space, by defining any set $O \subset S$ to be open in the **relative topology** $\mathbb{T}|_S$ iff it is the same as the intersection of some open set $\widetilde{O} \in \mathbb{T}$ with S, i.e. $O \in \mathbb{T}|_S \iff O = \widetilde{O} \cap S$ for some $\widetilde{O} \in \mathbb{T}$. The relative topology never ceases to surprise me, because $S \subset \mathbb{X}$ may be neither open nor closed in $(\mathbb{X}, \mathbb{T})$ yet it is both open and closed in $(S, \mathbb{T}|_S)$.

EXERCISE 1.9.1 *Let $(\mathbb{X}, d)$ be a metric space, let $\mathbb{T}_d(\mathbb{X})$ be the associated natural topology and let $S \subset \mathbb{X}$. Show that the natural topology on the metric space $(S, d|_S)$ is the same as the relative topology $\mathbb{T}|_S$, i.e. that $\mathbb{T}_{d|_S}(S) = \mathbb{T}_d(\mathbb{X})|_S$.*

EXERCISE 1.9.2 *Verify that the discrete topology on a space $\mathbb{X}$ is same as the natural topology associated with the metric space $(\mathbb{X}, d)$ where $d(x, y) = 1$ whenever $x, y \in \mathbb{X}$, with $x \neq y$, and $d(x, x) = 0$.*

Topology generated by a basis

Let $\{O_i : i \in \mathcal{I}\}$ be a collection of subsets of a space $\mathbb{X}$. Then the smallest topology $\mathbb{T}$ on $\mathbb{X}$ such that $O_i \in \mathbb{T}$ for all $i \in \mathcal{I}$ is called the topology **generated by the basis** $\{O_i : i \in \mathcal{I}\}$. It can be proved (by you) that

$$\mathbb{T} = \left\{ O \subset \mathbb{X} : O = \bigcup_{i \in \mathcal{J}} O_i, \text{ for some } \mathcal{J} \subset \mathcal{I} \right\};$$

that is, the open sets of $\mathbb{T}$ are exactly those that can be written as unions of members of the basis. Of course the sets in the basis, the individual O_i, are open in $\mathbb{T}$.

EXERCISE 1.9.3 *Let $\mathbb{T}$ denote the topology for the space $\mathbb{X}$ generated by a basis $\{O_i \subset \mathbb{X} : i \in \mathcal{I}\}$. Show that $\mathbb{T} = \cap \left\{ \widetilde{\mathbb{T}} : \widetilde{\mathbb{T}} \text{ is a topology for } \mathbb{X}, O_i \in \widetilde{\mathbb{T}} \text{ for all } i \in \mathcal{I} \right\}$.*

It turns out to be very useful to have a **countable** basis for a topological space. In Chapter 5 we will describe certain **ergodic** properties associated with fractals, which are established by first showing that they hold for each member of a countable basis for $\mathbb{R}^n$. So we note the following:

THEOREM 1.9.4 *A countable basis for $\mathbb{R}^n$ is provided by the set of all open balls with rational radii and centres at rational points.*

PROOF   See [70], p. 192, Exercise 3.   □

In the following exercise we introduce a useful collection of decompositions of a closed rectangle in $\mathbb{R}^2$. These will be used in Chapter 2 to describe 'pixel functions'.

EXERCISE 1.9.5   *Show that a countable basis for* $(\square^{interior}, d_{euclidean})$ *is provided by the interiors of the set of squares*

$$\left\{\square_{w,h}^{W,H} : w \in \{1, 2, \ldots, W\}, h \in \{1, 2, \ldots, H\}, W \in \mathbb{N}, H \in \mathbb{N}\right\}$$

where

$$\square^{interior} = \{(x, y) \in \mathbb{R}^2 : 0 < x < 1, 0 < y < 1\}$$

and

$$\square_{w,h}^{W,H} = \left\{(x, y) \in \mathbb{R}^2 : \frac{w-1}{W} \leq x < \frac{w}{W}, \frac{h-1}{H} \leq y < \frac{h}{H}\right\},$$

$$\square_{W,h}^{W,H} = \left\{(x, y) \in \mathbb{R}^2 : \frac{W-1}{W} \leq x \leq 1, \frac{h-1}{H} \leq y < \frac{h}{H}\right\},$$

$$\square_{w,H}^{W,H} = \left\{(x, y) \in \mathbb{R}^2 : \frac{w-1}{W} \leq x < \frac{w}{W}, \frac{H-1}{H} \leq y \leq 1\right\},$$

for all $h \in \{1, \ldots, H-1\}$, $w \in \{1, 2, \ldots, W-1\}$ *and*

$$\square_{W,H}^{W,H} = \left\{(x, y) \in \mathbb{R}^2 : \frac{W-1}{W} \leq x \leq 1, \frac{H-1}{H} \leq y \leq 1\right\}.$$

Product topology

Let $\{(\mathbb{X}_i, \mathbb{T}_i)\}_{n=1}^{\infty}$ be an infinite sequence of topological spaces. Let $\mathbb{X} = \mathbb{X}_1 \times \mathbb{X}_2 \times \cdots$ denote the space whose points are sequences of the form $x = \{x_n \in \mathbb{X}_n : n = 1, 2, \ldots\}$. Then the **product topology** for the space $\mathbb{X}$ is defined as the topology that is generated by sets of the form

$$O = O_1 \times O_2 \times \cdots$$

where $O_n \in \mathbb{T}_n$ for all $n = 1, 2, \ldots$ and for only *finitely* many values of n is it true that $O_n \neq \mathbb{X}_n$. Similarly we define the product topology on the finite product space $\mathbb{X} = \mathbb{X}_1 \times \mathbb{X}_2 \times \cdots \times \mathbb{X}_N$ to be the topology generated by sets of the form $O_1 \times O_2 \times \cdots \times O_N$, where now the only constraint is that $O_n \in \mathbb{T}_n$ for all $n = 1, 2, \ldots, N$.

The case that interests us is where $\mathbb{X}_n = \mathcal{A}$ for all $n = 1, 2, \ldots$ and $\mathbb{T}_n = \mathbb{T}_{discrete}(\mathcal{A})$ is the discrete topology on the alphabet $\mathcal{A}$. In this case we note that

$$\mathcal{A}^{\infty} := \mathcal{A} \times \mathcal{A} \times \cdots = \Omega_{\mathcal{A}}.$$

In general, if $\mathbb{X}$ is a space then we write $\mathbb{X}^{\infty}$ to denote the product space $\mathbb{X} \times \mathbb{X} \times \cdots$. Thus we obtain, in a very simple way, the product topology $\mathbb{T}_{product}(\Omega_{\mathcal{A}})$ on code space. We have the following observations, which we leave as either an exercise or an act of faith for the reader.

T H E O R E M 1.9.6 *The product topology on code space is the same as the natural topology associated with the metric d_Ω (and with the metric $d_{|\mathcal{A}|}$). That is,*

$$\left(\Omega_\mathcal{A}, \mathbb{T}_{product}(\Omega_\mathcal{A})\right) = \left(\Omega_\mathcal{A}, \mathbb{T}_{d_{|\mathcal{A}|}}\right) = \left(\Omega_\mathcal{A}, \mathbb{T}_{d_\Omega}\right).$$

Henceforth we refer to the natural topology on code space $\Omega_\mathcal{A}$ as *the* topology on code space, and it should be assumed that this topology is the one meant when no other assertion is made. It turns out that there is a wonderful basis for the product topology on code space. To describe it we need the following definition.

D E F I N I T I O N 1.9.7 A **cylinder set** of the code space $\Omega_\mathcal{A}$ is a subset of $\Omega_\mathcal{A}$ that can be written in the form

$$\mathcal{C}(\sigma) := \{\omega \in \Omega_\mathcal{A} : \omega_n = \sigma_n \text{ for all } n = 1, 2, \ldots, |\sigma|\},$$

for some $\sigma \in \Omega'_\mathcal{A}$.

We will also refer to $\mathcal{C}(\sigma)$ as a cylinder subset of $\Omega_\mathcal{A}$. Notice that the set of cylinder subsets of $\Omega_\mathcal{A}$ is addressed by the code space $\Omega'_\mathcal{A}$.

E X E R C I S E 1.9.8 *Show that for each $m = 1, 2, \ldots$*

$$\Omega_\mathcal{A} = \bigcup \{\mathcal{C}(\sigma) : \sigma \in \Omega'_\mathcal{A}, |\sigma| = m\}.$$

Cylinder sets are used in the construction of fractals. Indeed, this is one reason why we introduced $\Omega'_\mathcal{A}$. We mention the following because of its relevance to the existence of certain invariant measures on fractals.

T H E O R E M 1.9.9 *A countable basis for $(\Omega_\mathcal{A}, \mathbb{T}_{product})$ is the set of all cylinder sets $\{\mathcal{C}(\sigma) \subset \Omega_\mathcal{A} : \sigma \in \Omega'_\mathcal{A}\}$.*

P R O O F A basis for $(\Omega_\mathcal{A}, \mathbb{T}_{product})$ is, from the definition of $\mathbb{T}_{product} = \mathbb{T}_{product}(\Omega_\mathcal{A})$, the set of sets that can be written in the form

$$O_1 \times O_2 \times \cdots \times O_N \times \mathcal{A} \times \mathcal{A} \times \cdots$$

for some finite integer N, where O_k can be written as $\{a_{k,1}, a_{k,2}, \ldots, a_{k,n_k}\} \subset \mathcal{A}$ with $n_k \in \{1, 2, \ldots, |\mathcal{A}|\}$ for $k = 1, 2, \ldots, N$. (Note that some of the O_k may be equal to $\mathcal{A}$, and recall that $|\mathcal{A}|$ is finite.) It follows that every set in $\mathbb{T}_{product}(\Omega_\mathcal{A})$ can be written as a union of sets of the form

$$\bigcup_{\substack{k_1=1,\ldots,n_1 \\ k_2=1,\ldots,n_2 \\ \vdots \\ k_N=1,\ldots,n_N}} \{a_{1,k_1}\} \times \{a_{2,k_2}\} \times \cdots \times \{a_{N,k_N}\} \times \mathcal{A} \times \mathcal{A} \times \cdots.$$

But this last expression is the same as

$$\bigcup_{\substack{k_1=1,\ldots,n_1 \\ k_2=1,\ldots,n_2 \\ \vdots \\ k_N=1,\ldots,n_N}} \mathcal{C}(a_{1,k_1} a_{2,k_2} \cdots a_{N,k_N}),$$

i.e. it is a union of cylinder sets. So every set in $\mathbb{T}_{product}(\Omega_A)$ can be written as a union of cylinder sets, which is obviously countable. $\qquad\square$

Identification topologies

Let $(\mathbb{X}, \mathbb{T})$ be a topological space, say a Hausdorff space. Let $x_1, x_2 \in \mathbb{X}$, with $x_1 \neq x_2$. Define a new topology $\widetilde{\mathbb{T}}$ on $\mathbb{X}$ as follows: remove from $\mathbb{T}$ all those sets that contain either x_1 or x_2 but not both x_1 and x_2; then $\widetilde{\mathbb{T}}$ consists of the sets that remain. It is readily verified that $\widetilde{\mathbb{T}}$ is a topology. But it is no longer a Hausdorff topology, for there is no open set that contains x_1 but not x_2. According to the topology $\widetilde{\mathbb{T}}$ the set $\{x_1, x_2\}$ behaves like a single point in the sense that whenever $O \in \widetilde{\mathbb{T}}$ we have: $x_1 \in O \iff \{x_1, x_2\} \subset O$.

EXAMPLE 1.9.10 Let $\mathbb{X} = \{x_1, x_2, x_3, x_4\}$ and Let $\mathbb{T} = \{\varnothing, \mathbb{X}, \{x_1, x_2, x_3\}, \{x_1, x_2, x_4\}, \{x_1, x_3, x_4\}, \{x_2, x_3, x_4\}, \{x_1, x_2\}, \{x_1, x_3\}, \{x_1, x_4\}, \{x_2, x_3\}, \{x_2, x_4\}, \{x_3, x_4\}, \{x_1\}, \{x_2\}, \{x_3\}, \{x_4\}\}$. Then $\widetilde{\mathbb{T}} = \{\varnothing, \mathbb{X}, \{x_1, x_2, x_3\}, \{x_1, x_2\}, \{x_3, x_4\}, \{x_3\}, \{x_4\}\}$. In this case we have started with the discrete topology on $\mathbb{X}$ and have ended up with a new topology $\widetilde{\mathbb{T}}$. It looks quite like the discrete topology on $\widetilde{\mathbb{X}} = \{\{x_1, x_2\}, x_3, x_4\}$. Notice how the topology $\widetilde{\mathbb{T}}$ is coarser than $\mathbb{T}$, that is, $\widetilde{\mathbb{T}} \subset \mathbb{T}$.

DEFINITION 1.9.11 Let $f : \mathbb{X} \to \mathbb{Y}$ be a mapping from a topological space $(\mathbb{X}, \mathbb{T})$ to a space $\mathbb{Y}$. Let $\mathbb{T}_f(\mathbb{X}) = \mathbb{T}_{f:\mathbb{X}\to\mathbb{Y}}(\mathbb{X})$ be the topology on $\mathbb{Y}$ specified by: $O \subset \mathbb{X}$ is open iff $O \in \mathbb{T}$ and $f^{-1}f(O) = O$. Then $\mathbb{T}_f(\mathbb{X})$ is called the **identification topology on** $\mathbb{X}$ induced by $f : \mathbb{X} \to \mathbb{Y}$.

Here we use the notation $f^{-1}f$ to denote the mapping obtained by first applying f and then applying f^{-1}. We might also have written $f^{-1}(f(O))$ or $f^{-1} \circ f(O)$. We prove now that Definition 1.9.11 is a good one.

PROOF Let $\mathbb{T}_f(\mathbb{X})$ denote the set of subsets of $\mathbb{X}$ specified in the definition. We need to demonstrate that it is a topology for $\mathbb{X}$.

(i) Since $\mathbb{X} \in \mathbb{T}$ and $f^{-1}f(\mathbb{X}) = \mathbb{X}$ it follows that $\mathbb{X} \in \mathbb{T}_f(\mathbb{X})$. Since $\varnothing \in \mathbb{T}$ and $f^{-1}f(\varnothing) = \varnothing$ it follows that $\varnothing \in \mathbb{T}_f(\mathbb{X})$.

(ii) Suppose that $\{O_\alpha \in \mathbb{T}_f(\mathbb{X}) : \alpha \in \mathcal{I}\}$ is a collection of sets in $\mathbb{T}_f(\mathbb{X})$. Then $O_\alpha \in \mathbb{T}$ for all $\alpha \in \mathcal{I}$, and so $\bigcup_{\alpha \in \mathcal{I}} O_\alpha \in \mathbb{T}$. Using Exercise 1.3.2(i), (iii), $f^{-1}f(\bigcup_{\alpha \in \mathcal{I}} O_\alpha) = f^{-1}(\bigcup_{\alpha \in \mathcal{I}} f(O_\alpha)) = \bigcup_{\alpha \in \mathcal{I}} f^{-1}(f(O_\alpha)) = \bigcup_{\alpha \in \mathcal{I}} O_\alpha$. It follows that $\bigcup_{\alpha \in \mathcal{I}} O_\alpha \in \mathbb{T}_f(\mathbb{X})$.

(iii) Suppose that $O_1, O_2 \in \mathbb{T}_f(\mathbb{X})$. Then $O_1, O_2 \in \mathbb{T}$ and so $O_1 \cap O_2 \in \mathbb{T}$. It remains to prove that $f^{-1}f(O_1 \cap O_2) = O_1 \cap O_2$. This follows at once from Exercise 1.3.2(iv) provided that $f(O_1 \cap O_2) = f(O_1) \cap f(O_2)$. But from Exercise 1.3.2(ii) we know that $f(O_1 \cap O_2) \subset f(O_1) \cap f(O_2)$. So the proof is complete if we can show that $f(O_1 \cap O_2) \supset f(O_1) \cap f(O_2)$. Suppose that $y \in f(O_1) \cap f(O_2)$; then there exists $x_1 \in O_1$ such that $f(x_1) = y$ and $x_2 \in O_2$ such that $f(x_2) = y$. But since $f^{-1}f(O_1) = O_1$ we must have $f^{-1}(y) = f^{-1}f(x_1) \subset O_1$ and similarly $f^{-1}(y) = f^{-1}f(x_2) \subset O_2$. So $f^{-1}(y) \subset O_1 \cap O_2$. It follows upon applying f to both sides that $y \in f(O_1 \cap O_2)$. $\qquad\square$

EXERCISE 1.9.12 *In Example 1.9.10 choose* $\mathbb{Y} = \{\{x_1, x_2\}, x_3, x_4\}$ *and define* $f : \mathbb{X} \to \mathbb{Y}$ *by* $f(x_1) = \{x_1, x_2\}$, $f(x_2) = \{x_1, x_2\}$, $f(x_3) = x_3$, $f(x_4) = x_4$. *Verify that the identification topology on* $\mathbb{X}$ *induced by* $f : \mathbb{X} \to \mathbb{Y}$ *is exactly* $\widetilde{\mathbb{T}}$.

DEFINITION 1.9.13   Let $f : \mathbb{X} \to \mathbb{Y}$ be a mapping from a topological space $(\mathbb{X}, \mathbb{T})$ onto a space $\mathbb{Y}$. Let $\mathbb{T}_f(\mathbb{Y})$ $(= \mathbb{T}_{f:\mathbb{X}\to\mathbb{Y}}(\mathbb{Y}))$ be the topology on $\mathbb{X}$ specified by: $O \subset \mathbb{Y}$ is open iff $f^{-1}(O) \in \mathbb{T}$. Then $\mathbb{T}_f(\mathbb{Y})$ is called the **identification topology on** $\mathbb{Y}$ induced by $f : \mathbb{X} \to \mathbb{Y}$.

The proof that $\mathbb{T}_f(\mathbb{Y})$ is indeed a topology is similar but easier than the proof (above) that $\mathbb{T}_f(\mathbb{X})$ is a topology, and we leave it to the reader.

EXAMPLE 1.9.14   In Example 1.9.10 choose $\mathbb{Y} = \{\{x_1, x_2\}, x_3, x_4\}$ and define $f : \mathbb{X} \to \mathbb{Y}$ by $f(x_1) = \{x_1, x_2\}$, $f(x_2) = \{x_1, x_2\}$, $f(x_3) = x_3$, $f(x_4) = x_4$. Then $\mathbb{T}_f(\mathbb{X}) = \{\varnothing, \mathbb{X}, \{x_1, x_2, x_3\}, \{x_1, x_2\}, \{x_3, x_4\}, \{x_3\}, \{x_4\}\}$ while $\mathbb{T}_f(\mathbb{Y}) = \{\varnothing, \mathbb{Y}, \{\{x_1, x_2\}, x_3\}, \{\{x_1, x_2\}\}, \{x_3, x_4\}, \{x_3\}, \{x_4\}\}$.

Identification topologies are rather simple in the case of finite sets of points, but they become decidedly interesting in the case of non-denumerable spaces. For example, let $\mathbb{X} = [0, 1] \subset \mathbb{R}$ and $\mathbb{Y} = S^1$, the circle in $\mathbb{R}^2$ of radius 1 centred at the origin, let $\mathbb{T}$ be the natural topology on $\mathbb{R}^2$ and let $f : (\mathbb{X}, \mathbb{T}) \to \mathbb{Y}$ be defined by

$$f(x) = (\cos 2\pi x, \; \sin 2\pi x). \qquad (1.9.1)$$

Then the two points $x_1 = 0$ and $x_2 = 1$ in $[0, 1]$ are mapped onto the single point $P := (1, 0) \in \mathbb{R}^2$. In the identification topology on S^1 induced by the mapping $f : [0, 1] \to S^1$, the point P is an element of each of the many open sets that consist of arcs of the circle that contain P but do not terminate at P and do not contain the points that define their extent. Indeed, the identification topology on S^1 induced by f is just the natural topology as a subset of $\mathbb{R}^2$. But the corresponding open sets in $[0, 1]$, which contain both the points x_1 and x_2, are of the form $[0, a) \cup (b, 1]$ where $a, b \in (0, 1)$. See Figure 1.20(i).

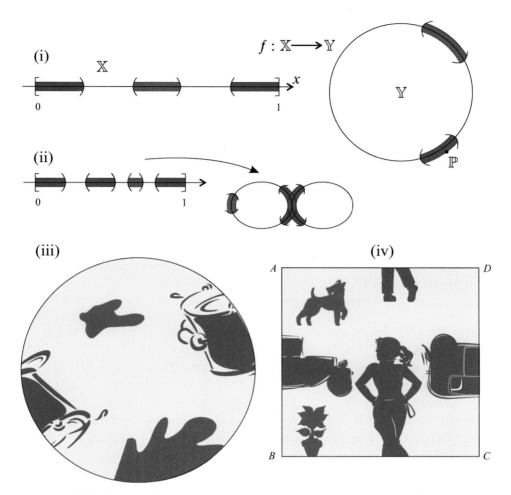

Figure 1.20 This figure illustrates some open sets (purple and green) and some sets that are not open (red) in various identification topologies. In (i) the identification topology is induced by a mapping f from the real closed interval [0, 1] to a circle; see Equation (1.9.1). Neither purple subinterval of [0, 1] is open *on its own* (in the induced topology) but their union is open. The single green subinterval is also open.

In (ii) the transformation is from the interval [0, 1] onto a sideways figure-eight in $\mathbb{R}^2$; see Equation (1.9.2). None of the three purple subintervals is open, nor any pairwise union of them, but the union of all three is open. The image of this union is the purple X-shaped segment of the sideways figure eight.

In (iii) is shown a model of the projective plane; it consists of a disk centred at the origin, with opposite points on its circular boundary identified (via an appropriate mapping from the disk onto itself minus half its circular boundary). The two purple regions comprise a single open set (a bucket) in the identification topology, but neither on its own is open. The red region represents a set that includes part of the circle but since none of its points expands across the opposite side it cannot represent an open set.

In (iv) each point in the side AD of the filled square $ABCD$ is identified with the point 'vertically below it' in BC. Each point on AB is then identified with the opposite point (through the centre of the square) on DC. The purple car, which becomes inverted as it 'drives through the barrier DC to emerge through AB', represents an open set, as does the purple girl and the green dog. The red region, however, does not represent an open set because although it touches BC the bottom of the flower-pot does not extend below AD.

A related example is obtained by changing the definition of $f : [0, 1] \to S^1$ from Equation (1.9.1) to

$$f(x) = (\sin 2\pi x, \cos 4\pi x). \tag{1.9.2}$$

See Figure 1.20(ii).

EXERCISE 1.9.15 *Let* $f : \mathbb{X} \to \mathbb{Y}$ *be a mapping from a topological space* $(\mathbb{X}, \mathbb{T})$ *to a space* $\mathbb{Y}$. *Let* $C \subset \mathbb{Y}$. *Show that* C *is closed in the identification topology on* $\mathbb{Y}$ *induced by* $f : \mathbb{X} \to \mathbb{Y}$ *iff* $f^{-1}(C)$ *is closed in the topology* $\mathbb{T}$.

The following theorem tells us that f is almost but not quite a homeomorphism with respect to the identification topologies it induces because in general f is not one-to-one as a point map.

THEOREM 1.9.16 *Let* $f : \mathbb{X} \to \mathbb{Y}$ *be a mapping from a topological space* $(\mathbb{X}, \mathbb{T})$ *to a space* $\mathbb{Y}$. *Then, as a mapping from subsets of* $\mathbb{X}$ *to subsets of* $\mathbb{Y}$, *restricted to* $\mathbb{T}_f(\mathbb{X})$, f *is one-to-one from* $\mathbb{T}_f(\mathbb{X})$ *onto* $\mathbb{T}_f(\mathbb{Y})$.

PROOF We show first that f maps from $\mathbb{T}_f(\mathbb{X})$ into $\mathbb{T}_f(\mathbb{Y})$. Let $O \in \mathbb{T}_f(\mathbb{X})$. Then $f(O)$ is in $\mathbb{T}_f(\mathbb{Y})$ because $f^{-1}(f(O)) = O$ and $O \in \mathbb{T}$.

Next we show that $f : \mathbb{T}_f(\mathbb{X}) \to \mathbb{T}_f(\mathbb{Y})$ is onto. Suppose that $\widetilde{O} \in \mathbb{T}_f(\mathbb{Y})$. Then $f^{-1}(\widetilde{O}) \in \mathbb{T}$ and $f^{-1}f(f^{-1}(\widetilde{O})) = f^{-1}(\widetilde{O})$, so $f^{-1}(\widetilde{O})$ is in $\mathbb{T}_f(\mathbb{X})$. And $f(f^{-1}(\widetilde{O})) = \widetilde{O}$ since $f \circ f^{-1}$ is the identity map.

Finally we show that $f : \mathbb{T}_f(\mathbb{X}) \to \mathbb{T}_f(\mathbb{Y})$ is one-to-one. Suppose that $A, B \in \mathbb{T}_f(\mathbb{X})$ and $f(A) = f(B)$. Then applying f^{-1} to both sides we obtain $f^{-1}(f(A)) = f^{-1}(f(B))$. But $A = f^{-1}(f(A))$ since $A \in \mathbb{T}_f(\mathbb{X})$, and $B = f^{-1}(f(B))$ since $B \in \mathbb{T}_f(\mathbb{X})$. So $A = B$. $\square$

We will be particularly interested in identification topologies on code space $\Omega_\mathcal{A}$ that are associated with mappings from $\Omega_\mathcal{A}$ onto subspaces of $\mathbb{R}^n$ such as fractals. For example, a general theorem in Section 4.14 implies in particular that *the natural topology on* $[0, 1] \subset \mathbb{R}$ *is the identification topology induced by the continuous mapping* $f : (\Omega_\mathcal{A}, \mathbb{T}_\Omega) \to [0, 1]$ defined as follows. Take $\mathcal{A}$ to be $\{0, 1, 2, \ldots, N - 1\}$ and set

$$f(\sigma) = \sum_{n=1}^{\infty} \frac{\sigma_n}{N^n} \quad \text{for all } \sigma \in \Omega_\mathcal{A}. \tag{1.9.3}$$

This tells us that the real interval can indeed be thought of, from a topological point of view, as being code space 'joined to itself' at those pairs of points, namely addresses, $\varpi, \omega \in \Omega_\mathcal{A}$, of the form

$$\varpi = \sigma_1 \sigma_2 \cdots \sigma_{M-1} \sigma_M \overline{0} \quad \text{and} \quad \omega = \sigma_1 \sigma_2 \cdots \sigma_{M-1} (\sigma_M - 1) \overline{(N-1)}$$

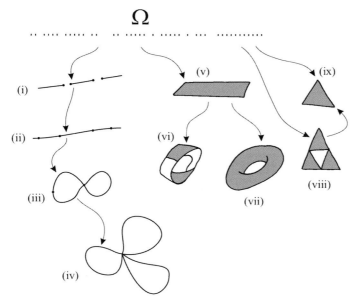

Figure 1.21   Illustration of various spaces with identification topologies induced by functions on a code space Ω. (i) A set of intervals; (ii) a line segment; (iii) a figure eight; (iv) three loops meeting at a point; (v) a filled rectangle in $\mathbb{R}^2$; (vi) a Möbius strip; (vii) a torus; (viii) a triangle with a triangular hole; (ix) a filled triangle. Note that (viii) cannot be obtained as an identification topology induced by a function whose domain is the filled triangle (ix) − why?

for $\sigma_m \in \{0, 1, 2, \ldots, N - 1\}$, $m \in \{1, 2, \ldots, M - 1\}$, and for $\sigma_M \in \{1, 2, \ldots, N - 1\}$, $M \in \{1, 2, 3, \ldots\}$. The reason is that these are exactly the points that are identified by f, i.e. $f(\varpi) = f(\omega)$.

E X E R C I S E 1.9.17   *Prove that the identification topology on* $[0, 1] \subset \mathbb{R}$ *induced by* $f : (\Omega_{\mathcal{A}}, \mathbb{T}_\Omega) \to [0, 1]$ *as defined in Equation (1.9.3) is the natural topology.*

Many natural topologies on interesting 'smooth' objects in $\mathbb{R}^n$ are in fact identification topologies induced by mappings from code space to the objects. Examples include intervals, disks, Möbius strips, a model for the projective plane and so on, as illustrated in Figure 1.21. But to us the most remarkable and fascinating realization is that the natural topologies of diverse fractals are induced by mappings from code space; see Chapter 4. This relates to our theme that code space is somehow protoplasmic, the stem cell material of fractal geometry, the meristem of plant growth.

1.10   Some key topological invariants

In this section we follow the theme of looking at properties that are invariant under transformations. Such properties are called topological because they are invariant under homeomorphisms. We also continue to describe properties of code spaces.

DEFINITION 1.10.1   Let $\mathbb{X}$ be a topological space. Then $\mathbb{X}$ is said to be **perfect** iff it is equal to the set of its accumulation points.

For example, the space $\mathbb{R}^n$ is perfect in the natural topology, and so is $[0, 1] \subset \mathbb{R}$; but $[0, 1] \cup \{2\} \subset \mathbb{R}$ is not perfect because 2 is not a limit of any sequence of points in $[0, 1]$.

THEOREM 1.10.2   *When $|\mathcal{A}| > 1$, the code space Ω_A is perfect.*

PROOF   The natural topology is implied. Let $\sigma \in \Omega_A$ be given. Then we can choose $\omega \in \Omega_A$ in such a way that $\omega_n \neq \sigma_n$ for $n = 1, 2, \ldots$ Now define a sequence $\{\alpha_n \in \Omega_A\}_{n=1}^{\infty}$ by $(\alpha_n)_m$ (i.e. the mth component of α_n) $= \sigma_m$ for $m = 1, 2, \ldots, n$ and $(\alpha_n)_m = \omega_m$ for $m = n + 1, n + 2, \ldots$ Then it is easy to see that $\alpha_p \neq \alpha_q$ for all $p, q \in \{1, 2, \ldots\}$ with $p \neq q$, and that $\lim_{n \to \infty} \alpha_n = \sigma$. Hence σ is an accumulation point of Ω_A.   □

EXERCISE 1.10.3   *Show that if $f : \mathbb{X} \to \mathbb{Y}$ is a homeomorphism then $\mathbb{X}$ is perfect iff $\mathbb{Y}$ is perfect.*

DEFINITION 1.10.4   Let $\mathbb{X}$ be a topological space. Then $\mathbb{X}$ is said to be **connected** iff the only two subsets of $\mathbb{X}$ that are both open and closed are $\mathbb{X}$ and $\varnothing$. A subset $S \subset \mathbb{X}$ is said to be **connected** iff the space S with the relative topology is connected. S is said to be **disconnected** iff it is not connected. S is said to be **totally disconnected** iff the only nonempty connected subsets of S are those that contain single points.

EXERCISE 1.10.5   *Let $\mathbb{X}$ be a space. Show that $(\mathbb{X}, \mathbb{T}_{discrete})$ is totally disconnected.*

EXERCISE 1.10.6   *Show that $(\Omega_A \cup \Omega'_A, d_\Omega)$ is totally disconnected in the natural topology.*

DEFINITION 1.10.7   Let $\mathbb{X}$ be a topological space. Let $S \subset \mathbb{X}$. Then S is said to be **pathwise connected** iff whenever $x, y \in S$ there is a continuous mapping $f : [0, 1] \subset \mathbb{R} \to S$ such that $x, y \in f([0, 1])$.

Each property, of being connected, disconnected, totally disconnected or pathwise connected, is invariant under homeomorphism. They are topological properties. For example, if $f : \mathbb{X} \to \mathbb{Y}$ is a homeomorphism between topological spaces and $S \subset \mathbb{X}$ then S is a connected subset of $\mathbb{X}$ iff $f(S)$ is a connected subset of $\mathbb{Y}$. See Figure 1.22.

If $\mathbb{X}$ is pathwise connected then it is connected. But the converse is not true. For example, let $g(x) = \sin(x/(10 - x))$, let G_g denote the graph of $g : [0, 10) \subset$

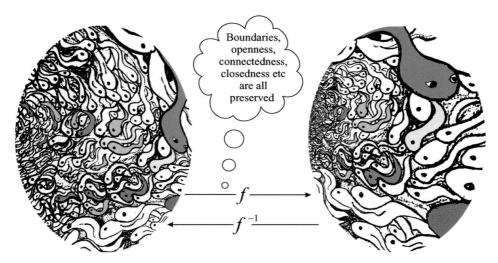

Figure 1.22 Homeomorphisms preserve topological properties such as being connected, being the boundary of a subset, being the interior of a subset, being an accumulation point of a subset and so on. Here the action of a certain homeomorphism acting on an elliptical subspace of $\mathbb{R}^2$ is illustrated by showing how it acts upon various coloured subsets. See Chapter 2 for more precision on what it means for a transformation to act upon a picture.

$\mathbb{R} \to \mathbb{R}^2$ and let $L := \{(10, y) \in \mathbb{R}^2 : -1 \le y \le 1\}$ be a line segment. Then the subset $S = L \cup G_g \subset \mathbb{R}^2$, illustrated in Figure 1.23, is connected but not pathwise connected (see [70], p. 141): there does not exist any curve, homeomorphic to $[0, 1] \subset \mathbb{R}$, that passes through both the points $(0, 0)$ and $(10, 0) \in S$. Clearly, by following G_g one can find a curve that connects $(0, 0)$ to a point lying arbitrarily close to $(10, 0)$. But one cannot find a curve that 'crosses the divide'.

In Figure 1.24 we have illustrated variants of the previous example. Part (i) of the figure shows the graph G_l of a piecewise linear function $l : (0, 1] \to [0, 1]$. The set $S = G_l \cup L_1$, where $L_1 := \{(0, y) \in \mathbb{R}^2 : 0 \le y \le 1\}$, is a connected but not pathwise connected subset of $\mathbb{R}^2$.

We note that S has the following property:

$$S = w_1(S) \cup w_2(S),$$

where the transformations $w_1, w_2 : \mathbb{R}^2 \to \mathbb{R}^2$ are given by

$$w_1(x, y) = (0.7x, -y + 1),$$
$$w_2(x, y) = (0.3x + 0.7, x). \tag{1.10.1}$$

We have taken the origin of coordinates to be at the lower left-hand corner of Figure 1.24(i) and the width of S to be one unit. The transformation w_1 shrinks the x-coordinates of S by a factor 0.7 and reflects the result in the line $y = 0.5$, so that $w_1(S)$ is all S minus the line segment $L_2 := \{(x, \frac{1}{3}(10x - 7)) : 0.7 \le x \le 1.0\}$.

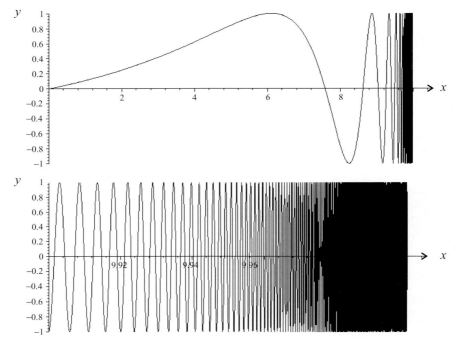

Figure 1.23   The graph of sin($x/(10 - x)$) for $0 \leq x < 10$ and for $0 \leq 9.9 < 10$. This graph, completed with a line segment parallel to the y-axis at $x = 10$, makes a connected set that is not pathwise connected.

The transformation w_2 maps S onto L_2. (In IFS theory it is known that the attractor of an IFS of two strictly contractive maps on $\mathbb{R}^2$ is either connected or totally disconnected; see Chapter 4). In the present case one map is not strictly contractive, and the attractor is neither connected nor totally disconnected.

In Figure 1.24(ii) a further variant of the connected-but-not-pathwise-connected type is illustrated. This time the figure is made of four transformations of itself: can you spot the transformations? Now the set is quite a bit more broken up; it is not pathwise connected at a countable infinity of places.

In Figure 1.24(iii) we zoom to a comb-shaped part of the curve.

See also Figure 1.25.

DEFINITION 1.10.8   Let S be a subset of a topological space $\mathbb{X}$. Then the **boundary** of S is the set of points in $\mathbb{X}$ such that every neighbourhood of x contains a point in S and one in $\mathbb{X}\backslash S$.

The boundary of the open disk $\{(x, y) \in \mathbb{R}^2 : x^2 + y^2 < 1\}$ as a subset of $\mathbb{R}^2$ is the circle of radius 1, centred at the origin. The boundary of the set $\mathbb{R}\backslash\{x = 0\}$ as a subset of $\mathbb{R}$ is the point $x = 0$, and as a subset of $\mathbb{R}^2$ it is $\mathbb{R}$. The boundary of a closed set is always contained in the set.

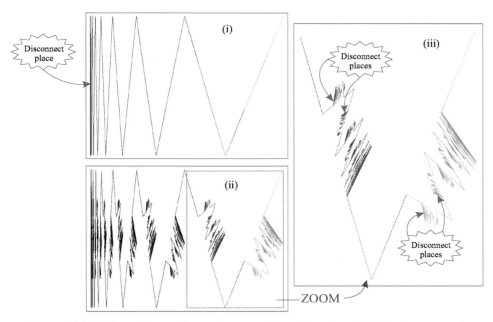

Figure 1.24   (i) The set illustrated here is homeomorphic to that in Figure 1.23. The **disconnect place**, where the set is pathwise disconnected, is indicated by the red arrow. Note that the set is a union of two transformed copies of itself, according to Equations (1.10.1). (ii) The situation is pretty bad here; each squiggly bit is pathwise disconnected as in (i) and, moreover, it is a transformed copy of the whole set, so that actually there are infinitely many disconnect places. Can you work out how (ii) is the union of four transformed copies of itself? (iii) Part of the set in (ii) in shown magnified, revealing more disconnect places.

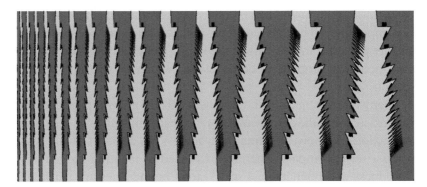

Figure 1.25   Part of Figure 1.24(ii), magnified, with colours red and yellow demarking the regions 'above' and 'below' the curve.

The boundary of an open set is the empty set. If $f : \mathbb{X} \to \mathbb{Y}$ is a homeomorphism then ∂S is the boundary of a set $S \subset \mathbb{X}$ iff $f(\partial S)$ is the boundary of $f(S) \subset \mathbb{Y}$. That is, the concept of a boundary is a topological one. Note that a topological space has an empty boundary.

EXERCISE 1.10.9   *Show that the boundary of $\Omega'_{\mathcal{A}}$ in $(\Omega_{\mathcal{A}} \cup \Omega'_{\mathcal{A}}, d_{\Omega})$ is $\Omega_{\mathcal{A}}$.*

DEFINITION 1.10.10   Let S be a subset of a topological space $\mathbb{X}$. Then the **interior** of S is the set of points each of which belongs to an open set contained in S.

The interior of the closed interval $[0, 1] \subset \mathbb{R}$ is the open interval $(0, 1)$. The interior of the open ball

$$B(x_0, \epsilon) := \{ d(x_0, x) < \epsilon : x \in \mathbb{X} \}$$

in the metric space $(\mathbb{X}, d)$ is the open ball itself, where $\epsilon > 0$. The interior of $\overline{B(x_0, \epsilon)}$ is also $B(x_0, \epsilon)$. The interior of an open set is the set itself. If $f : \mathbb{X} \to \mathbb{Y}$ is a homeomorphism then S°, say, is the interior of a set $S \subset \mathbb{X}$ iff $f(S^{\circ})$ is the interior of $f(S)$. That is, the concept of the interior of a set is a topological one.

EXERCISE 1.10.11   *Show that in $(\Omega_{\mathcal{A}} \cup \Omega'_{\mathcal{A}}, d_{\Omega})$ the interior of $\Omega'_{\mathcal{A}}$ is $\Omega'_{\mathcal{A}}$ and the interior of $\Omega_{\mathcal{A}}$ is empty.*

1.11   Compact sets and spaces

Over time, some mathematical concepts become clearly established as being of key importance. They are concepts that can be expressed concisely, occur often and are powerful ingredients of theorems. Continuity is such a concept, and so is the topological property of compactness, which we shall introduce in the present section.

Many fractal objects with which we will deal are compact, and indeed owe their very existence to the compactness of the spaces in which we seek them. So here we are anxious not only to define compactness but also to provide ways of knowing when a set is compact. Therefore we need to mention sequential compactness, closedness and boundedness in $\mathbb{R}^n$, total boundedness in metric spaces and the compactness of code spaces.

Let $\mathbb{X}$ be a space. A sequence $\{y_n\}_{n=1}^{\infty} \subset \mathbb{X}$ is called a **subsequence** of the sequence $\{x_n\}_{n=1}^{\infty} \subset \mathbb{X}$ iff there is an increasing sequence of positive integers $\{n_k\}_{k=1}^{\infty} \subset \mathbb{R}$ such that $x_{n_k} = y_k$ for all $k = 1, 2, \ldots$ We may write $\{x_{n_k}\}_{k=1}^{\infty}$ to denote this subsequence.

Let $\mathbb{X}$ be a topological space. A collection of sets $\{O_i \subset \mathbb{X} : i \in \mathcal{I}\}$ is called a **cover** for or **covering** of $S \subset \mathbb{X}$ iff every point in S lies in at least one of the O_i.

That is, $S \subset \cup \{O_i \subset \mathbb{X} : i \in \mathcal{I}\}$. The collection of sets $\{O_i \subset \mathbb{X} : i \in \mathcal{I}\}$ is called an **open covering** of S iff it is a cover for S and each of the sets O_i is open. A cover is called a **finite covering** iff it consists of finitely many sets.

DEFINITION 1.11.1   A topological space $\mathbb{X}$ is **compact** iff every open cover of $\mathbb{X}$ contains a finite cover of $\mathbb{X}$. $\mathbb{X}$ is said to be **sequentially compact** iff every infinite sequence $\{x_n\}_{n=1}^{\infty} \subset \mathbb{X}$ contains a subsequence $\{x_{n_m}\}_{m=1}^{\infty}$ that converges to a point $x \in \mathbb{X}$. A subset $S \subset \mathbb{X}$ is said to be (sequentially) compact iff it is (sequentially) compact in the relative topology.

The property of being (sequentially) compact is invariant under homeomorphism and so is indeed a topological property. A simple source of compact sets is provided by the closed subsets of compact spaces.

THEOREM 1.11.2   *Let $\mathbb{X}$ be a compact space, and let $S \subset \mathbb{X}$ be closed. Then S is compact.*

PROOF   See [70], Theorem 2.11, p. 168.   □

Sequential compactness and compactness are equivalent in the case of metric spaces.

THEOREM 1.11.3   *When $\mathbb{X}$ is a metric space, a subset $S \subset \mathbb{X}$ is compact iff it is sequentially compact.*

PROOF   See [70], Theorem 5.9, p. 183.   □

A rich source of compact sets is the set of closed bounded subsets of $\mathbb{R}^n$, as the following theorem attests.

THEOREM 1.11.4   *Let $\mathbb{X}$ be a subspace of $\mathbb{R}^n$ with the natural topology. Then the following three properties are equivalent.*
 (i) $\mathbb{X}$ is compact.
 (ii) $\mathbb{X}$ is closed and bounded.
(iii) Each infinite subset of $\mathbb{X}$ has at least one accumulation point in $\mathbb{X}$.

PROOF   See [70], Corollary 5.1, p. 183.   □

One of the main ways of establishing that a metric space is compact involves the following concept.

DEFINITION 1.11.5   A metric space $(\mathbb{X}, d)$ is said to be **totally bounded** iff, for every given $\epsilon > 0$, there is a finite set of points $\{x_1, x_2, \ldots, x_L\}$ such that

$$\mathbb{X} = \bigcup \{B(x_l, \epsilon) : l = 1, 2, \ldots, L\}.$$

We have given no proofs of compactness results so far. But the proof of the following key theorem gives a good idea of how such proofs are constructed.

THEOREM 1.11.6 *Let* $(\mathbb{X}, d)$ *be a complete metric space. Then* $\mathbb{X}$ *is compact iff it is totally bounded.*

PROOF Suppose that $\mathbb{X}$ is totally bounded. Then, for some finite integer L,

$$\mathbb{X} = \bigcup \{B(y_l, 1) : l = 1, 2, \ldots, L\}$$

for some points $y_l \in \mathbb{X}$, for $l = 1, 2, \ldots, L$. Let $\{x_n\}_{n=1}^{\infty} \subset \mathbb{X}$ be an infinite sequence of points. Since there are infinitely many points in $\{x_n\}_{n=1}^{\infty}$, one of the $B(y_l, 1)$ must contain an infinite subsequence, which we denote by $\{x_{n_{1,k}}\}_{k=1}^{\infty}$. Let us call this ball $\mathbb{X}_1$. Then $\mathbb{X}_1$ is totally bounded because $\mathbb{X}$ is. So we can re-peat the same argument, this time applied to the infinite sequence $\{x_{n_{1,k}}\}_{k=1}^{\infty} \subset \mathbb{X}_1$ with balls of radius $\frac{1}{2}$. Then we find that one of these balls, which we will denote by $\mathbb{X}_2$, contains an infinite subsequence $\{x_{n_{2,k}}\}_{k=1}^{\infty} \subset \mathbb{X}_2$. We assume that $n_{1,k} < n_{2,k}$ with no loss of generality. We continue in this manner to obtain a **decreasing** sequence,

$$\mathbb{X}_1 \supset \mathbb{X}_2 \supset \mathbb{X}_3 \supset \cdots$$

where $\mathbb{X}_n$ is a ball of radius $1/2^{n-1}$. We also obtain the sequence of points $\{x_{n_{m,1}} \in \mathbb{X}_m : m = 1, 2, 3, \ldots\}$, where $n_{1,k} < n_{2,k} < n_{3,k} < \cdots$, which is a subsequence of $\{x_n\}_{n=1}^{\infty}$. Since the diameter of the $\mathbb{X}_m$ tends to zero as m tends to infinity it fol-lows that $\{x_{n_{m,1}}\}_{m=1}^{\infty}$ is a Cauchy sequence and, since $\mathbb{X}$ is complete, converges to a point $x \in \mathbb{X}$. So $\mathbb{X}$ is compact.

Conversely, suppose that $\mathbb{X}$ is compact but not totally bounded. Then for some $\epsilon > 0$ we can find an *infinite* sequence of points $\{y_l\}_{l=1}^{\infty}$ such that $d(y_l, y_m) > \epsilon$ whenever $l \neq m$. But since $\mathbb{X}$ is assumed to be compact, it must possess a convergent subsequence $\{y_{l_j}\}_{j=1}^{\infty}$. So we can find $s, t \in \{1, 2, \ldots\}$ such that $d(y_{l_s}, y_{l_t}) < \epsilon$, which is a contradiction. $\square$

One way in which fractals are constructed is by means of decreasing sequences of subsets. In Section 1.4 we claimed that the decreasing sequence of real closed intervals in Equation (1.4.3) converges to a point $x \in \mathbb{R}$. Here is the justification of that claim.

THEOREM 1.11.7 *Let* $(\mathbb{X}, d)$ *be a complete metric space and let* $\{C_n \subset \mathbb{X}\}_{n=1}^{\infty}$ *be a decreasing sequence of nonempty compact sets, that is*

$$C_1 \supset C_2 \supset C_3 \supset \cdots.$$

Then

$$C := \bigcap_{n=1}^{\infty} C_n$$

is a nonempty compact set.

PROOF C is compact, because if we have any open cover of C we can extend it to an open cover of C_1 by adding to it the open set $\mathbb{X} \backslash C$. Then this open cover contains a finite subcover of C_1. This subcover also covers C. If this subcover contains $\mathbb{X}\backslash C$ then we remove $\mathbb{X}\backslash C$ from the subcover. The resulting set of open sets continues to cover C and is now a finite subset of the original open covering of C. So we have found a finite subcover of the original open cover of C.

To show that C is nonempty, make an infinite sequence $\{x_n \in C_n\}_{n=1}^{\infty}$ by choosing one point from each of the C_n. This sequence must contain a convergent subsequence, and it is easy to show that the limit must belong to each of the C_n and hence must belong to C. □

Of great importance to us is the fact that code space is compact.

THEOREM 1.11.8   *The code space $\Omega = \Omega_A \cup \Omega'_A$ is compact.*

PROOF   The natural topology is implied, and it suffices to work with the metric d_Ω. We already know from Theorem 1.7.5 that Ω is complete, so we merely need to prove that Ω is totally bounded. It suffices to prove that Ω_A is totally bounded, because $\Omega_A \cup \Omega'_A$ can be embedded in $\Omega_{\widetilde{A}}$ where $|\widetilde{A}| = |A| + 1$, as in Equation (1.6.3). Let $\epsilon > 0$ be given. Choose m so that $2^{-m} < \epsilon$ and choose $L = |A|^m$. Then recall from Exercise 1.9.8 that $\Omega_A = \bigcup\{\mathcal{C}(\sigma) : \sigma \in \Omega_A, |\sigma| = m\}$, where we note that each cylinder has diameter less than $\epsilon/2$. Thus, Ω_A can be covered by 2^m balls each of radius ϵ, each centred on a point in a different cylinder set. □

1.12   The Hausdorff metric

In this section we develop and explore a wonderful metric, the Hausdorff metric. It measures the distances between nonempty compact subsets of a metric space. Later we will use the Hausdorff metric to describe the convergence of sequences of approximations to fractals.

In order to help form our intuition about how the Hausdorff metric works, we will explain it in several stages and explore some examples in detail. We also mention connections between optical processes and the Hausdorff metric. These connections lead us to speculate that in the future the metric may be computed by optical means.

This section also illustrates how we can build a new space of mathematical objects out of an underlying space. In the present case the underlying space is a complete metric space. The mathematical objects are the compact nonempty subsets of this space. A metric on the new space is derived from that on the underlying space. What properties of the new metric space are inherited from the original metric space?

The distance from a point to a set

To define the Hausdorff metric, first we need the concept of the distance from a point to a set.

THEOREM 1.12.1   *Let* $(\mathbb{X}, d)$ *be a complete metric space and let* $\mathbb{H}(\mathbb{X})$ *denote the space of nonempty compact subsets of* $\mathbb{X}$. *Let* $x \in \mathbb{X}$ *and let* $B \in \mathbb{H}(\mathbb{X})$. *Then there exists at least one point* $\widehat{b} = \widehat{b}(x) \in B$ *such that*

$$d(x, b) \geq d(x, \widehat{b}(x)) \quad \text{for all } b \in B.$$

PROOF   Fix $x \in \mathbb{X}$. Then the function $f : B \subset \mathbb{X} \to \mathbb{R}$ defined by

$$f(b) = d(x, b) \quad \text{for all } b \in B$$

is continuous and B is compact. Hence there exists at least one point in B where the value of f is a minimum. We denote such a point by $\widehat{b} \in B$. Notice that $\widehat{b}$ may change when x changes, so we write $\widehat{b} = \widehat{b}(x)$. $\square$

Theorem 1.12.1 enables us to make the following definition.

DEFINITION 1.12.2   Let $(\mathbb{X}, d)$ be a complete metric space. Let $\mathbb{H}(\mathbb{X})$ denote the space of nonempty compact subsets of $\mathbb{X}$. Then the **distance from a point** $x \in X$ **to** $B \in \mathbb{H}(\mathbb{X})$ is defined by

$$\mathcal{D}_B(x) := \min\{d(x, b) : b \in B\}.$$

We refer to $\mathcal{D}_B(x)$ as the **shortest-distance function** of the set B.

EXERCISE 1.12.3   *Let* $\mathbb{X} = \square = \{(x, y) \in \mathbb{R}^2 : 0 \leq x \leq 1, \, 0 \leq y \leq 1\}$. *Let* $d_{max}((x_1, y_1), (x_2, y_2)) = \max\{|x_1 - x_2|, |y_1 - y_2|\}$. *Let* $B = \{(x, y) \in \square : x^2 + y^2 = 0.25\}$. *Calculate* $\mathcal{D}_B((0.6, 0.8))$.

EXERCISE 1.12.4   *Show that*

$$\mathcal{D}_B(x) \leq d(x, y) + \mathcal{D}_B(y) \quad \text{for all } x, y \in \mathbb{X}.$$

Use this to show that $\mathcal{D}_B(x)$ *is a continuous function of* $x \in \mathbb{X}$.

EXERCISE 1.12.5   *Prove that if* $C, D \in \mathbb{H}(\mathbb{X})$ *with* $C \subset D$ *then* $\mathcal{D}_C(x) \geq \mathcal{D}_D(x)$ *for all* $x \in \mathbb{H}(\mathbb{X})$.

For given $d \geq 0$ we call the set of points

$$L_d := \{x \in \mathbb{X} : \mathcal{D}_B(x) = d\}$$

a **level set** of $\mathcal{D}_B(x)$. All points on L_d are at the same distance d from B. In $\mathbb{R}^2$ these level sets $\{L_d : d \geq 0\}$ may form a graceful family of curves, like patterns of ripples, shaped like B, produced by simultaneous disturbances on a water surface or like the wavefronts of light at successive equally spaced time intervals after tiny coherent light pulses are emitted by the points of B at an initial time.

We can imagine optical devices, based on the latter idea, that generate approximate level sets of $\mathcal{D}_B(x)$ when $B \subset \mathbb{R}^2$. For example, schematically, we can imagine a collection of light-emitting diodes organized in two dimensions to form a discrete model for B. We suppose that these diodes are turned on and off rapidly,

Figure 1.26   The space around a fern image is painted using different colours for different level curves of the shortest-distance function $\mathcal{D}_F(x, y)$. These level curves do not possess well-defined tangents at all their points. Also, the perturbation P_0, a small purple disk, makes no difference to the shortest-distance function close to the fern but modifies it further away.

while an array of ultra-fast and sensitive charge-coupled devices, something like the CCD chip in a digital camera, in the same plane as the diodes is used to 'photograph' the wavefront at different times.

In Figure 1.26 we show for comparison $\mathcal{D}_F(x)$ and $\mathcal{D}_{F \cup P_0}(x)$, where F is a fern-like subset of $\mathbb{R}^2$ and $P_0 \subset \mathbb{R}^2$ is a small disk. From left to right: the subset $F \subset \mathbb{R}^2$ and a small disk P_0; some level curves of $\mathcal{D}_F(x)$; some level curves of $\mathcal{D}_{F \cup P_0}(x)$ (the outermost contour, red, contains points equidistant from F and P_0); the same as the preceding image but more contours are shown. We see that $\mathcal{D}_F(x) = \mathcal{D}_{F \cup P_0}(x)$ whenever $\mathcal{D}_F(x)$ is sufficiently small but that P_0 provides a serious perturbation to the shortest-distance function at points sufficiently far away from F, in some directions. In Figure 1.27 we show a close-up of the level sets of $\mathcal{D}_F(x)$ in the vicinity of the subset $F \subset \mathbb{R}^2$. It is fascinating to imagine these lovely patterns at higher resolutions. In Figure 1.28 we show an artificial artistic work. It was made using the shortest-distance function associated with the euclidean metric. Four objects were drawn and coloured, then level sets of the shortest-distance function for the coloured points were computed and rendered.

Paths of steepest descent

In this subsection we continue to discuss shortest-distance functions.

Let $B \in \mathbb{H}(\mathbb{R}^2)$. At those points $(x, y) \in \mathbb{R}^2$ where the shortest-distance function $\mathcal{D}_B(x, y)$ is differentiable,

$$-\operatorname{grad} \mathcal{D}_B(x, y) = -\left(\frac{\partial \mathcal{D}_B}{\partial x}, \frac{\partial \mathcal{D}_B}{\partial y} \right)$$

is a vector pointing along the **path of steepest descent** from x to the nearest point on B. When the underlying metric is the euclidean metric, and the level sets of $\mathcal{D}_B(x, y)$ are differentiable curves, this vector is oriented perpendicular to the level set through the point (x, y). In this case, paths of steepest descent for

Figure 1.27 The shortest-distance function associated with a fern image is illustrated by variously coloured level curves, each corresponding to a different distance from the fern.

shortest-distance functions are found typically to be straight-line segments, as illustrated below in Exercise 1.12.6; more generally they may lie along geodesics.

EXERCISE 1.12.6   *In $\mathbb{R}^2$ let L_0 denote the line $y = -\frac{1}{4}$ and let F denote the point* $(0, \frac{1}{4})$. *Show that the level curves of $D_{L_0 \cup F}(x, y)$ have discontinuous gradients on the parabola P defined by $y = x^2$, and sketch the paths of steepest descent. Notice that F is the **focus** of the parabola, while L_0 is its **directrix**.*

As a slightly more complicated example, we consider the shortest-distance function $\mathcal{D}_P(x, y)$ of (part of) the parabola P in $\mathbb{R}^2$ arising in Exercise 1.12.6. P is defined by $(x_0, y_0) \in P$ iff

$$y_0 = x_0^2. \tag{1.12.1}$$

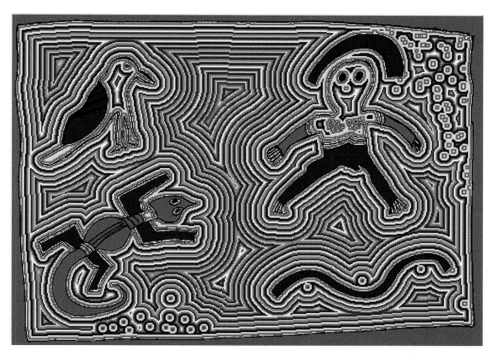

Figure 1.28 The level sets of a shortest-distance function provide a visually appealing way of filling up blank space in a drawing. Such patterns are used in aboriginal art. See for example Jack Jakamarra Ross *et al.*, 'Karrku Jukurrpa, 1996', acrylic on canvas, shown on p. 203 in Howard Morphy, *Aboriginal Art*, Phaidon Press, London, 1998.

Let $(x_1, y_1) \in \mathbb{R}^2$ be given and let (x_0, y_0) be the point of P closest to (x_1, y_1). From elementary coordinate geometry we know that (x_1, y_1) lies on the normal to the parabola at (x_0, y_0). At $(x_0, y_0) \in P$ the slope of the parabola is $dy/dx|_{(x_0, y_0)} = 2x_0$, so the slope of the normal to P at (x_0, y_0) is $-1/(2x_0)$. It follows that

$$y_1 - y_0 = \frac{-1}{2x_0}(x_1 - x_0). \tag{1.12.2}$$

At the point (x_1, y_1) on the level set (curve) L_d we must also have

$$(x_1 - x_0)^2 + (y_1 - y_0)^2 = d^2. \tag{1.12.3}$$

We now use Equations (1.12.1), (1.12.2) and (1.12.3) to express both x_1 and y_1 in terms of x_0 and d. We find, from consideration of the geometry, see Figure 1.29, that

$$\begin{cases} x_1 = x_0 \pm \dfrac{2x_0 d}{\sqrt[2]{1 + 4x_0^2}}, \\ y_1 = x_0^2 \mp \dfrac{d}{\sqrt[2]{1 + 4x_0^2}}, \end{cases}$$

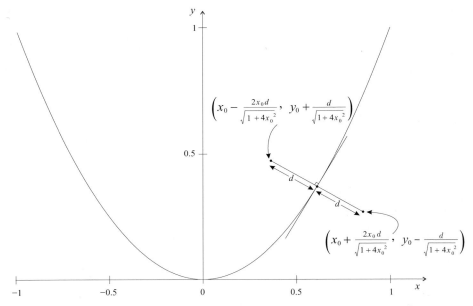

Figure 1.29 This illustrates the locations of points at (shortest) distance d from the point (x_0, y_0) on the parabola $y = x^2$.

where it is assumed that $d \geq 0$ and that x_1 and x_0 are either both positive or both negative. The upper sign corresponds to points outside the parabola while the lower sign corresponds to points inside the parabola. In the latter case we find that for x_1 to be positive when x_0 is positive we must have

$$1 - \frac{2d}{\sqrt[2]{1 + 4x_0^2}} \geq 0,$$

which implies that, when $d \geq \frac{1}{2}$, x_0 jumps from $\sqrt[2]{4d^2 - 1}$ to $-\sqrt[2]{4d^2 - 1}$ as (x_1, y_1) crosses from $x_1 > 0$ to $x_1 < 0$. Hence, while $\mathcal{D}_P(x, y)$ is continuous for all $(x, y) \in \mathbb{R}^2$, grad $\mathcal{D}_B(x, y)$ is discontinuous when $x = 0$ and $y > \frac{1}{2}$. This discontinuity is illustrated in Figure 1.30. These different renderings of $\mathcal{D}_P(x, y)$ show that elementary coordinate geometry may be colourful, beautiful, and mysterious.

EXERCISE 1.12.7   *Analyze $\mathcal{D}_P(x, y)$ when the underlying metric is*

$$d_{\max}((x_1, y_1), (x_2, y_2)) = \max\{|x_1 - x_2|, |y_1 - y_2|\}.$$

What do the level sets look like? Show that in this case $\mathcal{D}_P(x, y)$ has discontinuities where $|y - x^2| = |\sqrt[2]{y} - x|$ and make a sketch of this set of points. You will be delighted how neatly everything works out.

EXERCISE 1.12.8   *Analyze the shortest-distance function $\mathcal{D}_E(x, y)$ for the ellipse E defined by $(x_0, y_0) \in E$ iff $4x_0^2 + y_0^2 = 4$.*

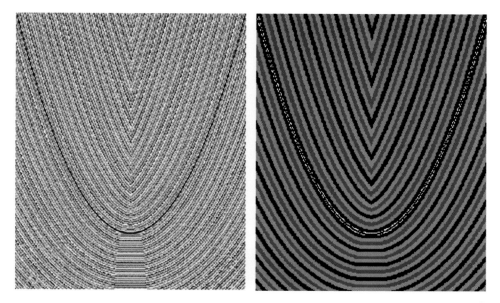

Figure 1.30 Approximate level sets for the shortest distance function for part of the parabola $y = x^2$. The exact level sets are differentiable curves close enough to the parabola. But inside the parabola, on the axis of symmetry, they have a discontinuous first derivative for $y > 0.5$.

The distance from one set to another

In this subsection we complete the definition of the Hausdorff metric.

DEFINITION 1.12.9   Let $(\mathbb{X}, d_{\mathbb{X}})$ be a metric space. Let $\mathbb{H}(\mathbb{X})$ denote the space of nonempty compact subsets of $\mathbb{X}$. The **distance from** $A \in \mathbb{H}(\mathbb{X})$ **to** $B \in \mathbb{H}(\mathbb{X})$ is defined by

$$\mathcal{D}_B(A) := \max\{\mathcal{D}_B(a) : a \in A\} \quad \text{for all } A, B \in \mathbb{H}(\mathbb{X}).$$

Again, this definition makes sense because $\mathcal{D}_B(x)$ is a continuous function of $x \in A$ and A is compact, so there must exist a point $\widehat{a} \in A$ such that $\mathcal{D}_B(\widehat{a}) \geq \mathcal{D}_B(a)$ for all $a \in A$.

In Figure 1.31 we illustrate a visual, 'optical', way of calculating and thinking about the function $\mathcal{D}_B : \mathbb{H}(\mathbb{X}) \to [0, \infty)$ when $\mathbb{X} = \mathbb{R}^2$. The top left panel illustrates the interaction between the shortest-distance functions for a fern-like subset of $\mathbb{R}^2$ and a square subset. The level sets of the shortest-distance function for the fern-like subset are coloured in various intensities of turquoise. Specifically, the level set L_d is coloured according to $red = 0$, $green = d$, $blue = d$, for $d = 0, 1, 2, \ldots, 255$. Superimposed upon this picture, in the red bitplane, is a picture of a square, coloured according to $red = 200$, $green = 0$, $blue = 0$. The result is that the brightest points on the square are those that are at the greatest distance from the fern. That is, each point $\widehat{a}$ on the square which is brightest,

Figure 1.31 Level sets of shortest-distance functions for a fern-like set and a square set. See the main text. If each red band in the lower left panel corresponds to one unit of distance, what is (approximately) the greatest shortest distance *from* the fern *to* the square, $\mathcal{D}_{square}(fern)$?

somewhere in the white part of the square, occurs where $\mathcal{D}_{fern}(square) = \mathcal{D}_{fern}(\widehat{a})$; see Figure 1.26. An optical device could in principle be used to find the brightest points.

In practice, some digital image processing effects can be seen in the top left panel of Figure 1.31; they are quantization bands associated with the printing and render this description even more approximate than it would otherwise be.

The bottom left panel in Figure 1.31 illustrates the shortest-distance functions for both the fern and the square, with the level sets of the latter represented in shades of red. See also Figure 1.33.

The following theorem provides a kind of triangle inequality for the function $\mathcal{D}_B(A)$.

THEOREM 1.12.10   *Let* $(\mathbb{X}, d_\mathbb{X})$ *be a metric space and* $\mathbb{H}(\mathbb{X})$ *denote the nonempty compact subsets of* $\mathbb{X}$. *Then*

$$\mathcal{D}_B(A) \leq \mathcal{D}_B(C) + \mathcal{D}_C(A) \quad \text{for all } A, B, C \in \mathbb{H}(\mathbb{X}).$$

PROOF   For any $a \in A$ we have

$$\begin{aligned}
\mathcal{D}_B(a) &= \min_{b \in B} d(a, b) \\
&\leq \min_{b \in B}(d(a, c) + d(c, b)) \quad \text{for all } c \in C, \\
&= d(a, c) + \min_{b \in B} d(c, b) \quad \text{for all } c \in C.
\end{aligned}$$

It follows that

$$\begin{aligned}
\mathcal{D}_B(a) &\leq \min_{c \in C} d(a, c) + \max_{c \in C} \min_{b \in B} d(c, b) \\
&= \mathcal{D}_C(a) + \mathcal{D}_B(C) \quad \text{for all } a \in A.
\end{aligned}$$

Now take the maximum over $a \in A$ on both sides of this equation. $\square$

EXERCISE 1.12.11   *Show that*

$$\mathcal{D}_A(B \cup C) = \max\{\mathcal{D}_A(B), \mathcal{D}_A(C)\} \quad \text{for all } A, B, C \in \mathbb{H}(\mathbb{X}).$$

Draw a picture to illustrate the content of this equation.

EXERCISE 1.12.12   *Show that*

$$\mathcal{D}_{A \cup B}(C) \leq \min\{\mathcal{D}_A(C), \mathcal{D}_B(C)\} \quad \text{for all } A, B, C \in \mathbb{H}(\mathbb{X}).$$

Draw a picture to illustrate the content of this equation.

Finally we are in a position to define the Hausdorff metric.

THEOREM 1.12.13   *Let* $(\mathbb{X}, d_\mathbb{X})$ *be a metric space and* $\mathbb{H}(\mathbb{X})$ *denote the nonempty compact subsets of* $\mathbb{X}$. *Let*

$$d_{\mathbb{H}(\mathbb{X})}(A, B) := \max\{\mathcal{D}_B(A), \mathcal{D}_A(B)\} \quad \text{for all } A, B \in \mathbb{H}(\mathbb{X}).$$

Then $(\mathbb{H}(\mathbb{X}), d_{\mathbb{H}(\mathbb{X})})$ *is a metric space.*

PROOF   We write $d_{\mathbb{H}(\mathbb{X})} = d_\mathbb{H}$. We will demonstrate with reference to Definition 1.5.1 that $d_\mathbb{H}$ is indeed a metric on the space $\mathbb{H}(\mathbb{X})$. (i) $d_\mathbb{H}(A, B) = \max\{\mathcal{D}_B(A), \mathcal{D}_A(B)\} = \max\{\mathcal{D}_A(B), \mathcal{D}_B(A)\} = d_\mathbb{H}(B, A)$. (ii) and (iii) Notice that $d_\mathbb{H}(A, B)$ equals either $\mathcal{D}_B(A)$ or $\mathcal{D}_A(B)$. Hence, using the compactness of A and B and the continuity of $d(x, y)$, it then follows that $d_\mathbb{H}(A, B) = d(\widehat{a}, \widehat{b})$ for some $\widehat{a} \in A$ and $\widehat{b} \in B$. It then follows that $0 \leq d_\mathbb{H}(A, B) < \infty$. Suppose that $A \neq B$. Then, without loss of generality, we can assume that there exists a point $a \in A$ such that $a \notin B$. Hence $\mathcal{D}_B(A) = \max\{\mathcal{D}_B(a) : a \in A\} > 0$ and so $d_\mathbb{H}(A, B) > 0$. (iv) From Theorem 1.12.10 we have that $\mathcal{D}_B(A) \leq \mathcal{D}_C(A) + \mathcal{D}_B(C)$ and that $\mathcal{D}_A(B) \leq \mathcal{D}_A(C) + \mathcal{D}_C(B)$. Hence $d_{\mathbb{H}}(A, B) \leq \max\{\mathcal{D}_C(A) + \mathcal{D}_B(C), \mathcal{D}_A(C) + \mathcal{D}_C(B)\} \leq \max\{\mathcal{D}_C(A), \mathcal{D}_A(C)\} + \max\{\mathcal{D}_B(C), \mathcal{D}_C(B)\} = d_{\mathbb{H}}(A, C) + d_{\mathbb{H}}(C, B)$. $\qquad\square$

DEFINITION 1.12.14 The metric $d_{\mathbb{H}} = d_{\mathbb{H}(\mathbb{X})}$ is called the **Hausdorff metric**. The quantity $d_{\mathbb{H}}(A, B)$ is called the **Hausdorff distance** between the points $A, B \in \mathbb{H}(\mathbb{X})$.

We remark as an aside that it is possible to define a type of 'distance' between any pair of bounded subsets of a metric space by replacing the maximum and minimum operators by supremum and infimum operators, which are defined as follows. When $S \subset \mathbb{R}$ is a bounded set then $\inf S = \max\{x \in \mathbb{R} : x \leq s$ for all $s \in S\}$, and similarly $\sup S = \min\{x \in \mathbb{R} : x \geq s$ for all $s \in S\}$. But the result is not a metric, in general. For example the 'distance' between an open set O and its closure $\overline{O}$ is zero but it is not true in general that $O = \overline{O}$.

The following theorem provides a characteristic but at first sight somewhat suprising property of the Hausdorff distance. It will be most useful later on.

THEOREM 1.12.15 *Let $(\mathbb{X}, d_{\mathbb{X}})$ be a metric space and $\mathbb{H}(\mathbb{X})$ denote the nonempty compact subsets of $\mathbb{X}$. Then*

$$d_{\mathbb{H}}(A \cup B, C \cup D) \leq \max\{d_{\mathbb{H}}(A, C), d_{\mathbb{H}}(B, D)\}$$

for all $A, B, C, D \in \mathbb{H}(\mathbb{X})$.

PROOF First we verify the claim in Exercise 1.12.11: we have

$$\begin{aligned}
\mathcal{D}_A(B \cup C) &= \max_{x \in B \cup C} \min_{a \in A} d(a, x) \\
&= \max\left\{\max_{b \in B} \min_{a \in A} d(a, b), \max_{c \in C} \min_{a \in A} d(a, c)\right\} \\
&= \max\{\mathcal{D}_A(B), \mathcal{D}_A(C)\}.
\end{aligned}$$

It follows that

$$\mathcal{D}_{A \cup B}(C \cup D) = \max\{\mathcal{D}_{A \cup B}(C), \mathcal{D}_{A \cup B}(D)\}. \tag{1.12.4}$$

Now we verify the claim in Exercise 1.12.12: we have

$$\begin{aligned}
\mathcal{D}_{A \cup B}(C) &= \max_{c \in C} \min_{x \in A \cup B} d(c, x) = \max_{c \in C} \min\left\{\min_{a \in A} d(c, a), \min_{b \in B} d(c, b)\right\} \\
&\leq \min\left\{\max_{c \in C} \min_{a \in A} d(c, a), \max_{c \in C} \min_{b \in B} d(c, b)\right\} \\
&= \min\{\mathcal{D}_A(C), \mathcal{D}_B(C)\}.
\end{aligned}$$

It follows that

$$\mathcal{D}_{A \cup B}(C) \leq \mathcal{D}_A(C) \quad \text{and} \quad \mathcal{D}_{A \cup B}(D) \leq \mathcal{D}_B(D).$$

Substituting from the latter pair of equations into the right-hand side of Equation (1.12.4) we obtain

$$\mathcal{D}_{A \cup B}(C \cup D) \leq \max\{\mathcal{D}_A(C), \mathcal{D}_B(D)\}.$$

It follows that

$$\mathcal{D}_{C \cup D}(A \cup B) \leq \max\{\mathcal{D}_C(A), \mathcal{D}_D(B)\}.$$

Hence

$$
\begin{aligned}
d_{\mathbb{H}}(A \cup B, C \cup D) &= \max\{\mathcal{D}_{A \cup B}(C \cup D), \mathcal{D}_{C \cup D}(A \cup B)\} \\
&\leq \max\big\{\max\{\mathcal{D}_A(C), \mathcal{D}_B(D)\}, \ \max\{\mathcal{D}_C(A), \mathcal{D}_D(B)\}\big\} \\
&\leq \max\{\mathcal{D}_A(C), \mathcal{D}_B(D), \ \mathcal{D}_C(A), \mathcal{D}_D(B)\} \\
&= \max\big\{\max\{\mathcal{D}_A(C), \mathcal{D}_C(A)\}, \ \max\{\mathcal{D}_D(B), \mathcal{D}_B(D)\}\big\} \\
&= \max\{d_{\mathbb{H}}(A, C), d_{\mathbb{H}}(B, D)\}.
\end{aligned}
$$

□

The metric space $(\mathbb{H}(\mathbb{X}), d_{\mathbb{H}})$ inherits properties from the underlying metric space $(\mathbb{X}, d)$. For example, in Section 1.13, we show that if $(\mathbb{X}, d)$ is complete then $(\mathbb{H}(\mathbb{X}), d_{\mathbb{H}})$ is complete. Also, if $(\mathbb{X}, d)$ is compact then $(\mathbb{H}(\mathbb{X}), d_{\mathbb{H}})$ is compact and, under certain conditions, when $(\mathbb{X}, d)$ is connected then $(\mathbb{H}(\mathbb{X}), d_{\mathbb{H}})$ is connected. The inheritance of completeness is of particular importance to us because it leads to beautiful simple proofs of the existence of many fractals and superfractals.

EXERCISE 1.12.16   *Let* $A = \{(x, y) \in \mathbb{R}^2 : x^2 + y^2 = 1\}$ *and* $B = \{(x, y) \in \mathbb{R}^2 : y = 0, 0 \leq x \leq 1\}$. *Compute* $d_{\mathbb{H}}(A, B)$ *when the underlying metric is the euclidean metric.*

EXERCISE 1.12.17   *Suppose that* $A \subset B$. *Show that* $d_{\mathbb{H}}(A, B) = \mathcal{D}_A(B)$.

EXERCISE 1.12.18   *Estimate the Hausdorff distance* $d_{\mathbb{H}}(A, B)$ *between the two sets* A *and* B, *which look like leaves, in Figure 1.32. Assume that the underlying metric is* $d_{\max}$. *Mark on the figure a pair of points* $\widehat{a}, \widehat{b}$ *such that* $d(\widehat{a}, \widehat{b}) = d_{\mathbb{H}}(A, B)$.

EXERCISE 1.12.19   *Let* $(\mathbb{X}, d)$ *be a metric space and let*

$$\widetilde{d_{\mathbb{H}}}(A, B) = \mathcal{D}_B(A) + \mathcal{D}_A(B) \quad \text{for all } A, B \in \mathbb{H}(\mathbb{X}).$$

Prove that $(\mathbb{H}(\mathbb{X}), \widetilde{d_{\mathbb{H}}})$ *is a metric space.*

Dilations of sets

In this subsection we explore an alternative characterization of the Hausdorff metric that has an 'optical' interpretation.

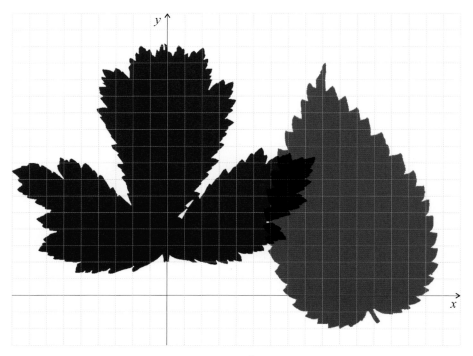

Figure 1.32 See Exercise 1.12.18: a pair of points $\hat{a}, \hat{b}$, one on each leaf, whose distance apart is equal to the Hausdorff distance between the leaves is to be located.

Given $C \in \mathbb{H}(\mathbb{X})$ and $r \in [0, \infty]$ we define the **set C dilated by** r to be

$$\mathcal{B}_C(r) = \{x \in \mathbb{X} : \mathcal{D}_C(x) \leq r\} = \cup \{\overline{B(x, r)} : x \in C\}.$$

That is, $\mathcal{B}_C(r)$ is obtained by taking the union of all closed balls of radius r centred at points of C. Clearly $\mathcal{B}_C : [0, \infty] \to \mathbb{S}(\mathbb{X})$ (the set of subsets of $\mathbb{X}$) and $\mathcal{B}_C(r)$ is an increasing family of subsets (i.e. $r_1 < r_2 \implies \mathcal{B}_C(r_1) \subset \mathcal{B}_C(r_2)$) with $\mathcal{B}_C(0) = C$ and $\mathcal{B}_C(\infty) = \mathbb{X}$. We refer to these subsets as **dilations** of C. We can characterize the Hausdorff distance in terms of dilations in the following manner.

THEOREM 1.12.20   *Let $(\mathbb{X}, d)$ be a complete metric space and let $(\mathbb{H}(\mathbb{X}), d_{\mathbb{H}})$ denote the corresponding space of compact nonempty subsets that has the Hausdorff metric. Then, for given $C, D \in \mathbb{H}(\mathbb{X})$, $d_{\mathbb{H}}(C, D)$ is the minimum value of r such that the dilation of C by r contains D and the dilation of D by r contains C.*

PROOF   We leave this as an exercise, or else see [9].   □

Notice in particular that

$$D \subset \mathcal{B}_C(d_{\mathbb{H}}(C, D)) \quad \text{and} \quad C \subset \mathcal{B}_D(d_{\mathbb{H}}(C, D)) \quad \text{for all } C, D \in \mathbb{H}(\mathbb{X}).$$
$$(1.12.5)$$

We will use this observation below in the proof of Theorem 1.13.2.

One reason why we are interested in characterizing the Hausdorff distance in terms of dilations is that, at least in the case of $\mathbb{R}^2$ with the euclidean metric, dilations of bounded sets may be computed by means of optical algorithms; see for example [93]. In the future, it may be possible to compute rapidly Hausdorff distances between images using optical computation.

It is not suprising that optical algorithms can be used to compute dilations. If you are shortsighted then, roughly speaking, dots at a fixed distance from the eye, close to the optical axis, are blurred to become, upon the retina, small disks of some radius ρ. A viewed flat object at the same distance, in a plane perpendicular to the optical axis, treated as a collection of dots, is similarly blurred, yielding upon the retina the dilation by ρ of the object.

Indeed, suppose we represent bounded subsets of $\mathbb{R}^2$ as black pictures against a white background. Suppose we 'look at' these pictures from various distances d with eyes or a camera of fixed resolving power. Then the effective dilation ρ of the pictures becomes greater when we look at them from further away. Let $\widehat{d}(A, B)$ denote the smallest distance from the plane at which the pictures of two subsets A, B are indistinguishable. Then roughly speaking $\widehat{d}(A, B) = f(d_{\mathbb{H}}(A, B))$ for all A, $B \in \mathbb{H}(\mathbb{X})$, where $f : [0, \infty) \to [0, \infty)$ is a monotone increasing function.

EXERCISE 1.12.21   *Prove that $\mathcal{B}_C(r) \subset \mathbb{X}$ is compact for all $C \in \mathbb{H}(\mathbb{X})$ and $r \geq 0$.*

We illustrate the application of dilations to the computation of the Hausdorff distance between pairs of compact sets with the aid of Figure 1.33. This illustrates the shortest-distance functions for three (nonempty compact) sets: a green fern, a red Sierpinski triangle and a blue square. *Equivalently it represents increasing families of dilated sets.* The outer boundaries of an increasing family of dilations of the green fern are illustrated in shades of green (in the green bitplane). The outer boundaries of an increasing family of dilations of the red Sierpinski triangle are illustrated in shades of red and boundaries of successive dilations of the blue square are represented in shades of blue. Figure 1.33 is the image that results when the red, green and blue bitplanes are superimposed.

Now imagine that each coloured band in Figure 1.33 represents one unit of distance. Then, by counting blue bands out from the square until the fern is engulfed, that is, reading off the minimum value of r such that $\mathcal{B}_{square}(r) \supset fern$ we find that $\mathcal{D}_{square}(fern) \simeq 8$. Similarly, by counting out the green bands from the fern until the square is engulfed, we obtain $\mathcal{D}_{fern}(square) \simeq 3$. Hence $d_{\mathbb{H}}(fern, square) \simeq 8$. In a similar manner we find that $\mathcal{D}_{square}(Sierpinski) \simeq 4.5$ and $\mathcal{D}_{Sierpinski}(square) \simeq 6.5$, so that $d_{\mathbb{H}}(square, Sierpinski) \simeq 6.5$. Also, we find that $\mathcal{D}_{fern}(Sierpinski) \simeq 3.5$, $\mathcal{D}_{Sierpinski}(fern) \simeq 9$ and $d_{\mathbb{H}}(fern, Sierpinski) \simeq 9$. The triangle inequality tells us that

$$d_{\mathbb{H}}(fern, Sierpinski) \leq d_{\mathbb{H}}(fern, square) + d_{\mathbb{H}}(square, Sierpinski),$$

which in the present case reads $9 \leq 8 + 6.5$.

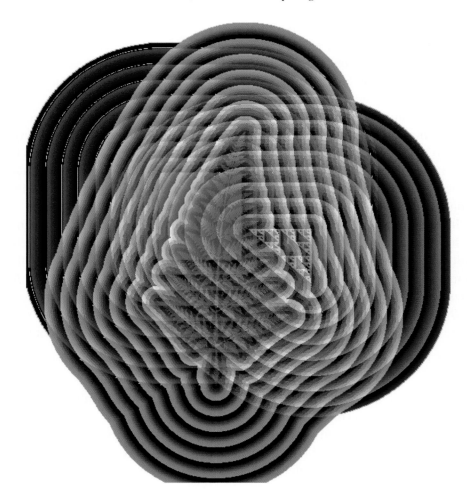

Figure 1.33 Interacting level sets of shortest-distance functions. The level sets are for a fern, a Sierpinski triangle and a square. See the main text.

EXERCISE 1.12.22   *Show that*

$$d_{\mathbb{H}}(C \cup D, E \cup F) \leq \max\{d_{\mathbb{H}}(C, E), d_{\mathbb{H}}(D, F)\} \quad \text{for all } C, D, E, F \in \mathbb{H}(\mathbb{X}).$$

EXERCISE 1.12.23   *Show that*

$$A \subset \mathcal{B}_B(\mathcal{D}_B(A)) \quad \text{for all } A, B \in \mathbb{H}(\mathbb{X}).$$

The furthest-distance function

In this subsection we discuss optimization problems associated with the Hausdorff distance.

In applications of the Hausdorff metric on $\mathbb{H}(\mathbb{R}^2)$ to pattern matching and fractal approximation, we are led to consider the minimization of $d_{\mathbb{H}}(A, B)$, where the 'target' set A is held fixed and the set B depends upon certain parameters. The goal is to adjust the parameters so that B is as close as possible to A. For example, as illustrated in Figure 1.34, A might be a set that looks like a leaf, B_0 might represent the silhouette of another leaf and B might represent B_0 translated by the vector (x, y), namely

$$B = B(x, y) = \{(x + x_0, y + y_0) : (x_0, y_0) \in B_0\} \quad \text{for all } (x, y) \in \mathbb{R}^2.$$

In this case the Hausdorff distance from $B(x, y)$ to A is a function $f : \mathbb{R}^2 \to [0, \infty)$ defined by

$$f(x, y) = d_{\mathbb{H}}(A, B(x, y)) \quad \text{for all } (x, y) \in \mathbb{R}^2. \tag{1.12.6}$$

We wish to search for the minimum value of f and the locations (x, y) at which this minimum is achieved.

There are many approaches to optimization problems that might be applied, but the problem is that finding the Hausdorff distance is computationally expensive and we do not have neat formulas or approximations for $f(x, y)$ with which to work. Also, we might wish to compare A with many different leaves B_0. How do we *start* to think about developing efficient algorithms to approach this type of problem?

Further insight into the behaviour of the Hausdorff distance on $\mathbb{H}(\mathbb{R}^2)$, apropos this question, is provided by the furthest-distance function. Let $x \in \mathbb{X}$ and $A \subset \mathbb{H}(\mathbb{X})$. Then

$$d_{\mathbb{H}}(\{x\}, A) = \max\left\{\mathcal{D}_A(\{x\}), \mathcal{D}_{\{x\}}(A)\right\} = \mathcal{D}_{\{x\}}(A).$$

We refer to $\mathcal{F}_A(x) := \mathcal{D}_{\{x\}}(A)$ as the **furthest-distance function** for A. We have

$$\mathcal{F}_A(x) = \max\{d(x, a) : a \in A\}.$$

Imagine that $A \subset \mathbb{R}^2$ represents a leaf and that an (infinitesimally small) ant is located at a point $x \in \mathbb{R}^2$. Then the path of steepest descent for $\mathcal{F}_A(x)$, which cuts the level sets of $\mathcal{F}_A(x)$ at right angles, where they are differentiable curves, represents the route to be followed by the ant to decrease the Hausdorff distance most rapidly. Then, usually, by following a path of steepest descent of $\mathcal{F}_A(x)$ the ant will arrive at some 'central' point on the leaf such that the Hausdorff distance between the leaf and the ant is a minimum. This is in contrast to what happens when the ant follows a path of steepest descent of $\mathcal{D}_A(x)$, which may lead the ant to a point on the boundary of the leaf nearest to its starting point. When driving to a city, the distance to the city specified on road signs is often the distance to

Figure 1.34 How do we adjust parameters such as x and y, which represent the positions of members of a set $B = B(x, y)$, so as to minimize the Hausdorff distance $d_{\mathbb{H}}(A, B(x, y))$?

the city centre rather than to the boundary of the city. But when driving to France from Italy, the distances quoted on road signs are to the border.

EXERCISE 1.12.24 *Figure 1.35 shows some level sets for both the shortest-distance function (in red) and the furthest-distance function (in black) associated with a line segment L. The number on a contour gives the value of the corresponding distance function. Neatly draw a few paths of steepest descent, for both $\mathcal{D}_L(x)$ and $\mathcal{F}_L(x)$. Where do the paths of steepest descent for $\mathcal{F}_L(x)$ terminate?*

Given the two functions $\mathcal{D}_A(x)$ and $\mathcal{F}_A(x)$ for some fixed $A \in \mathbb{H}(\mathbb{X})$, the following theorem provides a simple upper bound to $d_{\mathbb{H}}(A, B)$. This upper bound can be evaluated using only extrema of the two functions for $x \in B \in \mathbb{H}(\mathbb{X})$. This makes it easier to compare approximate distances for different values of B.

THEOREM 1.12.25 *Let $(\mathbb{X}, d)$ be a metric space and let $(\mathbb{H}(\mathbb{X}), d_{\mathbb{H}})$ denote the space of compact nonempty subsets of $\mathbb{X}$ together with the Hausdorff metric. Then*

$$d_{\mathbb{H}}(A, B) \leq \max \left\{ \max_{x \in B} \mathcal{D}_A(x), \min_{x \in B} \mathcal{F}_A(x) \right\} \quad \text{for all } A, B \in \mathbb{H}(\mathbb{X}).$$

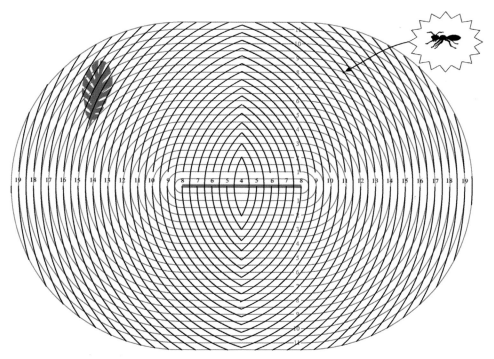

Figure 1.35 Level sets for the shortest-distance function $\mathcal{D}_L(x)$ (in red) and the furthest-distance function $\mathcal{F}_L(x)$ (in black) for the set $L \subset \mathbb{R}^2$. The set L is the line segment, shown in green, at the centre. The underlying metric is the euclidean distance. Contours are labelled with corresponding distances. This array of level sets can be used to estimate the Hausdorff distance from a set B that is overlaid on the contours. Imagine that the ant is very small, relative to the spacing of the contours, and located as indicated by the arrow. Then you can estimate $d_{\mathbb{H}}(ant, L)$ very accurately! Can you find an upper bound for $d_{\mathbb{H}}(leaf, L)$?

PROOF   This follows from

$$d_{\mathbb{H}}(A, B) = \max\{\mathcal{D}_A(B), \mathcal{D}_B(A)\},$$

where $\mathcal{D}_A(B) = \max_{x \in B} \mathcal{D}_A(x)$ and

$$\mathcal{D}_B(A) = \max_{a \in A} \min_{b \in B} d(a, b) \leq \min_{b \in B} \max_{a \in A} d(a, b) = \min_{x \in B} \mathcal{F}_A(x).$$

$\square$

Thus, the Hausdorff distance from A to B is bounded by the larger of the maximum value (on B) of the shortest-distance function $\mathcal{D}_A(x)$ and the minimum value (on B) of the furthest-distance function $\mathcal{F}_A(x)$.

EXERCISE 1.12.26   *Use Figure 1.35 to obtain an upper bound for $d_{\mathbb{H}}(L, leaf)$; see the figure caption.*

EXERCISE 1.12.27   *In Figure 1.36 trace the paths of shortest descent for both ants, with respect to $\mathcal{D}_L(x)$ and $\mathcal{F}_L(x)$.*

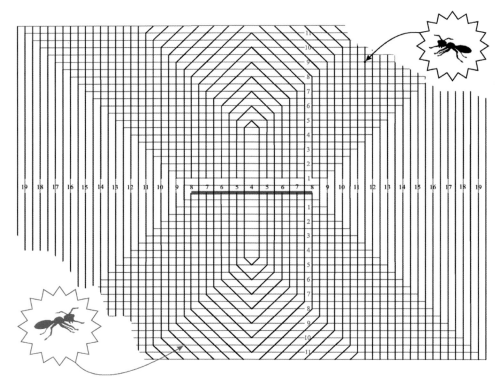

Figure 1.36 This figure is similar to Figure 1.35, but here the underlying metric is $d_{\max}$. Sketch some paths of steepest descent. What route will each ant follow to decrease $d_{\mathbb{H}}(ant, L)$ as rapidly as possible? Where will each ant end up?

EXERCISE 1.12.28 *Find the minimum value of $f(x, y)$ in Equation (1.12.6) and a value of (x, y) at which it occurs, when $A = \{(x, y) \in \mathbb{R}^2 : x^2 + y^2 = 1, x \geq 0\}$ and $B = \{(x, y) \in \mathbb{R}^2 : x^2 + y^2 = \frac{1}{2}\} \cup \{(0, y) \in \mathbb{R}^2 : -\frac{1}{2} \leq y \leq -\frac{1}{4}\}$.*

Hausdorff distances on code spaces

Here we consider distance functions on $\mathbb{H}(\Omega)$, the set of nonempty compact subsets of a code space. The code space may be the metric space (Ω, d_Ω), where d_Ω is defined in Equation (1.6.1), or $(\Omega, d_{|\mathcal{A}|})$, where $d_{|\mathcal{A}|}$ is defined in Equation (1.6.6), or more generally (Ω, d_ξ), where $\xi : \Omega \to \mathbb{R}^2$ is an embedding and

$$d_\xi(\sigma_1, \sigma_2) = |\xi(\sigma_1) - \xi(\sigma_2)| \quad \text{for all } \sigma_1, \sigma_2 \in \Omega,$$

as in Theorem 1.5.5.

When the underlying metric is obtained by embedding, as in the case of $d_{|\mathcal{A}|}$ and more generally d_ξ, it is possible to make 'pictures' of the associated shortest-distance functions and to think quite geometrically and 'optically' about the metric, as illustrated in Figures 1.37 and 1.38.

In Figure 1.37 we consider the code space (Ω, d_ξ), where $\Omega = \Omega_{\{0,1,2,3\}}$ and the embedding function $\xi : \Omega_{\{0,1,2,3\}} \to \mathbb{R}^2$ is defined by

$$\xi(\sigma) = \lim_{n\to\infty} f_{\sigma_1} \circ f_{\sigma_2} \circ \cdots \circ f_{\sigma_n}(x_0, y_0) \quad \text{for all } \sigma \in \Omega,$$

for some fixed $(x_0, y_0) \in \mathbb{R}^2$. Here

$$f_\omega(x, y) = \left(\frac{x}{3} + \frac{2}{3}\left[\frac{\omega}{2}\right], \frac{y}{3} + \frac{(1 - (-1)^\omega)}{3} \right)$$

for all $(x, y) \in \mathbb{R}^2$, $\omega \in \{0, 1, 2, 3\}$, where $[x]$ denotes the greatest integer less than or equal to the real number x. We defer until Chapter 4 a proof that ξ is indeed an embedding function and a more precise discussion of such embeddings. What matters here is that the embedded set $\xi(\Omega)$ looks like the set of green points in the bottom left panel of Figure 1.37. What you cannot see is that each small green rectangle represents many more green rectangles organized in the same sort of way as those green rectangles that you can see, and so on. $\xi(\Omega)$ is in fact of the form $C \times C \subset \mathbb{R}^2$, where $C \subset \mathbb{R}$ is a classical Cantor set.

The top right panel of Figure 1.37 represents the embedded set $\xi(S)$, where $S \subset \Omega$. The top left panel shows level sets of $\mathcal{D}_{\xi(S)}(x)$ for $x \in \square \subset \mathbb{R}^2$. Of course, this top left panel is not a picture of the level sets of $\mathcal{D}_S(\sigma)$ for $\sigma \in \Omega$, because most points x on level sets of $\mathcal{D}_{\xi(S)}(x)$ do not correspond to points in Ω. But the level sets of $\mathcal{D}_{\xi(S)}(x)$ accurately describe the distances from points in Ω to points in S for all $x \in R_\xi$, the range of ξ, because

$$\mathcal{D}_{\xi(S)}(\xi(\sigma)) = \mathcal{D}_S(\sigma) \quad \text{for all } \sigma \in \Omega.$$

The function $\mathcal{D}_{\xi(S)} : \mathbb{R}^2 \to [0, \infty)$ is a continuous extension of $\mathcal{D}_{\xi(S)} : R_\xi \subset \mathbb{R}^2 \to [0, \infty)$ to all $\mathbb{R}^2$.

The bottom right panel shows the level sets of $\mathcal{D}_{\xi(S)}(x)$, for $x \in \square$, superimposed on $\xi(\Omega)$. Since this panel contains in principle the points of both $\xi(S)$ and $\xi(\Omega)$, we can use it to estimate $\mathcal{D}_{\xi(S)}(\Omega)$ and hence, since $\mathcal{D}_\Omega(\xi(S)) = 0$, $d_{\mathbb{H}}(\Omega, S)$. But our purpose, of course, is not so much to do this as it is to enable us to think geometrically and visually about code-space metrics.

In Figure 1.38 the level sets of $\mathcal{D}_{\xi(S)}(x)$, in the right-hand panel, may be compared with the level sets of $\mathcal{D}_{\xi(\Omega)}(x)$, in the left-hand panel, for $x \in \square \subset \mathbb{R}^2$. In this example $\Omega = \Omega_{\{0,1\}} \cup \Omega'_{\{0,1\}}$, where $\Omega_{\{0,1\}}$ and $\Omega'_{\{0,1\}}$ are the code spaces defined in Section 1.4, and a different embedding function $\xi : \Omega \to \mathbb{R}^2$ is used, similar to the one used in Figure 1.15. The points of $\xi(\Omega'_{\{0,1\}})$ are situated at the branch points, also called nodes, of a tree-like structure in $\mathbb{R}^2$ much the same as that seen in Figure 1.15 and the points of $\xi(\Omega_{\{0,1\}})$ are located on the canopy of the tree-like structure. Although it does not appear to be so, the canopy $\xi(\Omega_{\{0,1\}})$

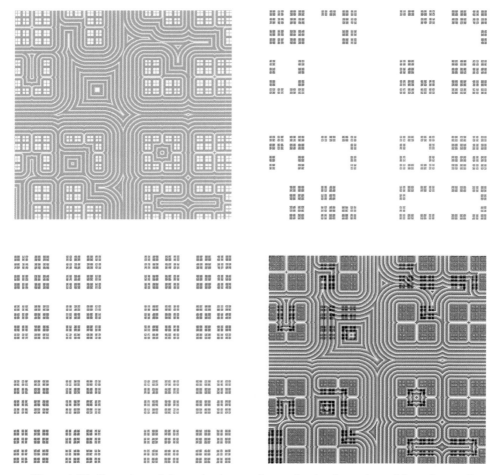

Figure 1.37 Illustrations relating to the shortest-distance function for a subset of the codespace $\Omega = \Omega_{\{0,1,2,3\}}$. The bottom left panel shows an embedded set $\xi(\Omega)$ of the code space Ω in $\mathbb{R}^2$, where $\xi : \Omega \to \mathbb{R}^2$ is the embedding function. The top right panel shows the embedded set $\xi(S)$ of a compact subset $S \subset \Omega$. The top left panel shows level sets of $\mathcal{D}_{\xi(S)}(x)$. The bottom right panel shows the level sets of $\mathcal{D}_{\xi(S)}(x)$ superimposed on $\xi(\Omega)$. Assuming that the width of each band of level sets is one unit, can you estimate $\mathcal{D}_{\xi(S)}(\Omega)$?

is totally disconnected. You can deduce the locations of some of the points of $\xi(\Omega'_{\{0,1\}})$ because they are at the centres of concentric circles formed by level sets. Comparison between the two images in Figure 1.38 enables us, as in the previous example, to estimate $d_{\mathbb{H}}(S, \Omega)$ by making use of the fact that $\mathcal{D}_{\xi(S)}(\xi(\sigma)) = \mathcal{D}_S(\sigma)$ for all $\sigma \in \Omega$.

When the underlying metric is d_Ω it is hard to make illustrations similar to Figures 1.37 and 1.38, because generally there exist large sets of equidistant points in the metric space (Ω, d_Ω). For example, when $\Omega = \Omega_{\{0,1\}}$ there exists a set containing 2^m points, each of which is at a distance $1/2^m$ from all the other points

Figure 1.38 Can you spot the differences and thereby estimate the Hausdorff distance between the two embedded sets? This figure shows the level sets of the shortest-distance function for embeddings of the code space $\Omega_{\{0,1\}} \cup \Omega'_{\{0,1\}}$ and a subset of the code space. The embedding function here is similar to the one used in Figure 1.15.

in the set, for all $m = 1, 2, \ldots$ This implies that there does not exist an embedding $\xi : \Omega_{\{0,1\}} \to \mathbb{R}^n$ such that $d_\Omega(\sigma, \omega) = d_{euclidean}(\xi(\sigma), \xi(\omega))$ for all $\sigma, \omega \in \Omega_{\{0,1\}}$ and all $n = 1, 2, \ldots$; if the latter were the case then there would exist in $\mathbb{R}^n$ a set containing more than $n + 1$ points, each of which is at unit euclidean distance from all the other points in the set. The latter statement is not true, as demonstrated in Exercise 1.5.17. See also Figure 1.12.

In this sense we can think of the space $(\Omega_\mathcal{A}, d_\Omega)$ as being very high dimensional, whereas we can think of $(\Omega_\mathcal{A}, d_{|\mathcal{A}|})$ as being contained in a one-dimensional space. Recall that $d_{|\mathcal{A}|}$ is defined in Equation (1.6.6) by means of an embedding of Ω in $\mathbb{R}$. Despite this difference, remember that, as asserted in Theorem 1.9.6, the natural topology on (Ω, d_Ω) is the same as the natural topology on $(\Omega, d_{|\mathcal{A}|})$.

EXERCISE 1.12.29 *Show that in the code space $(\Omega_{\{0,1\}}, d_\Omega)$ the furthest-distance function $\mathcal{F}_{\Omega_{\{0,1\}}}(\sigma)$ of $\Omega_{\{0,1\}}$ is constant for all $\sigma \in \Omega_{\{0,1\}}$. Show too that in $(\Omega_{\{0,1\}}, d_{|\mathcal{A}|})$ we have $\mathcal{F}_{\Omega_{\{0,1\}}}(\sigma) = \max\{\xi(\sigma), \xi(111\cdots) - \xi(\sigma)\}$, where ξ is the embedding function defined in Equation (1.6.6).*

EXERCISE 1.12.30 *Prove that in the metric space $(\Omega_{\{0,1\}}, d_\Omega)$ there exists a set containing 2^m points, each of which is at a distance $1/2^m$ from all the other points in the set, for all $m = 1, 2, \ldots$*

1.13   The metric spaces $(\mathbb{H}(\mathbb{X}), d_\mathbb{H})$, $(\mathbb{H}(\mathbb{H}(\mathbb{X})), d_{\mathbb{H}(\mathbb{H})})$, ...

In this section we investigate some properties of the space $(\mathbb{H}(\mathbb{X}), d_\mathbb{H})$ that it inherits from the underlying metric space $(\mathbb{X}, d)$. The space $(\mathbb{H}(\mathbb{X}), d_\mathbb{H})$ is a very natural setting in which to study fractal sets. As we will see in Chapter 4, sequences of approximations to fractal sets may be described as Cauchy sequences of points in $(\mathbb{H}(\mathbb{X}), d_\mathbb{H})$. Thus the existence of limits of such sequences, the fractals themselves, depends upon the the completeness of the space $(\mathbb{H}(\mathbb{X}), d_\mathbb{H})$. Similarly, the existence of superfractal sets depends upon the completeness of the space $(\mathbb{H}(\mathbb{H}(\mathbb{X})), d_{\mathbb{H}(\mathbb{H})})$, as we will see in Chapter 5. So we begin by showing that $(\mathbb{H}(\mathbb{X}), d_\mathbb{H})$ inherits the property of completeness from the space $(\mathbb{X}, d)$.

The completeness of $\mathbb{H}(\mathbb{X})$

The statements and proofs of Theorems 1.13.1 and 1.13.2 follow closely [9], p. 34, Lemma 7.2 and p. 35, Theorem 7.1.

THEOREM 1.13.1   *(Extension lemma)   Let* $(\mathbb{X}, d_\mathbb{X})$ *be a complete metric space and let* $\{A_n \in \mathbb{H}(\mathbb{X})\}_{n=1}^\infty$ *be a Cauchy sequence in* $(\mathbb{H}(\mathbb{X}), d_\mathbb{H})$. *Suppose that* $\{x_{n_j} \in A_{n_j}\}_{j=1}^\infty$ *is a Cauchy sequence in* $(\mathbb{X}, d_\mathbb{X})$, *where* $\{n_j\}_{j=1}^\infty$ *is an increasing sequence of positive integers. Then there exists a Cauchy sequence* $\{x_n \in A_n\}_{n=1}^\infty$ *in* $(\mathbb{X}, d_\mathbb{X})$ *for which* $\{x_{n_j} \in A_{n_j}\}_{j=1}^\infty$ *is a subsequence.*

PROOF   Let $n_0 = 0$. For each $j \in \{1, 2, 3, \ldots\}$ and $n \in \{n_{j-1} + 1, \ldots, n_j\}$ choose $x_n \in A_n$ such that $\mathcal{D}_{A_n}(x_{n_j}) = d_\mathbb{X}(x_n, x_{n_j})$. Then $\{x_{n_j} \in A_{n_j}\}_{j=1}^\infty$ is a subsequence of $\{x_n \in A_n\}_{n=1}^\infty$. To show that the latter is a Cauchy sequence let $\epsilon > 0$ be given. There is an integer $N_1 > 0$ such that whenever $n_k, n_l \geq N_1$ we have $d_\mathbb{X}(x_{n_k}, x_{n_l}) \leq \epsilon/3$. Also, there is an integer $N_2 > 0$ such that whenever $m, n \geq N_2$ we have $d_\mathbb{H}(A_n, A_m) \leq \epsilon/3$.

So we assume that $n_k, n_l \geq N_1$ and that $m, n \geq N_2$. Then we note that, by the triangle inequality,

$$d_\mathbb{X}(x_m, x_n) \leq d_\mathbb{X}(x_m, x_{n_k}) + d_\mathbb{X}(x_{n_k}, x_{n_l}) + d_\mathbb{X}(x_{n_l}, x_n).$$

Let k, l be such that $m \in \{n_{k-1} + 1, \ldots, n_k\}$, $n \in \{n_{l-1} + 1, \ldots, n_l\}$ and let $m, n \geq \max\{N_1, N_2\}$. Then $d_\mathbb{X}(x_m, x_{n_k}) = \mathcal{D}_{A_m}(x_{n_k}) \leq \mathcal{D}_{A_m}(A_{n_k}) \leq d_\mathbb{H}(A_n, A_{n_k}) \leq \epsilon/3$; similarly, $d_\mathbb{X}(x_{n_l}, x_n) \leq \epsilon/3$. Since we also have $d_\mathbb{X}(x_{n_k}, x_{n_l}) \leq \epsilon/3$ it follows that $d_\mathbb{X}(x_m, x_n) \leq \epsilon$ for all $m, n \geq \max\{N_1, N_2\}$.   □

The following result provides not only a general condition under which $(\mathbb{H}(\mathbb{X}), d_\mathbb{H})$ is complete but also a characterization of the limits of Cauchy sequences in $\mathbb{H}(\mathbb{X})$.

THEOREM 1.13.2   *Let* $(\mathbb{X}, d_\mathbb{X})$ *be a complete metric space. Then* $(\mathbb{H}(\mathbb{X}), d_\mathbb{H})$ *is a complete metric space. Moreover, if* $\{A_n \in \mathbb{H}(\mathbb{X})\}_{n=1}^\infty$ *is a Cauchy sequence then* $A := \lim_{n\to\infty} A_n$ *can be characterized as*

$$A = \{x \in \mathbb{X} : \text{there is a Cauchy sequence } \{x_n \in A_n\}_{n=1}^{\infty} \text{ that converges to } x\}.$$

(1.13.1)

Proof   Let $\{A_n \in \mathbb{H}(\mathbb{X})\}_{n=1}^{\infty}$ and let A be defined by Equation (1.13.1). We prove that (i) $A \neq \varnothing$; (ii) A is closed and hence complete; (iii) for given $\epsilon > 0$ there is an N such that for $n \geq N$ we have $A \subset \mathcal{B}_{A_n}(\epsilon)$; (iv) A is totally bounded and hence by (ii) is compact; (v) $A = \lim_{n\to\infty} A_n$.

Proof of (i): We establish the existence of a Cauchy sequence $\{x_n \in A_n\}_{n=1}^{\infty}$ in $\mathbb{X}$. We can select an increasing sequence of positive integers $\{N_n\}_{n=1}^{\infty}$ such that $d_{\mathbb{H}}(A_n, A_m) \leq 1/2^i$ for $m, n > N_i$. Choose $x_{N_1} \in A_{N_1}$. Then since $d_{\mathbb{H}}(A_{N_1}, A_{N_2}) \leq 1/2$ we can find $x_{N_2} \in A_{N_2}$ such that $d_{\mathbb{X}}(x_{N_1}, x_{N_2}) < 1/2$. Now use an inductive argument to show that we can find an infinite sequence $\{x_{n_j} \in A_{n_j}\}_{j=1}^{\infty}$ such that $d_{\mathbb{X}}(x_{N_j}, x_{N_{j+1}}) < 1/2^j$. Then it readily follows that $\{x_{n_j} \in A_{n_j}\}_{j=1}^{\infty}$ is a Cauchy sequence. Now use Theorem 1.13.1 to yield the existence of a convergent sequence $\{x_n \in A_n\}_{n=1}^{\infty}$. Since $\mathbb{X}$ is complete the limit exists and, by the definition of A, Equation (1.13.1), it belongs to A.

Proof of (ii): To show that A is closed, suppose that $\{a_i \in A\}_{i=1}^{\infty}$ converges to a point $a \in \mathbb{X}$. We need to show that $a \in A$. Hence we can find an increasing sequence of integers $\{N_n\}_{n=1}^{\infty}$ such that $d_{\mathbb{X}}(a_{N_n}, a) < 1/n$. Also, since $a_i \in A$ it follows from the definition of A that there is a sequence $\{a_{i,n} \in A_n\}_{n=1}^{\infty}$ that converges to a_i for each i. And so we can find an increasing sequence of integers $\{M_n\}_{n=1}^{\infty}$ such that $d_{\mathbb{X}}(a_{N_n, M_n}, a_{N_n}) < 1/n$. It follows that $d_{\mathbb{X}}(a_{N_n, M_n}, a) < 2/n$. Hence the sequence $\{x_{N_n} = a_{N_n, M_n} \in A_{N_n}\}_{n=1}^{\infty}$ is a Cauchy sequence convergent to a. By Theorem 1.13.1 it can be extended to a sequence $\{x_n \in A_n\}_{n=1}^{\infty}$ convergent to a and, by the definition of A, Equation (1.13.1), it follows that a belongs to A.

Proof of (iii): Let $\epsilon > 0$. Then there exists N such that $n, m \geq N$ implies that $d_{\mathbb{H}}(A_n, A_m) \leq \epsilon$ and, as in Equation (1.12.5), $A_m \subset \mathcal{B}_{A_n}(\epsilon)$. Let $a \in A$ and let $\{a_m \in A_m\}_{m=1}^{\infty}$ be a sequence that converges to a. Then we must have $a_m \in \mathcal{B}_{A_n}(\epsilon)$ whenever $n, m \geq N$. But $\mathcal{B}_{A_n}(\epsilon)$ is closed because A_n is compact. So $a \in \mathcal{B}_{A_n}(\epsilon)$ whenever $n \geq N$ and therefore $A \subset \mathcal{B}_{A_n}(\epsilon)$ for all $n \geq N$.

Proof of (iv): Suppose that A is not totally bounded. Then for some $\epsilon > 0$ we can find a sequence of points $\{x_i \in A\}_{i=1}^{\infty}$ such that $d_{\mathbb{X}}(x_i, x_j) \geq \epsilon$ whenever $i \neq j$. From (iii) we have that $A \subset \mathcal{B}_{A_n}(\epsilon/3)$ for some large enough n. It follows that for each x_i we can find a corresponding $y_i \in A_n$ such that $d_{\mathbb{X}}(x_i, y_i) \leq \epsilon/3$. Since A_n is compact some subsequence $\{y_{i_j}\}_{j=1}^{\infty}$ of $\{y_i\}_{i=1}^{\infty}$ converges. So we can find points y_{j_1} and y_{j_2} such that $d_{\mathbb{X}}(y_{j_1}, y_{j_2}) < \epsilon/3$. But then it follows that

$$d_{\mathbb{X}}(x_{j_1}, x_{j_2}) \leq d_{\mathbb{X}}(x_{j_1}, y_{j_1}) + d_{\mathbb{X}}(y_{j_1}, y_{j_2}) + d_{\mathbb{X}}(y_{j_2}, x_{j_2}) < \epsilon.$$

This is a contradiction. So A is totally bounded. Since A is also complete, by (ii), it must be compact.

Proof of (v): From (iv) we have that $A \in \mathbb{H}(\mathbb{X})$. Let $\epsilon > 0$ and choose N so large that $n, m \geq N$ implies $d_{\mathbb{H}}(A_n, A_m) \leq \epsilon/2$ and $A_m \subset \mathcal{B}_{A_n}(\epsilon/2)$. Let $n \geq N$ and $y \in A_n$. There exists an increasing sequence of integers greater than n, $\{N_i\}_{i=1}^{\infty}$, such that for $m, k \geq N_j$, $A_m \subset \mathcal{B}_{A_k}(\epsilon/2^{j+1})$. Note that $A_n \subset \mathcal{B}_{A_{N_1}}(\epsilon/2)$. Since $y \in A_n$ there is a point $x_{N_1} \in A_{N_1}$ such that $d_{\mathbb{X}}(y, x_{N_1}) \leq \epsilon/2$. Since $x_{N_2} \in A_{N_2}$ there is a point $x_{N_2} \in A_{N_2}$ such that $d_{\mathbb{X}}(x_{N_1}, x_{N_2}) \leq \epsilon/2^2$. Continuing in this manner we may show by induction that there is a sequence $\{x_{N_j} \in A_{N_j}\}_{j=1}^{\infty}$ such that $d_{\mathbb{X}}(x_{N_j}, x_{N_{j+1}}) \leq \epsilon/2^{j+1}$. It follows that $\{x_{N_j} \in A_{N_j}\}_{j=1}^{\infty}$ is a Cauchy sequence that converges to a point $x \in A$ and that $d(y, x_{N_j}) \leq \epsilon$ for all j. The latter implies that $d(y, x) \leq \epsilon$. Hence $A_n \subset \mathcal{B}_A(\epsilon)$ for all $n \geq N$. But, by (iii), $A \subset \mathcal{B}_{A_n}(\epsilon)$ for all sufficiently large n. It follows that $d_{\mathbb{H}}(A_n, A) \leq \epsilon$ for all sufficiently large n. Hence $A = \lim_{n \to \infty} A_n$. $\qquad\qquad\qquad\square$

A simple example of a Cauchy sequence of points in $\mathbb{H}(\mathbb{X})$ is $\{\mathcal{B}_A(1/n)\}_{n=1}^{\infty}$ for $A \in \mathbb{H}(\mathbb{X})$. Clearly $\{\mathcal{B}_A(1/n)\}_{n=1}^{\infty}$ converges to A, whether or not $\mathbb{X}$ is complete.

Figure 1.39 shows a sequence of images that represents a Cauchy sequence of compact subsets of $\mathbb{R}^3$. Read the images from left to right and from top to bottom. The intensity of green represents the z-component of the set. The base of each image is taken to lie on the x-axis. In such cases we can infer the existence of the limiting fractal fern from the existence of the Cauchy sequence and the completeness of $\mathbb{R}^3$.

EXERCISE 1.13.3   *Show that if $(\mathbb{X}, d_{\mathbb{X}})$ is a compact metric space then $(\mathbb{H}(\mathbb{X}), d_{\mathbb{H}})$ is a compact metric space. Hint: Assume that $\mathbb{X}$ is nonempty. Define $A_n = \mathbb{X}$ for all $n = 1, 2, \ldots$ Then $\{A_n \in \mathbb{H}(\mathbb{X})\}_{n=1}^{\infty}$ is a Cauchy sequence that converges to $\mathbb{X}$. Now look back at the proof of Theorem 1.13.2.*

EXERCISE 1.13.4   *Show that $\mathbb{H}(\mathbb{R})$ is pathwise connected.*

EXERCISE 1.13.5   *In Figure 1.40 we show the first four generations of shield subsets of $\mathbb{R}^2$. Let A_n denote the union of the boundaries of the 2^{n-1} shields belonging to the nth generation. Show that $\{A_n\}_{n=1}^{\infty}$ converges in the Hausdorff metric to a line segment.*

The space $(\mathbb{H}(\mathbb{H}(\mathbb{X})), d_{\mathbb{H}(\mathbb{H})})$

It is at first sight amazing. But it is true. The space $\mathbb{H}(\mathbb{H}(\mathbb{X}))$ is highly nontrivial: it is fascinating, rich, at least as interesting as is $\mathbb{H}(\mathbb{X})$ relative to $\mathbb{X}$ and it has significant applications to superfractal sets.

As we showed in Theorem 1.12.13, the condition that $(\mathbb{X}, d)$ is a metric space implies that $(\mathbb{H}(\mathbb{X}), d_{\mathbb{H}})$ is a metric space. It follows that $(\mathbb{H}(\mathbb{H}(\mathbb{X})), d_{\mathbb{H}(\mathbb{H})})$ is also a metric space, where $\mathbb{H}(\mathbb{H}(\mathbb{X}))$ is the space of compact subsets of the set of compact subsets of the metric space $(\mathbb{X}, d)$ and $d_{\mathbb{H}(\mathbb{H})}$ is the Hausdorff metric on $\mathbb{H}(\mathbb{H}(\mathbb{X}))$

Figure 1.39   These images represent a sequence of compact subsets of $\mathbb{R}^3$ that converges in the Hausdorff metric. The intensity of green represents the z-component of the set, which in each case lies in a plane parallel to $z = 0$. The base of each image lies on the x-axis.

implied by the Hausdorff metric $d_{\mathbb{H}}$ on $\mathbb{H}(\mathbb{X})$. That is, for all $\alpha, \beta \in \mathbb{H}(\mathbb{H}(\mathbb{X}))$,

$$d_{\mathbb{H}(\mathbb{H})}(\alpha, \beta) = \max\left\{\mathcal{D}_{\alpha}^{\mathbb{H}}(\beta), \mathcal{D}_{\beta}^{\mathbb{H}}(\alpha)\right\}$$

where

$$\mathcal{D}_{\alpha}^{\mathbb{H}}(\beta) = \max_{B \in \beta} \min_{A \in \alpha} d_{\mathbb{H}}(A, B).$$

We summarise the basic inheritance properties of $(\mathbb{H}(\mathbb{H}(\mathbb{X})), d_{\mathbb{H}(\mathbb{H})})$ in the following theorem.

THEOREM 1.13.6   *Let* $(\mathbb{X}, d)$ *be a metric space. Then* $(\mathbb{H}(\mathbb{H}(\mathbb{X})), d_{\mathbb{H}(\mathbb{H})})$ *is a metric space. If* $(\mathbb{X}, d)$ *is complete then* $(\mathbb{H}(\mathbb{H}(\mathbb{X})), d_{\mathbb{H}(\mathbb{H})})$ *is complete. If* $(\mathbb{X}, d)$ *is compact then* $(\mathbb{H}(\mathbb{H}(\mathbb{X})), d_{\mathbb{H}(\mathbb{H})})$ *is compact.*

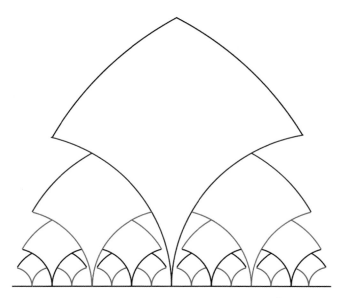

Figure 1.40   The first four generations of shield subsets of $\mathbb{R}^2$. If A_n denotes the union of the boundaries of the 2^{n-1} shields belonging to the nth generation then $\{A_n\}$ converges to a line segment in the Hausdorff metric on $\mathbb{R}^2$. If α_n denotes the set of boundaries of the nth generation shields, to which set of sets does the sequence $\{\alpha_n\}_{n=1}^{\infty}$ converge, in the metric $d_{\mathbb{H}(\mathbb{H})}$? See also Figure 1.9.

Figure 1.41 illustrates, in its four panels, four different related points in $\mathbb{H}(\mathbb{H}(\mathbb{R}^2))$. These points may be imagined to belong to a sequence of similarly constructed points and to converge to a set of subsets of $\mathbb{R}^2$, which, taken together, constitute a single point in $\mathbb{H}(\mathbb{H}(\mathbb{R}^2))$. The first point, represented in the top left panel of Figure 1.41, contains four sets that look like leaves (green). We will refer to these sets as leaf sets. In the same way, let us refer to calyx sets (mauve) and flower sets (yellows, dark purple and pale mauve). *Furthermore, from time to time elsewhere in this book we will use a similar abbreviated nomenclature to describe sets represented by parts of images.* Then we can say that the points represented successively in the other panels of Figure 1.41 contain more and more, smaller and smaller, copies of leaf sets, calyx sets and flower sets. We may suppose that the point in $\mathbb{H}(\mathbb{H}(\mathbb{R}^2))$ to which the sequence converges is $\{\{x\} \subset \mathbb{R}^2 : x \in \blacktriangle\}$ where $\blacktriangle$ denotes a certain filled triangle. Then any neighbourhood, however small, of any such $x \in \blacktriangle$ would contain a set of sets that contains at least one minute leaf set, at least one minute calyx set and at least one minute flower set, all belonging to a point in the sequence.

EXERCISE 1.13.7   *Calculate $d_{\mathbb{H}(\mathbb{H})}(\alpha, \beta)$ for the case when the underlying space is $(\mathbb{R}^2, d_{euclidean})$, $\alpha = \{A, B\}$ and $\beta = \{C, D\}$, where $A = \{(x, y) \in \mathbb{R}^2 : x^2 + y^2 = 1, x \geq 0\}$, $B = \{(x, y) \in \mathbb{R}^2 : x^2 + y^2 = \frac{1}{2}\}$, $C = \{(0, y) \in \mathbb{R}^2 : -2 \leq y \leq -1\}$ and $D = \square$. Compare this distance with $d_{\mathbb{H}}(A \cup B, C \cup D)$.*

Figure 1.41 Four points in $\mathbb{H}(\mathbb{H}(\mathbb{R}^2))$ are represented here. The first point consists of four leaf sets (green), a calyx set (mauve), and a flower set (yellows, dark purple and pale mauve). The points represented in the other panels contain more and more, smaller and smaller, copies of leaf sets, calyx sets and flower sets. Assume that these points belong to a sequence of points in $\mathbb{H}(\mathbb{H}(\mathbb{R}^2))$ in the implied progression, which converges to the set of all singleton sets $\{x\}$, where x belongs a filled triangle. Then any neighbourhood, however small, of any such $\{x\}$ would contain a set of sets that contains at least one minute leaf set, at least one minute calyx set and at least one minute flower set, all belonging to a single point in the sequence.

A significant difference between the relationship of $\mathbb{H}(\mathbb{H}(\mathbb{X}))$ to $\mathbb{H}(\mathbb{X})$ and the relationship of $\mathbb{H}(\mathbb{X})$ to $\mathbb{X}$ is that if $\alpha \in \mathbb{H}(\mathbb{H}(\mathbb{X}))$ is a finite set of sets then $\bigcup_{A \in \alpha} A \in \mathbb{H}(\mathbb{X})$; it is not true in general that if $A \in \mathbb{H}(\mathbb{X})$ is a finite set of points then $\bigcup_{a \in A} a \in \mathbb{X}$. That is, we can often 'project' from $\mathbb{H}(\mathbb{H}(\mathbb{X}))$ to $\mathbb{H}(\mathbb{X})$ in a way that cannot analogously be used to link $\mathbb{H}(\mathbb{X})$ to $\mathbb{X}$. This leads us to the comparisons in Theorems 1.13.8 and 1.13.9 below. Theorem 1.13.8 asserts that the metric $d_{\mathbb{H}(\mathbb{H})}$ is a 'stronger' metric than $d_{\mathbb{H}}$.

THEOREM 1.13.8   *Let $(\mathbb{X}, d)$ be a metric space. Let $\alpha, \beta \in \mathbb{H}(\mathbb{H}(\mathbb{X}))$ be such that*

$$\{a \in A : A \in \alpha\}, \{b \in B : B \in \beta\} \in \mathbb{H}(\mathbb{X}).$$

Then

$$d_{\mathbb{H}}(\{a \in A : A \in \alpha\}, \{b \in B : B \in \beta\}) \leq d_{\mathbb{H}(\mathbb{H})}(\alpha, \beta).$$

PROOF   Firstly, we note that

$$
\begin{aligned}
\mathcal{D}_A(\{b \in B : B \in \beta\}) &= \max_{\{b \in B : B \in \beta\}} \mathcal{D}_A(b) \\
&= \max_{B \in \beta} \max_{b \in B} \mathcal{D}_A(b) \\
&= \max_{B \in \beta} \mathcal{D}_A(B).
\end{aligned}
$$

Secondly, we note that

$$
\begin{aligned}
\mathcal{D}_{\{a \in A : A \in \alpha\}}(B) &= \max_{b \in B} \min_{\{a \in A : A \in \alpha\}} d(a, b) \\
&= \max_{b \in B} \min_{A \in \alpha} \min_{a \in A} d(a, b) \\
&\leq \min_{A \in \alpha} \max_{b \in B} \min_{a \in A} d(a, b) \\
&= \min_{A \in \alpha} \mathcal{D}_A(B).
\end{aligned}
$$

It follows that

$$
\begin{aligned}
\mathcal{D}_{\{a \in A : A \in \alpha\}}(\{b \in B : B \in \beta\}) &= \max_{B \in \beta} \mathcal{D}_{\{a \in A : A \in \alpha\}}(B) \\
&\leq \max_{B \in \beta} \min_{A \in \alpha} \mathcal{D}_A(B).
\end{aligned}
$$

Hence

$$
\begin{aligned}
& d_{\mathbb{H}}(\{a \in A : A \in \alpha\}, \{b \in B : B \in \beta\}) \\
& = \max \left\{ \mathcal{D}_{\{a \in A : A \in \alpha\}}(\{b \in B : B \in \beta\}), \mathcal{D}_{\{b \in B : B \in \beta\}}(\{a \in A : A \in \alpha\}) \right\} \\
& \leq \max \left\{ \max_{B \in \beta} \min_{A \in \alpha} \mathcal{D}_A(B), \max_{A \in \alpha} \min_{B \in \beta} \mathcal{D}_B(A) \right\}.
\end{aligned}
$$

But

$$
\begin{aligned}
& d_{\mathbb{H}(\mathbb{H})}(\alpha, \beta) \\
& = \max \left\{ \mathcal{D}_\alpha^{\mathbb{H}}(\beta), \mathcal{D}_\beta^{\mathbb{H}}(\alpha) \right\} \\
& = \max \left\{ \max_{B \in \beta} \min_{A \in \alpha} \max\{\mathcal{D}_A(B), \mathcal{D}_B(A)\}, \max_{A \in \alpha} \min_{B \in \beta} \max\{\mathcal{D}_A(B), \mathcal{D}_B(A)\} \right\} \\
& \geq \max \left\{ \max_{B \in \beta} \min_{A \in \alpha} \mathcal{D}_A(B), \max_{A \in \alpha} \min_{B \in \beta} \mathcal{D}_B(A) \right\}.
\end{aligned}
$$

This completes the proof. $\square$

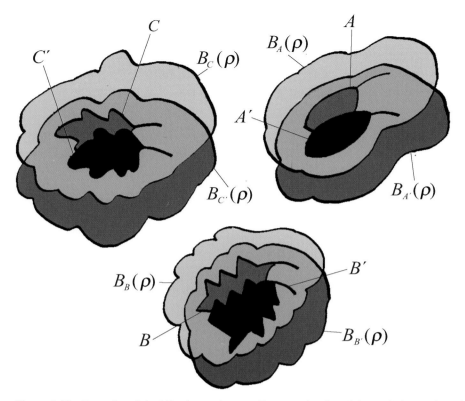

Figure 1.42 Illustration of the falling leaves theorem. For an explanation of the symbols, see the main text immediately before Theorem 1.12.20.

Theorem 1.13.8 enables us, in some cases, to think about approximation in $\mathbb{H}(\mathbb{X})$ in terms of approximation in $\mathbb{H}(\mathbb{H}(\mathbb{X}))$. According to Theorem 1.13.6, a sequence $\{\alpha_n \in \mathbb{H}(\mathbb{H}(\mathbb{X}))\}_{n=1}^{\infty}$ converges to a point $\alpha \in \mathbb{H}(\mathbb{H}(\mathbb{X}))$ iff for each $A \in \alpha$ there is a sequence of sets $A_n \in \alpha_n$ such that $\{A_n \in \mathbb{H}(\mathbb{X})\}_{n=1}^{\infty}$ converges to A. So in the case of $\mathbb{R}^2$ we may for example discuss the possible convergence of a sequence of approximations to a tree set in terms of sequences of sets that contain sequences of leaf sets that converge to leaf sets, sequences of foliage sets that converge to foliage sets and a sequence of trunk sets that converges to a trunk set. We obtain a richer view of convergence in the Hausdorff metric.

Falling leaves theorem

Leaves fall from the sky, the sun is setting, and the shadows of three leaves float down a white wall. At one instant t the set of leaf shadows is represented by $\alpha = \{A, B, C\} \subset \mathbb{H}(\mathbb{H}(\mathbb{R}^2))$ while at a later instant t' it is represented by $\alpha' = \{A', B', C'\} \subset \mathbb{H}(\mathbb{H}(\mathbb{R}^2))$. Here A' and A represent the shadows of a given leaf, B' and B represent the shadows of the second leaf and C' and C represent the shadows of a third leaf; see Figure 1.42.

We will suppose that the leaf shadows A, B, C are disjoint. The following theorem tells us that when t and t' are sufficiently close the Hausdoff distance between the union of the shadows at time t and the union of the shadows at time t' is the same as the distance $d_{\mathbb{H}(\mathbb{H})}(\alpha, \alpha')$. We have framed this result for three leaves, but you will easily see how it is true for any finite set of leaves.

THEOREM 1.13.9 *(Falling leaves theorem)   If $A, B, C \in \mathbb{H}(\mathbb{X})$ are disjoint and $A', B', C' \in \mathbb{H}(\mathbb{X})$ are such that $d_{\mathbb{H}}(A, A')$, $d_{\mathbb{H}}(B, B')$ and $d_{\mathbb{H}}(C, C')$ are all sufficiently small then*

$$d_{\mathbb{H}(\mathbb{H})}(\{A, B, C\}, \{A', B', C'\}) = d_{\mathbb{H}}(A \cup B \cup C, A' \cup B' \cup C').$$

PROOF   We can suppose that

$$d_{\mathbb{H}}(A, A') < \min\left\{\tfrac{1}{2}d_{\mathbb{H}}(A, B), \tfrac{1}{2}d_{\mathbb{H}}(A, C)\right\},$$

$$d_{\mathbb{H}}(B, B') < \min\left\{\tfrac{1}{2}d_{\mathbb{H}}(B, A), \tfrac{1}{2}d_{\mathbb{H}}(B, C)\right\}$$

and

$$d_{\mathbb{H}}(C, C') < \min\left\{\tfrac{1}{2}d_{\mathbb{H}}(C, A), \tfrac{1}{2}d_{\mathbb{H}}(C, B)\right\}.$$

Now we start from the triangle inequality $d_{\mathbb{H}}(A, B') \geq d_{\mathbb{H}}(A, B) - d_{\mathbb{H}}(B, B')$ and find that

$$d_{\mathbb{H}}(A, B') \geq d_{\mathbb{H}}(A, B) - \tfrac{1}{2}d_{\mathbb{H}}(B, A) = \tfrac{1}{2}d_{\mathbb{H}}(A, B) > d_{\mathbb{H}}(A, A').$$

Similarly we find that $d_{\mathbb{H}}(A, C') > d(A, A')$. It follows that

$$d_{\mathbb{H}}(A, A') = \min\{d_{\mathbb{H}}(A, A'), d_{\mathbb{H}}(A, B'), d_{\mathbb{H}}(A, C')\}. \qquad (1.13.2)$$

Also,

$$d_{\mathbb{H}}(B, B') = \min\{d_{\mathbb{H}}(B, A'), d_{\mathbb{H}}(B, B'), d_{\mathbb{H}}(B, C')\}$$

and

$$d_{\mathbb{H}}(C, C') = \min\{d_{\mathbb{H}}(C, A'), d_{\mathbb{H}}(C, B'), d_{\mathbb{H}}(C, C')\}.$$

Thus

$$\mathcal{D}^{\mathbb{H}}_{\{A,B,C\}}(\{A', B', C'\}) = \max_{F' \in \alpha'} \ \min_{G \in \alpha} d_{\mathbb{H}}(F', G)$$
$$= \max\{d_{\mathbb{H}}(A, A'), d_{\mathbb{H}}(B, B'), d_{\mathbb{H}}(C, C')\}.$$

It now follows that $\mathcal{D}^{\mathbb{H}}_{\{A',B',C'\}}(\{A, B, C\}) = \mathcal{D}^{\mathbb{H}}_{\{A,B,C\}}(\{A', B', C'\})$ and hence that

$$d_{\mathbb{H}(\mathbb{H})}(\{A, B, C\}, \{A', B', C'\}) = \max\{d_{\mathbb{H}}(A, A'), d_{\mathbb{H}}(B, B'), d_{\mathbb{H}}(C, C')\}.$$
$$(1.13.3)$$

But from Theorem 1.12.20 the expression on the right-hand side of Equation (1.13.3) is the same as $d_{\mathbb{H}}(A \cup B \cup C, A' \cup B' \cup C')$.   □

The above result depends upon the natural pairing of the leaves from the sets α and α' in the calculation of the Hausdorff distances. A is paired with A', as in Equation (1.13.2), B with B', and C with C'. But as the gap between the instants becomes larger this pairing up breaks down and it is likely that

$$d_{\mathbb{H}(\mathbb{H})}(\alpha, \beta) > d_{\mathbb{H}}(A \cup B \cup C, A' \cup B' \cup C').$$

EXERCISE 1.13.10   *Suppose that $\{A_n \in \mathbb{H}(\mathbb{X})\}_{n=1}^{\infty}$ converges to $A \in \mathbb{H}(\mathbb{X})$ in the metric $d_{\mathbb{H}}$. Then does $\{\mathbb{H}(A_n) \in \mathbb{H}(\mathbb{H}(\mathbb{X}))\}_{n=1}^{\infty}$ converge to $\mathbb{H}(A) \in \mathbb{H}(\mathbb{H}(\mathbb{X}))$ in the metric $d_{\mathbb{H}(\mathbb{H})}$?*

EXERCISE 1.13.11   *In Figure 1.40 we showed the first four generations of shield subsets of $\mathbb{R}^2$. Let α_n denote the set of the boundaries of the 2^{n-1} shields belonging to the nth generation. Show that $\{\alpha_n\}_{n=1}^{\infty}$ converges in the Hausdorff metric to the set of sets $\alpha = \{\{x\} : x \in [0, 1]\}$.*

Other metric spaces

We have seen how, starting from a metric space $(\mathbb{X}, d)$, we may form new metrics $d_{\mathbb{H}}$ and $d_{\mathbb{H}(\mathbb{H})}$ and new spaces $\mathbb{H}(\mathbb{X})$ and $\mathbb{H}(\mathbb{H}(\mathbb{X}))$ and how the important properties of completeness and compactness are inherited. We will discover later that there are many other such hierarchical constructions of spaces, of more elaborate mathematical objects, with similar inheritance properties. In the next chapter, where we introduce measures, we will mention a space $\mathbb{P}(\mathbb{X})$ of measures. We will show how, with appropriate straightforward conditions, we can define a metric $d_{\mathbb{P}} = d_{\mathbb{P}(\mathbb{X})}$ on $\mathbb{P}(\mathbb{X})$ such that it too is complete and even compact.

With this machinery in place we can go on a construction spree. We can form metric spaces such as $(\mathbb{P}(\mathbb{H}(\mathbb{X})), d_{\mathbb{P}(\mathbb{H}(\mathbb{X}))})$, $(\mathbb{P}(\mathbb{P}(\mathbb{X})), d_{\mathbb{P}(\mathbb{P}(\mathbb{X}))})$, $(\mathbb{H}(\mathbb{P}(\mathbb{X})), d_{\mathbb{H}(\mathbb{P}(\mathbb{X}))})$ and even, for example, $(\mathbb{P}^L(\mathbb{H}^M(\mathbb{X}^N)), d_{\mathbb{P}^L(\mathbb{H}^M(\mathbb{X}^N))})$. We will discover that despite the initial appearance of a Baroque elaborateness these spaces are entirely natural and, like collections of multiscale, many-layered, natural objects, from skies full of clouds to seas full of protozoa, they too contain rich and beautiful objects, for example when $\mathbb{X}$ is $\mathbb{R}^2$ or real projective space. It is in these spaces that we will find superfractals.

1.14   Fractal dimensions

In the literature there are many different definitions of a theoretical quantity called the fractal dimension of a subset of $\mathbb{X} \subset \mathbb{R}^n$. A mathematically convenient definition is the Hausdorff dimension. This is always well defined. Its numerical value often but not always coincides with the values provided by other definitions, when they apply.

The following two definitions are discussed in [34], pp. 25 *et seq.*

DEFINITION 1.14.1   Let $S \subset \mathbb{X}$, $\delta > 0$ and $0 \leq s < \infty$. Let

$$H_\delta^s(S) = \inf \left\{ \sum_{i=1}^{\infty} |U_i|^s : \{U_i\} \text{ is a } \delta\text{-cover of } S \right\},$$

where $|U_i|^s$ denotes the sth power of the diameter of the set U_i, and where a δ-cover of S is a covering of S by subsets of $\mathbb{X}$ of diameter less than δ. Then the s-dimensional Hausdorff measure of the set S is defined to be

$$H^s(S) = \lim_{\delta \to 0} H_\delta^s(S).$$

The limit exists but may be infinite, since $H_\delta^s(S)$ increases as δ decreases. Moreover $H^s(S)$ is non-increasing as s increases from zero to infinity. For any $s < t$ we have $H_\delta^s(S) \geq H_\delta^t(S)$, which implies that if $H_\delta^t(S)$ is positive then $H^s(S)$ is infinite. Thus there is a unique value, given by the following definition.

DEFINITION 1.14.2   The **Hausdorff dimension** or **fractal dimension** of the set $S \subset \mathbb{X}$ is defined to be

$$\dim_H S = \inf\{s \,|\, H^s(S) = 0\}.$$

There is much written about fractal dimensions in many sources, including *Fractals Everywhere* [9]. It is important to read Mandelbrot's book [64] to understand his vision of why fractal dimension is important. Other useful references are [34] and [96].

CHAPTER 2

Transformations of points, sets, pictures and measures

2.1 Introduction

There are many types of transformation, not just mathematical ones; see Figure 2.1. In this chapter, however, we consider two important types of mathematical transformation. **Projective** transformations are remarkable because our sight depends upon them. **Möbius** transformations are remarkable because of their beauty. For these reasons among others we use these two families of transformations to describe fractal sets, measures and pictures.

An important goal of fractal geometry is to describe images in terms of transformations that in some way leave the images unaltered. For us an image is a set, measure or mathematical picture.

How does a transformation on $\mathbb{R}^2$ act upon a picture? To answer this we begin in Section 2.2 by defining mathematical pictures. Then we explain the meaning of $f(\mathfrak{P})$, where $\mathfrak{P}$ is a picture and f is a transformation. We discover practical problems that derive from the question 'Where do pictures come from?' For example, in the process of constructing a digital picture, how does one decide on the colour of a pixel? The need for a model for pictures that is consistent with transformation and discretization provides a motivation to model pictures using measures. An alternative approach to modelling pictures, using fractal tops, is described in Chapter 4.

How does a transformation on $\mathbb{R}^2$ act upon a measure? To answer this we begin by introducing measure theory in Section 2.3. We will do this both intuitively and rigorously, with an emphasis on the interpretation of measures in terms of pictures. Then we define and illustrate $f(\mu)$ where μ is a measure and $f : \mathbb{X} \to \mathbb{X}$ is a transformation. We conclude Section 2.3 with the definition of an invariant measure of a transformation and with examples of pictures of invariant measures relating to projective and Möbius transformations on $\mathbb{R}^2$.

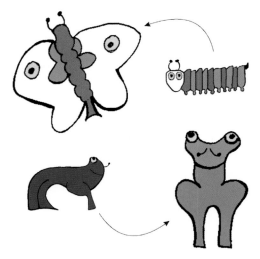

Figure 2.1   There are many types of transformation, not just mathematical ones. For example we have the following definition: '**transformation** n. 2 *Zool.* a change of form at metamorphosis, esp. of insects, amphibia, etc.' (*The Concise Oxford Dictionary*, Clarendon Press, Oxford, 1990)

We formulate invariance properties of sets, measures and pictures under transformation in terms of fixed-point properties of transformations acting on appropriate spaces. This motivates us in Section 2.4 to discuss fixed-point theorems and to add to our collection a new metric space $(\mathbb{P}(\mathbb{X}), d_{\mathbb{P}})$ whose elements are measures. These concepts will be used in Chapters 4 and 5 to construct fractal sets, measures and pictures.

But the central question which we need to answer and which we pursue in this chapter is how, specifically, do Möbius and projective transformations deform space and, consequently, change or leave unaltered aspects of sets, pictures and measures? How do these transformations not only affect the locations of points within a picture but also, when they act upon a picture treated as a measure, alter contrast and brightness? In order to understand these questions better and so be able to model images with fractals, we will explore the geometry of Möbius and projective transformations in a detailed and specific way.

Since transformations on real spaces relate to transformations on code spaces, we conclude this chapter with a section on transformations on code space, our 'meristem' or 'formative tissue'. The relationship between transformations on code spaces and transformations on sets, pictures and measures is a key theme of this book.

Another theme of this chapter is that sets, measures and pictures founded in $\mathbb{R}^2$ may be complicated but even so can have invariance properties under geometrically simple transformations. Such invariances can in principle be used to reduce the amount of information needed to describe apparently complicated sets, measures and pictures. Until the end of this chapter, code space is off the stage while we develop the theme of transformations. Then in Chapter 3 we start to combine the two themes.

Structure of this chapter

In Section 2.2 we define pictures and digital pictures and explain the action of transformations upon them. You might now like to glance ahead at Figures 2.2–2.6 to get a feel for the content of this section. We conclude Section 2.2 by illustrating the concept of a picture that is invariant under a transformation.

We start Section 2.3 by explaining in an intuitive and visual manner what a measure is. Again you might like to glance ahead, at Figure 2.10. Then we introduce fields and σ-algebras of subsets of a space $\mathbb{X}$. Upon these we define and construct measures, and we introduce the space $\mathbb{P}(\mathbb{X})$ of normalized Borel measures upon a metric space $(\mathbb{X}, d)$. Then we explain how continuous transformations act on measures and give examples of transformations acting upon pictures of measures. We conclude this section by explaining what it means for a measure to be invariant under a transformation.

Then in Section 2.4 we consider fixed points. When does a transformation $f : \mathbb{H}(\mathbb{X}) \to \mathbb{H}(\mathbb{X})$ possess a fixed point? We are interested because a fixed point of f is a set that is unchanged when f is applied to it. We introduce a metric $d_{\mathbb{P}}$ on $\mathbb{P}(\mathbb{X})$. In the right circumstances $(\mathbb{P}(\mathbb{X}), d_{\mathbb{P}})$ is a compact metric space, another remarkable example of inheritance. Contraction mappings on $(\mathbb{P}(\mathbb{X}), d_{\mathbb{P}})$ possess unique fixed points, yielding measures unchanged by transformations. Also, since often the space $\mathbb{P}(\mathbb{X})$ will be linear and convex, the Schauder–Tychenoff fixed-point theorem applies and ensures the existence of invariant measures in broad circumstances. We will need these ideas in the later chapters.

After Sections 2.2, 2.3 and 2.4 we will be in a position to discuss the actions of Möbius and projective transformations on sets, measures and pictures founded on $\mathbb{R}^2$. Actually, the underlying space upon which Möbius transformations act is the Riemann sphere, which is equivalent to $\mathbb{R}^2 \cup \{\infty\}$ where ∞ is an additional point called 'the point at infinity'. The underlying space upon which projective transformations act is $\mathbb{RP}^2$, which is equivalent to $\mathbb{R}^2 \cup \{L_\infty\}$ where L_∞ is an additional straight line, the 'line at infinity'. To explain these transformations we need first to understand in a geometrical way how linear transformations in $\mathbb{R}^2$ and $\mathbb{R}^3$ behave. This is considered in Section 2.5. Here we assume a basic knowledge of linear spaces and linear transformations but include a brief review of two-dimensional linear algebra as a reminder and as a way of introducing our notation. The main result that we need is that an invertible linear transformation in $\mathbb{R}^3$ can always be expressed as the composition of rescalings along three perpendicular directions, a possible reflection and a rotation. This is very useful!

We describe Möbius transformations in Section 2.6. Möbius transformations can be represented by linear transformations in $\mathbb{C}^2$ and so may be expressed with eight real parameters, which explains why they are efficient carriers of information. What intrigues us is that they map the set of straight lines and circles into itself while at the same time preserving angles, yet at the same time they can effect huge distortions. This really is remarkable: how can the nature of $\mathbb{R}^2 \cup \{\infty\}$ be such that this is possible? Realization of the nature of Möbius transformations was a key idea behind the discovery of non-euclidean geometry, which eluded geometers for nearly two thousand years.

We describe projective transformations in Section 2.7. Any projective transformation can expressed in terms of a linear transformation in three dimensions and can be represented using nine real parameters. When you view a picture on a flat plane, such as the screen of a modern television or movie screen, from two different positions, the relationship between the two images upon the retina of one eye will be provided by a projective transformation. Indeed the actual differences between the images, the distortions in going from one to the other, can be quite extreme. But the vision system compensates for such projective transformations. This observation motivated mathematicians of an earlier era to study projective geometry most intently, to discover what it is, mathematically, that is left unchanged by projective transformations. We recall some of their results in this section. But our goal in later chapters is to exploit these transformations by using finite collections of them to describe completely certain sets, pictures and measures. We discuss some transformations on code space in Section 2.8.

2.2   Transformations of pictures

Definition of a picture

DEFINITION 2.2.1   We define a **picture function** $\mathfrak{P}$ to be a function

$$\mathfrak{P} : D_{\mathfrak{P}} \subset \mathbb{R}^2 \to \mathfrak{C},$$

where $\mathfrak{C}$ is a **colour space** and $D_{\mathfrak{P}}$ is called the **domain** of the **picture**. The value of $\mathfrak{P}(x)$ gives the **colour** of the picture at the point $x \in D_{\mathfrak{P}}$. We denote the space of all pictures with colour space $\mathfrak{C}$ by $\Pi = \Pi_{\mathfrak{C}}$.

Throughout this book we usually suppose that the colour space $\mathfrak{C}$ is a subset of $\mathbb{R}^3$ such as

$$\mathfrak{C} = [0, \infty)^3 \subset \mathbb{R}^3, \quad \mathfrak{C} = [0, 255]^3 \subset \mathbb{R}^3 \quad \text{or} \quad \mathfrak{C} = \{0, 1, \dots, 255\}^3 \subset \mathbb{R}^3.$$

When $\mathfrak{C} \subset \mathbb{R}^3$ the components of a point $c = (c_1, c_2, c_3) \in \mathfrak{C}$ may be called the **colour components**, with c_1 named the **red** component, c_2 named the **green** component and c_3 named the **blue** component.

But there are other possibilities, corresponding to different models for images: for example c_1 might represent intensity, c_2 saturation and c_3 hue, with appropriate ranges of values. The points in the space $\mathfrak{C}$ could be simply one dimensional, corresponding to **intensities** in **greyscale** pictures. Or they may have more than three dimensions. For example, in applications to the high-quality image printing industry a four-dimensional colour space is used, whose axes are cyan, magenta, yellow and black.

There are diverse possible choices for the domain $D_\mathfrak{P}$ of the picture function $\mathfrak{P}$; it may be a line segment, a curve, an open ball, a closed rectangle or any other subset of $\mathbb{R}^2$. It may represent the region that is yellow in a watercolour of a flower, the part of a piece of photographic paper on which a photo has been developed, the retina of your eye, the painted region of an artist's canvas, the screen of a computer or a patch of vision in your mind's eye.

In some cases we assume that

$$D_\mathfrak{P} = \square = \{(x, y) \in \mathbb{R}^2 : x_L \leq x \leq x_H, y_L \leq y \leq y_H\}$$

where $(x_L, y_L) \in \mathbb{R}^2$ is called the **lower left corner** and $(x_H, y_H) \in \mathbb{R}^2$ is called the **upper right corner** of the (domain of the) picture. In the absence of other information, for mathematical purposes we take $(x_L, y_L) = (0, 0)$ and $(x_H, y_H) = (1, 1)$.

The domain of a picture function is an important part of its definition. The characteristic function of a subset $S \subset \mathbb{R}^2$,

$$\chi_S(x) = \begin{cases} 1 & \text{if } x \in S, \\ 0 & \text{if } x \notin S, \end{cases}$$

may be treated as a picture function that represents S. Another representation of S is provided by the picture function $\mathfrak{P}_S$ with domain S and constant value, say $\mathfrak{P}_S(x) = 1$, for all $x \in S$. With the aid of χ_S or $\mathfrak{P}_S$ we can embed classical geometrical objects such as circles, lines or triangles in the space of picture functions.

We will usually refer to a picture function as a **picture**; our intention is that it should be clear from the context whether we mean a picture function or just a picture, as on the pages in this book. We refer to a picture as a geometer might refer to a triangle, meaning either a concrete image or the abstract mathematical entity.

Pictures are available to us in various forms. They may be defined explicitly, in much the same way as a parabola or a sphere is defined, by reference to mathematical algorithms and formulas, including for example the kinds of expressions produced by and interpreted by computer graphics software. They may be piecewise constant functions, such as digital pictures (see below), defined using arrays of data obtained from devices such as scanners and digital cameras that focus, sample, filter and interpolate real-world scenes. But for us they are always, in the end, mathematical entities.

We have defined pictures as having their domains in $\mathbb{R}^2$. But it is easy to see how this definition may be extended to pictures with domains in other fundamentally two-dimensional spaces such as a spherical shell or a projective plane.

Transformations of pictures

Let $f : D_f \subset \mathbb{R}^2 \rightarrow \mathbb{R}^2$ be a *one-to-one* transformation and let $\mathfrak{P} : D_\mathfrak{P} \subset \mathbb{R}^2 \rightarrow \mathfrak{C}$ be a picture with $D_\mathfrak{P} \subset D_f$. Then we define

$$f(\mathfrak{P}) : D_{f(\mathfrak{P})} \subset \mathbb{R}^2 \rightarrow \mathbb{R}^2$$

to be the **picture $\mathfrak{P}$ transformed by** f, or equivalently, the **transformation f applied to the picture $\mathfrak{P}$**, where

$$D_{f(\mathfrak{P})} = f(D_\mathfrak{P})$$

and

$$f(\mathfrak{P})(x) = \mathfrak{P}(f^{-1}(x)) \quad \text{for all } x \in D_{f(\mathfrak{P})}.$$

We also denote $f(\mathfrak{P})$ by $f \circ \mathfrak{P}$. Note that when $f : \mathbb{R}^2 \rightarrow \mathbb{R}^2$ the picture $f(\mathfrak{P})$ is always well defined and $f : \Pi \rightarrow \Pi$. Figure 2.2 shows the pictures produced when three different transformations f_1, f_2, $f_3 : \mathbb{R}^2 \rightarrow \mathbb{R}^2$ are applied to $\mathfrak{P}$, a picture of a fish.

EXERCISE 2.2.2 *Why have we restricted transformations of pictures to being one-to-one?*

EXERCISE 2.2.3 *Where in the real world do we see interesting transformations of pictures? Some sources are mirrors, uneven glass, the distortions produced by water in a fish tank or by hot rising air, and the reflections in shiny metal surfaces such as the surface of a ball bearing. Name some other sources.*

Invariant sets and pictures

A set $S \subset \mathbb{X}$ is said to be invariant under a transformation $f : \mathbb{X} \rightarrow \mathbb{X}$ iff

$$f^{-1}(S) = S.$$

We will refer to such a set S as an **invariant set** of the transformation f. Note that this implies

$$f(S) = S,$$

but the converse is not true unless f is one-to-one.

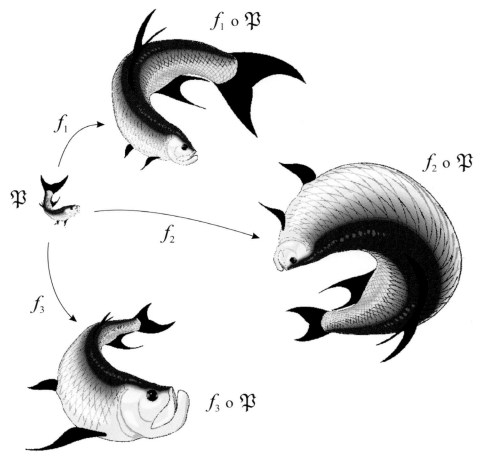

Figure 2.2 A picture function $\mathfrak{P}$ representing a fish, and three different Möbius transformations of it, $f_1 \circ \mathfrak{P}$, $f_2 \circ \mathfrak{P}$ and $f_3 \circ \mathfrak{P}$.

Similarly, a picture $\mathfrak{P}$ is said to be invariant under a *one-to-one* transformation $f : \mathbb{R}^2 \to \mathbb{R}^2$ iff

$$f(\mathfrak{P}) = \mathfrak{P}.$$

We will refer to such a picture $\mathfrak{P}$ as an **invariant picture** of the transformation f.

For example, Figure 2.3 shows a picture that is invariant under the transformation defined by $f(x, y) = (-x, y)$ and Figure 2.4 shows a picture that is invariant under the transformation $f(x, y) = (x, -y)$. Figure 2.5 illustrates a set and a picture that are invariant under the same transformation $R_\theta : \mathbb{R}^2 \to \mathbb{R}^2$, a clockwise rotation through $\theta = 36°$.

Figure 2.3   This picture is invariant under a familiar type of transformation on $\mathbb{R}^2$, a reflection. This invariance *partly* defines the picture.

Figure 2.4   This picture shows a mathematically perfect reflection. But photographs of real swans on real water are not exactly invariant under reflection.

Figure 2.5   Both pictures here are invariant under the rotation transformation $R_{36°}$. The left-hand picture also represents an invariant set.

There are many instances of sets and pictures that are invariant under transformations. In graphic design and art the transformations under which a picture is invariant may be referred to as its **symmetries**. Wallpaper pictures, pictures of flowers and architectural motifs may be invariant under translational and/or rotational transformations.

As a more complicated example, Figure 2.6 illustrates a set $S \subset \mathbb{R}^2$ that is invariant under the Möbius transformation (Section 2.6) $\mathcal{M} = \widehat{\mathcal{M}}_\rho \circ R_\theta \circ \widehat{\mathcal{M}}_\rho$, where $\widehat{\mathcal{M}}_\rho : \mathbb{C} \to \mathbb{C}$ is defined by

$$\widehat{\mathcal{M}}_\rho(z) = \frac{\rho z}{1 + (\rho - 1)z} \tag{2.2.1}$$

for values of $\rho > 1$. This transformation obeys $\widehat{\mathcal{M}}_\rho(0) = 0$ and $\widehat{\mathcal{M}}_\rho(1) = 1$. The visible part of the invariant set is $S \cap \{(x, y) \in \mathbb{R}^2 : -2 \leq x, y \leq 2\}$. R_θ denotes a rotation through angle θ about the origin.

An even more complicated example of an invariant picture is illustrated in Figure 2.7. In this case the transformation $f : \square \to \square$, where $\square = \{(x, y) \in \mathbb{R}^2 : 0 \leq x, y < 1\}$, is defined by

$$f(x, y) = \begin{cases} \left(\frac{1}{2}x, 2y - 1\right) & \text{when } \frac{1}{2} \leq y < 1, \\ \left(\frac{1}{2}x + \frac{1}{2}, 2y\right) & \text{when } 0 \leq y < \frac{1}{2}. \end{cases} \tag{2.2.2}$$

Elaborate sets and pictures that are invariant under simple transformations, those whose formulas may be described explicitly in a succinct manner involving less than say sixteen free parameters, can be produced in various ways. In Chapter 3 we show how new pictures generated by groups of simple transformations, such as Möbius transformations and projective transformations, acting on a given picture

Figure 2.6 Invariant sets of simple transformations may be elaborate. This figure shows part of an invariant set for the Möbius transformation $\mathcal{M}_\rho : \mathbb{R}^2 \to \mathbb{R}^2$, defined via Equation (2.2.1) .

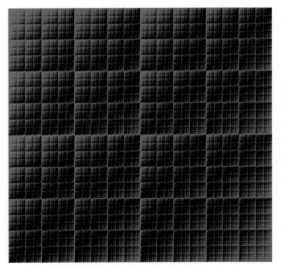

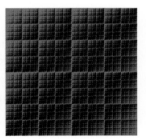

Figure 2.7 Two views of a picture that is exactly invariant under the transformation defined in Equation (2.2.2). The colours of the diadic rational points $(k/2^9, l/2^9)$ are plotted in the left-hand panel for $k, l = 0, 1, \ldots, 511$. The colours of the points $(m/2^8, n/2^8)$ for $m, n = 0, 1, \ldots, 256$ are plotted in the right-hand panel, which is thus a precise subsample of the left-hand panel.

may be used to define invariant pictures. Such families of transformations may be produced by autonomous differential equations that model physical systems. Indeed, phase portraits associated with autonomous systems in two dimensions can be thought of as invariant sets for appropriate transformations.

EXERCISE 2.2.4 *Show that if $g : \mathbb{R}^2 \to \mathbb{R}^2$ is invertible and if the picture $\mathfrak{P}$ is invariant under the rotation R_θ then the picture $\widetilde{\mathfrak{P}} := g(\mathfrak{P})$ is invariant under the transformation $\widetilde{g} := g \circ R_\theta \circ g^{-1}$.*

EXERCISE 2.2.5 *Define a transformation $f : \Omega_{\{0,1,2\}} \to \Omega_{\{0,1,2\}}$ by the expression $f(\sigma_1\sigma_2\sigma_3 \cdots) = 2\sigma_1\sigma_2\sigma_3 \cdots$ for all $\sigma = \sigma_1\sigma_2\sigma_3 \cdots \in \Omega_{\{0,1,2\}}$. Find an invariant set for f.*

Digital pictures

Let $W, H \in \mathbb{N} = \{1, 2, 3, \ldots\}$. Suppose that $\mathfrak{C}$ is a discrete space, such as $\{0, 1, \ldots, 255\}^3$, and that the picture function $\mathfrak{P} : \square \subset \mathbb{R}^2 \to \mathfrak{C}$ is constant on each rectangular region in a $W \times H$ array of rectangular regions $\square_{w,h}$, each of the same width and height,

$$(\square_{w,h}) := \{\square_{w,h} : w = 1, 2, \ldots, W; h = 1, 2, \ldots, H\}$$

such that

$$\square = \bigcup \square_{w,h}$$

and

$$\square_{w,h} \cap \square_{w',h'} = \varnothing \quad \text{whenever } (w, h) \neq (w', h').$$

Then $\mathfrak{P}$ is called a **digital picture**.

We will suppose that the array of rectangles $(\square_{w,h})$ is organized similarly to the elements of a matrix but flipped and transposed, so that $\square_{1,1}$ is in the lower left corner of $\square$ and $\square_{W,H}$ is in the upper right corner of $\square$, as illustrated in Figure 2.8. Each rectangle may be open, closed or partly open and partly closed, as indicated. We may write $\square_{w,h}^{W,H}$ to denote $\square_{w,h}$ more precisely. Exercise 1.9.5 provides a canonical set of choices for $\square_{w,h}^{W,H}$.

The picture $\mathfrak{P}_{w,h} : \square_{w,h} \subset \mathbb{R}^2 \to \mathfrak{C}$, where $\mathfrak{P}_{w,h}$ is the restriction of the digital picture $\mathfrak{P}$ to the rectangle $\square_{w,h}$, is called a **pixel function** or, simply, a **pixel**. The constant value of $\mathfrak{P}_{w,h}(x) \in \mathfrak{C}$ for $x \in \square_{w,h}$ is called the colour of the pixel $\mathfrak{P}_{w,h}$.

We will denote a typical digital image, as described here, by $\mathfrak{P}_{W \times H}$ and a typical pixel as $\mathfrak{P}_{w,h}$ or more specifically as $\mathfrak{P}_{w,h}^{W,H}$. We call W the **width** of the digital image and H the **height** of the digital image, in pixel units. We refer to $\min\{W, H\}$ as the **resolution** of the digital picture $\mathfrak{P}_{W \times H}$.

Let $f : \mathbb{R}^2 \to \mathbb{R}^2$ and let $\mathfrak{P}_{W \times H}$ be a digital picture. Then in general $f \circ \mathfrak{P}_{W \times H}$ is not a digital picture. So the set of digital images is not mapped into itself under

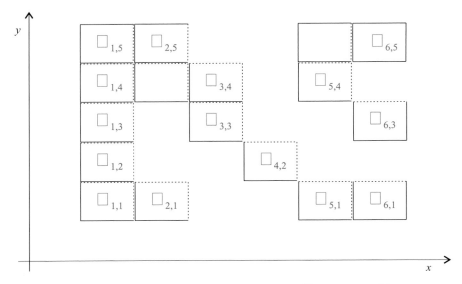

Figure 2.8 This illustrates the organization of the pixel domains $\square_{w,h}$ in an array $(\square_{w,h})$ constituting a digital image. Here $W = 4$ and $H = 3$.

general transformations. This difficulty is related to the question: where do digital pictures come from?

Digitization

Suppose that we are given a picture $\mathfrak{P} \in \Pi_\mathfrak{C}$ and we wish to convert it into a digital picture $\mathfrak{P}_{W \times H}$. How do we decide which colour to assign to each pixel $\mathfrak{P}_{w,h}$?

If the picture $\mathfrak{P}$ is assumed to belong to a class of functions which are suitably smooth or regular in some way, so that they do not vary too wildly over the domain of a pixel, then it may be easy to select a typical value of $\mathfrak{P}(x)$ for $x \in \square_{w,h}$ and for each $\square_{w,h} \in (\square_{w,h})$, discretize this value and thus define $\mathfrak{P}_{w,h}$ and $\mathfrak{P}_{W \times H}$.

In fractal geometry, particularly, we are concerned with pictures $\mathfrak{P}$ that may be very complicated. The problem is of the following nature: the rectangle $\square_{w,h}$ contains a vast collection of points, a countless infinity of them, each coloured in one of many possible colours, and we have to select a single representative colour for the pixel $\mathfrak{P}_{w,h}$. This idea is illustrated in Figure 2.9. Some kind of colour selection procedure is needed, perhaps based on averaging. But unless we make assumptions about the nature of the picture, for example concerning the type of function it is, thereby providing some mathematical cohesion between the colours of nearby points, then it is difficult to define a colour selection algorithm in such a way as to satisfy our intuitions about how digital pictures of different resolutions should relate to one another. We want to be able to capture the idea that we can

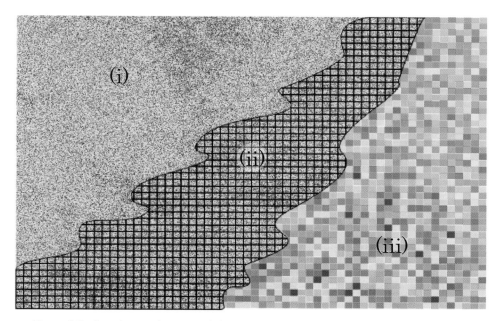

Figure 2.9 A mathematical picture is defined by assigning one colour to each point in a domain. How does one select the colour of a pixel to make a digital picture, an approximation to the original picture, that satisfies our intuitions? In (i) we envisage part of a mathematical picture that consists of an uncountable collection of points in the euclidean plane, each with its own colour. In (ii) part of the picture has been overlaid by a grid of domains or pixels. In (iii) each pixel has been assigned the colour of one point within it, yielding a digital picture. Vastly many different pictures many result!

increase the resolution without limit, revealing more and more intricate detail as we do so. But we do not want to prejudice the kinds of picture that we are trying to find by making assumptions about the function class they may belong to, in effect saying something about what they 'look like' before we have seen them.

A different model for images is provided by measure theory. This model yields a consistent way of defining digital pictures at different resolutions. Yet another type of model is provided by fractal tops, which we will encounter in Chapter 4.

2.3 Transformations of measures

At a first reading, if you do not already have some familiarity with measures, you could now read only the intuitive introduction to measures given below, study the figures in this section and then go straight on to Section 2.4. Try not to get bogged down at this point.

Intuitive description of some measures

Firstly we present an intuitive idea of a measure on a subset $\square \subset \mathbb{R}^2$. Then we introduce a beautiful formalism, measure theory, that captures this intuition.

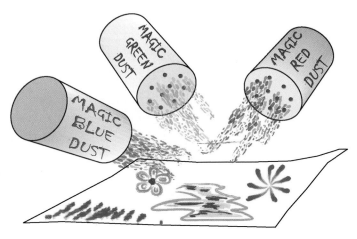

Figure 2.10 Measures are mysterious in some ways. Intuitively you may think of a measure $v = (v_{red}, v_{green}, v_{blue}) \in M^3(\square)$ as being the result of distributing magic luminous dust from three shakers upon $\square$. Although individual grains of dust have no brightness, the cumulative effect of countless infinities of grains within a subset $S \subset \square$ may provide a total amount of brightness of the red, green and blue components of light emitted by S, yielding $v(S)$. But S needs to be a subset of $\square$ of a special type, called 'measurable'.

Let us suppose two things. First, that we have a certain magical luminous powder. This powder glows. It has the total brightness of one candle. But it is also infinitely fine, so that its individual particles emit no light. Second, suppose that the luminous powder can be attached to the points of the euclidean plane.

All the powder could be attached to a single point $x_0 \in \square \subset \mathbb{R}^2$. Then x_0 would glow with the brightness of one candle. Or all the powder could be attached uniformly to a line segment $L \subset \square$. Then the line segment would glow with the brightness of one candle. But if we were to look more and more closely at one part of the line then the brightness of the observed part would steadily diminish towards zero. Or again, the powder could be distributed on $\square$ unevenly, some of it concentrated on the forms of clouds and some of it filling in the shape of the moon, say, to produce a picture of varying brightness but emitting in total the light of one candle.

Similarly we may model arbitrary coloured images in this intuitive manner. We have illustrated this idea in Figure 2.10.

However, we cannot describe a measure by means of a picture function $\mathfrak{P} : \square \subset \mathbb{R}^2 \to \mathfrak{C}$. For example, if all the luminous powder were concentrated on a single point $x_0 \in \square$ then we would have to take $\mathfrak{P}(x) = 0$ for all $x \in \square$ with $x \neq x_0$. Digitized versions of such a picture function would tend to take the value zero *everywhere*. A similar problem would occur if the powder were attached uniformly to the points of a line segment $L \subset \square$.

Figure 2.11   A measure with red, green and blue components has here been digitized to produce digital pictures with two different resolutions. The digital picture on the left has one quarter the resolution of that on the right. In principle, with uniform lighting both pictures should reflect the same number of red, green and blue photons per second, from each little pixel domain corresponding to the lower-resolution image. In practice, printing and saturation effects make this only an approximate statement. Nonetheless, try moving back from the two pictures until you cannot tell them apart.

You may at this point be tempted to try to describe the distribution of luminous powder generally with the aid of picture functions that involve 'densities' and 'delta functions'. But this is not generally possible either: to see why, imagine that all the powder were attached to a non-denumerable totally disconnected subset of $\square$, or to the set of points in $\square$ whose coordinates are pairs of numbers having binary expansions that contain more zeros than ones asymptotically. Neither of these methods by which powder is attached to points in $\mathbb{R}^2$ can be described in the usual way by densities or delta functions.

However, given any distribution of the luminous powder on $\square$, we can in fact imagine how it may be used to deduce a corresponding digital picture in a consistent manner. To each pixel $\mathfrak{P}_{w,h}$ we simply assign a value equal to the total brightness of the powder lying upon the rectangle $\square_{w,h}$, some number such as 0.01 candlepower. In this way, if a single pixel is treated as being made of smaller pixels, the sum of the brightness of the smaller pixels is equal to the brightness of the larger one, component by component.

In Figure 2.11 we show two digital pictures, one at one quarter the resolution of the other, both with the same domain $\square \subset \mathbb{R}^2$ corresponding to the same measure. Limitlessly, new detail in the image is revealed as the resolution is increased, because the underlying measure is a fractal one, as we shall understand better in Chapter 5. See also Figure 2.12.

Figure 2.12 Here a measure has been digitized at lower and lower resolutions and printed, from left to right, in such a way that the pixels are all the same size. So all images emit the same amount of red, green and blue light. Once a pixel reaches full intensity in any component, however, it becomes saturated.

A more extraordinary fact, making the subject of measure theory depart from intuition in some deep and mysterious way, is this: you might simply decide to say that the measure of any subset of $\square \subset \mathbb{R}^2$ is the 'amount' of luminous powder that is attached to the points of the subset. Surely this will give us a consistent description of brightness so that, when we break up a set into several parts, the sum of the measures of the parts is equal to the measure of the whole? This is not possible. We have to restrict the class of sets to those that we can suppose to have known amounts of luminous powder attached to them. Loosely we call such subsets 'measurable'. *Not all subsets of $\square$ can be measurable, if a consistent picture is to emerge.* This extraordinary mathematical fact is deep and inspires us to explore the true magical nature of the euclidean plane.

The spaces on which measures may be defined: fields and σ-algebras

In order to define measures we need first to discuss the types of collections of subsets with which they may be associated.

DEFINITION 2.3.1   Let $\mathbb{X}$ be a space. Let $\mathcal{F}(\mathbb{X})$ be a nonempty collection of subsets of $\mathbb{X}$ with these properties:

(i) if $\mathcal{O}_1, \mathcal{O}_2 \in \mathcal{F}(\mathbb{X})$ then $\mathcal{O}_1 \cup \mathcal{O}_2 \in \mathcal{F}(\mathbb{X})$;
(ii) if $\mathcal{O} \in \mathcal{F}(\mathbb{X})$ then $\mathbb{X} \backslash \mathcal{O} \in \mathcal{F}(\mathbb{X})$.

The collection of subsets $\mathcal{F}(\mathbb{X})$ is called a **field** on $\mathbb{X}$. If, moreover,

(iii) whenever $\mathcal{O}_i \in \mathcal{F}(\mathbb{X})$ for all $i = 1, 2, \ldots$ we have $\bigcup_{i=1}^{\infty} \mathcal{O}_i \in \mathcal{F}(\mathbb{X})$

then $\mathcal{F}(\mathbb{X})$ is called a σ-**algebra** on $\mathbb{X}$.

A σ-algebra may also be called a σ-**field**. The focus of our interest is on σ-algebras but it is convenient to work with fields and then extend the results to the corresponding σ-algebras. For example, the pixel field described in Exercise 2.3.5 below does not contain rectangles whose dimensions are irrational numbers, yet we are certainly going to want to discuss the 'mass' contained in such rectangles.

Given any space $\mathbb{X}$ one can always find at least one σ-algebra on $\mathbb{X}$, namely $\mathbb{S}(\mathbb{X})$, the space of all subsets of $\mathbb{X}$. Indeed, $\mathbb{S}(\mathbb{X})$ is the largest σ-algebra on $\mathbb{X}$ since it contains all other σ-algebras on $\mathbb{X}$. The pair of sets $\{\mathbb{X}, \varnothing\}$ is also a σ-algebra on $\mathbb{X}$, as is the set of sets $\{\mathbb{X}, S, \mathbb{X}\backslash S, \varnothing\}$ where $S \subset \mathbb{X}$.

One way to construct a σ-algebra is as follows. Start with any collection $\mathbb{G}$ of subsets of $\mathbb{X}$. Let $\mathcal{F}'(\mathbb{X})$ be the set of all sets that can be described by finite-length expressions (that make sense) involving the space $\mathbb{X}$ together with the sets of $\mathbb{G}$, using unions and complements. So if $\mathbb{G}_1, \mathbb{G}_2 \in \mathbb{G}$ then examples of members of $\mathcal{F}'(\mathbb{X})$ are:

$$\mathbb{G}_1, \quad \mathbb{G}_2, \quad \mathbb{G}_1 \cup \mathbb{G}_2, \quad (\mathbb{X}\backslash\mathbb{G}_1) \cup \mathbb{G}_2, \quad ((\mathbb{X}\backslash\mathbb{G}_2) \cup \mathbb{G}_2) \cap \mathbb{G}_1, \quad \ldots$$

You can verify that $\mathcal{F}'(\mathbb{X})$ is a field by checking that if the sets $O, O_1, O_2 \subset \mathbb{X}$ are defined by such expressions then so are the expressions in Definition 2.3.1(i), (ii). We call $\mathcal{F}'(\mathbb{X})$ the **field generated by** $\mathbb{G}$.

EXERCISE 2.3.2 *Prove that you generate exactly the same set $\mathcal{F}'(\mathbb{X})$ if you allow intersections, as well as unions and complements, in the above description.*

Let $\mathcal{F}_{\mathbb{G}}(\mathbb{X})$ denote the intersection of all the σ-algebras on $\mathbb{X}$ that contain $\mathbb{G}$. Then it is straightforward to verify that $\mathcal{F}_{\mathbb{G}}(\mathbb{X})$ is also a σ-algebra. It is called the σ-algebra **generated by** $\mathbb{G}$.

EXERCISE 2.3.3 *Prove that $\mathcal{F}_{\mathbb{G}}(\mathbb{X})$ is a σ-algebra.*

It is convenient to think of $\mathcal{F}_{\mathbb{G}}(\mathbb{X})$ as the smallest σ-algebra on $\mathbb{X}$ that contains the field $\mathcal{F}'(\mathbb{X})$.

EXERCISE 2.3.4 *Prove that if the set of generators $\mathbb{G}$ is finite then $\mathcal{F}'(\mathbb{X}) = \mathcal{F}_{\mathbb{G}}(\mathbb{X})$.*

In Figure 2.13 we illustrate the σ-algebra generated by two subsets of $\square$.

EXERCISE 2.3.5 *Let*

$$\mathbb{G}_{pixels} = \left\{ \square_{w,h}^{W,H} : w = 1, 2, \ldots, W; h = 1, 2, \ldots, H; W \in \mathbb{N}; H \in \mathbb{N} \right\}$$

*denote the set of all domains of all pixels of all digital images with domain $\square \subset \mathbb{R}^2$ and having lower left corner $(0, 0)$ and upper right corner $(1, 1)$. Then let us call $\mathcal{F}'(\square)$ the **pixel field for** $\square \subset \mathbb{R}^2$. Similarly, let us call $\mathcal{F}_{\mathbb{G}_{pixels}}(\square)$ the **pixel σ-algebra for** $\square \subset \mathbb{R}^2$. Show that the area of each element of $\mathcal{F}'(\square)$ is a rational*

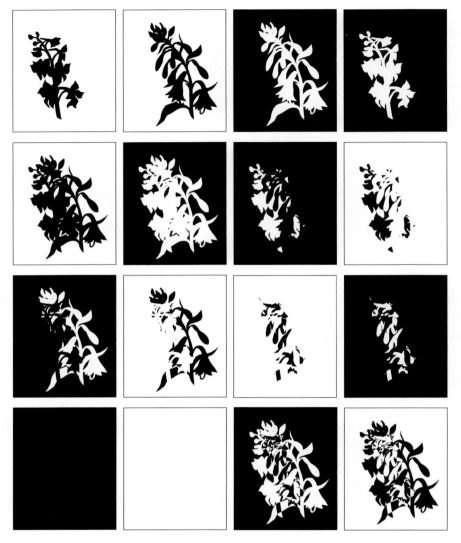

Figure 2.13 This represents the sets in a σ-algebra $\mathcal{F}_{\mathbb{G}}(\square)$ of subsets of $\square \subset \mathbb{R}^2$. It is generated by a pair of sets $\mathbb{G}$, top left, each of which looks like the silhouette of a flower.

number. Show that $S \in \mathcal{F}_{\mathbb{G}_{pixels}}(\square)$, where

$$S = \big\{(x, y) : 0 < x, y < \sqrt[4]{2}\big\}.$$

Show that the area of S is $\sqrt{2}$. Conclude that S is not in the pixel field but that it is in the pixel σ-algebra.

EXERCISE 2.3.6   *Recall that a cylinder set of the code space Ω_A is a set that can be written in the form*

$$\mathcal{C}(\sigma) := \{\omega \in \Omega_A : \omega_n = \sigma_n \text{ for all } n = 1, 2, \ldots, |\sigma|\},$$

for some $\sigma \in \Omega'_A$. Let $\mathcal{F}'(\Omega_A)$ and $\mathcal{F}(\Omega_A)$ denote respectively the field and the σ-algebra generated by the cylinder subsets of Ω_A. Let $\mathcal{A} = \{0, 1\}$. Show that the set

$$S := \{\sigma \in \Omega_A : \sigma_n = 1 \text{ when } n \text{ is an odd integer}\}$$

is in $\mathcal{F}(\Omega_A)$ but not in $\mathcal{F}'(\Omega_A)$.

An important σ-algebra for digital pictures is that generated by the open subsets of $\mathbb{R}^2$, because it is preserved by continuous transformations.

DEFINITION 2.3.7   Let $(\mathbb{X}, \mathbb{T}(\mathbb{X}))$ be a topological space. Then the set of **Borel subsets** of $\mathbb{X}$ is the σ-algebra $\mathcal{F}_{\mathbb{T}(\mathbb{X})}(\mathbb{X})$ generated by the open subsets $\mathbb{T}(\mathbb{X})$ of $\mathbb{X}$. We denote the Borel subsets of $\mathbb{X}$ by $\mathcal{B}(\mathbb{X})$, namely

$$\mathcal{B}(\mathbb{X}) := \mathcal{F}_{\mathbb{T}(\mathbb{X})}(\mathbb{X}).$$

EXERCISE 2.3.8   *Show that the pixel σ-algebra for $\square \subset \mathbb{R}^2$ is the same as the set of the Borel subsets of $\mathcal{B}(\square)$, where the underlying topology is the natural topology induced by the euclidean metric. That is,*

$$\mathcal{B}(\square) = \mathcal{F}_{\mathbb{G}_{pixels}}(\square).$$

See also Exercise 1.9.5.

Definition of a measure

DEFINITION 2.3.9   A **measure** on a space $\mathbb{X}$ is a function $\nu : \mathcal{F}(\mathbb{X}) \to [0, \infty)$, where $\mathcal{F}(\mathbb{X})$ is a field, with

$$\sum_{n=1}^{\infty} \nu(\mathcal{O}_n) = \nu\left(\bigcup_{n=1}^{\infty} \mathcal{O}_n\right) \tag{2.3.1}$$

whenever $\{\mathcal{O}_n \in \mathcal{F}(\mathbb{X}) : n = 1, 2, \ldots\}$ is a sequence such that $\bigcup_{n=1}^{\infty} \mathcal{O}_n \in \mathcal{F}(\mathbb{X})$ and

$$\mathcal{O}_n \cap \mathcal{O}_m = \varnothing$$

for all $n, m \in \mathbb{N}$ with $n \neq m$. In other texts a measure as defined here may be called a **finite** measure. Sometimes, too, it may be referred to as a **positive** measure.

When $\nu(\mathbb{X}) = 1$ the measure ν may be called a **normalized** measure or a **probability** measure on $\mathbb{X}$. When $\mathcal{F}(\mathbb{X}) = \mathcal{B}(\mathbb{X})$, the Borel σ-algebra, the measure ν may be called a **Borel measure**. We denote the set of Borel measures on $\mathbb{X}$ by $\mathbb{M}(\mathbb{X})$ and the set of normalized Borel measures on $\mathbb{X}$ by $\mathbb{P}(\mathbb{X})$.

Let $(\mathbb{X}, d)$ be a metric space and let μ be a Borel measure. Then the **support** of μ is the set of points $x \in \mathbb{X}$ such that $\mu(\mathcal{O}_x) > 0$ whenever $\mathcal{O}_x$ is an open set that contains x.

Examples of measures

EXAMPLE 2.3.10   Suppose that $\mathbb{X}$ consists of an array of 50×70 points and that Figure 2.13 represents a σ-algebra $\mathcal{F}(\mathbb{X})$. (That is, suppose that each panel in Figure 2.13 represents an array of 50×70 dots, some black and the others white and that the black dots within the panel represent the members of $\mathbb{X}$ in the corresponding subset.) Then define $\nu(B)$ to be the number of black dots in the set B, for each $B \in \mathcal{F}(\mathbb{X})$. Then $\nu : \mathcal{F}(\mathbb{X}) \to \{0, 1, \dots, 3500\}$ is an example of a measure. In this case $\nu(\mathbb{X}) = 3500$.

EXERCISE 2.3.11   *Assign a value $\nu(B) \in \{0, 1, \dots, 3500\}$ to each of the panels B in Figure 2.13 so as to define a measure on the σ-algebra $\mathcal{F}(\mathbb{X})$ in Example 2.3.10.*

EXAMPLE 2.3.12   Let $\mathcal{B}(\square)$ denote the set of Borel subsets of $\square \subset \mathbb{R}^2$. Let $x_0 \in \square$. Define $\delta_{x_0} : \mathcal{B}(\square) \to [0, \infty)$ by

$$\delta_{x_0}(B) = \begin{cases} 1 & \text{if } x_0 \in B, \\ 0 & \text{if } x_0 \notin B. \end{cases}$$

Then δ_{x_0} is a normalized Borel measure on $\square$. That is, $\delta_{x_0} \in \mathbb{P}(\square)$.

EXAMPLE 2.3.13   Let $\mathcal{F}'(\Omega_A)$ denote the field generated by the cylinder subsets of the code space Ω_A, as in Exercise 2.3.6. Let $\mathcal{A} = \{1, 2, \dots, N\}$. Let $p_1 \geq 0, p_2 \geq 0, \dots, p_N \geq 0$ and $p_0 + p_1 + \cdots + p_N = 1$. Then there is a unique measure ν' on the field $\mathcal{F}'(\Omega_A)$ generated by the cylinder subsets $\{\mathcal{C}(\sigma) : \sigma \in \Omega'\}$ such that

$$\nu'(\mathcal{C}(\sigma)) = p_{\sigma_1} p_{\sigma_2} \cdots p_{\sigma_{|\sigma|}} \quad \text{for all } \sigma \in \Omega'.$$

Let us say that the length of the cylinder subset $\mathcal{C}(\sigma)$ is $|\sigma|$. Then it is straightforward to show that any element of $\mathcal{F}'(\Omega_A)$ may be written as a union of cylinder subsets of the same length, and also that distinct cylinder subsets of the same length are disjoint. It follows that the measure of any element of the field $\mathcal{F}'(\Omega_A)$ can be written as the sum of numbers of the form $p_{\sigma_1} p_{\sigma_2} \cdots p_{\sigma_{|\sigma|}}$ for some fixed value of $|\sigma|$.

EXAMPLE 2.3.14   Let $\rho : \square \subset \mathbb{R}^2 \to [0, \infty)$ be a continuous positive function. It describes a surface over $\square$. Using standard integration, one can in principle evaluate the integral

$$\nu'(\square_{w,h}) := \int \int_{\square_{w,h}} \rho(x, y) dx dy.$$

This yields the volume between the part of the surface lying 'vertically above' the rectangle $\Box_{w,h}$ and the xy-plane. In a similar manner one can evaluate $v'(S)$ for any set S in the pixel field $\mathcal{F}'(\Box)$. Then it can be shown (straightforwardly) that $v' : \mathcal{F}'(\Box) \to [0, \infty)$ is a measure. In this case the measure is indeed described by the **density function** $\rho(x, y)$. This same construction of a measure works equally well if the density function is piecewise continuous and its discontinuities occur on sets that are not too complicated, for example boundaries of domains of pixels.

Digital pictures of measures

Given any vector of measures,

$$v_\rho = (v_1, v_2, v_3),$$

where v_i is a measure on the pixel field $\mathcal{F}'(\Box)$ for $i = 1, 2, 3$, we can uniquely specify a corresponding digital picture $\mathfrak{P}_{W \times H}$ for each $W, H \in \mathbb{N}$. The members of the resulting family of digital pictures will be consistent with one another in this sense: the value of any pixel $\mathfrak{P}_{w,h}$ will be greater than or equal to the sum of the values of any set of pixels from a family whose domains are disjoint and whose union is contained in $\Box_{w,h}$, while $\mathfrak{P}_{w,h}$ will be less than or equal to the sum of the values of any set of pixels from a family whose union contains $\Box_{w,h}$.

However, in order to be able to discuss what happens when we transform a measure, say under a continuous transformation, we need to know the value of the measure not just on the pixel field but also on the pixel σ-algebra, because pixels are not transformed into pixels by general transformations. Luckily, the following theorem tells us two wonderful things, that once we have a measure on a field we can extend it uniquely to the σ-algebra generated by the field and also how to evaluate the measure on the σ-algebra using only its values on the field.

THEOREM 2.3.15   *Let $\mathbb{X}$ be a space, let $\mathcal{F}'(\mathbb{X})$ be a field on $\mathbb{X}$ and let $\mathcal{F}(\mathbb{X})$ be the smallest σ-algebra on $\mathbb{X}$ that contains $\mathcal{F}'(\mathbb{X})$. Let $v' : \mathcal{F}'(\mathbb{X}) \to [0, \infty)$ be a measure. Then there exists a unique measure $v : \mathcal{F}(\mathbb{X}) \to [0, \infty)$ such that*

$$v(B) = v'(B) \quad \text{for all } B \in \mathcal{F}'(\mathbb{X}).$$

Moreover,

$$v(A) = \inf \left\{ \sum_{n=1}^{\infty} v(B_n) : A \subset \bigcup_{n=1}^{\infty} B_n, \, B_n \in \mathcal{F}'(\mathbb{X}) \text{ for all } n = 1, 2, \ldots \right\}$$

for all $A \in \mathcal{F}(\mathbb{X})$.

The notation $\inf S$, where $S \subset \mathbb{R} \cup \{-\infty, +\infty\}$, means the largest number in $\mathbb{R} \cup \{-\infty, +\infty\}$ that is less than or equal to all the numbers in the set S. For example, $\inf \mathbb{R} = -\infty$, $\inf(1, 2] = 1$ and $\inf\{1/n : n = 1, 2, \ldots\} = 0$. Similarly the notation sup S, where $S \subset \mathbb{R} \cup \{-\infty, +\infty\}$, is defined to be the smallest number in $\mathbb{R} \cup \{-\infty, +\infty\}$ that is greater than or equal to all the numbers in the set S.

PROOF   The proof of Theorem 2.3.15 is beautiful and subtle, and it may be found in most books on measure theory; see for example [31], Theorem 5, p. 180. The key steps are the following. (i) Define a function, called an **outer measure**, $\nu^0 : \mathbb{S}(\mathbb{X}) \to [0, \infty)$, by

$$\nu^0(S) = \inf \left\{ \sum_{n=1}^{\infty} \nu'(B_n) : S \subset \bigcup_{n=1}^{\infty} B_n, B_n \in \mathcal{F}'(\mathbb{X}) \text{ for } n = 1, 2, \dots \right\}$$

for all $S \in \mathbb{S}(\mathbb{X})$.

(ii) Show that D is a σ-algebra and that $\nu^0 : D \subset \mathbb{S}(\mathbb{X}) \to [0, \infty)$ is a measure, where $D := \{ S \in \mathbb{S}(\mathbb{X}) : \nu^0(S \cap T) + \nu^0((\mathbb{X} \backslash S) \cap T)) = \nu^0(T) \text{ for all } T \in \mathbb{S}(\mathbb{X}) \}$.

(iii) Show that $\mathcal{F}'(\mathbb{X}) \subset D$ and that ν^0 agrees with ν' on $\mathcal{F}'(\mathbb{X})$.

(iv) Define ν to be ν^0 restricted to $\mathcal{F}(\mathbb{X})$. Note that $\mathcal{F}(\mathbb{X}) \subset D$.

(v) Check uniqueness.   □

EXAMPLE 2.3.16   Let $\rho : \square \subset \mathbb{R}^2 \to [0, \infty)$ be a continuous (or piecewise continuous) density function, as discussed in Example 2.3.14. Then Theorem 2.3.15 tells us that there exists a unique measure ν_ρ on the pixel σ-algebra that agrees with the measure ν', defined in Example 2.3.14, on the pixel field $\mathcal{F}'(\square)$. Since, as in Exercise 2.3.8, $\mathcal{F}_{\mathbb{G}_{pixels}}(\square) = \mathcal{B}(\square)$, ν_ρ is a Borel measure.

EXAMPLE 2.3.17   There exists a unique measure ν on the code space σ-algebra $\mathcal{F}(\Omega_A)$ that agrees with the measure ν' on the code space field $\mathcal{F}'(\Omega_A)$, as described in Example 2.3.13. Since the cylinder sets generate the natural topology on Ω_A, the measure ν is actually a Borel measure.

EXAMPLE 2.3.18   Any digital picture defines a (vector of) Borel measure(s) in the following manner. Let α denote the area of the domain of the pixel $\mathfrak{P}_{w,h}$. The area of each pixel in the digital picture $\mathfrak{P}_{W \times H}$ is the same. Then we define a piecewise-constant density function by

$$\rho(x) = \frac{1}{\alpha} \mathfrak{P}_{w,h}(x) \quad \text{when } x \in \square_{w,h}, \text{ for all } w \in \{1, 2, \dots, W\},$$

$$h \in \{1, 2, \dots, H\}.$$

We now use this measure to define a Borel measure on $\mathcal{B}(\square)$ as in Examples 2.3.14 and 2.3.16. Then if we make a digital picture $\widetilde{\mathfrak{P}}_{W \times H}$ of this measure, we will have $\widetilde{\mathfrak{P}}_{W \times H} = \mathfrak{P}_{W \times H}$. The advantage of converting a digital picture into a Borel measure is that it can then be manipulated by continuous transformations and digitized, in a consistent manner.

A **measurable set** is one to which a measure may be assigned, that is, a member of a field or σ-algebra. Usually, for us, this will mean a Borel set. (Imagine that we could distribute one candlepower of luminous powder on a subset of $\square$ that is not a Borel set. How would one make digital pictures of the resulting glowing thing?)

In Chapter 4 we will discover a multitude of interesting measures on $\square$. For now we need to know that there exist diverse measures on $\square$, that they define arrays of pictures, one for each W and H, namely digital pictures, and that they behave as nicely as picture functions under continuous transformations, as Theorem 2.3.19 below shows. Throughout this book we give many examples of digital pictures of measures.

Transformations of measures

Let us first describe intuitively what we would like to happen when a transformation is applied to a measure. Suppose that we are given a normalized Borel measure v on $\square \subset \mathbb{R}^2$, which we imagine to be a luminous picture in the euclidean plane. Perhaps it is embedded in infinitely thin flat material, like the skin of a vast balloon. Let $f : \mathbb{R}^2 \to \mathbb{R}^2$ be a continuous transformation. Then we may think of f as deforming, stretching, shrinking and folding the luminous material. Regions that are stretched will tend to become less bright, regions that are compressed will become brighter and parts that are folded on top of one another will have a brightness that is the sum of the brightnesses of the parts. Figure 2.14 illustrates this idea. This is how we would like to think of the continuous transformation of a measure. We want the result to be a new luminous picture, that is, another Borel measure.

This inspires us to define below the action of a transformation on a measure in a certain obvious sort of way. But is the resulting transformed measure indeed always a Borel measure? Does the transformation process damage the underlying σ-algebra? No, wonderfully, it does not.

Theorem 2.3.19 defines the continuous transformation of a Borel measure and assures us that we obtain a new Borel measure. The key ideas are that the Borel sets are generated by the open sets and that the inverse images of open sets, under continuous transformation, are open sets.

THEOREM 2.3.19   *Let $v \in \mathbb{M}(\mathbb{X})$ be a Borel measure and let $f : \mathbb{X} \to \mathbb{X}$ be continuous. Then there exists on $\mathbb{X}$ a unique Borel measure $\mu \in \mathbb{M}(\mathbb{X})$ such that*

$$\mu(B) = v(f^{-1}(B)) \quad \text{for all } B \in \mathcal{B}(\mathbb{X}).$$

We denote this measure μ by $f(v)$ and also by $f \circ v$.

DEFINITION 2.3.20   The measure $f(v)$ is called the **transformation of the measure** v by the function f or the transformation f applied to the measure v.

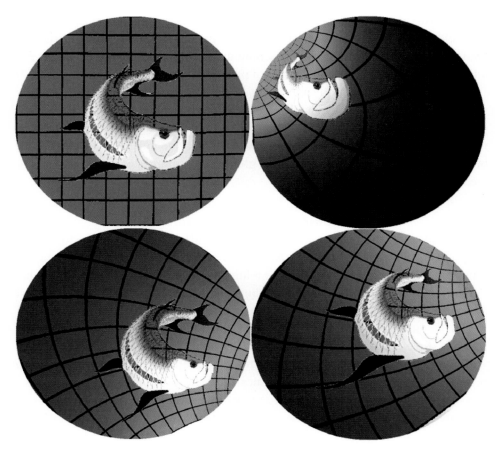

Figure 2.14 Each image represents a coloured measure on a disk in $\mathbb{R}^2$. Each pair of measures is related by a Möbius transformation that maps the disk to itself. So the total brightness of each image is in principle the same, as is the total brightness of each curvaceous square.

When $v = (v_1, v_2, v_3) \in \mathbb{M}(\mathbb{X})^3$ we define

$$f(v) = (f(v_1), f(v_2), f(v_3)) \in \mathbb{M}(\mathbb{X})^3.$$

We may refer to a transformation of a measure where we mean a **transformation of a vector of measures**.

PROOF OF THEOREM 2.3.19   We will show that (i) $f \circ v$ is defined on $\mathcal{F}'$, the field generated by the open subsets of $\mathbb{X}$; (ii) $f \circ v$ is a measure on $\mathcal{F}'$, that is, it obeys Equation (2.3.1) in Definition 2.3.9; (iii) $f \circ v$ is a measure on $\mathcal{F}_{\mathbb{T}(\mathbb{X})}$, the smallest σ-algebra that contains $\mathcal{F}'$. In fact (iii) follows immediately from Theorem 2.3.15 once (i) and (ii) are established. So we need only to prove (i) and (ii).

Proof of (i): Suppose that $S \in \mathcal{F}'$. Then S can be written as a finite expression (that makes sense) involving unions and complements (with respect to $\mathbb{X}$) of a finite number of open sets, say $S = E(O_1, O_2, \ldots, O_N)$ where $O_1, O_2, \ldots, O_N$ are open sets. Then, since $f^{-1}(V \cup W) = f^{-1}(V) \cup f^{-1}(W)$ and $f^{-1}(\mathbb{X} \backslash V) = \mathbb{X} \backslash f^{-1}(V)$ whenever $V, W \in \mathbb{X}$, as you learnt in Exercise 1.3.2, it follows that $f^{-1}(S) = E(f^{-1}(O_1), f^{-1}(O_2), \ldots, f^{-1}(O_N))$. Furthermore, because f is continuous it follows that each of the sets $f^{-1}(O_1), f^{-1}(O_2), \ldots,$ and $f^{-1}(O_N)$ is open. Hence $f^{-1}(S)$ is a finite expression that makes sense, involving unions and complements of a finite number of open sets, and so belongs to $\mathcal{F}'$.

Proof of (ii): Let us suppose that $\{\mathcal{O}_n \in \mathcal{F}'(\mathbb{X}) : n = 1, 2, \ldots\}$ is a sequence such that $\bigcup_{n=1}^{\infty} \mathcal{O}_n \in \mathcal{F}'(\mathbb{X})$ and

$$\mathcal{O}_n \cap \mathcal{O}_m = \varnothing$$

for all $n, m \in \mathbb{N}$ with $n \neq m$. Then we need to show that

$$\sum_{n=1}^{\infty} \nu(f^{-1}(\mathcal{O}_n)) = \nu \left(\bigcup_{n=1}^{\infty} f^{-1}(\mathcal{O}_n) \right). \tag{2.3.2}$$

But $\bigcup_{n=1}^{\infty} \mathcal{O}_n \in \mathcal{F}'(\mathbb{X})$ implies that $f^{-1}(\bigcup_{n=1}^{\infty} \mathcal{O}_n) \in \mathcal{F}'(\mathbb{X})$ by (i). Moreover $f^{-1}(\bigcup_{n=1}^{\infty} \mathcal{O}_n) = \bigcup_{n=1}^{\infty} f^{-1}(\mathcal{O}_n)$ so $\bigcup_{n=1}^{\infty} f^{-1}(\mathcal{O}_n) \in \mathcal{F}'(\mathbb{X})$. Also, that $\mathcal{O}_n \cap \mathcal{O}_m = \varnothing$ for all $m \neq n$ implies $f^{-1}(\mathcal{O}_n \cap \mathcal{O}_m) = f^{-1}(\mathcal{O}_n) \cap f^{-1}(\mathcal{O}_m) = \varnothing$ for all $m \neq n$. So $\{f^{-1}(\mathcal{O}_n) \in \mathcal{F}'(\mathbb{X}) : n = 1, 2, \ldots\}$ is a sequence such that $\bigcup_{n=1}^{\infty} f^{-1}(\mathcal{O}_n) \in \mathcal{F}'(\mathbb{X})$ and $f^{-1}(\mathcal{O}_n) \cap f^{-1}(\mathcal{O}_m) = \varnothing$ for all $m \neq n$. Since ν is a measure on $\mathcal{F}'(\mathbb{X})$ it follows that Equation (2.3.2) holds as desired. $\qquad \square$

Figure 2.15 illustrates transformations of a Borel measure. The top left panel represents a vector (red, green and blue) ν of Borel measures on $\square \subset \mathbb{R}^2$. The other three panels represent three different projective transformations of the measure. Projective transformations map straight lines into straight lines and quadrilaterals into quadrilaterals. In particular, the total amount of light (component by component) given off by the points inside each quadrilateral should be the same. See also Figures 2.14 and 2.16.

EXERCISE 2.3.21 *Let $\nu \in \mathbb{M}(\mathbb{X})$ be a Borel measure and let $f : \mathbb{X} \to \mathbb{X}$ be continuous. Show that*

$$(f \circ \nu)(\mathbb{X}) = \nu(\mathbb{X})$$

and hence that $f : \mathbb{P}(\mathbb{X}) \to \mathbb{P}(\mathbb{X})$.

EXERCISE 2.3.22 *In our discussion of transformations of pictures we restricted our attention to one-to-one transformations but we do not do so in the case of transformations of measures. Why is this?*

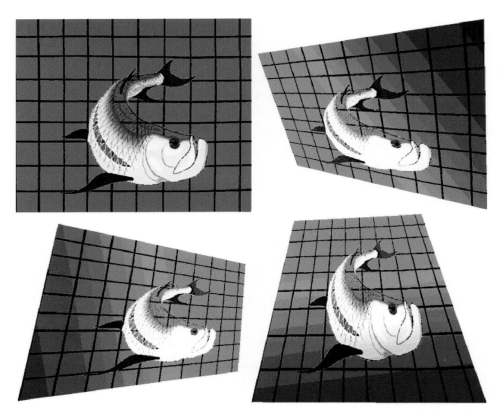

Figure 2.15 These four images illustrate transformations of a Borel measure. The top left image represents a vector (red, green, blue) v of Borel measures on $\square \subset \mathbb{R}^2$. The other three images represent three different projective transformations of the measure. The transformed measures tend to be brighter where the space is compressed and less bright where the space is stretched. But in these images, where a colour should have a value greater than 255 it is assigned the value 255, and we say that the colour is **saturated**. The colours are also **quantized**, that is, they only take certain discrete values, producing jumps in intensity rather than a smooth gradation.

Invariant measures

DEFINITION 2.3.23   Let $(\mathbb{X}, d)$ be a metric space and let $f : \mathbb{X} \to \mathbb{X}$ be continuous. A measure $\mu \in \mathbb{M}(\mathbb{X})$ is said to be invariant under f iff

$$\mu(B) = \mu(f^{-1}(B)) \quad \text{for all Borel sets } B \in \mathcal{B}(\mathbb{X}). \qquad (2.3.3)$$

Such a measure μ is called an **invariant measure** of the transformation f.

Notice that Equation (2.3.3) is equivalent to

$$f(\mu) = \mu.$$

EXAMPLE 2.3.24   The measure δ_{x_0} defined in Example 2.3.12 is invariant under the transformation $f : \square \to \square$ defined by $f(x) = \frac{1}{2}(x - x_0)$.

EXAMPLE 2.3.25   A measure $\mu \in \mathbb{M}([0, 1] \subset \mathbb{R})$ can be defined with density $\rho(x) = (\sqrt{x(1 - x)})^{-1}$. That is,

$$\mu(B) = \int_B \frac{dx}{\sqrt{x(1 - x)}} \quad \text{for all Borel sets } B \in \mathcal{B}(\mathbb{R}^2).$$

This measure is invariant under the transformation $f : [0, 1] \to [0, 1]$ defined by $f(x) = 4x(1 - x)$.

In Figure 2.17 we show two pictures of (vectors of Borel) measures that are invariant under transformations that map $\mathbb{R}^2$ into itself. The left-hand panel is a picture of a measure that is invariant under any rotation $R_\theta : \mathbb{R}^2 \to \mathbb{R}^2$, where θ is a multiple of $36°$, about the origin, which corresponds to the centre of the picture. The right-hand panel is a picture of a measure that is invariant under any Möbius rotation

$$\mathcal{M} = \mathcal{M}_a \circ R_\theta \circ \mathcal{M}_a^{-1}, \tag{2.3.4}$$

where, in complex notation with $z = x + iy$,

$$\mathcal{M}_a(z) = \frac{a - z}{1 - \overline{a}z} \quad \text{for all } (x, y) \in \mathbb{R}^2$$

and $a = (-0.25, 0.15) \in \mathbb{C}$ is the centre of the rotation. The transformation $\mathcal{M}_a$ maps the circle of radius 1 centred at the origin into itself, while mapping the point a to the origin. Such transformations are discussed in Section 2.6.

An example that illustrates a closely related picture and measure, both of which are invariant under the Möbius rotation in Equation (2.3.4), is illustrated in Figure 2.18. Clearly the same transformation may possess many different invariant pictures and invariant measures. Similarly Figures 2.19 and 2.20 contrast (parts of) pictures and measures that are invariant under the same transformation as that illustrated in Figure 2.6.

EXERCISE 2.3.26   *Show that if $f, g : \mathbb{X} \to \mathbb{X}$ are both continuous and the measure $\mu \in \mathbb{M}(\mathbb{X})$ is invariant under f then the measure $g(\mu)$ is invariant under $g \circ f \circ g^{-1}$.*

2.4   Fixed points and fractals

DEFINITION 2.4.1   Let $\mathbb{X}$ be a space and let $f : \mathbb{X} \to \mathbb{X}$ be a transformation. Then a point $a \in \mathbb{X}$ such that

$$f(a) = a$$

is called a **fixed point** of the transformation f.

Let $\mathbb{X}$ be a metric space, or a topological space, so that $\mathbb{H}(\mathbb{X})$ is defined. Then an invariant set $A \in \mathbb{H}(\mathbb{X})$ of a transformation $f : \mathbb{X} \to \mathbb{X}$ is a fixed point of $f : \mathbb{H}(\mathbb{X}) \to \mathbb{H}(\mathbb{X})$ because it obeys

$$f(A) = A.$$

An invariant measure $\mu \in \mathbb{P}(\mathbb{X})$ for a transformation $f : \mathbb{X} \to \mathbb{X}$ is similarly a fixed point of $f : \mathbb{P}(\mathbb{X}) \to \mathbb{P}(\mathbb{X})$ because

$$f(\mu) = \mu.$$

Also, an invariant picture $\mathfrak{P} \in \Pi$ of a *one-to-one* transformation $f : \mathbb{R}^2 \to \mathbb{R}^2$ is a fixed point of $f : \Pi \to \Pi$. So in our search for an understanding of when sets, pictures and measures may be invariant under transformations it is natural to consider conditions relating to the existence of fixed points.

Contraction mapping theorem

DEFINITION 2.4.2   Let $(\mathbb{X}, d)$ be a metric space. A transformation $f : \mathbb{X} \to \mathbb{X}$ is said to be **Lipschitz** with **Lipschitz constant** $l \in \mathbb{R}$ iff

$$d(f(x), f(y)) \leq l \cdot d(x, y) \quad \text{for all } x, y \in \mathbb{X}.$$

A transformation $f : \mathbb{X} \to \mathbb{X}$ is called **contractive** iff it is Lipschitz with Lipschitz constant $l \in [0, 1)$. A Lipschitz constant $l \in [0, 1)$ is also called a **contraction factor**. A contractive transformation is also called a **contraction mapping**.

We may write $Lip_l(\mathbb{X})$ to denote the set of Lipschitz transformations $F : \mathbb{X} \to \mathbb{X}$ with Lipschitz constant $l \geq 0$.

The following theorem, for all its formal elegance, is of great practical importance to us. We will use it over and over again to construct fractal sets, pictures, measures and superfractals.

THEOREM 2.4.3 (*Contraction mapping theorem*)   *Let $\mathbb{X}$ be a complete metric space. Let $f : \mathbb{X} \to \mathbb{X}$ be a contraction mapping with contraction factor l. Then f has a unique fixed point $a \in \mathbb{X}$. Moreover, if x_0 is any point in $\mathbb{X}$ and we have $x_n = f(x_{n-1})$ for $n = 1, 2, 3, \ldots$ then*

$$d(x_0, a) \leq \frac{d(x_0, x_1)}{1 - l} \tag{2.4.1}$$

and

$$\lim_{n \to \infty} x_n = a.$$

PROOF   The proof of this theorem is an enjoyable exercise. Start by showing that $\{x_n\}_{n=0}^{\infty}$ is a Cauchy sequence. Let $a \in \mathbb{X}$ be the limit of this sequence. Then use the continuity of f to yield $a = f(a)$.   □

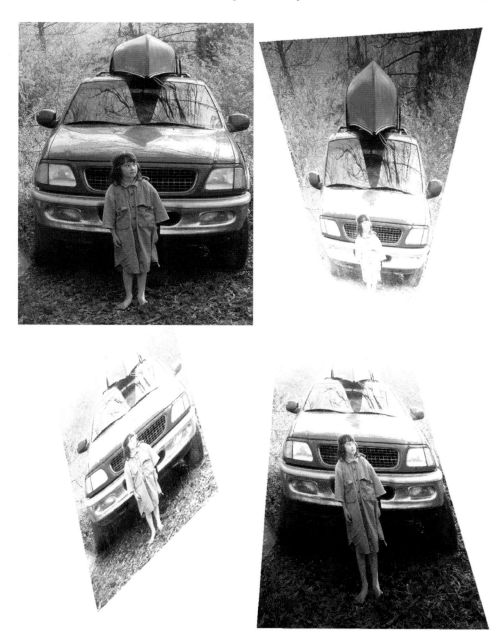

Figure 2.16 Projective transformations of a digital photograph, treated as a measure. Colour saturation effects can be seen here.

Figure 2.17 Two examples of pictures of measures that are invariant under transformations. The measure represented by the picture on the left is invariant by a rotation through 36°. The measure represented inside the disk, in the right-hand panel, is invariant under a Möbius rotation, as in Equation (2.3.4). The picture fades where it expands and brightens where it contracts.

Equation (2.4.1) tells us an upper bound for the distance from x_0 to the fixed point a that involves only $d(x_0, f(x_0))$ and l. We will use this bound in Chapters 4 and 5 to help construct fractal approximations to given sets, pictures and measures.

EXAMPLE 2.4.4 The transformation $f : \mathbb{R} \to \mathbb{R}$ defined by $f(x) = \frac{2}{3} + \frac{1}{3}x$ is a contraction mapping in the euclidean metric with contractivity factor $l = \frac{1}{3}$. Let $x_0 = 0$. Then

$$x_n = 1 - \frac{1}{3^n}$$

and the fixed point is the limit of the sequence $0, \frac{2}{3}, \frac{8}{9}, \frac{26}{27}, \ldots$, namely $a = 1$. In this case

$$d(x_0, 1) = 1 \leq \frac{d(x_0, x_1)}{1 - l} = 1.$$

EXAMPLE 2.4.5 Let $f : \Omega_{\{0,1\}} \to \Omega_{\{0,1\}}$ be defined by $f(\sigma) = 01\sigma$. Then

$$d_{\Omega}(f(\sigma), f(\omega)) \leq \frac{1}{2^3} d(\sigma, \omega)$$

for all $\sigma, \omega \in \Omega_{\{0,1\}}$. Let us choose $x_0 = \overline{0}$. Then $x_n = 010101 \cdots 01\overline{0}$ and it follows that $a = \overline{01} \in \Omega_{\{0,1\}}$ is the unique fixed point.

There are many different examples and applications of the contraction mapping theorem, involving diverse transformations and spaces. But we are primarily interested in fixed points of transformations on spaces such as $\mathbb{H}(\mathbb{X})$, $\mathbb{P}(\mathbb{X})$ and Π.

Contractive transformations on $(\mathbb{H}(\mathbb{X}), d_{\mathbb{H}})$ **and the existence of fractal sets**

In this subsection we show how contractive transformations on an underlying space $\mathbb{X}$ can be used as building blocks to construct contractive transformations on $\mathbb{H}(\mathbb{X})$. The fixed points of such contractive transformations on $\mathbb{H}(\mathbb{R}^2)$ are examples of the fractal sets that we shall explore in Chapter 4. For now it is important to understand the dependence on transformations on the underlying space, in order to help guide and motivate our investigation of Möbius and projective transformations acting on $\mathbb{R}^2$ later in this chapter.

The next theorem tells us that the property of $f : \mathbb{X} \to \mathbb{X}$ of being contractive is inherited by $f : \mathbb{H}(\mathbb{X}) \to \mathbb{H}(\mathbb{X})$.

THEOREM 2.4.6 *Let* $f : \mathbb{X} \to \mathbb{X}$ *be a contractive transformation on the metric space* $(\mathbb{X}, d)$ *with contractivity factor* l. *Then* $f : \mathbb{H}(\mathbb{X}) \to \mathbb{H}(\mathbb{X})$ *is a contractive transformation on the metric space* $(\mathbb{H}(\mathbb{X}), d_{\mathbb{H}})$ *with contractivity factor* l.

PROOF   Let $A, B \in \mathbb{H}(\mathbb{X})$. Then

$$\mathcal{D}_{f(A)}(f(B)) = \max_{a \in A} \min_{b \in B}\{d(f(a), f(b))\} \le l \max_{a \in A} \min_{b \in B}\{d(a, b)\} = l\mathcal{D}_A(B).$$

It follows that

$$d_{\mathbb{H}}(f(A), f(B)) = \max\{\mathcal{D}_{f(A)}(f(B)), \mathcal{D}_{f(B)}(f(A))\}$$
$$\le l \max\{\mathcal{D}_A(B), \mathcal{D}_B(A)\} = l d_{\mathbb{H}}(A, B).$$

$\square$

EXAMPLE 2.4.7   Let $f(x) = \frac{2}{3} + \frac{1}{3}x$ as in Example 2.4.4. Then f is a contractive transformation on $(\mathbb{H}(\mathbb{R}), d_{\mathbb{H}})$ with contraction factor $\frac{1}{3}$. Its unique fixed point is the nonempty compact set $A = \{1\}$. Also, if $A_0 \in \mathbb{H}(\mathbb{R})$ and $A_n = f(A_{n-1})$ for $n = 1, 2, 3, \ldots$ then we must have $\lim_{n\to\infty} A_n = A$. For example, the sequence of closed intervals $\left[0, \frac{1}{2}\right], \left[\frac{2}{3}, \frac{5}{6}\right], \left[\frac{8}{9}, \frac{17}{18}\right], \ldots$ converges in the Hausdorff metric to $\{1\}$.

It is clear that if $f : \mathbb{X} \to \mathbb{X}$ is a contractive transformation on a complete metric space with unique fixed point $a \in \mathbb{X}$ then $f : \mathbb{H}(\mathbb{X}) \to \mathbb{H}(\mathbb{X})$ is a contractive transformation on a complete metric space with unique fixed point $A = \{a\}$. It might appear that we have not gained much, with all our elaboration and inheritance. But actually we have achieved the start of a beautiful constructive theory for deterministic fractal sets, the first hint of which is provided by the following theorem. This theory, based on ideas in a visionary book, entitled *Fractals: Form, Chance, and Dimension*, by Benoit B. Mandelbrot, see [63], was first analyzed and presented in a general mathematical framework by John Hutchinson in [48]. See also [4] and [44].

Figure 2.18 A picture (left) and a picture of a measure (right) both of which are invariant under the Möbius rotation defined in Equation (2.3.4). Möbius transformations are discussed in Section 2.6.

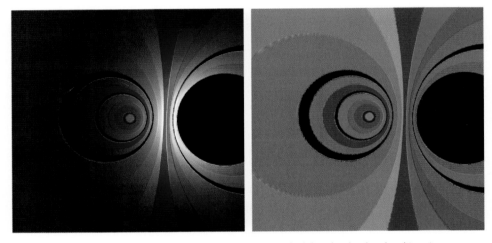

Figure 2.19 These pictures illustrate an invariant picture (right) and a closely related invariant measure (left) for the same Möbius transformation as that discussed in Figure 2.6. See also Figure 2.20.

THEOREM 2.4.8 *Let* $f_n : \mathbb{H}(\mathbb{X}) \to \mathbb{H}(\mathbb{X})$ *be a contractive transformation on* $(\mathbb{H}(\mathbb{X}), d_{\mathbb{H}})$ *with contractivity factor* l_n, *for* $n = 1, 2, \ldots, N$ *for some finite positive integer* N. *Then* $\mathcal{F} : \mathbb{H}(\mathbb{X}) \to \mathbb{H}(\mathbb{X})$ *defined by*

$$\mathcal{F}(B) = f_1(B) \cup f_2(B) \cup \cdots \cup f_N(B) \quad \text{for all } B \in \mathbb{H}(\mathbb{X})$$

is a contractive transformation with contractivity factor $l = \max\{l_n : n = 1, 2, \ldots, N\}$.

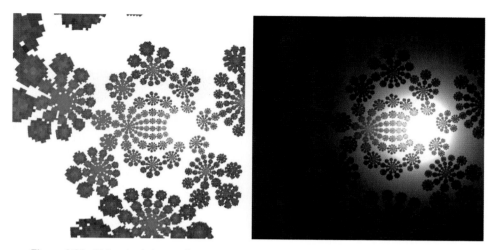

Figure 2.20 This pair of pictures illustrates part of an invariant picture, on the left, and a closely related invariant measure, on the right, for the Möbius transformation in Equation (2.2.1). See also Figures 2.6 and 2.19.

PROOF   We prove the result for the case $N = 2$. By Theorem 1.12.15,

$$d_{\mathbb{H}}(A_1 \cup A_2, A_3 \cup A_4) \leq \max\{d_{\mathbb{H}}(A_1, A_3), \ d_{\mathbb{H}}(A_2 \cup A_4)\}$$

for all $A_1, A_2, A_3, A_4 \in \mathbb{H}(\mathbb{X})$. It follows that

$$d_{\mathbb{H}}(f_1(A) \cup f_2(A), \ f_1(B) \cup f_2(B)) \leq \max\{d_{\mathbb{H}}(f_1(A), f_1(B)), \ d_{\mathbb{H}}(f_2(A), f_2(B))\}$$
$$\leq \max\{l_1, l_2\}d_{\mathbb{H}}(A, B)$$

for all $A, B \in \mathbb{H}(\mathbb{X})$.   □

Notice that a contractive transformation $f_n : \mathbb{H}(\mathbb{X}) \to \mathbb{H}(\mathbb{X})$ need not derive from a contractive transformation $f_n : \mathbb{X} \to \mathbb{X}$. For example we may define $f_1 : \mathbb{H}(\mathbb{X}) \to \mathbb{H}(\mathbb{X})$ by

$$f_1(B) = B_0 \quad \text{for all } B \in \mathbb{H}(\mathbb{X})$$

for some fixed $B_0 \in \mathbb{H}(\mathbb{X})$; then $f_1 : \mathbb{H}(\mathbb{X}) \to \mathbb{H}(\mathbb{X})$ is contractive but does not correspond to any transformation on $\mathbb{X}$ unless $B_0 = \{b\}$ for some $b \in \mathbb{X}$.

DEFINITION 2.4.9   Let $\mathbb{X}$ be a complete metric space, and let the transformations $f_n : \mathbb{X} \to \mathbb{X}$ be contractions. Then the unique fixed point of $\mathcal{F} : \mathbb{H}(\mathbb{X}) \to \mathbb{H}(\mathbb{X})$ is called the **set attractor**, or **fractal set**, associated with $\{\mathbb{X}; f_1, f_2, \ldots, f_N\}$.

We will develop the theory in later chapters. We will refer to $\{\mathbb{X}; f_1, f_2, \ldots, f_N\}$ as an **iterated function system** or **IFS**. These terms were first introduced in [4]. We will say that 'an IFS is contractive' if the functions $f_1, f_2, \ldots, f_N$ are contractions.

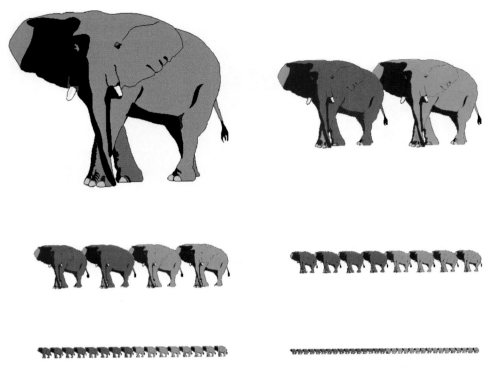

Figure 2.21   This figure shows a sequence of sets $A_0, A_1, A_2, A_3, A_4, A_5 \subset \square \subset \mathbb{R}^2$ obtained by successive application of the transformation $F : \mathbb{H}(\square) \to \mathbb{H}(\square)$. Here A_0 is a set that looks like an elephant and $A_n = F(A_{n-1}) = f_1(A_{n-1}) \cup f_2(A_{n-1})$ for $n = 1, 2, 3, 4, 5$ where f_1, f_2 are contractive with contractivity factor 0.5. The sequence of sets is converging towards the unique fixed point of F, in this case a straight line segment.

Figure 2.21 illustrates a sequence of sets in $\mathbb{H}(\square)$ produced by successive application of the operator $\mathcal{F}$ in the case where $N = 2$ and f_1 and f_2 are contractive transformations that shrink $\square$ by a linear factor 2.

Figure 2.22 shows an example of a fractal set, or set attractor, that is the fixed point of a transformation $\mathcal{F} : \mathbb{H}(\mathbb{X}) \to \mathbb{H}(\mathbb{X})$ constructed using two contractive transformations on $\mathbb{R}^2$.

EXERCISE 2.4.10   *Consider the case where* $\mathbb{X} = \mathbb{R}$, $f(x) = \frac{2}{3} + \frac{1}{3}x$ *and* $g(x) = \frac{1}{3}x$. *Describe the unique fixed point of* $\mathcal{F} : \mathbb{H}(\mathbb{R}) \to \mathbb{H}(\mathbb{R})$ *in this case.*

The metric spaces $(\mathbb{P}(\mathbb{X}), \widehat{d_{\mathbb{P}}})$ and $(\mathbb{P}(\mathbb{X}), d_{\mathbb{P}})$

If your understanding of measures is new and frail you might wish to skip forward to Section 2.5 after briefly surveying the rest of the material in Section 2.4.

Here we define two metrics on $\mathbb{P}(\mathbb{X})$, $d_{\mathbb{P}} = d_{\mathbb{P}(\mathbb{X})}$ and $\widehat{d_{\mathbb{P}}} = \widehat{d}_{\mathbb{P}(\mathbb{X})}$. The resulting metric spaces $(\mathbb{P}(\mathbb{X}), d_{\mathbb{P}})$ and $(\mathbb{P}(\mathbb{X}), \widehat{d_{\mathbb{P}}})$ are complete. Furthermore, both metrics

Figure 2.22 This picture represents, in red and mauve, the unique fixed point of a transformation $\mathcal{F} : \mathbb{H}(\mathbb{R}^2) \to \mathbb{H}(\mathbb{R}^2)$ promised by Theorem 2.4.8. Here $\mathcal{F}(A) := f_1(A) \cup f_2(A)$ for all $A \in \mathbb{H}(\mathbb{R}^2)$, where $f_1, f_2 : \mathbb{R}^2 \to \mathbb{R}^2$ are strictly contractive. Can you see how the fixed point is made of two transformations of *itself*?

are such that if $f : \mathbb{X} \to \mathbb{X}$ is a contraction mapping then $f : \mathbb{P}(\mathbb{X}) \to \mathbb{P}(\mathbb{X})$ is a contraction mapping with respect to both the metrics $d_{\mathbb{P}}$ and $\widehat{d}_{\mathbb{P}}$. That is, in both cases contractivity is inherited. So an analogous situation regarding the construction of contractive transformations and the existence of fixed points in $(\mathbb{H}(\mathbb{X}), d_{\mathbb{H}})$, discussed above, applies both to $(\mathbb{P}(\mathbb{X}), d_{\mathbb{P}})$ and to $(\mathbb{P}(\mathbb{X}), \widehat{d}_{\mathbb{P}})$. This relates to the construction of fractal measures; this topic is developed in later chapters, particularly Chapter 4.

We begin by establishing the space $(\mathbb{P}(\mathbb{X}), \widehat{d}_{\mathbb{P}})$. The metric $\widehat{d}_{\mathbb{P}}$ is quite easy to understand on the basis of what has been encountered so far in this book.

THEOREM 2.4.11   *Let $(\mathbb{X}, d)$ be a compact metric space. Let $\widehat{d}_{\mathbb{P}} : \mathbb{P}(\mathbb{X}) \times P(X) \to [0, \infty)$ be defined by*

$$\widehat{d}_{\mathbb{P}}(\mu, \nu) = \inf\{r \geq 0 : \mu(A) \leq \nu(\mathcal{B}_A(r)) \text{ for all } A \in \mathcal{B}(\mathbb{X})\} \quad \text{for all } \mu, \nu \in \mathbb{P}(\mathbb{X}).$$

$$(2.4.2)$$

Then $\widehat{d}_{\mathbb{P}}$ is a metric on $\mathbb{P}(\mathbb{X})$ and the metric space $(\mathbb{P}(\mathbb{X}), \widehat{d}_{\mathbb{P}})$ is complete.

PROOF   Recall that $\mathcal{B}_A(r)$ denotes the dilation of the set A by $r \geq 0$, whereas $\mathcal{B}(\mathbb{X})$ denotes the Borel subsets of $\mathbb{X}$. It is straightforward to show that $\widehat{d_\mathbb{P}}$ is a metric, and we leave this as an exercise. Completeness is proved in [9] based on [35], Theorem 9.1. See also [81], p. 160, and references given therein.   □

DEFINITION 2.4.12   Let $(\mathbb{X}, d)$ be a compact metric space. The metric $\widehat{d_\mathbb{P}}$ is called the **uniform Prokhorov metric** on $\mathbb{P}(\mathbb{X})$.

This metric was introduced to fractal geometry by Falconer [34], because of its scaling property. John Hutchinson pointed it out to me and suggested the name. It is important to note that although $(\mathbb{P}(\mathbb{X}), \widehat{d_\mathbb{P}})$ inherits from $(\mathbb{X}, d)$ the property of completeness, provided that the former space is *compact*, it does not in general inherit the property of compactness. For example, an infinite sequence of points in $(\mathbb{P}([0, 1]), \widehat{d_\mathbb{P}})$ is

$$\left\{ \frac{n-1}{n} \delta_0 + \frac{1}{n} \lambda_{[0,1]} \right\}_{n=1}^{\infty},$$

but the Prokhorov distance between each pair of distinct points in this sequence is unity, so this sequence contains no subsequence that is a Cauchy sequence. Here $\lambda_{[0,1]}$ denotes the uniform Borel measure of total mass unity on the real interval $[0, 1]$.

EXERCISE 2.4.13   *Show that* $\widehat{d_\mathbb{P}}(\mu, \nu) = 1$ *where* μ *is the uniform distribution of unit total mass on* $[0, 1] \times [0, 1] \subset \mathbb{R}^2$ *and* $\nu = \delta_{(1,1)}$ *is a unit mass located at the point* $(1, 1) \in \mathbb{R}^2$.

EXERCISE 2.4.14   *Verify that* $\widehat{d_\mathbb{P}}(\mu, \nu) = \widehat{d_\mathbb{P}}(\nu, \mu)$. *Hint: Look at what happens when you replace a set by its complement in Equation (2.4.2).*

Next we describe the metric space $(\mathbb{P}(\mathbb{X}), d_\mathbb{P})$. The metric $d_\mathbb{P}$ has the spectacular advantage over $\widehat{d_\mathbb{P}}$ that it admits the inheritance of compactness. It has the disadvantage that it involves measure-theoretic integration, which we do not develop in this book; see [9] for a gentle formal presentation. Also, it is weaker than $\widehat{d_\mathbb{P}}$. This latter weakness is also a strength, because $d_\mathbb{P}$ provides contractivity in situations where $\widehat{d_\mathbb{P}}$ does not.

The metric $d_\mathbb{P}$ depends on the evaluation of **integrals** such as $\int_\mathbb{X} h d\mu$, where $\mu \in \mathbb{P}(\mathbb{X})$ and $h : \mathbb{X} \to \mathbb{R}$ is continuous. Here you will not go far off course if your intuition is guided by the following fact: *if the Borel measure μ is defined by a continuous density function $\rho : \mathbb{X} \to \mathbb{R}$, that is, $\mu(B) = \int_B \rho(x) dx$ for all Borel sets B, then $\int_\mathbb{X} h d\mu = \int_\mathbb{X} h(x)\rho(x) dx$.*

THEOREM 2.4.15   *Let $(\mathbb{X}, d)$ be a compact metric space. Let $d_\mathbb{P} : \mathbb{P}(\mathbb{X}) \times \mathbb{P}(\mathbb{X}) \to [0, \infty)$ be defined by*

$$d_\mathbb{P}(\mu, \nu) = \sup_{h:\mathbb{X} \to \mathbb{R}} \left\{ \int_\mathbb{X} h d\mu - \int_\mathbb{X} h d\nu : h \in Lip_1(\mathbb{X}) \right\} \quad \text{for all } \mu, \nu \in \mathbb{P}(\mathbb{X}).$$

(2.4.3)

Then $d_\mathbb{P}$ is a metric on $\mathbb{P}(\mathbb{X})$ and the metric space $(\mathbb{P}(\mathbb{X}), d_\mathbb{P})$ is compact.

PROOF   It is straightforward to verify that $d_\mathbb{P}$ is a metric. You might also be able to verify that the natural topology corresponding to the metric $d_\mathbb{P}$ is exactly the same as what is called the **weak*** ('weak-star') **topology** on $\mathbb{P}(\mathbb{X})$, where a basis for the weak* topology is the set of 'balls' of measures $Ball(a, b, h) :=$ $\{\mu \in \mathbb{P}(\mathbb{X}) : a < \int_\mathbb{X} h d\mu < b\}$ for all $a < b \in \mathbb{R}$ and all continuous $h : \mathbb{X} \to \mathbb{R}$. By Alaoglu's theorem, [30], p. 424, the weak* topology on $\mathbb{P}(\mathbb{X})$ is compact when $\mathbb{X}$ is compact, so $(\mathbb{P}(\mathbb{X}), d_\mathbb{P})$ is a compact metric space.   □

DEFINITION 2.4.16   Let $(\mathbb{X}, d)$ be a compact metric space. The metric $d_\mathbb{P}$ defined in Equation (2.4.3) is called the **Monge–Kantorovitch metric** on $\mathbb{P}(\mathbb{X})$.

EXERCISE 2.4.17   *Show that $d_\mathbb{P}(\mu, \nu) = \frac{1}{2}$ when μ is the uniform distribution of unit total mass on $[0, 1] \times [0, 1] \subset \mathbb{R}^2$ and $\nu = \delta_{(1,1)}$ is a unit mass located at the point $(1, 1) \in \mathbb{R}^2$.*

The following theorem tells us that the metric $\widehat{d_\mathbb{P}}$ is stronger than $d_\mathbb{P}$. What it does not tell us is that it is almost too strong.

THEOREM 2.4.18   *Let $(\mathbb{X}, d)$ be a compact metric space. Then*

$$d_\mathbb{P}(\mu, \nu) \leq \widehat{d_\mathbb{P}}(\mu, \nu) \quad \text{for all } \mu, \nu \in \mathbb{P}(\mathbb{X}).$$

PROOF   This is proved in [19]. You might like to try to prove it for yourself.   □

Contractive transformations on $(\mathbb{P}(\mathbb{X}), \widehat{d_\mathbb{P}})$ and $(\mathbb{P}(\mathbb{X}), d_\mathbb{P})$ and the existence of fractal measures

Throughout this section we assume that $(\mathbb{X}, d)$ is a compact metric space. When dealing with contraction mappings acting on an underlying space such as $\mathbb{R}^2$, we leave it up to you, gentle reader, to remember to work in a big closed ball of the space such that the transformations map this ball into itself.

We have previously shown that if a transformation f has the property of being contractive on a compact metric space $\mathbb{X}$ then this property is inherited when f acts on $\mathbb{H}(\mathbb{X})$. We have promised that this has spectacular consequences, which we shall see in Chapter 4. In this section we show that the same sort of infectious inheritance applies with regard to the action of f on $\mathbb{P}(\mathbb{X})$. This will lead us, in Chapter 4, to a constructive theory for those visually elusive beautiful mathematical objects, 'deterministic' fractal measures.

The following theorem tells us that the property of f of being a contraction mapping is indeed inherited from $(\mathbb{X}, d)$ to $(\mathbb{P}(\mathbb{X}), \widehat{d}_\mathbb{P})$ and to $(\mathbb{P}(\mathbb{X}), d_\mathbb{P})$.

THEOREM 2.4.19 *Let $(\mathbb{X}, d)$ be a compact metric space. Let $f : \mathbb{X} \to \mathbb{X}$ be a contractive transformation with contractivity factor $l \geq 0$. Then $f : \mathbb{P}(\mathbb{X}) \to \mathbb{P}(\mathbb{X})$ is a contractive transformation with contractivity factor l, with respect to both the metrics $\widehat{d}_\mathbb{P}$ and $d_\mathbb{P}$.*

PROOF   For all $\mu, \nu \in \mathbb{P}(\mathbb{X})$ we have, assuming $l > 0$ for brevity,

$$
\begin{aligned}
\widehat{d}_\mathbb{P}(f(\mu), f(\nu)) &= \inf\{r \geq 0 : \mu(f^{-1}(A)) \leq \nu(f^{-1}(\mathcal{B}_A(r))) \text{ for all } A \in \mathcal{B}(\mathbb{X})\} \\
&\leq \inf\{r \geq 0 : \mu(f^{-1}(A)) \leq \nu(\mathcal{B}_{f^{-1}(A)}(r/l))) \text{ for all } A \in \mathcal{B}(\mathbb{X})\} \\
&\leq l \inf\{r \geq 0 : \mu(\widetilde{A}) \leq \nu(\mathcal{B}_{\widetilde{A}}(r)) \text{ for all } \widetilde{A} = f^{-1}(A), A \in \mathcal{B}(\mathbb{X})\} \\
&\leq l \inf\{r \geq 0 : \mu(\widetilde{A}) \leq \nu(\mathcal{B}_{\widetilde{A}}(r)) \text{ for all } \widetilde{A} \in \mathcal{B}(\mathbb{X})\} \\
&= l\widehat{d}_\mathbb{P}(\mu, \nu),
\end{aligned}
$$

where we have used the observation that $f^{-1}(\mathcal{B}_A(r)) \supset \mathcal{B}_{f^{-1}(A)}(r/l)$.

Also, we have

$$
\begin{aligned}
d_\mathbb{P}(f(\mu), f(\nu)) &= \sup_{h:\mathbb{X}\to\mathbb{R}} \left\{ \int_\mathbb{X} h d(f \circ \mu) - \int_\mathbb{X} h d(f \circ \nu) : h \in Lip_1(\mathbb{X}) \right\} \\
&= \sup_{h:\mathbb{X}\to\mathbb{R}} \left\{ \int_\mathbb{X} h \circ f d\mu - \int_\mathbb{X} h \circ f d\nu : h \in Lip_1(\mathbb{X}) \right\} \\
&= l \sup_{h:\mathbb{X}\to\mathbb{R}} \left\{ \int_\mathbb{X} \frac{1}{l} h \circ f d\mu - \frac{1}{l} \int_\mathbb{X} h \circ f d\nu : h \in Lip_1(\mathbb{X}) \right\} \\
&\leq l \sup_{\widetilde{h}:\mathbb{X}\to\mathbb{R}} \left\{ \int_\mathbb{X} \widetilde{h} d\mu - \int_\mathbb{X} \widetilde{h} d\nu : \widetilde{h} \in Lip_1(\mathbb{X}) \right\} \\
&= l \, d_\mathbb{P}(\mu, \nu),
\end{aligned}
$$

where we have used the observation that

$$
\left| \frac{1}{l} h \circ f(x) - \frac{1}{l} h \circ f(y) \right| \leq |h(x) - h(y)| \leq d(x, y) \quad \text{when } h \in Lip_1(\mathbb{X});
$$

see also Hutchinson [48].   □

EXERCISE 2.4.20   *Let $f(x) = \frac{2}{3} + \frac{1}{3}x$ as in Exercises 2.4.14 and 2.4.17 above. Then f is a contractive transformation on both $(\mathbb{P}(\mathbb{X}), \widehat{d}_\mathbb{P})$ and $(\mathbb{P}(\mathbb{X}), d_\mathbb{P})$ with contractivity factor $\frac{1}{3}$. Its unique fixed point is δ_1, the measure that assigns unit mass to the point $x = 1$. Also, if $\mu_0 \in \mathbb{P}(\mathbb{X})$ and $\mu_n = f(\mu_{n-1})$ for $n = 1, 2, 3, \ldots$ then we must have $\lim_{n\to\infty} \mu_n = \delta_1$. For example, the sequence of measures*

$$
2\lambda_{[0,\frac{1}{2}]}, \quad 6\lambda_{[\frac{2}{3},\frac{5}{6}]}, \quad 18\lambda_{[\frac{8}{9},\frac{17}{18}]}, \quad \ldots
$$

converges in both metrics, $\widehat{d}_\mathbb{P}$ and $d_\mathbb{P}$, to δ_1.

It is clear that if $f : \mathbb{X} \to \mathbb{X}$ is a contractive transformation on a compact metric space $\mathbb{X}$ with unique fixed point $a \in \mathbb{X}$ then $f : \mathbb{P}(\mathbb{X}) \to \mathbb{P}(\mathbb{X})$ is a contractive transformation on the complete metric space $\mathbb{P}(\mathbb{X})$ with unique fixed point δ_a. It might again appear, as in the case where $f : \mathbb{H}(\mathbb{X}) \to \mathbb{H}(\mathbb{X})$, that we have not gained much with all our elaboration and inheritance. But we have: the construction that follows is the starting point for the theory of deterministic fractal measures.

Given that we have a set of continuous transformations $\{f_n : \mathbb{P}(\mathbb{X}) \to \mathbb{P}(\mathbb{X}) : n = 1, 2, \ldots, N\}$ and a set of probabilities $p_1, p_2, \ldots, p_N > 0$, with $p_1 + p_2 + \cdots + p_N = 1$, we can define a new transformation $\mathcal{F} : \mathbb{P}(\mathbb{X}) \to \mathbb{P}(\mathbb{X})$, where

$$\mathcal{F}(\mu) = p_1 f_1(\mu) + p_2 f_2(\mu) + \cdots + p_N f_N(\mu) \quad \text{for all } \mu \in \mathbb{P}(\mathbb{X}). \quad (2.4.4)$$

Notice that the transformations $f_n : \mathbb{P}(\mathbb{X}) \to \mathbb{P}(\mathbb{X})$ in this theorem need not derive from transformations $f_n : \mathbb{X} \to \mathbb{X}$. For example, we may define $f_1 : \mathbb{P}(\mathbb{X}) \to \mathbb{P}(\mathbb{X})$ by

$$f_1(\mu) = \omega \quad \text{for all } \mu \in \mathbb{P}(\mathbb{X}),$$

for some fixed $\omega \in \mathbb{P}(\mathbb{X})$.

THEOREM 2.4.21 *Let $f_n : \mathbb{P}(\mathbb{X}) \to \mathbb{P}(\mathbb{X})$ be a Lipschitz transformation on $(\mathbb{P}(\mathbb{X}), d_\mathbb{P})$, with Lipschitz constant l_n for $n = 1, 2, \ldots, N$, for some finite positive integer N. Then the transformation $\mathcal{F} : \mathbb{P}(\mathbb{X}) \to \mathbb{P}(\mathbb{X})$ defined in Equation (2.4.4) is Lipschitz with respect to $\widehat{d}_\mathbb{P}$, with Lipschitz constant*

$$\widehat{l} = \max\{l_1, l_2, \ldots, l_N\},$$

and Lipschitz with respect to $d_\mathbb{P}$ with Lipschitz constant

$$\bar{l} = p_1 l_1 + p_2 l_2 + \cdots + p_N l_N.$$

In particular, $\mathcal{F} : (\mathbb{P}(\mathbb{X}), \widehat{d}_\mathbb{P}) \to (\mathbb{P}(\mathbb{X}), \widehat{d}_\mathbb{P})$ is a contraction mapping when $\widehat{l} < 1$ and $\mathcal{F} : (\mathbb{P}(\mathbb{X}), d_\mathbb{P}) \to (\mathbb{P}(\mathbb{X}), d_\mathbb{P})$ is a contraction mapping when $\bar{l} < 1$.

PROOF   We prove the result for $N = 2$. We have, for all $\mu, \nu \in \mathbb{P}(\mathbb{X})$, that

$\widehat{d}_\mathbb{P}(\mathcal{F}(\mu), \mathcal{F}(\nu))$
$\quad = \inf\{r \geq 0 : \mathcal{F}(\mu)(A) \leq \mathcal{F}(\nu)(\mathcal{B}_A(r)) \text{ for all } A \in \mathcal{B}(\mathbb{X})\}$
$\quad = \inf\{r \geq 0 : p_1 f_1(\mu)(A) + p_2 f_2(\mu)(A) \leq p_1 f_1(\nu)(\mathcal{B}_A(r)) + p_2 f_2(\nu)(\mathcal{B}_A(r))$
$\qquad\qquad\qquad\qquad\qquad\qquad\qquad\qquad\qquad\qquad \text{for all } A \in \mathcal{B}(\mathbb{X})\}$
$\quad \leq \inf\{r \geq 0 : p_1 f_1(\mu)(A) \leq p_1 f_1(\nu)(\mathcal{B}_A(r)) \text{ and } p_2 f_2(\mu)(A) \leq p_2 f_2(\nu)(\mathcal{B}_A(r))$
$\qquad\qquad\qquad\qquad\qquad\qquad\qquad\qquad\qquad\qquad \text{for all } A \in \mathcal{B}(\mathbb{X})\}$
$\quad \leq \max\{l_1, l_2\}\, \widehat{d}_\mathbb{P}(\mu, \nu).$

Also,

$$d_{\mathbb{P}}(\mathcal{F}(\mu), \mathcal{F}(\nu))$$

$$= \sup_{h:\mathbb{X}\to\mathbb{R}} \left\{ \int_{\mathbb{X}} h\, d\mathcal{F}(\mu) - \int_{\mathbb{X}} h\, d\mathcal{F}(\nu) : h \in Lip_1(\mathbb{X}) \right\}$$

$$= \sup_{h:\mathbb{X}\to\mathbb{R}} \left\{ p_1\left(\int_{\mathbb{X}} h\, df_1(\mu) - \int_{\mathbb{X}} h\, df_1(\nu) \right) + p_2\left(\int_{\mathbb{X}} h\, df_2(\mu) - \int_{\mathbb{X}} h\, df_2(\nu) \right) : \right.$$

$$\left. h \in Lip_1(\mathbb{X}) \right\}$$

$$\leq \sup_{h_i:\mathbb{X}\to\mathbb{R}, i\in\{1,2\}} \left\{ p_1\left(\int_{\mathbb{X}} h_1\, df_1(\mu) - \int_{\mathbb{X}} h_1\, df_1(\nu) \right) \right.$$

$$\left. + p_2\left(\int_{\mathbb{X}} h_2\, df_2(\mu) - \int_{\mathbb{X}} h_2\, df_2(\nu) \right) : h_1, h_2 \in Lip_1(\mathbb{X}) \right\}$$

$$= p_1 d_{\mathbb{P}}(f_1(\mu), f_1(\nu)) + p_2 d_{\mathbb{P}}(f_2(\mu), f_2(\nu))$$

$$\leq (p_1 l_1 + p_2 l_2) d_{\mathbb{P}}(\mu, \nu).$$

$\square$

Generally speaking we call the fixed points of the transformation $\mathcal{F}$: $\mathbb{P}(\mathbb{X}) \to \mathbb{P}(\mathbb{X})$ fractal measures or **measure attractors**.

An example of a sequence of measures $\mu_0, \mu_1, \mu_2, \mu_3, \mu_4, \mu_5 \in \mathbb{P}(\square)$ converging towards a fixed point $\mu_\infty \in \mathbb{P}(\square)$ is shown in Figure 2.23; μ_0 is represented by a green rectangle and $\mu_n = \mathcal{F}(\mu_{n-1})$ for $n = 1, 2, 3, 4, 5$. The transformation $\mathcal{F} : \mathbb{P}(\square) \to \mathbb{P}(\square)$ is defined as in Equation (2.4.4), where f_1, f_2 are contractive projective transformations. The picture of μ_∞ represents the fixed point of $\mathcal{F}$. As in other digital pictures of measures, colour intensity values above 255 are replaced by 255.

A picture of part of a measure that is a fixed point of $\mathcal{F} : \mathbb{P}(\mathbb{X}) \to \mathbb{P}(\mathbb{X})$ with $N = 2$, as promised by Theorem 1.13.9, is illustrated in Figure 2.24.

Finally we note that quite generally, even when the f_n are not contractive, the transformation $\mathcal{F} : \mathbb{P}(\mathbb{X}) \to \mathbb{P}(\mathbb{X})$ possesses at least one fixed point.

THEOREM 2.4.22 *Let $\mathbb{X}$ be a compact metric space. Let $f_n : \mathbb{X} \to \mathbb{X}$ be continuous for $n = 1, 2, \ldots, N$ in Equation (2.4.4). Then there exists a measure $\mu \in \mathbb{P}(\mathbb{X})$ that is invariant for $\mathcal{F}$.*

PROOF   This follows from the Schauder–Tychenoff fixed-point theorem; see [30], p. 456. It uses the fact that $\mathbb{P}(\mathbb{X})$ is a convex, compact, subset of a normed linear space, namely the space of signed Borel measures on $\mathbb{X}$, that is, the space of continuous linear functions from $C(\mathbb{X})$ into $\mathbb{R}$. Moreover the continuity of $f_n : \mathbb{X} \to \mathbb{X}$ for $n = 1, 2, \ldots, N$ implies that $\mathcal{F} : \mathbb{P}(\mathbb{X}) \to \mathbb{P}(\mathbb{X})$ is continuous with respect to the weak* topology, which implies that it is continuous in the metric $d_{\mathbb{P}}$. $\square$

Two examples of measures $\mu \in \mathbb{P}(\square \subset \mathbb{R}^2)$ that are fixed points of a continuous transformation that is not contractive are illustrated in Figure 2.25.

Figure 2.23 Approximate pictures of the measures μ_0, μ_1, μ_2, μ_3, μ_4, μ_5, μ_6 and μ_∞ obtained by recursive application of a transformation $\mathcal{F} : \mathbb{P}(\square) \to \mathbb{P}(\square)$, as in Equation (2.4.4), where f_1, f_2 are projective transformations.

2.5 Linear and affine transformations in two and three dimensions

In this section we describe the behaviour of linear transformations on $\mathbb{R}^2$ and $\mathbb{R}^3$. These linear transformations are fundamental to the description of Möbius and projective transformations. We want to know how linear transformations on $\mathbb{R}^2$ act on points, sets, pictures, and measures.

As a reminder and to set some notation in place, we begin by recalling a few details from linear algebra. But generally we assume familiarity with vector spaces and linear transformations, including such concepts as inner products and eigenvectors, eigenvalues, adjoints, transposes, and inverses of matrices.

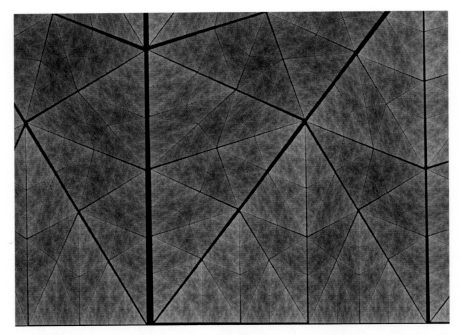

Figure 2.24   Picture of part of a measure that is the unique fixed point of a transformation f : $\mathbb{P}(\mathbb{X}) \to \mathbb{P}(\mathbb{X})$ of the form in Equation (2.4.4) with $N = 2$ as promised by Theorem 2.4.21. The fixed point is made of two transformed copies of itself, and we are looking at a close-up. But you can see transformed replicas of the fixed point, a triangular object, all over the image.

Figure 2.25   Pictures of two different measures on $\square$ that are invariant under the continuous 'expanding' transformation $f(x, y) = (\min\{2x, 2 - 2x\}, \min\{2y, 2 - 2y\})$.

Linear algebra

Recall that $\mathbb{R}, \mathbb{R}^2, \mathbb{R}^3, \ldots$ and $\mathbb{C}, \mathbb{C}^2, \mathbb{C}^3, \ldots$ are examples of finite-dimensional **linear spaces**. (The notation $\mathbb{C}$ signifies the complex plane.) Any pair of points in a linear space can be added to produce a new point in the space. Also any point in a linear space can be multiplied by any **scalar** to produce a new point in the space. When the linear space is one of $\mathbb{R}, \mathbb{R}^2, \mathbb{R}^3, \ldots$ the customary set of scalars is $\mathbb{R}$. When the linear space is one of $\mathbb{C}, \mathbb{C}^2, \mathbb{C}^3, \ldots$ the scalars may be either $\mathbb{R}$ or $\mathbb{C}$. The operations of addition and multiplication by a scalar are consistent with one another. A linear space is also called a **vector space**.

If $V = V(\mathbb{F})$ and $W = W(\mathbb{F})$ are linear spaces, both with the same set of scalars $\mathbb{F}$, then $f : V \to W$ is called a **linear transformation** iff

$$f(\alpha x_1 + \beta x_2) = \alpha f(x_1) + \beta f(x_2)$$

for all $\alpha, \beta \in \mathbb{F}$ and all $x_1, x_2 \in V$.

Let $M_N(\mathbb{F})$ denote the set of $N \times N$ matrices whose entries belong to $\mathbb{F}$, for $N = 1, 2, 3, \ldots$ To any linear transformation $f : \mathbb{R}^2 \to \mathbb{R}^2$ there corresponds a unique matrix

$$A := \begin{pmatrix} a & b \\ c & d \end{pmatrix} \in M_2(\mathbb{R}) \tag{2.5.1}$$

such that

$$(f(x, y))^T = A \begin{pmatrix} x \\ y \end{pmatrix} = \begin{pmatrix} a & b \\ c & d \end{pmatrix} \begin{pmatrix} x \\ y \end{pmatrix} = \begin{pmatrix} ax + by \\ cx + dy \end{pmatrix}$$

for all $(x, y) \in \mathbb{R}^2$, where $a, b, c, d \in \mathbb{R}$. That is,

$$f(x, y) = (ax + by, cx + dy).$$

The superscript T denotes the transpose of the vector or matrix to which it is applied. Note that $f(x, y) \in \mathbb{R}^2$ is a coordinate pair, which we treat here as a row vector of length 2.

Conversely, each matrix $A \in M_2(\mathbb{R})$ represents a unique linear transformation on $\mathbb{R}^2$. The linear transformation is said to be represented by the corresponding matrix. In general we will not distinguish between matrices and the linear transformations that they represent. But you should keep an eye on the domains of the transformations.

If both $f_1 : \mathbb{R}^2 \to \mathbb{R}^2$ and $f_2 : \mathbb{R}^2 \to \mathbb{R}^2$ are linear transformations then so is $f_1 \circ f_2$. Moreover, if

$$f_1 = \begin{pmatrix} a_1 & b_1 \\ c_1 & d_1 \end{pmatrix} \quad \text{and} \quad f_2 = \begin{pmatrix} a_2 & b_2 \\ c_2 & d_2 \end{pmatrix}, \tag{2.5.2}$$

then $f_1 \circ f_2$ is represented by the matrix product

$$f_1 \cdot f_2 = \begin{pmatrix} a_1 & b_1 \\ c_1 & d_1 \end{pmatrix} \begin{pmatrix} a_2 & b_2 \\ c_2 & d_2 \end{pmatrix} = \begin{pmatrix} a_1 a_2 + b_1 c_2 & a_1 b_2 + b_1 d_2 \\ c_1 a_2 + d_1 c_2 & c_1 b_2 + d_1 d_2 \end{pmatrix}. \qquad (2.5.3)$$

Note that the linear transformation $f : \mathbb{R}^2 \to \mathbb{R}^2$ is invertible iff $(ad - bc) \neq 0$, and that in this case

$$f^{-1} = (ad - bc)^{-1} \begin{pmatrix} d & -b \\ -c & a \end{pmatrix}.$$

Remarks similar to those above apply to linear transformations on $\mathbb{R}^n$ and on $\mathbb{C}^n$.

EXERCISE 2.5.1   *Repeat the above discussion for the case of the linear space* $\mathbb{R}^3$.

EXERCISE 2.5.2   *Find the inverse of the matrix*

$$\begin{pmatrix} 6 & 5 & 4 \\ 0 & 3 & 2 \\ 0 & 0 & 1 \end{pmatrix}.$$

EXERCISE 2.5.3   *An **affine transformation** $g : \mathbb{R}^2 \to \mathbb{R}^2$ is one that can be expressed in the form $g = f + t$ for all $x \in \mathbb{R}^2$, where $f : \mathbb{R}^2 \to \mathbb{R}^2$ is a linear transformation and $t \in \mathbb{R}^2$. Show that if $g_1 : \mathbb{R}^2 \to \mathbb{R}^2$ and $g_2 : \mathbb{R}^2 \to \mathbb{R}^2$ are affine transformations on $\mathbb{R}^2$ then so is $g_1 \circ g_2$. Let $g_1 = f_1 + t_1$ and $g_2 = f_2 + t_2$, where f_1, f_2 are given by the 2×2 matrices in Equation (2.5.2). Let $t_i = (h_i, k_i)$ for $i = 1, 2$. Show that if g_i is represented by the matrix*

$$\begin{pmatrix} a_i & b_i & h_i \\ c_i & d_i & g_i \\ 0 & 0 & 1 \end{pmatrix}$$

for $i = 1, 2$ then $g_1 \circ g_2$ is represented by the matrix $g_1 \cdot g_2$.

Geometrical behaviour

The following theorem tells us that any invertible linear transformation on $\mathbb{R}^3$ consists of three rescalings, one in each of three perpendicular directions, followed by a rotation of the whole space. The rotation may include a reflection. By a rescaling in a particular direction we mean that, for each vector that represents a point in that space, the component in that direction is multiplied by a constant positive factor while the components in perpendicular directions are unaltered. If the factor is of magnitude less than unity then we say that the space has contracted or shrunk in that direction, while if the factor is greater than unity it has expanded.

THEOREM 2.5.4   *Every invertible linear transformation $f : \mathbb{R}^n \to \mathbb{R}^n$ can be represented as the product of an orthogonal transformation and a symmetric linear transformation.*

PROOF   We choose $n = 3$ but the same proof works for any $n \in \mathbb{N}$. Let f^T denote the transpose or adjoint of f. Then it is readily verified that $f^T \cdot f$ is symmetric, that is, self-adjoint. It follows that there exists a rectangular coordinate system, with orthogonal unit vectors $\psi_1, \psi_2, \psi_3 \in \mathbb{R}^3$, eigenvectors of $f^T \cdot f$, with corresponding strictly positive eigenvalues $\lambda_1, \lambda_2, \lambda_3$, such that

$$f^T \cdot f = \begin{pmatrix} \psi_1^T & \psi_2^T & \psi_3^T \end{pmatrix} \begin{pmatrix} \lambda_1 & 0 & 0 \\ 0 & \lambda_2 & 0 \\ 0 & 0 & \lambda_3 \end{pmatrix} \begin{pmatrix} \psi_1 \\ \psi_2 \\ \psi_3 \end{pmatrix}.$$

The strict positivity of the λ_i follows from the inner product

$$(x, f^T \cdot fx) = (fx, fx) \geq 0$$

and the fact that f is invertible. It follows that

$$f^T \cdot f = h \cdot h$$

where

$$h = \begin{pmatrix} \psi_1^T & \psi_2^T & \psi_3^T \end{pmatrix} \begin{pmatrix} \sqrt{\lambda_1} & 0 & 0 \\ 0 & \sqrt{\lambda_2} & 0 \\ 0 & 0 & \sqrt{\lambda_3} \end{pmatrix} \begin{pmatrix} \psi_1 \\ \psi_2 \\ \psi_3 \end{pmatrix}.$$

Thus

$$f = \left(f^T\right)^{-1} \cdot (h \cdot h) = \left(\left(f^T\right)^{-1} \cdot h\right) \cdot h.$$

Now it is remarkable but true that

$$q := \left(f^T\right)^{-1} \cdot h$$

is an orthogonal transformation. Indeed,

$$q \cdot q^T = \left(f^T\right)^{-1} \cdot h \cdot \left(\left(f^T\right)^{-1} \cdot h\right)^T$$
$$= \left(f^T\right)^{-1} \cdot h \cdot h \cdot f^{-1} = \left(f^T\right)^{-1} \cdot f^T \cdot f \cdot f^{-1} = I,$$

where

$$I = \begin{pmatrix} 1 & 0 & 0 \\ 0 & 1 & 0 \\ 0 & 0 & 1 \end{pmatrix}.$$

□

Theorem 2.5.4 tells us exactly how a linear transformation acts on a picture in terms of scalings and rotations. For example, Figure 2.26 illustrates the application

Figure 2.26 An invertible linear transformation applied to a picture, lower right, produces the same result, left, as rescaling in two perpendicular directions, upper right, followed by a rotation.

of the linear transformation

$$\begin{pmatrix} 0.18 & 0.31 \\ -0.49 & 0.42 \end{pmatrix} = \begin{pmatrix} 0.6 & 0.8 \\ -0.8 & 0.6 \end{pmatrix} \begin{pmatrix} 1 & -0.3 \\ -0.3 & 1 \end{pmatrix}$$

to a picture of a leaf, lower right, within a circular domain in $\mathbb{R}^2$; the symmetrical transformation

$$\begin{pmatrix} 1 & -0.3 \\ -0.3 & 1 \end{pmatrix}$$

rescales the picture in two perpendicular directions, producing a distorted leaf within an elliptical domain, upper right, and then the rotation

$$\begin{pmatrix} 0.6 & 0.8 \\ -0.8 & 0.6 \end{pmatrix}$$

produces the final transformed picture, on the left.

EXERCISE 2.5.5   *Show that the transformation*

$$\begin{pmatrix} 1 & -0.3 \\ -0.3 & 1 \end{pmatrix}$$

rescales by factors 1.3 *and* 0.7 *in two perpendicular directions. What are these two directions?*

Notice that, in the special case where the linear transformation $f : \mathbb{R}^n \to \mathbb{R}^n$ is symmetric, the scaling factors referred to in Theorem 2.5.4 are just the eigenvalues of f.

How does a linear transformation $f : \mathbb{R}^2 \to \mathbb{R}^2$ act on a measure $\mu \in \mathbb{P}(\mathbb{R}^2)$? Suppose that f is represented by the matrix A in Equation (2.5.1) and that it is invertible. Then f scales areas by the constant factor $|ad - bc|$. So $f \circ \mu$ assigns mass $\mu(B)$ to $f(B)$, while the area of $f(B)$ is $|ad - bc|$ times the area of B, for all Borel sets $B \in \mathcal{B}(\mathbb{R}^2)$. It follows that

$$\frac{\text{mass of } f(B)}{\text{area of } f(B)} = \frac{\mu(B)}{|ad - bc|} \tag{2.5.4}$$

for all $B \in \mathcal{B}(\mathbb{R}^2)$ with nonzero area. So in special cases where μ can be described by a continuous density function $\rho(x)$, $f \circ \mu$ can be described by the density function $\widetilde{\rho}(x) = |ad - bc|^{-1} \rho(f^{-1}(x))$.

For example, suppose that we have a vector of measures that represents a cartoon, that is, a picture composed of regions of constant colour separated by smooth one-dimensional boundaries. Then if we compare digital pictures, at the same resolution, of the measure before and after a linear transformation has been applied to it, the general effects will be: (i) the picture is altered as described by Theorem 2.5.4; and (ii) the brightness is changed by a constant factor. One point we are making here is that linear and affine transformations act uniformly on many types of picture.

Clearly any invertible linear transformation $f : \mathbb{R}^2 \to \mathbb{R}^2$ always fixes the point $a = (0, 0)$. It follows that the set $\{a\} \in \mathbb{H}(\mathbb{R}^2)$, the measure $\delta_a \in \mathbb{P}(\mathbb{R}^2)$ and any picture $\mathfrak{P} \in \Pi$ with $D_{\mathfrak{P}} = \mathbb{R}^2$ and $\mathfrak{P}(x) = c$ for some constant $c \in \mathfrak{C}$ are all invariant under f. Another example of a set, picture and measure each of which is invariant under a rotational linear transformation is illustrated in Figure 2.27. The

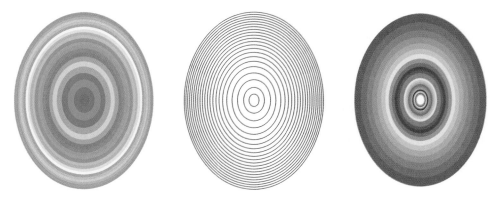

Figure 2.27   From left to right, an invariant picture, an invariant set and a picture of an invariant measure of a linear transformation of the form in Equation (2.5.5). Notice that the measure is uniform within each elliptical annulus, where it is represented by a single shade of grey.

linear transformation is of the form

$$\begin{pmatrix} \lambda_1 & 0 \\ 0 & \lambda_2 \end{pmatrix} \begin{pmatrix} \cos\theta & \sin\theta \\ -\sin\theta & \cos\theta \end{pmatrix} \begin{pmatrix} \lambda_1^{-1} & 0 \\ 0 & \lambda_2^{-1} \end{pmatrix} = \begin{pmatrix} \cos\theta & \lambda_1\lambda_2^{-1}\sin\theta \\ -\lambda_1^{-1}\lambda_2\sin\theta & \cos\theta \end{pmatrix}$$

(2.5.5)

where $\lambda_1, \lambda_2 > 0$. Other examples that are related to wallpaper patterns and Figures 3.68, 3.69 and 3.75 may be constructed.

EXERCISE 2.5.6   *Show that the ellipse* $\{(\lambda_1^{-1}\cos\theta, \lambda_2^{-1}\sin\theta) : 0 \leq \theta \leq 360°\}$ *is an invariant set for the linear transformation* $f : \mathbb{R}^2 \to \mathbb{R}^2$ *in Equation (2.5.5) but that a measure* $\mu \in \mathbb{P}(\mathbb{R}^2)$ *that assigns unit mass to the ellipse and then assigns this mass in proportion to arc length around the ellipse is not invariant for* f.

Affine geometry

The most general **affine** transformation $\mathcal{A} : \mathbb{R}^2 \to \mathbb{R}^2$ consists of a linear transformation $A : \mathbb{R}^2 \to \mathbb{R}^2$ followed by a translation. That is, it can be written as

$$(\mathcal{A}(x, y))^T = A\begin{pmatrix} x \\ y \end{pmatrix} + \begin{pmatrix} e \\ f \end{pmatrix} = \begin{pmatrix} a & b \\ c & d \end{pmatrix}\begin{pmatrix} x \\ y \end{pmatrix} + \begin{pmatrix} e \\ f \end{pmatrix},$$

where A is a linear transformation and $a, b, c, d, e, f \in \mathbb{R}$ are parameters. It is readily verified that $\mathcal{A}$ is invertible iff A is invertible, that is, iff $ad - bc \neq 0$. The inverse of an invertible affine transformation is also an invertible affine transformation. Furthermore, the composition of two affine transformations is an affine transformation.

An affine transformation acts on sets, pictures and measures in essentially the same way as does a linear transformation. If the affine transformation $\mathcal{A}$ has a fixed point a and we make the change of coordinates $x' = x - a$, that is, we change the origin of coordinates to the point a, then $\mathcal{A}(x') = Ax'$.

The basic properties of affine transformations are that they (i) map straight lines into straight lines, (ii) preserve ratios of distances between points on straight lines and (iii) map parallel straight lines into parallel straight lines, triangles into triangles and interiors of triangles into interiors of triangles.

To state what is obvious to you: all these properties can be interpreted as applying to pictures. For example, if three points *in a picture* lie on a straight line then the corresponding points after affine transformation lie on a straight line *in the transformed picture*. Similarly, ratios of distances along straight lines in pictures are preserved and parallel lines in pictures are transformed to parallel lines. It is well worth looking at Figure 2.26 to see these statements in practice.

The set of invertible affine transformations acting on the euclidean plane $\mathbb{R}^2$ provides an example of a geometry.

THEOREM 2.5.7 *(Fundamental theorem of affine geometry)   Let P, Q and R and P′, Q′ and R′ be two sets of three non-collinear points in $\mathbb{R}^2$. Then there is a unique affine transformation $f : \mathbb{R}^2 \to \mathbb{R}^2$ that maps P, Q and R to P′, Q′ and R′ respectively.*

PROOF   This is a good exercise. Hint: Start by choosing $P = (0, 0)$, $Q = (1, 0)$ and $R = (0, 1)$. See [25], p. 67.   □

EXERCISE 2.5.8   *Various affine transformations of a picture of a flower are illustrated in Figure 2.28. What properties can you **observe** to be common to all the flower pictures? What differences can you see?*

EXERCISE 2.5.9   *Find the unique affine transformation promised by Theorem 2.5.7 when $P = (0, 0)$, $Q = (1, 0)$, $R = (0, 1)$, $P′ = (1, 0)$, $Q′ = (0, 1)$, $R′ = (0, 0)$.*

A **translation** is an affine transformation in which the linear part is the identity.

A **similitude** is an affine transformation in which the scalings by the linear part, as promised by Theorem 2.5.4, are all of the same magnitude. In two dimensions a similitude has the property that it maps circles into circles and straight lines into straight lines. The most general invertible similitude $\mathcal{A} : \mathbb{R}^2 \to \mathbb{R}^2$ that uses proper rotation, i.e. does not include a reflection, can be written in the form

$$(\mathcal{A}(x, y))^T = \begin{pmatrix} \lambda \cos \theta & -\lambda \sin \theta \\ \lambda \sin \theta & \lambda \cos \theta \end{pmatrix} \begin{pmatrix} x \\ y \end{pmatrix} + \begin{pmatrix} e \\ f \end{pmatrix}, \qquad (2.5.6)$$

where $\lambda > 0$, $\theta \in [0, 2\pi)$ and e, $f \in \mathbb{R}$.

A two-dimensional **shear transformation** is an affine transformation which possesses a set of fixed points that lie on a straight line. The linear part of a shear transformation possesses an eigenvalue equal to unity, and the corresponding (first) eigenvector direction is the same as that of the line of fixed points. Any line that

Figure 2.28   Each pair of pictures here is related by an affine transformation. What properties do all the pictures have in common?

intersects the fixed line is rotated about the intersection point. Lines of points in the second eigenvector direction are mapped onto themselves, being simply stretched or shrunk, according to the second eigenvalue. See Figure 2.29.

EXERCISE 2.5.10   *Show that an example of a shear transformation $\mathcal{S} : \mathbb{R}^2 \to \mathbb{R}^2$ is given by*

$$(\mathcal{S}(x, y))^T = \begin{pmatrix} 1 & 0 \\ 2 & 3 \end{pmatrix} \begin{pmatrix} x \\ y \end{pmatrix} + \begin{pmatrix} 4 \\ 5 \end{pmatrix}.$$

Identify the line of fixed points and the two eigenvector directions.

EXERCISE 2.5.11   *Find the unique shear transformation $\mathcal{S} : \mathbb{R}^2 \to \mathbb{R}^2$ such that $\mathcal{S}(1, 1) = (1, 1)$, $\mathcal{S}(2, 1) = (2, 1)$ and $\mathcal{S}(a_1, a_2) = (a_1', a_2')$, where (a_1, a_2), $(a_1', a_2') \in \mathbb{R}^2$ are such that neither point lies on the straight line through $(2, 1)$ and $(1, 1)$.*

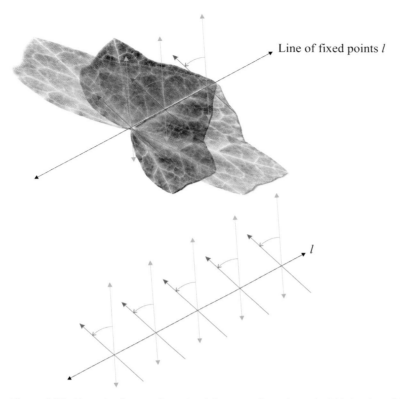

Line of fixed points *l*

l

Figure 2.29   Example of a two-dimensional shear transformation: a leaf (darker image) is transformed to produce the lighter image. Parallel lines through the line of fixed points *l* are rotated about their intersections with *l*; lines through *l* in the direction of the second eigenvector are stretched or shrunk with zero rotation. It can be seen that such transformations are affine.

EXERCISE 2.5.12   *Write down a formula for a similitude* $\mathcal{A} : \mathbb{R}^2 \to \mathbb{R}^2$ *for which the determinant of the linear part is negative.*

EXERCISE 2.5.13   *Show that any affine transformation* $\mathcal{A} : \mathbb{R}^2 \to \mathbb{R}^2$ *can be written as a composition of a translation T, a similitude S and a shear transformation* $\mathcal{F}$, *according to*

$$\mathcal{A} = \mathcal{F} \circ \mathcal{S} \circ \mathcal{T}.$$

This is particularly useful for the interactive adjustment, of an affine transformation $\mathcal{A}$ *applied to a given picture* $\mathfrak{P}_1$, *to make the picture* $\mathcal{A}(\mathfrak{P}_1)$ *look as close as possible to a second picture* $\mathfrak{P}_2$, *as illustrated in Figure 2.30, where we describe the 'move-three-points' algorithm. This type of manipulation of image segments may be used interactively to find affine transformations that approximately map a segment of a picture to a segment of a picture, when applying the collage theorem; see Chapter 4.*

2.6 Möbius transformations

Möbius transformations are specified by eight real parameters. They are geometrically simple and cheap to describe, communicate and compute. Small sets of them may be used to represent apparently complex images, as we will show in Chapters 4 and 5. So here we start to explain what they are, how they act on points, sets, pictures and measures and what sorts of sets, pictures and measures are invariant under them.

Definition of a Möbius transformation

Möbius transformations have the quite extraordinary property that they map the set of all circles and straight lines onto the set of all circles and straight lines while, typically, substantially distorting other shapes. In addition, they preserve angles and the orientation of angles.

Various Möbius transformations applied to a picture of a cyclist riding a bike are illustrated in Figure 2.31. Notice how the rims of the wheels are all nearly circular and how corresponding angles in the bike frames are all the same. But the tyres themselves are distorted and the relative sizes of the two wheels vary from bike to bike. Also, the straight lines in the bike frame are mapped onto arcs of circles.

Some other illustrations are shown in Figure 2.2, where the three large fish are each related to the small fish by a Möbius transformation. Notice how the eyes of all the fish are round, how angles are preserved and how different yet fish-like all the fish look. See also Figure 2.32.

A Möbius transformation is a mapping $\mathcal{M} : \widehat{\mathbb{R}^2} \to \widehat{\mathbb{R}^2}$, where $\widehat{\mathbb{R}^2} = \mathbb{R}^2 \cup \{\infty\}$ denotes the **extended real plane** and ∞ is called **the point at infinity.** Both the domain and the range of a Möbius transformation include ∞ because, as we explain in the next subsection, this point can be handled in a consistent manner, resulting in a continuous, one-to-one, onto, invertible transformation. A Möbius transformation may be represented by a formula such as

$$\mathcal{M}(x, y) = \left(\frac{3x + 4y}{5x^2 + 5y^2}, \frac{4x - 3y}{5x^2 + 5y^2} \right). \tag{2.6.1}$$

This maps the unit circle $\mathcal{C}$ centred at the origin $\mathcal{O} = (0, 0)$ onto itself, maps the interior $\mathcal{D}$ of the unit disk centred at $\mathcal{O}$ onto the region outside $\mathcal{C}$, maps $\mathcal{O}$ to ∞ and ∞ to $\mathcal{O}$ and involves both a reflection in the y-axis and a rotation about $\mathcal{O}$. The behaviour at $\mathcal{O}$ and ∞ may be deduced by using continuity and taking limits.

The most general Möbius transformation $\mathcal{M} : \widehat{\mathbb{R}^2} \to \widehat{\mathbb{R}^2}$ may be expressed in terms of eight real parameters $a_R, a_I, b_R, b_I, c_R, c_I, d_R, d_I \in \mathbb{R}^2$, which are constrained only by the condition that

$$\text{either} \quad a_R d_R - a_I d_I - b_R c_R + b_I c_I \neq 0 \quad \text{or} \quad a_I d_R + a_R d_I - b_R c_I - b_I c_R \neq 0.$$
$$\tag{2.6.2}$$

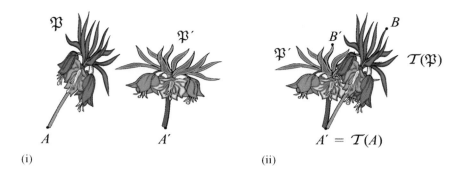

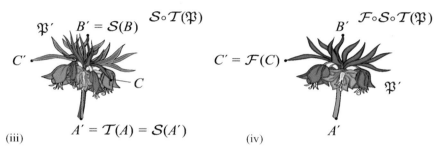

Figure 2.30 This illustrates the 'move-three-points' algorithm, which is defined as follows. (i) Identify a pair of points $A \in \mathfrak{P}$ and $A' \in \mathfrak{P}'$. (ii) Apply to $\mathfrak{P}$ the unique translation $\mathcal{T}$ such that $\mathcal{T}(A) = A'$. Identify a second pair of points $B \in \mathcal{T}(\mathfrak{P})$ and $B' \in \mathfrak{P}'$. (iii) Apply to $\mathcal{T}(\mathfrak{P})$ the unique similitude $\mathcal{S}$ such that $\mathcal{S}(A') = A'$ and $\mathcal{S}(B) = B'$. Identify a third pair of points $C \in \mathcal{S} \circ \mathcal{T}(\mathfrak{P})$ and $C' \in \mathfrak{P}'$. (iv) Apply to $\mathcal{S} \circ \mathcal{T}(\mathfrak{P})$ the unique shear transformation $\mathcal{F}$ such that $\mathcal{F}(A') = A'$, $\mathcal{F}(B') = B'$ and $\mathcal{F}(C') = C'$.

The general formula is

$$\mathcal{M}(x, y) = \left(\frac{A(x, y)}{C(x, y)}, \frac{B(x, y)}{C(x, y)} \right), \tag{2.6.3}$$

where

$$A(x, y) = (a_R x - a_I y + b_R)(c_R x - c_I y + d_R)$$
$$+ (a_R y + a_I x + b_I)(c_I x + c_R y + d_I),$$

$$B(x, y) = (a_R y + a_I x + b_I)(c_R x - c_I y + d_R)$$
$$- (a_R x - a_I y + b_R)(c_I x + c_R y + d_I)$$

and

$$C(x, y) = (c_R x - c_I y + d_R)^2 + (c_R y + c_I x + d_I)^2.$$

In order to evaluate expressions where both the numerator and denominator may vanish, limits must be taken. But these formulas are best handled using complex notation.

We identify $\mathbb{R}^2$ with the complex plane $\mathbb{C}$ in the obvious way, mapping the point $(x, y) \in \mathbb{R}^2$ to the point $x + iy = z \in \mathbb{C}$ where $i = \sqrt{-1}$. If we write

$$a = a_R + ia_I, \quad b = b_R + ib_I, \quad c = c_R + ic_I, \quad d = d_R + id_I,$$

then the condition in Equation (2.6.2) becomes

$$ad - bc \neq 0$$

and our transformation $\mathcal{M} : \widehat{\mathbb{C}} \to \widehat{\mathbb{C}}$, where $\widehat{\mathbb{C}} = \mathbb{C} \cup \{\infty\}$ is known as the **extended complex plane**, becomes quite simply

$$\mathcal{M}(z) = \frac{az + b}{cz + d}. \tag{2.6.4}$$

In this representation we have $\mathcal{M}(\infty) = a/c$ and $\mathcal{M}(-d/c) = \infty$ if $c \neq 0$, and $\mathcal{M}(\infty) = \infty$ if $c = 0$.

EXERCISE 2.6.1 *Verify that Equations (2.6.3) and (2.6.4) are equivalent.*

With the aid of Equation (2.6.4) it is readily verified that the composition of two Möbius transformations is also a Möbius transformation; indeed,

$$f_1 \circ f_2(z) = \frac{(a_1a_2 + b_1c_2)z + (a_1b_2 + b_1d_2)}{(c_1a_2 + d_1c_2)z + (c_1b_2 + d_1d_2)}. \tag{2.6.5}$$

Does this look familiar? Compare it with Equation (2.5.3). This means that we can use the matrix operations of complex 2×2 matrices to compose and invert Möbius transformations.

EXERCISE 2.6.2 *Write down the Möbius transformation $\mathcal{M}$ in Equation (2.6.1) in complex notation. Then use matrix operations to find formulas for $\mathcal{M}^{-1}$ and $\mathcal{M} \circ \mathcal{M}$.*

It is easy for you to check that if $c \neq 0$ then the Möbius transformation in Equation (2.6.4) can be written in the form

$$\mathcal{M}(z) = \mathcal{M}_1 \circ \mathcal{M}_2 \circ \mathcal{M}_3(z)$$

where

$$\mathcal{M}_1(z) = \frac{bc - ad}{c}z + \frac{a}{c}, \quad \mathcal{M}_2(z) = \frac{1}{z}, \quad \mathcal{M}_3(z) = cz + d.$$

Both $\mathcal{M}_1$ and $\mathcal{M}_3$ are similitudes, which map the set of all **generalized circles**, namely the set of circles and straight lines, onto itself. The transformation $\mathcal{M}_2$ is an **inversion** that also maps the set of generalized circles onto itself, as we now show.

Figure 2.31 Various Mobius transformations have been applied to a picture of a person on a bike. What properties do all of the resulting bikes have in common?

Any generalized circle $\mathcal{C} \subset \widehat{\mathbb{C}}$ can be expressed in the form

$$\mathcal{C} = \left\{ z \in \widehat{\mathbb{C}} : \frac{|z - z_0|}{|z - z_1|} = \gamma \right\} \tag{2.6.6}$$

for some pair of points $z_0, z_1 \in \mathbb{C}$ and some $\gamma > 0$, as in Exercise 2.6.3 below. The inversion $\mathcal{M}_2(z) = 1/z$ maps the generalized circle $\mathcal{C}$ into the set

$$\widetilde{\mathcal{C}} = \left\{ z \in \widehat{\mathbb{C}} : \frac{|z - z_0^{-1}|}{|z - z_1^{-1}|} = \gamma \frac{|z_1|}{|z_0|} \right\}, \tag{2.6.7}$$

which is also a generalized circle. Here we have assumed $z_0, z_1 \neq 0$ for simplicity.

EXERCISE 2.6.3 *Verify that any generalized circle $\mathcal{C} \subset \widehat{\mathbb{C}}$ can be expressed as in Equation (2.6.6). Hint: $(x - x_0)^2 + (y - y_0)^2 = \gamma((x - x_1)^2 + (y - y_1)^2)$.*

The Riemann sphere

In order to understand how a Möbius transformation handles ∞ it is natural to model $\widehat{\mathbb{R}^2} := \mathbb{R}^2 \cup \{\infty\}$, or equivalently $\widehat{\mathbb{C}} = \mathbb{C} \cup \{\infty\}$, as the surface $\widehat{\mathbb{S}} \subset \mathbb{R}^3$

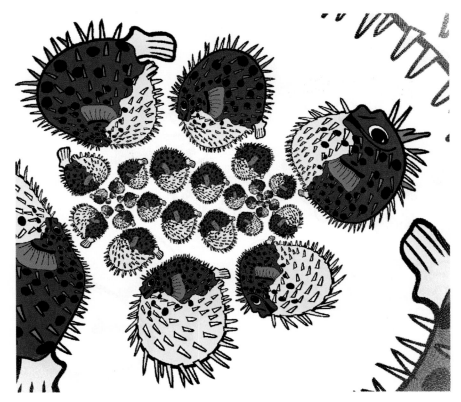

Figure 2.32   Various Möbius transformations have been applied to a picture of a puffer fish. Draw a circle through any three distinctive points on one fish, and another circle through the corresponding points on a second fish. Then if one of the circles goes through another distinctive point on the first fish, the second circle will go through the corresponding point on the second fish.

of a sphere of radius 1 centred at $(0, 0) \in \mathbb{R}^2$. The surface $\widehat{\mathbb{S}}$ is also called the **Riemann sphere**. The mapping between the plane and the sphere is achieved by **stereographic projection**; see Figure 2.33. The projection mapping $f : \widehat{\mathbb{R}^2} \to \widehat{\mathbb{S}}$ is readily found to be given by

$$f(x, y) = \left(\frac{2x}{x^2 + y^2 + 1}, \ \frac{2y}{x^2 + y^2 + 1}, \ \frac{x^2 + y^2 - 1}{x^2 + y^2 + 1} \right) \quad \text{for all } (x, y) \in \widehat{\mathbb{R}^2},$$

$$(2.6.8)$$

with inverse

$$f^{-1}(x', y', z') = \left(\frac{x'}{1 - z'}, \ \frac{y'}{1 - z'} \right) \quad \text{for all } (x', y', z') \in \widehat{\mathbb{S}}.$$

The point at infinity is mapped to the top of the sphere, $N = (0, 0, 1)$. All circles and straight lines in $\widehat{\mathbb{R}^2}$ correspond to circles on $\widehat{\mathbb{S}}$. Note that a circle on $\widehat{\mathbb{S}}$ is the

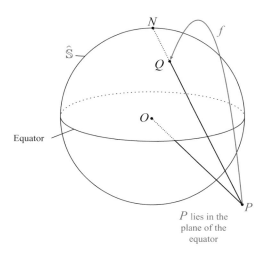

Figure 2.33 Illustration of a stereographic projection $f : \widehat{\mathbb{R}^2} \to \widehat{\mathbb{S}}$ between the extended plane and the surface of a sphere. The point P in the plane of the equator is mapped to the point Q where the straight line from P to the north pole N first meets $\widehat{\mathbb{S}}$. Circles and straight lines in the extended plane are mapped to circles on the sphere, and vice versa.

intersection of $\widehat{\mathbb{S}}$ with a plane in $\mathbb{R}^3$ that meets $\widehat{\mathbb{S}}$ in at least two points. Circles on $\widehat{\mathbb{S}}$ that go through $N = (0, 0, 1)$ correspond to straight lines in $\widehat{\mathbb{C}}$.

If we consider Möbius transformations acting on the sphere in place of the plane, we find that the point at infinity behaves exactly like all the other points on the sphere. Any Möbius transformation $f \circ \mathcal{M} \circ f^{-1}$ maps the sphere to itself in a one-to-one-onto continuous manner and can be expressed as a composition of rotations of the sphere and certain rescalings, corresponding to similitudes in the plane, each of which maps circles on the sphere to circles on the sphere. For example, the inversion $\mathcal{M}_2(z) = 1/z$ becomes simply a rotation of the sphere through $180°$ about the x-axis, i.e.

$$f \circ \mathcal{M}_2 \circ f^{-1}(x', y', z') = (x', -y', -z').\tag{2.6.9}$$

It is easy to see that the most general Möbius transformation for which ∞ is a fixed point is a similitude with a proper rotation; that is, it can be written in the form

$$\mathcal{M}(z) = \lambda e^{i\theta} z + t$$

for $\lambda > 0$, $\theta \in [0, 2\pi)$ and $t = (f + ig) \in \mathbb{C}$, which is equivalent to Equation (2.5.6). For example, when $\lambda > 1$ this transformation has two distinct fixed points, one of which is ∞, and it corresponds to a rotation of the sphere about the z-axis composed with a motion away from the south pole, following longitudinal great circles, towards the north pole. Indeed, with $t = 0$ and $\lambda > 1$, if we make the change of coordinates provided by the inversion that interchanges 0 and ∞, the transformation becomes

$$\widetilde{\mathcal{M}}(z) = \mathcal{M}_2 \circ \mathcal{M} \circ \mathcal{M}_2^{-1}(z) = \lambda^{-1} e^{-i\theta} z,$$

which is just like the original transformation except that λ is replaced by λ^{-1} and the direction of rotation is reversed.

A Möbius transformation that possesses two distinct fixed points either rotates points close to the fixed points in opposite directions, as illustrated in Figure 2.38, in which case it is called **loxodromic**, or else it does not rotate space about either fixed point. In the latter case it either expands points away from one fixed point and towards the other along arcs of generalized circles, in which case it is called **hyperbolic**, or else it is the identity transformation $\mathcal{M}(z) = z$.

The only other possible type of Möbius transformation is **parabolic** and possesses only one fixed point. This fixed point may be thought of as a limiting case of a family of hyperbolic transformations in which the two fixed points coalesce. As a consequence, a parabolic Möbius transformation behaves in a remarkable manner: some points are repelled and some are attracted by its fixed point.

Specifically, a parabolic Möbius transformation maps each circle tangent to a certain fixed line, through the fixed point, onto itself. Points are swept, along these circles, away from the fixed point on one side and towards it on the other side; the direction of this circling motion is clockwise on circles lying on one side of the fixed line and counterclockwise on circles lying on the other side.

An example of a parabolic transformation acting on a picture within a disk is illustrated in Figure 2.34; in this case the fixed point is at the top of the disk and the fixed line is tangent to the disk. Notice how colourful picture matter is maintained within each crescent, swept away from one side of the fixed point towards the other.

EXERCISE 2.6.4　*Verify Equation (2.6.9).*

We can define a metric $d_{Riemann}$ on $\mathbb{R}^2 \cup \{\infty\}$, or equivalently the Riemann sphere $\widehat{\mathbb{S}}$, by

$$d_{Riemann}((x_1, y_1), (x_2, y_2))$$
$$= \text{shortest distance between } f(x_1, y_1) \text{ and } f(x_2, y_2) \text{ on } \widehat{\mathbb{S}},$$

where f is given in Equation (2.6.8). Then $(\mathbb{R}^2 \cup \{\infty\}, d_{Riemann})$ is a compact metric space. The natural topology associated with $d_{Riemann}$ is such that any Möbius transformation $\mathcal{M} : \mathbb{R}^2 \cup \{\infty\} \to \mathbb{R}^2 \cup \{\infty\}$ is continuous.

Figure 2.34 Illustration of a **parabolic** Möbius transformation acting on a picture. The unique fixed point is at the top of the disk. Colourful picture material within each crescent, defined by adjacent pairs of circles, is swept round while staying within its allotted crescent. You should study carefully the two pictures, 'before' and 'after', to be sure you confirm this effect. The fixed point is both repulsive and attractive.

Fundamental theorem of Möbius transformations

THEOREM 2.6.5   *Let z_1, z_2, z_3 and w_1, w_2, w_3 be two sets of distinct points in the extended complex plane $\widehat{\mathbb{C}} = \mathbb{C} \cup \{\infty\}$. Then there exists a unique Möbius transformation that maps z_1 to w_1, z_2 to w_2 and z_3 to w_3.*

PROOF   This is a good exercise. Hint: Start by choosing $z_1 = 0$, $z_2 = 1$ and $z_3 = i$. See [25], p. 242.   □

One consequence of Theorem 2.6.5 is that there are many Möbius transformations that map any given generalized circle to another given generalized circle. In particular, there are many Möbius transformations that map the circle $\mathcal{C} = \{z \in \mathbb{C} : |z| = 1\}$ onto itself. Indeed, they are given by

$$\mathcal{M}_{a,\theta}(z) := \frac{(z - a)e^{i\theta}}{1 - \overline{a}z}, \qquad (2.6.10)$$

where $a \in \mathbb{C}$, with $a \neq 1$, and $0 \leq \theta < 2\pi$. It is readily verified that $\mathcal{M}_a$ takes three distinct points on $\mathcal{C}$, such as 1, i and -1, to points on $\mathcal{C}$. Notice that $\mathcal{M}_{a,\theta}(0) = -ae^{i\theta}$, so that $\mathcal{M}_{a,\theta}$ maps the interior of the circle $\mathcal{C}$ to itself when $|a| < 1$ but turns the circle 'inside out' when $|a| > 1$; examples of this type of transformation applied to a flower picture are shown in Figure 2.35 and to a fish measure in Figure 2.14. For $a \neq 0$, the fixed points of $\mathcal{M}_{a,\theta}(z)$ lie on the circle $|z| = 1$. For $0 < |a| \leq 1$ and $a \neq 1$, each member of this family of transformations is either parabolic or hyperbolic.

Figure 2.35 A Möbius transformation of the form given in Equation (2.6.10) has been applied to the picture on the left. Notice the big buds and the curved stems in the transformed picture on the right.

Another interesting family of Möbius transformations is given by

$$\widehat{\mathcal{M}}_\rho(z) = \frac{\rho z}{1 + (\rho - 1)z}, \qquad (2.6.11)$$

where $\rho \in \mathbb{C}$ with $\rho \neq 0, 1$. $\widehat{\mathcal{M}}_\rho(z)$ has two distinct fixed points, $z = 0$ and $z = 1$. It behaves like the similitude ρz near $z = 0$ and like the similitude $1 + \rho^{-1}(1 - z)$ near $z = 1$: if we rotate the coordinates through $180°$ about the point halfway between the two fixed points, by means of the transformation $t(z) = 1 - z$, we find that

$$\widehat{\mathcal{M}}_{1/\rho}(z) = t \circ \widehat{\mathcal{M}}_\rho \circ t^{-1}(z).$$

The transformations of this family are always either loxodromic or hyperbolic.

Two examples of the transformation in Equation (2.6.11) applied to the circular picture containing a flower in the left-hand panel of Figure 2.35 are shown in Figure 2.36. For the right-hand panel in Figure 2.36, $\rho = -0.35 - 0.1i$ and both the domain and the visible part of the range correspond to $\{x + iy : -1 \leq x, y \leq +1\}$. The white disk is the image of the exterior of the original disk. For the left-hand panel in Figure 2.36, $\rho = 0.3 - 0.2i$ and both the domain and the visible part of the range correspond to $\{x + iy : -2 \leq x, y \leq +2\}$.

EXERCISE 2.6.6 *Find the unique Möbius transformation $\mathcal{M}(z)$ that maps ∞ to 1, 0 to $-i$ and -1 to -1. Show that this transformation maps the upper half-plane to the interior of the circle of radius 1 centred at $z = 0$.*

Figure 2.36 Möbius transformations of the form given in Equation (2.6.11) have been applied to the left-hand picture in Figure 2.35. In each of these transformed pictures the disk containing the flower has been inverted and one of the blue petals has been stretched out to infinity.

Invariant points, sets, measures and pictures for Möbius transformations

Sets, pictures and measures which are invariant under certain Möbius transformations that are essentially rotations are illustrated in Figures 2.6 and 2.17–2.20.

Another type of invariant set is illustrated in Figure 2.37 and is associated with a transformation of the form in Equation (2.6.11): the invariance occurs because of an underlying group structure, to be explained in Chapter 3. This example illustrates clearly that although Möbius transformations map generalized circles into generalized circles, they do not preserve ellipses! A similar type of invariant picture is shown in Figure 2.38. Both these examples are interesting because the only associated invariant measures consist of point masses at the centres of the two spirals. These centres are the fixed points of the transformations. The transformations sweep all other finite measures along spiral paths away from one fixed point and in towards the other.

2.7 Projective transformations

Projective transformations in two dimensions are specified by nine real parameters. They are geometrically simple and may be efficiently described, communicated and computed. Small sets of them may be used to represent apparently complex images, as we will show in Chapters 4 and 5. So in this section we start to explain what they are and how they transform sets, pictures and measures. What sorts of sets, pictures and measures do they leave invariant?

We begin straight away by introducing projective transformations informally. Examples of projective transformations are illustrated in Figures 2.39 and 2.40.

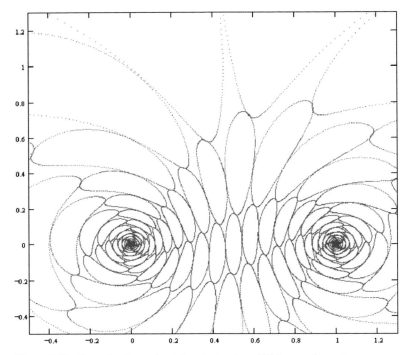

Figure 2.37   Example of a set that is invariant under a Möbius transformation.

Figure 2.39 shows a photograph of a framed picture, taken from directly in front, and next to it a photograph of the same picture taken from an oblique position. The picture on the right is to a very good approximation a translation and rotation of a **perspectivity** of the picture on the left. A perspectivity is a transformation $f : E_1 \to E_2$ between two planes $E_1, E_2 \subset \mathbb{R}^3$ that is defined, with the aid of a point $O \in \mathbb{R}^3$ with $O \notin E_1 \cup E_2$, by $f(P) = l(OP) \cap E_2$ for each $P \in E_1$, where $l(OP)$ denotes the line through O and P. The original two photographs in Figure 2.39 actually lie in different planes, defined by the photographic plane of the camera at the two instants the photos were taken, but in Figure 2.39 the second plane has been rigidly translated and rotated so as to position the two pictures side by side.

Perspective transformations, as well as translations and rotations, are carried out by the mental part of the human visual system to enable obliquely viewed pictures to seem in the mind's eye as though they are not distorted. For example, if you watch television from close up and to one side, you will not be aware, for more than a few moments, of the distortion, a significant optical perspectivity between the image on the screen and the one on your retina. Similarly, if you move a photograph around in front of you, or view it from different angles, you will continue to see the 'same' picture, not lots of different perspectivities of it.

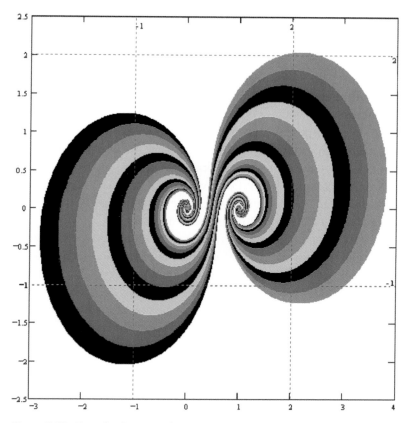

Figure 2.38   Example of a picture that is invariant under a Möbius transformation of the form of $\widehat{\mathcal{M}}_\rho(z)$ in Equation (2.6.11). Note the invariance with respect to rotation by $180°$ about the midpoint. Although they are mainly so small as to be invisible, the spirals about $z = 0$ and $z = 1$ contain infinitely many whirls. What do pictures of invariant measures for $\widehat{\mathcal{M}}_\rho(z)$ look like, in general?

Roughly, general **projective** transformations in two dimensions are those that are obtained by composing perspectivities. If you look at the right-hand picture in Figure 2.39 from an oblique position, or photograph it, the result on your retina or on the focal plane of the camera will be a projective transformation of the original framed picture. Moreover, in general these transformed pictures do not look like any picture that you could see by looking at the original framed picture from various positions. The reason is that the set of projective transformations strictly contains the set of perspective transformations; 'most' projective transformations are not perspectivities, nor are they rigid transformations of them.

Four different projective transformations of a picture are illustrated in Figure 2.40. Notice how they look quite odd, compared with perspective transformations. Notice also that points that lie on straight lines in the original picture are mapped to points that lie on straight lines in each of the transformed pictures.

This latter property is also true for affine transformations. The set of projective transformations includes the set of affine transformations.

The projective plane

There are several different useful representations or models for the space, called the **projective plane**, on which projective transformations act. They include: (i) $\mathbb{R}^2 \cup L_\infty$, which consists of a plane together with an extra line of points; (ii) $\mathbb{RP}^2$, whose points are lines through the origin in $\mathbb{R}^3$; (iii) a spherical shell with opposite points, relative to the centre of the sphere, identified; (iv) a filled disk with opposite points on its boundary identified. As we will see later, straightforward one-to-one invertible transformations between these spaces allow us to convert from one representation of the projective plane to another.

Because of our focus on pictures and because we want to develop our intuition about how projective transformations deform sets of points and measures in the plane, we begin by describing them in terms of the space $\mathbb{R}^2 \cup L_\infty$. This is the space in which we normally *see* projective transformations in action.

The space $\mathbb{R}^2 \cup L_\infty$ consists of the euclidean plane together with an extra set of points $L_\infty = \mathbb{R} \cup \{\infty\}$, which we call the **line at infinity**. Thus a point in $\mathbb{R}^2 \cup L_\infty$ may be denoted by $(x, y) \in \mathbb{R}^2$, $x \in \mathbb{R}$ or ∞, depending on whether it belongs to $\mathbb{R}^2$, $\mathbb{R}$ or $\{\infty\}$ respectively.

In $\mathbb{R}^2 \cup L_\infty$ we define a straight line L other than L_∞ in the usual way, representing it by means of a formula of the form $lx + my + n = 0$, where $l, m, n \in \mathbb{R}$ and $(l, m) \neq (0, 0)$, but we include on L a point belonging to L_∞; specifically

$$L = \begin{cases} \{(x, y) \in \mathbb{R}^2 : lx + my + n = 0\} \cup \{-m/l\} & \text{when } l \neq 0, \\ \{(x, y) \in \mathbb{R}^2 : lx + my + n = 0\} \cup \{\infty\} & \text{when } l = 0. \end{cases}$$

Since we can deduce the component of L that lies on L_∞ directly from the formula $lx + my + n = 0$ we will not usually make specific reference to this component. We will say simply 'L *is the line given by* $lx + my + n = 0$'.

A typical projective transformation maps all $\mathbb{R}^2$ minus one straight line L_D onto all $\mathbb{R}^2$ minus one straight line L_R; it maps the missing line L_D one-to-one onto L_∞ and it maps L_∞ one-to-one onto L_R. It does so in such a way that, from the right point of view, the line L_∞ behaves just like all other lines in the domain and range of the projective transformation. This is analogous to the way in which a typical Möbius transformation maps $\mathbb{R}^2$ minus one point z_D onto $\mathbb{R}^2$ minus one point z_R, the point z_D to the point at infinity and the point at infinity to z_R. By looking at Möbius transformations acting on the Riemann sphere, we saw that there was nothing special about the point at infinity. Analogously, in the case of projective transformations we will find that there is nothing special about L_∞, the line at infinity; but to see this we have to go to $\mathbb{RP}^2$, which we do in a later subsection.

Figure 2.39   A framed picture (on the left) has been photographed (on the right) from an oblique position. The result is a **perspective** transformation, plus some distortions of the colour.

Figure 2.40   Examples of projective transformations that are not perspective transformations. No ordinary photograph of the original framed picture could produce these transformed pictures. What properties do these four pictures have in common?

An example of a projective transformation on $\mathbb{R}^2 \cup L_\infty$

An example of a projective transformation $\mathcal{P} : \mathbb{R}^2 \cup L_\infty \to \mathbb{R}^2 \cup L_\infty$ is given by

$$P(x, y) = \left(\frac{6.5x}{-2x + 3y + 15}, \frac{6y}{-2x + 3y + 15} \right) \qquad (2.7.1)$$

for all $(x, y) \in \mathbb{R}^2 \backslash L_D, \mathcal{P}(L_D) = L_\infty$ and $\mathcal{P}(L_\infty) = L_R$, where

$$L_D \quad \text{is given by} \quad 2x - 3y - 15 = 0,$$

$$L_R \quad \text{is given by} \quad y - 2 - \frac{4}{6.5}x = 0.$$

Actually the formula in Equation (2.7.1) tells us all we need to know. L_D is just the straight line along which the denominators vanish, and L_R is the set of points in $\mathbb{R}^2$ that are not in the range of $\mathcal{P}$, that is, points for which the denominators in the inverse transformation vanish. The latter is given by

$$\mathcal{P}^{-1}(x, y) = \left(\frac{\frac{1}{6}x}{\frac{1}{45}x - \frac{2}{65}y + \frac{1}{15}}, \frac{\frac{2}{13}y}{\frac{1}{45}x - \frac{2}{65}y + \frac{1}{15}} \right).$$

The top left picture in Figure 2.40 corresponds to the application of $\mathcal{P}$ to the framed picture on the left in Figure 2.39, which has corners at, say, the points $(0, 0)$, $(1.2, 0)$, $(1.2, -1)$ and $(0, -1)$. These points are mapped by $\mathcal{P}$ onto the points $(0, 0)$, $(0.6, 0)$, $(0.8, -0.6)$ and $(0, -0.5)$ respectively.

EXERCISE 2.7.1   *Find the inverse of the projective transformation*

$$P(x, y) = \left(\frac{6.5x + y + 1}{-2x + 3y + 15}, \frac{2x - 6y + 2}{-2x + 3y + 15} \right).$$

Identify the set of points L_R mapped by $\mathcal{P}$ onto the line $L_D - 2x + 3y + 15 = 0$.

The dance of the points

Here we define projective transformations on $\mathbb{R}^2 \cup L_\infty$ in specific terms.

The most general projective transformation

$$\mathcal{P} : \mathbb{R}^2 \cup L_\infty \to \mathbb{R}^2 \cup L_\infty$$

may be expressed in terms of nine real parameters, constants $a, b, c, d, e, f,$ $g, h, l \in \mathbb{R}$, constrained only by the condition

$$\det P = a(dj - fh) - b(cj - fg) + e(ch - dg) \neq 0,$$

where P is the invertible 3×3 matrix

$$P = \begin{pmatrix} a & b & e \\ c & d & f \\ g & h & j \end{pmatrix}. \qquad (2.7.2)$$

The basic general formula is

$$P(x, y) = \left(\frac{ax + by + e}{gx + hy + j}, \frac{cx + dy + f}{gx + hy + j} \right) \quad \text{for all } (x, y) \in \mathbb{R}^2$$

with $(x, y) \notin L_D$, where L_D is given by $gx + hy + j = 0$ if $g \neq 0$ or $h \neq 0$, and $L_D = L_\infty$ if $g = h = 0$.

But to where on L_∞ does P specifically map the points of L_D, and to where does it map the points of L_∞? To answer these questions briefly, we restrict attention to the case where all the coefficients $a, b, c, d, e, f, g, h, j$ are nonzero. We define

$$P(x, y) = \frac{ax + by + c}{dx + ey + f} \in L_\infty \quad \text{for all } (x, y) \in L_D,$$

where the point $\infty \in L_\infty$ is assigned to the unique point $(\widehat{x}, \widehat{y}) \in L_D$, with $d\widehat{x} + e\widehat{y} + f = 0$. We define

$$P(x) = \left(\frac{ax + b}{gx + h}, \frac{dx + e}{gx + h} \right) \in \mathbb{R}^2 \quad \text{for all } x \in L_\infty \backslash \{\infty\}$$

and $P(\infty) = (a/g, d/g)$.

The general case can be deduced, as the above formulas were, by working in $\mathbb{RP}^2$, as described in Definition 2.7.15 *et seq.*

The whole system works consistently, in such a way that the composition of two projective transformations is also a projective transformation, each projective transformation possesses an inverse that is itself a projective transformation and so on.

Thus we see how a projective transformation may choreograph an elegant dance on $\mathbb{R}^2 \cup L_\infty$: 'most' points in $\mathbb{R}^2$ are mapped to new points in $\mathbb{R}^2$ but some seemingly disappear, leaving the dance floor so to speak, having been mapped to L_∞. If the transformation is applied again, these points will reappear in $\mathbb{R}^2$, in elegant continuous proximity once again with the points to which they had been near before they left the floor. There is this capability, this extra flexibility, compared with say affine transformations, to take some points out of the picture without losing track of them. This allows projective transformations to achieve some of their most beautiful moves, such as being able to map a circle onto a parabola or the vertices of any quadrilateral to the vertices of any other quadrilateral. Indeed, any invertible affine transformation on $\mathbb{R}^2$ is just the restriction to $\mathbb{R}^2$ of some projective transformation that maps L_∞ to L_∞; see Exercise 2.7.22.

EXERCISE 2.7.2   *Write down two specific projective transformations $P_1 : \mathbb{R}^2 \cup L_\infty \to \mathbb{R}^2 \cup L_\infty$ and $P_2 : \mathbb{R}^2 \cup L_\infty \to \mathbb{R}^2 \cup L_\infty$ corresponding to 3×3 matrices P_1 and P_2 with strictly positive entries. Verify that $P_1 \circ P_2$ corresponds to the matrix $P_1 P_2$. In particular, check its action on sets of points that are mapped to and from L_∞. Show that everything works out consistently.*

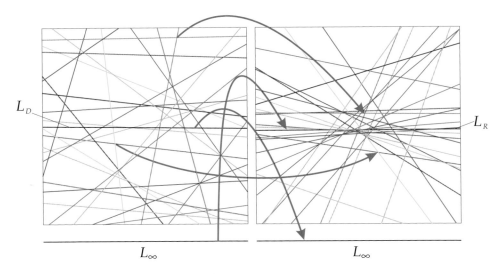

Figure 2.41   The dance of the lines in $\mathbb{R}^2 \cup L_\infty$ induced by a projective transformation is illustrated here. The set of lines in $\mathbb{R}^2 \cup L_\infty$ is mapped onto itself, but one line, L_D, is mapped to L_∞ while L_∞ is mapped to L_R. See also Figure 2.42.

The dance of the lines

It is quite easy to see that a projective transformation $\mathcal{P} : \mathbb{R}^2 \cup L_\infty \to \mathbb{R}^2 \cup L_\infty$ maps straight lines in $\mathbb{R}^2 \cup L_\infty$ to straight lines in $\mathbb{R}^2 \cup L_\infty$. For example, if (x, y) lies on the line given by $lx + my + n = 0$ but does not lie on the line given by $gx + hy + j = 0$, namely L_D, and we write $\mathcal{P}(x, y) = (X, Y)$ then it is readily verified that $LX + MY + N = 0$ where $(L, M, N) = (l, m, n)\mathcal{P}^{-1}$.

In fact it follows directly from the description below of $\mathcal{P}$ in terms of $\mathbb{RP}^2$ that $\mathcal{P}$ maps the set of straight lines in $\mathbb{R}^2 \cup L_\infty$ one-to-one onto itself.

So just as we can think of a projective transformation as describing a dance among the points of $\mathbb{R}^2 \cup L_\infty$, with points coming and going from L_∞, so too can we think of another dance, among the lines of $\mathbb{R}^2 \cup L_\infty$. But in this dance, only one line, L_D, may leave the floor and only one may return. See Figures 2.41 and 2.42. The latter figure reveals that the dance is very organized: lines that are parallel in the left-hand panel are mapped to lines that meet at the same point in the right-hand panel. The same transformation is used in both figures.

In Figure 2.43 we illustrate two different projective transformations applied to a picture of a beech tree leaf. Notice how the approximately straight lines of the veins are preserved.

Since a projective transformation maps straight lines into straight lines and points into points, it follows that it maps any figure of straight lines and their intersections into a figure of straight lines and their intersections. In thinking about this, notice that parallel straight lines intersect at a point, determined by their common direction, on L_∞.

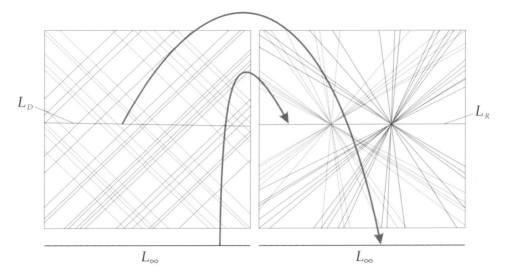

Figure 2.42   Lines that are parallel in the left-hand panel are mapped to lines which meet at the same point in the right-hand panel. The same transformation is used in Figures 2.41 and 2.42.

Figure 2.43   Original beech leaf, left, and two projective transformations of it. The tranformations preserve an ellipse that approximately surrounds the leaf and they are applied only to the picture inside the ellipse. Notice how straight lines approximately defined by the veins are preserved, while non-straightness is exaggerated.

A set of points $S \subset \mathbb{R}^2 \cup L_\infty$ is said to be **collinear** if there exists a straight line L in $\mathbb{R}^2 \cup L_\infty$ such that $S \subset L$.

EXERCISE 2.7.3   *Show that, in $\mathbb{R}^2 \cup L_\infty$, the three points $(1, 10)$, -1 and $\frac{1}{7}$ are not collinear but the three points $(1, 10)$, $(-1, -4)$ and $\frac{1}{7}$ are collinear.*

The following theorem tells us precisely how flexible projective transformations are. It is of particular importance to us, since it shows us one way to express the degrees of freedom of projective transformations, when we want to use them to map parts of pictures to parts of pictures.

THEOREM 2.7.4 *(Fundamental theorem of projective geometry)   In $\mathbb{R}^2 \cup L_\infty$ let A, B, C, D be a set of points, no three of which are collinear, and let A', B', C', D' be a second set of points no three of which are collinear. Then there exists a unique projective transformation $\mathcal{P} : \mathbb{R}^2 \cup L_\infty \to \mathbb{R}^2 \cup L$ that maps A to A', B to B', C to C' and D to D'.*

PROOF   See [25], p. 127.   □

In particular, a projective transformation will map any figure consisting of four distinct straight lines onto another figure consisting of four distinct straight lines. It is tempting to suppose that Theorem 2.7.4 asserts that there exists a projective transformation that not only maps the vertices of a quadrilateral α to the vertices of a quadrilateral β in any specified order but also maps the sides of the convex hull of α to the sides of the convex hull of β. This is not the case, as illustrated in Figure 2.44. The situation is somewhat analogous to the situation for Möbius transformations, where the interior of a filled-in circle may be mapped to the exterior of a filled-in circle.

Notice that there are twenty-four different projective transformations which map one set of four points to another set of four points when no three points in either set are collinear.

In the next subsection we discuss the mechanics of actually finding projective transformations that map one set of four points to another set of four points.

EXERCISE 2.7.5   *Let $(a, b) \in \mathbb{R}^2 \backslash \{(0, 0), (1, 0), (0, 1)\}$. Find the unique projective transformation $\mathcal{P} : \mathbb{R}^2 \cup L_\infty \to \mathbb{R}^2 \cup L_\infty$ such that*

$$\mathcal{P}(0, 0) = (0, 0), \quad \mathcal{P}(1, 0) = (1, 0), \quad \mathcal{P}(0, 1) = (0, 1), \quad \mathcal{P}(1, 1) = (a, b).$$

The dance of the conics

Suppose that we have a picture $\mathfrak{P}$ and a projective transformation $\mathcal{P}$. What does $\mathcal{P}$ do to $\mathfrak{P}$? Let us choose three distinctive non-collinear points A, B, C in the domain of $\mathfrak{P}$ and let A', B', C' denote their respective images under $\mathcal{P}$. Let $\mathcal{A}$

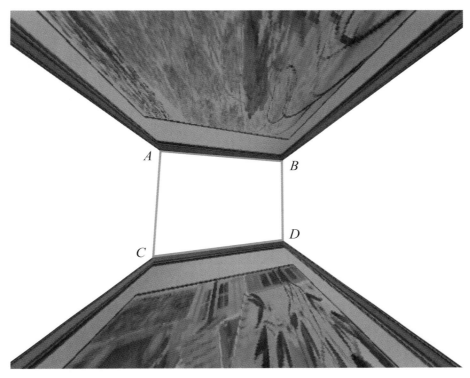

Figure 2.44   A projective tranformation may turn a rectangle 'inside out'. Here a picture that was framed in the quadrilateral $ABCD$ has been mapped by the unique projective transformation $\mathcal{P} : \mathbb{R}^2 \cup \{L_\infty\} \rightarrow \mathbb{R}^2 \cup \{L_\infty\}$ such that $\mathcal{P}(A) = A, \mathcal{P}(B) = B, \mathcal{P}(C) = D, \mathcal{P}(D) = C$. To where has the transformation mapped the line segment BC?

denote the unique affine transformation that maps the points A, B, C to the points $(0, 0)$, $(1, 0)$, $(0, 1)$ respectively. Let $\mathcal{A}'$ denote the unique affine transformation that maps the points A', B', C' to the points $(0, 0)$, $(1, 0)$, $(0, 1)$ respectively. Then

$$\mathcal{P} = \mathcal{A}'^{-1} \widehat{\mathcal{P}} \mathcal{A} \quad \text{where} \quad \widehat{\mathcal{P}} = \mathcal{A}' \mathcal{P} \mathcal{A}^{-1}.$$

$\widehat{\mathcal{P}}$ is a projective transformation that has $(0, 0)$, $(1, 0)$ and $(0, 1)$ as fixed points. It is readily verified that $\widehat{\mathcal{P}}$ belongs to the two-parameter family of projective transformations

$$\mathcal{P}_{a,b}(x, y) = \left(\frac{ax}{(a - 1)x + (b - 1)y + 1}, \frac{by}{(a - 1)x + (b - 1)y + 1} \right), \quad (2.7.3)$$

for $a, b \in \mathbb{R}$, $ab \neq 0$; see Figure 2.45. This two-parameter family of transformations must contain all the 'projective-but-not affine' aspects of $\mathcal{P}$. The central mystery of what a projective transformation does, which is fundamentally different

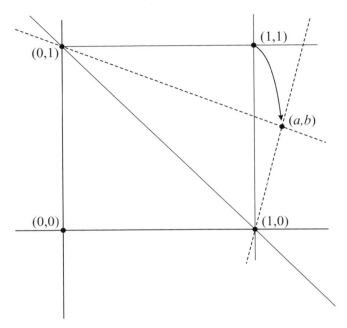

Figure 2.45 This illustrates the locations of the three fixed points $(0, 0)$, $(0, 1)$ and $(1, 0)$ of the canonical family of projective transformations $\mathcal{P}_{a,b}$ that take $(1, 1)$ to (a, b). This family of transformations leaves fixed the straight line passing through $(0, 0)$ and $(0, 1)$. It also leaves fixed the straight lines through $(0, 0)$ and $(1, 0)$ and through $(1, 0)$ and $(0, 1)$.

from what an affine transformation does, can be understood by considering how $\mathcal{P}_{a,b}$ acts on pictures.

EXERCISE 2.7.6 *In the above discussion, let D be a fourth point in the domain of $\mathfrak{P}$ such that no three of A, B, C, D are collinear, and let D' be a point in the domain of $\mathfrak{P}'$ such that no three of A', B', C', D' are collinear. Show that (a, b) can be choosen in such a way that $\mathcal{P}(D) = D'$. Thus devise a 'move-four-points' algorithm for adjusting projective transformations, analogous to the 'move-three-points' algorithm described in Figure 2.30.*

It is readily verified that

$$\mathcal{P}_{a,b}(0, 0) = (0, 0), \quad \mathcal{P}_{a,b}(1, 0) = (1, 0) \quad \text{and} \quad \mathcal{P}_{a,b}(0, 1) = (0, 1).$$

Moreover each transformation in the family maps each of the lines given by $x = 0$, $y = 0$ and $x + y = 1$ onto itself; it maps the line L_D given by

$$(a - 1)x + (b - 1)y + 1 = 0$$

to the line at infinity and L_∞ to the line L_R given by

$$\left(\frac{1}{a} - 1\right) x + \left(\frac{1}{b} - 1\right) y + 1 = 0.$$

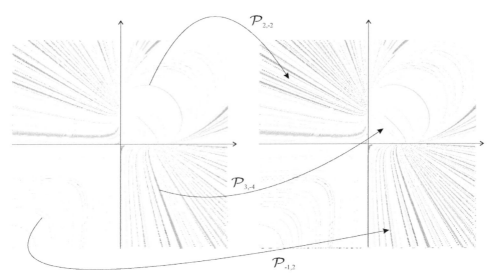

Figure 2.46   The dance of the conics! A projective transformation always maps conic sections into conic sections. Each of the two panels illustrates the family of conic sections $(x + y - 1)^2 + \gamma xy = 0$ where $\gamma \in \mathbb{R}$. Each projective transformation $\mathcal{P}_{a,b}$ in Equation (2.7.3) maps this family into itself.

Each member of this family of projective transformations has the remarkable property that it maps the family of conic sections $\{C_\gamma : \gamma \in \mathbb{R}\}$, where

$$C_\gamma := \{(x, y) \in \mathbb{R}^2 : (x + y - 1)^2 + \gamma xy = 0\}, \tag{2.7.4}$$

one-to-one onto itself. We now sketch the proof of this fact. Let $(x_0, y_0) \in \mathbb{R}^2$ and suppose that $(x_0, y_0) \in C_\gamma$. Let $\mathcal{P}_{a,b}(x_0, y_0) = (x, y)$. Then

$$(x_0, y_0) = \mathcal{P}_{a,b}^{-1}(x, y)$$
$$= \left(\frac{x/a}{(1/a - 1)x + (1/b - 1)y + 1}, \frac{y/b}{(1/a - 1)x + (1/b - 1)y + 1} \right)$$

and substituting into $(x_0 + y_0 - 1)^2 + \gamma x_0 y_0 = 0$, to formally eliminate x_0 and y_0, we obtain

$$(x + y - 1)^2 + \frac{\gamma xy}{ab} = 0,$$

from which it follows that

$$\mathcal{P}_{a,b}(C_\gamma) = C_{\gamma/(ab)}.$$

This completes the demonstration.

Figure 2.46 illustrates the family of conic sections given by Equation (2.7.4), and shows how some of them are mapped into others by members of the family of projective transformations $\mathcal{P}_{a,b}$.

Figure 2.47 The top left panel shows a picture $\mathfrak{P}$, in various colours, of parts of some of the conic sections $\mathcal{C}_\gamma$, defined by Equation (2.7.4), lying within the window $-3 \leq x \leq 3$, $-3 \leq y \leq 3$. The other three panels show, superimposed upon the original set of contours, the picture $\mathcal{P}_{a,b}(\mathfrak{P})$ superimposed upon $\mathfrak{P}$ for $(a, b) = (1.1, 1.1)$ (top right), $(a, b) = (0.9, 1.1)$ (bottom right) and $(a, b) = (1.1, 1.1)$ (bottom left). You can see quite clearly that $\mathcal{P}_{a,b}$ maps the underlying striated pattern onto itself, albeit, in these cases, that the colours are not preserved. The straight lines were added afterwards to show to where part of the boundary of the original picture is mapped.

In particular, if $ab = 1$ then $\mathcal{P}_{a,b}$ maps each conic section C_γ onto itself. So in this case, for example, the top left panel of Figure 2.47 represents an invariant picture for $\mathcal{P}_{a,b}$ because not only is the striated pattern preserved but the colours of the contours, before and after, are preserved too. Another example of a picture that is invariant under $\mathcal{P}_{a,b}$ when $ab = 1$ is shown in Figure 2.48.

EXERCISE 2.7.7 *Show that, when $\gamma = 1$, C_γ is the circle of radius 1 centred at* $(1, 1) \in \mathbb{R}^2$.

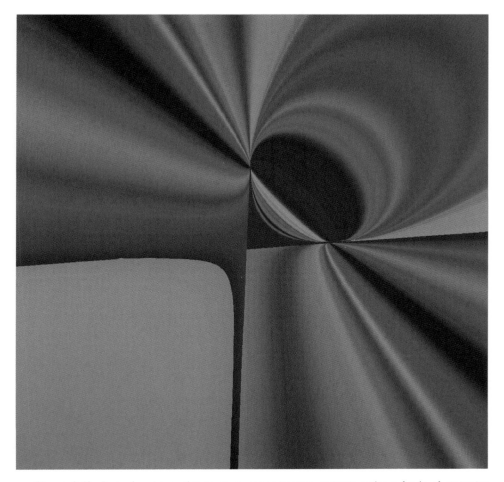

Figure 2.48 Part of a picture that is, to an approximation, invariant under a family of projective transformations.

When $ab = 1$ and $1 > a > 0$, the projective transformation $\mathcal{P}_{a,b}$ maps the disk D of radius 1 centred at $(1, 1)$ onto itself. Points are attracted towards the fixed points $(1, 0)$ and repelled by the fixed point $(0, 1)$. Any picture $\mathfrak{P}$ with domain D is transformed to a new picture with domain D. What is the relationship between $\mathfrak{P}$ and $\mathcal{P}_{a,b}(\mathfrak{P})$? We can think of $\mathcal{P}_{a,b}$ as sweeping lines of points around in a circle centred at $(0, 0)$ in such a way that the points on the lines follow elliptical paths, each concentric ellipse passing through the two fixed points $(1, 0)$ and $(0, 1)$. Straight lines are preserved, but so are these ellipses. This effect is illustrated in Figure 2.49, where we have chosen $a = 0.5$ and $b = 2.0$.

EXERCISE 2.7.8   *Compare the disk-preserving transformation illustrated in Figure 2.49 with that illustrated in Figure 2.35. What properties do the two transformations have in common?*

Figure 2.49   The projective transformation applied here maps a disk onto itself. Flowers in the picture are swept along elliptical paths away from one fixed point and towards the other. Where are the fixed points?

The way in which $\mathcal{P}_{a,b}$ maps the family of conic sections $\{C_\gamma : \gamma \in \mathbb{R}\}$ onto itself illustrates general properties of projective transformations in relation to conic sections.

DEFINITION 2.7.9   A non-degenerate **conic section** is a set of points $C \subset \mathbb{R}^2 \cup L_\infty$ given by an equation of the form

$$Ax^2 + Bxy + Cy^2 + Fx + Gy + H = 0, \quad \text{where } A, B, C, F, G, H \in \mathbb{R} \tag{2.7.5}$$

with $A \neq 0$, or $B \neq 0$ or $C \neq 0$, such that C does not contain a straight line or a single point. If Equation (2.7.5) describes an ellipse or circle then C includes no points on L_∞. If Equation (2.7.5) describes a hyperbola then C includes the two points on L_∞ at which the asymptotes of the hyperbola intersect L_∞. If Equation (2.7.5) describes a parabola then C includes the point at which the axis of the parabola intersects L_∞.

It is readily verified that if C is a non-degenerate conic section and $\mathcal{P} : \mathbb{R}^2 \cup L_\infty \to \mathbb{R}^2 \cup L_\infty$ is a projective transformation then $\mathcal{P}(C)$ is also a non-degenerate conic section.

The following theorem tells us not only that we can find a projective transformation that maps any given non-degenerate conic section onto any other one but also that we can do so in such a way that any three distinct points on the first conic can be mapped onto any three distinct points on the other conic, in any order.

THEOREM 2.7.10   *Let $C, C' \subset \mathbb{R}^2 \cup L_\infty$ be non-degenerate conic sections. Let three distinct points $A, B, C \in C$ and three distinct points $A', B', C' \in C'$ be* *given. Then there exists a projective transformation* $\mathcal{P} : \mathbb{R}^2 \cup L_\infty \to \mathbb{R}^2 \cup L_\infty$ *such that* $\mathcal{P}(C) = C'$ *and* $\mathcal{P}(A) = A', \mathcal{P}(B) = B', \mathcal{P}(C) = C'$.

PROOF   This follows from [25], p. 180, using the conventions adopted here regarding points on L_∞.   □

EXERCISE 2.7.11   *(i) Show that if* $\mathcal{P} : \mathbb{R}^2 \cup L_\infty \to \mathbb{R}^2 \cup L_\infty$ *is a projective transformation with a fixed point* $Q \in \mathbb{R}^2$ *and a fixed line* $l \subset \mathbb{R}^2 \cup L_\infty$ *through* Q, *for which* $Q \in l$, $\mathcal{P}(Q) = Q$ *and* $\mathcal{P}(l) = l$, *and if* C *is a non-degenerate conic section which is tangent to* l *at* Q, *then* $\mathcal{P}(C)$ *is a non-degenerate conic section which is tangent to* l *at* Q. *(ii) Use this result to deduce the formula in Equation (2.7.4) for a family of conic sections* $\{C_\gamma : \gamma \in \mathbb{R}\}$, *which is invariant under the family of projective transformations* $\{\mathcal{P}_{a,b} : a, b \in \mathbb{R}, \ ab \neq 0\}$ *in Equation (2.7.3). Is this family* $\{C_\gamma : \gamma \in \mathbb{R}\}$ *unique? Or can you find another nontrivial family of conic sections which contains conic sections different from those in the family* $\{C_\gamma : \gamma \in \mathbb{R}\}$ *and which is invariant under* $\mathcal{P}_{a,b}$ *for all* $a, b \in \mathbb{R}$ *with* $ab \neq 0$?

In order to complete our understanding of how our special canonical family of projective transformations $\{\mathcal{P}_{a,b} : a, b \in \mathbb{R}, \ ab \neq 0\}$ acts on pictures, we mention their behaviour in the vicinity of their fixed points.

The first derivative of $\mathcal{P}_{a,b}$ at the point (x, y) is the linear operator

$$\mathcal{P}'_{a,b}(x, y) = \begin{pmatrix} \dfrac{\partial}{\partial x}\left(\dfrac{ax}{(a-1)x + (b-1)y + 1}\right) & \dfrac{\partial}{\partial y}\left(\dfrac{ax}{(a-1)x + (b-1)y + 1}\right) \\[4mm] \dfrac{\partial}{\partial x}\left(\dfrac{by}{(a-1)x + (b-1)y + 1}\right) & \dfrac{\partial}{\partial y}\left(\dfrac{by}{(a-1)x + (b-1)y + 1}\right) \end{pmatrix}$$

$$= \frac{1}{((a-1)x + (b-1)y + 1)^2} \begin{pmatrix} a(b-1)y + a & -(b-1)a \\ -(a-1)b & b(a-1)x + b \end{pmatrix}.$$

$\mathcal{P}'_{a,b}(x, y)$ governs the local behaviour of $\mathcal{P}_{a,b}(x, y)$ in the vicinity of its fixed points. At the fixed points, we find

$$\mathcal{P}'_{a,b}(0, 0) = \begin{pmatrix} a & 0 \\ 0 & b \end{pmatrix}, \qquad \mathcal{P}'_{a,b}(1, 0) = \begin{pmatrix} a^{-1} & (1-b)a^{-1} \\ 0 & ba^{-1} \end{pmatrix},$$

$$\mathcal{P}'_{a,b}(0, 1) = \begin{pmatrix} ab^{-1} & 0 \\ (1-a)b^{-1} & ab^{-1} \end{pmatrix}. \tag{2.7.6}$$

The eigenvectors of $\mathcal{P}'_{a,b}(0, 0)$ are directed along the coordinate axes, with eigenvalues a and b. So when, for example, $ab = 1$ with $0 < a < 1$ the fixed point $(0, 0)$ is **hyperbolic**: any point on the x-axis sufficiently close but not equal to $(0, 0)$ is transformed by $\mathcal{P}_{a,b}$ to a point even closer to $(0, 0)$, while any point on the y-axis sufficiently close but not equal to $(0, 0)$ is transformed by $\mathcal{P}_{a,b}$ to a point further away from $(0, 0)$. Thus, a hyperbolic fixed point is **attractive** in some directions

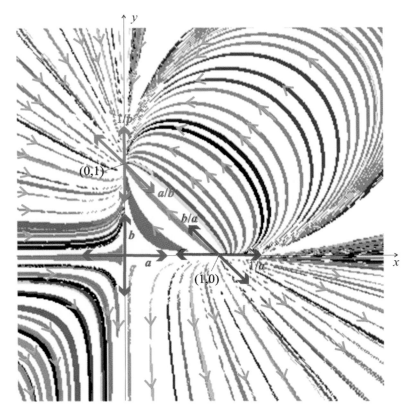

Figure 2.50   A vector field associated with the family of projective transformations $\mathcal{P}_{a,b}$. The crossed arrows represent the first-order linear approximations to $\mathcal{P}_{a,b}$ at the three fixed points, showing the eigenvalues and eigendirections. The grey arrows correspond to the case $ab = 1$ with $0 < a < 1$, and indicate the directions in which points are moved along the conic sections when $\mathcal{P}_{a,b}$ is applied.

and **repulsive** in others; upon iterative application of the transformation in the vicinity of a hyperbolic fixed point, many points follow orbits that may initially be drawn towards the fixed point but eventually are repelled by it.

The eigenvectors of $\mathcal{P}'_{a,b}(1, 0)$ are $(1, 0)^T$ and $(1, -1)^T$, with eigenvalues a^{-1} and ba^{-1} respectively. So for $ab = 1$ with $0 < a < 1$ the fixed point $(1, 0)$ is repulsive: any point sufficiently close but not equal to $(1, 0)$ is transformed by $\mathcal{P}_{a,b}$ to a point further away from $(1, 0)$.

The eigenvectors of $\mathcal{P}'_{a,b}(0, 1)$ are $(0, 1)^T$ and $(1, -1)^T$ with eigenvalues b^{-1} and ab^{-1} respectively. So for $ab = 1$ with $0 < a < 1$ the fixed point $(0, 1)$ is attractive: any point sufficiently close but not equal to $(0, 1)$ is transformed by $\mathcal{P}_{a,b}$ to a point nearer to $(0, 1)$. The situation is illustrated in Figure 2.50.

The linear operators in Equation (2.7.6) are useful in understanding how $\mathcal{P}_{a,b}$ transforms pictures. For example, suppose that $\mathfrak{P}$ is an invariant picture for $\mathcal{P}_{a,b}$. Then in the vicinity of a fixed point $\mathfrak{P}$ must look approximately like an invariant picture for the corresponding linear operator; that is, if we were to magnify $\mathfrak{P}$ in the vicinity of $(0, 0)$ we should see, within a finite fixed viewing window, what looks like part of an invariant picture for $\mathcal{P}'_{a,b}(0, 0)$.

EXERCISE 2.7.12 *What do the 'local' invariant pictures of the three linear transformations* $\mathcal{P}'_{a,b}(0, 0)$, $\mathcal{P}'_{a,b}(1, 0)$ *and* $\mathcal{P}'_{a,b}(0, 1)$ *corresponding to the invariant picture in Figure 2.48 look like?*

Projective transformations of measures

Let $\mu \in \mathbb{P}(\mathbb{R}^2 \cup L_\infty)$ be a measure that is absolutely continuous in the vicinity of a point $X \in \mathbb{R}^2$, with 'density' ρ in the vicinity of X. Let $\mathcal{P} : \mathbb{R}^2 \cup L_\infty \to \mathbb{R}^2 \cup L_\infty$ be a projective transformation such that $\mathcal{P}(X) \in \mathbb{R}^2$. Then $\mathcal{P}(\mu)$ is absolutely continuous in the vicinity of $\mathcal{P}(X)$, with density

$$\widetilde{\rho} = \frac{1}{|\det \mathcal{P}'(x, y)|} \rho, \tag{2.7.7}$$

where

$$|\det \mathcal{P}'(x, y)| = \left| \frac{\det P}{(gx + hy + j)^3} \right| \tag{2.7.8}$$

and P is the matrix, in Equation (2.7.2), which defines $\mathcal{P}$.

This remarkably simple formula, which is fun to verify, tells us that the factor by which brightness is scaled under projective transformation is constant on lines parallel to the line L_D mapped by $\mathcal{P}$ to the line at infinity. This effect is easy to see in Figures 2.15 and 2.16, which show pictures of projective transformations applied to measures. The lines of constant scaling factor are clearly seen.

Figure 2.16 shows another effect also. When the intensity value of a colour component at a pixel is scaled so that the result would be greater than 255, the result is set to the value 255; we say that the brightness becomes saturated. This saturation effect can cause colours of pixels to become distorted as they become brighter, because upon scaling one of the colour components may reach the value 255 before the others.

Figure 2.51 shows a picture that is invariant under a projective transformation $\widehat{\mathcal{P}} : \mathbb{R}^2 \cup L_\infty \to \mathbb{R}^2 \cup L_\infty$ of the form

$$\widehat{\mathcal{P}} = \mathcal{P}_{a,b} \mathcal{R} \mathcal{P}_{a,b}^{-1},$$

where $a = 1.5$, $ab = 1$ and $\mathcal{R}$ is a rotation through $\pi/5$. Figure 2.51 also shows a closely related picture of a vector of measures, each of which is invariant under $\widehat{\mathcal{P}}$.

EXERCISE 2.7.13 *Compare Figure 2.51 with Figure 2.17. How can you tell that Figure 2.51 does not represent a Möbius transformation?*

Figure 2.51 The picture on the right is invariant under a projective transformation $\widehat{\mathcal{P}} : \mathbb{R}^2 \cup L_\infty \to$ $\mathbb{R}^2 \cup L_\infty$ of the form $\widehat{\mathcal{P}} = \mathcal{P}\mathcal{R}\mathcal{P}^{-1}$, where $\mathcal{P}$ is a projective transformation and $\mathcal{R}$ is a rotation through $\pi/5$. On the left is a picture of a vector of measures, each of which is invariant under $\widehat{\mathcal{P}}$. How can you tell that $\widehat{\mathcal{P}}$ is not a Möbius transformation? See Exercise 2.7.13.

EXERCISE 2.7.14 *Simplify Equation (2.7.8) in the case where $\mathcal{P}|_{\mathbb{R}^2}$ is an affine transformation. Show that Equation (2.7.7) is consistent with Equation (2.5.4) when $\mathcal{P}|_{\mathbb{R}^2}$ is a linear transformation.*

Projective transformations on $\mathbb{RP}^2$

The most natural way to think mathematically and computationally about projective transformations is to represent them as acting on $\mathbb{RP}^2$. This greatly simplifies some aspects of understanding these transformations, though on its own it does not, in my experience, add much intuition to the specifics of how they act on pictures on $\mathbb{R}^2 \cup L_\infty$. This may be because the detailed way in which a picture is deformed, when it is mapped from a sphere to a plane, can be hard to imagine.

DEFINITION 2.7.15 The projective plane is denoted by $\mathbb{RP}^2$. It consists of the set of straight lines in $\mathbb{R}^3$ that pass through the origin of coordinates, $O = (0, 0, 0)$.

The points of $\mathbb{RP}^2$ are mathematical objects: each object consists of a set of points, namely a line, belonging to the underlying space $\mathbb{R}^3$.

EXERCISE 2.7.16 *Show that $x \in \mathbb{RP}^2$ iff there exists a point $l = (l_1, l_2, l_3) \in \mathbb{R}^3$, with $l \neq O$, such that*

$$x = \{(x_1, x_2, x_3) \in \mathbb{R}^3 : (x_1, x_2, x_3) = c \cdot (l_1, l_2, l_3) \text{ for some } c \in \mathbb{R}\}.$$

Any point $l = (l_1, l_2, l_3) \in \mathbb{R}^3$ with $l \neq O$ defines uniquely a corresponding point in $\mathbb{RP}^2$, namely the line through O and l. We denote this line by the same notation, $l = (l_1, l_2, l_3) \in \mathbb{RP}^2$. The only difference is the space to which it is asserted that the point belongs.

DEFINITION 2.7.17   A projective transformation $\widetilde{\mathcal{P}} : \mathbb{RP}^2 \to \mathbb{RP}^2$ is an invertible linear transformation $P : \mathbb{R}^3 \to \mathbb{R}^3$ treated as acting on the set of straight lines in $\mathbb{R}^3$ through O.

Thus $\widetilde{\mathcal{P}} : \mathbb{RP}^2 \to \mathbb{RP}^2$ is a projective transformation iff there exists an invertible linear transformation $P : \mathbb{R}^3 \to \mathbb{R}^3$ such that

$$\widetilde{\mathcal{P}}(l) = P(l) := \{P(x) : x \in l\}$$

for all $l \in \mathbb{RP}^2$.

EXERCISE 2.7.18   *Let* $l = (0.3, -1.2, 3.0) \in \mathbb{RP}^2$ *and* $m = (-0.33, 1.32, -3.3) \in \mathbb{RP}^2$. *Show that* $l = m$.

EXERCISE 2.7.19   *Show that the two linear transformations*

$$\begin{pmatrix} 1 & -1 & 3 \\ 2 & 0 & 1.1 \\ -10 & 0.56 & 0 \end{pmatrix} \quad and \quad \begin{pmatrix} 2.2 & -2.2 & 6.6 \\ 4.4 & 0 & 2.42 \\ -22 & 0.1232 & 0 \end{pmatrix}$$

define the same projective transformation $\widetilde{\mathcal{P}} : \mathbb{RP}^2 \to \mathbb{RP}^2$.

EXERCISE 2.7.20   *Show that any projective transformation* $\widetilde{\mathcal{P}} : \mathbb{RP}^2 \to \mathbb{RP}^2$ *maps* $\mathbb{RP}^2$ *one-to-one onto itself.*

The space $\mathbb{RP}^2$ may be imagined to look something like a pin cushion full of pins, except that the cushion itself consists of a single point, the origin, and the pins are infinitely long and infinitesimally thin; we imagine that each pin has been stuck through the cushion and out the other side, protruding to infinity in both directions. Each pin represents a single point in the projective plane. Following Theorem 2.5.4, a projective transformation produces a rescaling in three orthogonal directions of the space containing the pins, possibly followed by a reflection and/or a rotation. Some bundles of pins are squeezed more tightly while others are expanded, as illustrated in Figure 2.52.

We now describe the relationship between projective transformations acting on $\mathbb{RP}^2$ and projective transformations acting on $\mathbb{R}^2 \cup L_\infty$. Each point in $\mathbb{RP}^2$ is identified with a point in $\mathbb{R}^2 \cup L_\infty$ in a one-to-one onto manner. Each line in $\mathbb{R}^3$ through O that intersects the plane

$$\Pi_1 := \{(x, y, z) : z = 1\}$$

is identified by its point of intersection with Π_1. This represents 'most' of $\mathbb{RP}^2$ as a copy of $\mathbb{R}^2$. But some lines in $\mathbb{R}^3$ through O do not intersect Π_1, namely all the lines through O that lie in the plane

$$\Pi_0 := \{(x, y, z) : z = 0\}.$$

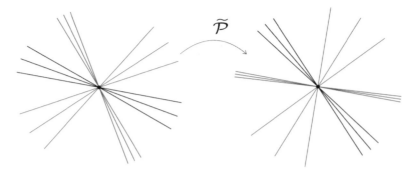

Figure 2.52 This diagram represents some bundles of lines *in three dimensions*, through the origin, before and after a linear transformation is applied. It is supposed that the lines are clustered around three orthogonal directions in which the linear transformation rescales space by constant factors, as in Theorem 2.5.4. The red lines on the left lie in a direction that is stretched by the transformation, and the same applies to the black lines. The blue lines lie in a direction that is compressed.

Furthermore, all except one of these lines intersect the line $\{(x, y) \in \Pi_0 : y = 1\}$. Accordingly, each line through O in the plane Π_0 is represented by its point of intersection with $\{(x, y) \in \Pi_0 : y = 1\}$. This leaves only the line $(1, 0, 0) \in \mathbb{RP}^2$ as so far unrepresented, and in fact it is represented by the point $\infty \in \mathbb{R}^2 \cup L_\infty$. In this way $\mathbb{RP}^2$ is represented using $\mathbb{R}^2 \cup \mathbb{R} \cup \{\infty\} = \mathbb{R}^2 \cup L_\infty$. The description of projective transformations, in the resulting new coordinate system, is exactly the one that we gave in the earlier subsection entitled 'The dance of the points'.

Specifically, the connection between the projective transformations $\widetilde{\mathcal{P}} : \mathbb{RP}^2 \to \mathbb{RP}^2$ and $\mathcal{P} : \mathbb{R}^2 \cup L_\infty \to \mathbb{R}^2 \cup L_\infty$ is provided by an invertible transformation $\mathcal{T} : \mathbb{RP}^2 \to \mathbb{R}^2 \cup L_\infty$ according to

$$\mathcal{P} = \mathcal{T} \circ \widetilde{\mathcal{P}} \circ \mathcal{T}^{-1},$$

where, for all $(l_1, l_2, l_3) \in \mathbb{RP}^2$,

$$\mathcal{T}(l_1, l_2, l_3) = \begin{cases} \left(\dfrac{l_1}{l_3}, \dfrac{l_2}{l_3} \right) \in \mathbb{R}^2 & \text{when } l_3 \neq 0, \\[2mm] \dfrac{l_1}{l_2} \in L_\infty & \text{when } l_2 \neq 0, l_3 = 0, \\[2mm] \infty \in L_\infty & \text{when } l_2 = 0, l_3 = 0. \end{cases}$$

The inverse transformation $\mathcal{T}^{-1} : \mathbb{R}^2 \cup L_\infty \to \mathbb{RP}^2$ is given by

$$\mathcal{T}^{-1}(X) = \begin{cases} (x_1, x_2, 1) \in \mathbb{RP}^2 & \text{when } X = (x_1, x_2) \in \mathbb{R}^2 \\ (x, 1, 0) \in \mathbb{RP}^2 & \text{when } X = x \in L_\infty, X \neq \infty, \\ (1, 0, 0) \in \mathbb{RP}^2 & \text{when } X = \infty, \end{cases}$$

for all $X \in \mathbb{R}^2 \cup L_\infty$. See also Figure 2.53.

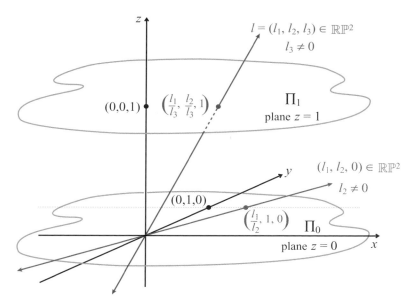

Figure 2.53   Construction of the mapping $\mathcal{T} : \mathbb{RP}^2 \to \mathbb{R}^2 \cup L_\infty$. The plane Π_1 represents $\mathbb{R}^2$.

EXERCISE 2.7.21   *Verify that $\mathcal{P} = \mathcal{T} \circ \widetilde{\mathcal{P}} \circ \mathcal{T}^{-1}$.*

EXERCISE 2.7.22   *Show that if $\mathcal{P} : \mathbb{R}^2 \cup L_\infty \to \mathbb{R}^2 \cup L_\infty$ is a projective transformation such that $\mathcal{P}(L_\infty) = L_\infty$ then $\mathcal{P}|_{\mathbb{R}^2}$, the restriction of $\mathcal{P}$ to $\mathbb{R}^2$, is an affine transformation.*

EXERCISE 2.7.23   *Show that if $\mathcal{P} : \mathbb{R}^2 \cup L_\infty \to \mathbb{R}^2 \cup L_\infty$ is such that it maps two distinct points on L_∞ to two distinct points on L_∞ then $\mathcal{P}(L_\infty) = L_\infty$.*

EXERCISE 2.7.24   *Show that a set $S \subset \mathbb{R}^2 \cup L_\infty$ is a straight line iff the set of straight lines defined by the set of points $\mathcal{T}^{-1}(S) \subset \mathbb{RP}^2$ defines a plane in $\mathbb{R}^3$. If S is the straight line in $\mathbb{R}^2 \cup L_\infty$ defined by $lx + my + n = 0$ with $n \neq 0$, what is the equation for the set of points in $\mathbb{R}^3$ that lie in the plane defined by $\mathcal{T}^{-1}(S)$?*

EXERCISE 2.7.25   *(i) Let $\mathbb{L}$ denote the set of straight lines in $\mathbb{X} = \mathbb{R}^2 \cup L_\infty$. Show that we can define an invertible mapping $Z : \mathbb{X} \to \mathbb{L}$ by*

$$Z(X) = \mathcal{T}\left(\{l \in \mathbb{RP}^2 : l \in \mathbb{R}^3; l \perp \mathcal{T}^{-1}(X)\}\right),$$

for all $X \in \mathbb{X}$, where $\perp$ means 'is perpendicular to' and $\mathcal{T}^{-1}(X)$ is treated as a line in $\mathbb{R}^3$. Describe the mapping $Z^{-1} : \mathbb{L} \to \mathbb{X}$.

(ii) Define $\widetilde{Z} : \mathbb{X} \cup \mathbb{L} \to \mathbb{X} \cup \mathbb{L}$ *by* $\widetilde{Z}(X) = Z(X)$ *when* $X \in \mathbb{X}$, $\widetilde{Z}(X) =$ $Z^{-1}(X)$ *when* $X \in \mathbb{L}$. *Show that* $\widetilde{Z}$ *is one-to-one and onto and has the following remarkable pair of properties: (a) if* $l_1, l_2 \in \mathbb{L}$, *with* $l_1 \neq l_2$, *intersect at the point* $p \in \mathbb{X}$ *then* $\widetilde{Z}(l_1), \widetilde{Z}(l_2) \in \mathbb{X}$ *are two distinct points that lie on the line* $\widetilde{Z}(p) \in \mathbb{L}$; *(b) if* $p_1, p_2 \in \mathbb{X}$, *with* $p_1 \neq p_2$, *lie on the line* l *then* $\widetilde{Z}(p_1), \widetilde{Z}(p_2) \in \mathbb{L}$ *are two distinct lines that intersect at the point* $\widetilde{Z}(l) \in \mathbb{X}$.

The mapping $\widetilde{Z}$ constructed in Exercise 2.7.25 is an example of a **duality** transformation. It can sometimes be used to convert the objects in a theorem that concerns straight lines, points and intersections in the projective plane into new objects, thereby yielding a new theorem.

Representation of the projective plane on a sphere and on a disk

Another way of representing $\mathbb{R}^2 \cup L_\infty$, or equivalently $\mathbb{RP}^2$, that reveals a natural topology for the projective plane is to describe each point $l \in \mathbb{RP}^2$ as the *pair of points* for which the corresponding line $l \in \mathbb{R}^3$ intersects the surface $\widehat{\mathbb{S}}$ of the sphere of radius 1 having its centre at O. In particular, a natural metric $d_{\mathbb{RP}^2}$ on $\mathbb{RP}^2$ is obtained by defining $d_{\mathbb{RP}^2}(l, l')$ to be the shortest distance, on $\widehat{\mathbb{S}}$, between the pair of points that represents l and and the pair of points that represents l'. The natural topology of $(\mathbb{RP}^2, d_{\mathbb{RP}^2})$ is the identification topology on $(\widehat{\mathbb{S}}, d_{euclidean})$ induced by mapping pairs of points lying on the same line through the centre to the same point. Clearly $(\mathbb{RP}^2, d_{\mathbb{RP}^2})$ is a compact metric space.

The behaviour of a projective transformation $\mathcal{P} : \mathbb{R}^2 \cup L_\infty \to \mathbb{R}^2 \cup L_\infty$ may be thought of in terms of the action, on the sphere $\widehat{\mathbb{S}}$, of the corresponding linear transformation $P : \mathbb{R}^3 \to \mathbb{R}^3$, as illustrated in Figure 2.54. It is clear, from this point of view, that any projective transformation is continuous with respect to the metric $d_{\mathbb{RP}^2}$.

In place of using two points on the spherical shell $\widehat{\mathbb{S}}$ to represent a single point of $\mathbb{RP}^2$ we can use just one of the points, say the one on the upper hemisphere. The only slight difficulty is that points on the equator, that is, on the intersection of $\widehat{\mathbb{S}}$ with the plane $z = 0$, are double points. So we must omit exactly half this circle. Then we see that we can represent $\mathbb{RP}^2$ by the set of points

$$\mathbb{S}_+ = \{(x, y, z) : x^2 + y^2 + z^2 = 1, z > 0\}$$
$$\cup \{(x, y, 0) : x^2 + y^2 = 1, y > 0\} \cup \{(1, 0, 0)\}.$$

Notice that we can define unique coordinates for $\mathbb{S}_+$ by using the points of

$$\mathbb{D}_+ := \{(x, y) \in \mathbb{R}^2 : (x, y, z) \in \mathbb{S}_+ \text{ for some } z \in \mathbb{R}^3\}.$$

The set $\mathbb{D}_+$ is just the orthogonal projection of $\mathbb{S}_+$ onto the xy-plane; it consists of the interior of the unit circle centred at O together with half the unit circle. The corresponding invertible mapping $T : \mathbb{R}^2 \cup L_\infty \to \mathbb{D}_+$ is illustrated in

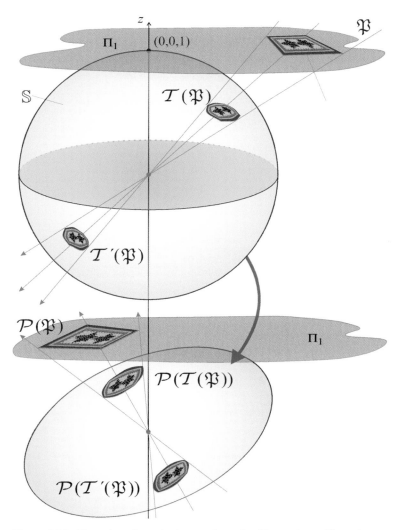

Figure 2.54 The action of a projective transformation $\mathcal{P}$ on a picture $\mathfrak{P}$ may be expressed in terms of how the corresponding linear transformation $P : \mathbb{R}^3 \to \mathbb{R}^3$ acts on the sphere $\widehat{\mathbb{S}}$. The plane Π_1 defined by $z = 1$ represents the space $\mathbb{R}^2$ in which $\mathfrak{P}$ lies. First $\mathfrak{P}$ is transformed into two pictures $\mathcal{T}(\mathfrak{P})$ and $\mathcal{T}'(\mathfrak{P})$ by central projection onto $\widehat{\mathbb{S}}$. Then a linear transformation $P : \mathbb{R}^3 \to \mathbb{R}^3$ is applied to the sphere and the two pictures on it, to yield two pictures on an ellipsoid. Finally, either of these pictures is projected back onto Π_1. It is always possible to choose the linear transformation P in such a way that the final result, back on Π_1, is $\mathcal{P}(\mathfrak{P})$.

Figure 2.55: it is defined by

$$T(x, y) = \left(\frac{x}{\sqrt{x^2 + y^2 + 1}}, \ \frac{y}{\sqrt{x^2 + y^2 + 1}} \right) \quad \text{for all } (x, y) \in \mathbb{R}^2, \quad (2.7.9)$$

while

$$T(x) = \left(\frac{x}{\sqrt{1 + x^2}}, \ \frac{1}{\sqrt{1 + x^2}} \right) \quad \text{for all } x \in L_\infty \backslash \{\infty\} \quad (2.7.10)$$

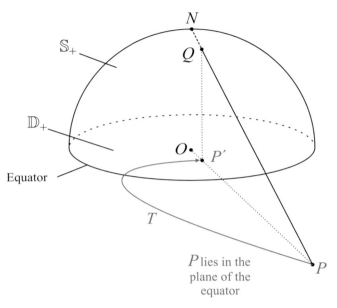

Figure 2.55   Construction of the invertible mapping $T : \mathbb{R}^2 \cup L_\infty \to \mathbb{D}_+$ described in the text, in Equations (2.7.9)–(2.7.11). Where are $T(L_\infty)$ and $T(\infty)$?

and

$$T(\infty) = (1, 0). \tag{2.7.11}$$

EXERCISE 2.7.26   *Verify that the inverse of the transformation T is given by*

$$T^{-1}(x, y) = \left(\frac{x}{\sqrt{1 - x^2 - y^2}}, \; \frac{y}{\sqrt{1 - x^2 - y^2}} \right)$$

for all $(x, y) \in \mathbb{D}_+$ such that $x^2 + y^2 < 1$. Calculate $T^{-1}\left(\frac{3}{5}, \frac{4}{5}\right)$.

In Figure 2.56 we show the result of applying the transformation $T : \mathbb{R}^2 \cup L_\infty \to \mathbb{D}_+$, defined in Equations (2.7.9)–(2.7.11), to pictures of periodic tilings of $\mathbb{R}^2$ by square tiles, where each tile has a white border and a black square in the middle. The sides of the tiles and of the black squares run parallel to the coordinate axes. Images of the straight lines formed by the boundaries of the tiles that are parallel to the x-axis seem to meet at a single point on $\mathbb{D}_+$. This meeting point is actually $T(\infty)$. Similarly, images of lines parallel to the y-axis meet at $T(0)$. Remember that the disk that represents $\mathbb{D}_+$ possesses only half its circular boundary.

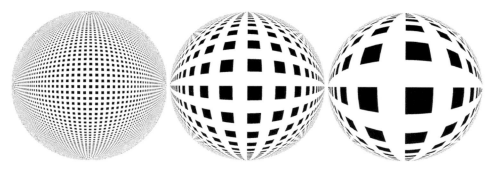

Figure 2.56 Here three regular arrays of square tiles, with their sides parallel to the coordinate axes, have been transformed by $T : \mathbb{R}^2 \to \mathbb{D}_+$ to produce three pictures. The different pictures correspond, from left to right, to successively larger tiles. Why do the images of parallel lines of tiles, in the plane, seem to converge to the same meeting point on $\mathbb{D}_+$?

We readily find that $T : \mathbb{R}^2 \to \mathbb{D}_+$ maps the straight line $y = c, c \in \mathbb{R}$, into the ellipse

$$x^2 + \left(1 + \frac{1}{c^2}\right) y^2 = 1.$$

This family of ellipses meets at the point $(1, 0) \in \mathbb{D}_+$. You should be able to spot illustrations of parts of this family of ellipses in Figure 2.56.

EXERCISE 2.7.27 *Show that $T : \mathbb{R}^2 \cup L_\infty \to \mathbb{D}_+$ maps the straight line given by*

$$lx + my + n = 0$$

into the conic section

$$(l^2 + n^2)x^2 + 2lmxy + (m^2 + n^2)y^2 - n^2 = 0.$$

Make a sketch of some of these ellipses for l and m fixed and several values of n.

In the left-hand panel of Figure 2.57 a regular array of pixels has been mapped by T onto $\mathbb{D}_+$. The right-hand panel shows the result of applying $T \circ L$ to the same array, where L is the linear transformation $L(x, y) = (2x, 2y)$. Notice how the line at infinity, represented by the boundary of the disk, remains fixed, the major axes of certain families of ellipses point to the same places and the picture material is squeezed out towards the boundary.

In place of looking at how a projective transformation $\mathcal{P} : \mathbb{R}^2 \cup L_\infty \to \mathbb{R}^2 \cup L_\infty$ acts upon a picture $\mathfrak{P}$ that has its domain in $\mathbb{R}^2$ we can instead look at how the conjugate transformation $\widetilde{\widetilde{\mathcal{P}}} : \mathbb{D}_+ \to \mathbb{D}_+$ defined by

$$\widetilde{\widetilde{\mathcal{P}}} := T\mathcal{P}T^{-1} \tag{2.7.12}$$

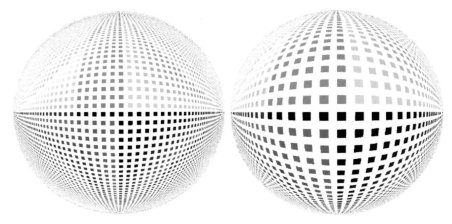

Figure 2.57　The left-hand panel shows the result of mapping an array of pixels onto $\mathbb{D}_+$. The right-hand panel shows the result of the same mapping after the dimensions of the domains of the pixels have been doubled.

acts upon a picture whose domain lies in $\mathbb{D}_+$. Specifically we find, for all $(x, y) \in \mathbb{D}_+$, that

$$\widetilde{\widetilde{\mathcal{P}}}(x, y) =$$
$$\left(\frac{(ax + by + eF(x, y)) \operatorname{sgn}(gx + hy + jF(x, y))}{\left\{ (ax+by+eF(x, y))^2 + (cx+dy+fF(x, y))^2 + (gx+hy+jF(x, y))^2 \right\}^{1/2}}, \right.$$
$$\left. \frac{(cx + dy + eF(x, y)) \operatorname{sgn}(gx + hy + jF(x, y))}{\left\{ (ax+by+eF(x, y))^2 + (cx+dy+fF(x, y))^2 + (gx+hy+jF(x, y))^2 \right\}^{1/2}} \right),$$

where

$$F(x, y) = \sqrt{1 - x^2 - y^2},$$

the underlying linear transformation is that given in Equation (2.7.2) and the function sgn is defined by

$$\operatorname{sgn}(x) = \begin{cases} +1 & \text{when } x \geq 0, \\ -1 & \text{when } x < 0. \end{cases}$$

Examples of the transformation $\widetilde{\widetilde{\mathcal{P}}} : \mathbb{D}_+ \to \mathbb{D}_+$, applied directly to pictures with domain $\mathbb{D}_+$, are illustrated in Figures 2.58–2.60. In each case the original picture is on the left, the transformed picture is on the right and the underlying linear transformation is the same, namely

$$\begin{pmatrix} 1.0 & -1.0 & 0.0 \\ 0.0 & 1.0 & 0.0 \\ 1.5 & 1.0 & -1.0 \end{pmatrix}.$$

Figure 2.58   The picture $\mathfrak{P}$ on the left represents, on $\mathbb{D}_+$, a regular array of pixels. The picture on the right shows the result of applying a tranformation $\widetilde{\widetilde{\mathcal{P}}} : \mathbb{D}_+ \to \mathbb{D}_+$ that is conjugate to a projective transformation. Notice how in the left-hand panel the pixels are squeezed towards the outer boundary of the disk which represents $\mathbb{D}_+$, but in the right-hand panel they are squeezed towards a smooth curve lying mainly in the interior of the disk.

Figure 2.59   A transformation $\widetilde{\widetilde{\mathcal{P}}} : \mathbb{D}_+ \to \mathbb{D}_+$, conjugate to a projective transformation acting on $\mathbb{R}^2 \cup L_\infty$, is applied to a picture whose domain is $\mathbb{D}_+$. Notice the lovely stretching lines and how the part of the picture on the boundary of $\mathbb{D}_+$ is mapped to two sides of an internal smooth curve.

We can see how a projective transformation acts on a picture $\mathfrak{P}$, especially in relation to L_∞, by comparing the pictures $\mathfrak{P}$, $T(\mathfrak{P})$, $\widetilde{\widetilde{\mathcal{P}}}(T(\mathfrak{P}))$ and $T^{-1}\big(\widetilde{\widetilde{\mathcal{P}}}(T(\mathfrak{P}))\big) = \mathcal{P}(\mathfrak{P})$. The relationship of $\mathfrak{P}$, $\mathcal{P}(\mathfrak{P})$ and L_∞ is conjugate to the relationship of $T(\mathfrak{P})$, $\widetilde{\widetilde{\mathcal{P}}}(T(\mathfrak{P}))$ and the boundary $\mathbb{D}_+$.

Figure 2.60 Here a projective transformation, represented as acting directly on $\mathbb{D}_+$, is applied to a picture, on the left, of a texture of foliage and branches. The result, on the right, is a very different looking kind of texture.

For example, in Figure 2.61 we illustrate the effect of the projective transformation associated with the linear transformation

$$P = \begin{pmatrix} -0.022998 & 0.118622 & -0.044233 \\ -0.001860 & 0.115182 & -0.048586 \\ -0.004235 & 0.237767 & -0.109326 \end{pmatrix}$$

acting on a picture $\mathfrak{P}$, in the top left panel, of nine cartoon trees. The tree at the centre is located in the vicinity of the origin. The top right panel shows the picture $T(\mathfrak{P})$; notice how the central tree is not much deformed but the other trees are squeezed against the boundary of $\mathbb{D}_+$. The picture $\widetilde{\widetilde{\mathcal{P}}}(T(\mathfrak{P}))$ is shown in the bottom right panel; the effect of $\widetilde{\widetilde{\mathcal{P}}}$ has been to reflect the picture $T(\mathfrak{P})$ about a horizontal line and then to displace the result, so that it seems to have slid off the disk across the top boundary of $\mathbb{D}_+$ and to have reappeared, with the orientation reversed, from across the bottom boundary. The bottom left panel shows $T^{-1}\big(\widetilde{\widetilde{\mathcal{P}}}(T(\mathfrak{P}))\big) = \mathcal{P}(\mathfrak{P})$; we can think of the picture $\mathfrak{P}$ as having been reflected in a horizontal line then slid off the euclidean plane, L_∞, at the top of the picture and slid from L_∞ back into view, with reversed orientation, at the bottom.

Figures 2.62 and 2.63 illustrate exactly the same sequence of transformations, but applied to different pictures. Each picture emphasizes different aspects of the same transformations. For example, notice how in Figure 2.62 the fish picture $T(\mathfrak{P})$ seems to nearly fill $\mathbb{D}_+$, while the flowers in the picture $\mathfrak{P}$ in Figure 2.63 are transformed by $\mathcal{P}$ to be closer together, no longer separated by the birds.

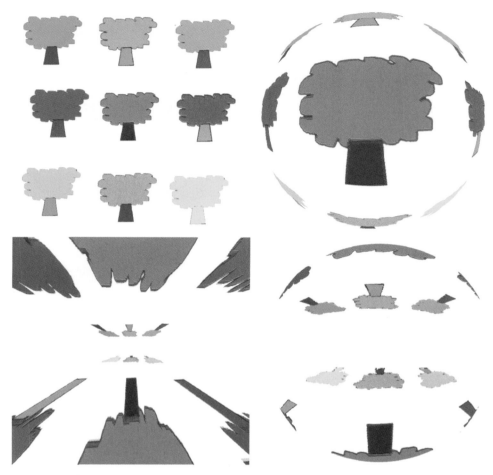

Figure 2.61 A picture $\mathfrak{P}$, top left, is mapped onto a disk, $\mathbb{D}_+$, at the top right. Then a transformation of the form given in Equation (2.7.12) is applied to produce the picture at the bottom right. This picture is mapped back onto the euclidean plane, yielding, at bottom left, a projective transformation $\mathcal{P}(\mathfrak{P})$ of the original picture $\mathfrak{P}$. Notice how the pictures on the right look something like the pictures on the left wrapped and stretched over a hemispherical shell – you can almost see the convexity of the hemisphere. The line at infinity corresponds, in the pictures on the right, to the boundaries of the disks.

The cross-ratio

Projective transformations do not in general preserve lengths, ratios of lengths or angles. But they always preserve cross-ratios.

DEFINITION 2.7.28 The **cross-ratio** of a sequence of four distinct collinear points $A, B, C, D \in \mathbb{R}^2 \cup L_\infty$ is a unique real number, which may be computed as follows. If $A, B, C, D \in \mathbb{R}^2$ then write

$$A = a \cdot (e_1, e_2) + (t_1, t_2), \quad B = b \cdot (e_1, e_2) + (t_1, t_2),$$
$$C = c \cdot (e_1, e_2) + (t_1, t_2), \quad D = d \cdot (e_1, e_2) + (t_1, t_2),$$

where $a, b, c, d \in \mathbb{R}$, $(e_1, e_2) \in \mathbb{R}^2$, $(t_1, t_2) \in \mathbb{R}^2$ and $e_1^2 + e_2^2 = 1$;

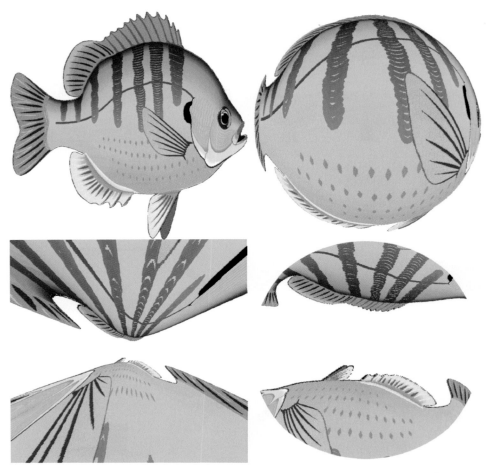

Figure 2.62 The two pictures on the left are related by a projective transformation $\mathcal{P} : \mathbb{R}^2 \cup L_\infty \to \mathbb{R}^2 \cup L_\infty$. Each picture on the right is the transform of the picture to its left under the transformation $T : \mathbb{R}^2 \cup L_\infty \to \mathbb{D}_+$. See also Figures 2.61 and 2.63. Where is the fish's eye?

then we have

$$\text{cross-ratio}(A, B, C, D) = \frac{c - a}{b - c} \frac{b - d}{d - a}.$$

When one of the points A, B, C, or D, lies on L_∞ the cross-ratio is given by the same formula in the limit as $a, b, c,$ or d respectively tends to infinity.

For example, if the four points lie on the x-axis with x-coordinates $a = 1$, $b = 2.3$, $c = -1$, $d = 10$ then the cross-ratio is

$$\frac{-1 - 1}{2.3 - (-1)} \frac{2.3 - 10}{10 - 1} = \frac{13.4}{29.7}.$$

In the limit as d tends to infinity the cross-ratio becomes

$$\frac{-1-1}{2.3-(-1)}(-1)=\frac{2}{3.3}.$$

EXERCISE 2.7.29 *Find the cross-ratio of the sequence of points* $(0,0)$, $(0.3, 0.4)$, $(1.2, 1.6)$, $(-1.2, -1.6)$.

THEOREM 2.7.30 *Let* $\mathcal{P} : \mathbb{R}^2 \cup L_\infty \rightarrow \mathbb{R}^2 \cup L_\infty$ *be a projective transformation and let* $A, B, C, D \in \mathbb{R}^2 \cup L_\infty$ *be a sequence of four distinct collinear points. Then*

$$\text{cross-ratio}(A, B, C, D) = \text{cross-ratio}(\mathcal{P}(A), \mathcal{P}(B), \mathcal{P}(C), \mathcal{P}(D)).$$

PROOF   See [25], p. 141.   □

EXERCISE 2.7.31 *Verify Theorem 2.7.30 explicitly when the four points* a, b, c, d *lie on the x-axis and* $\mathcal{P}$ *is an interesting projective transformation, which you choose.*

Does the cross-ratio correspond to some property of pictures you can somehow 'see', say in a picture that contains a line of equally spaced fence posts or four windows in a row on the front of a house? If the cross-ratio of four distinct collinear points, belonging to a picture of four copies of the same, is $\frac{4}{3}$ or one of the numbers $\{4, -3, \frac{1}{4}, -\frac{1}{3}, \frac{3}{4}\}$, depending upon the order in which the points are taken, then the answer, suitably qualified, may be positive, for these cross-ratios correspond to sets of points that can be transformed into a row of equally spaced points on the x-axis. The 'unique fourth-point theorem', [25], p. 141, tells us that if we know cross-ratio(A, B, C, D) and the locations of A, B and C then the location of D is uniquely determined. So, by looking at a picture containing three points $A, B, C \in \mathbb{R}^2$, can you locate by eye the point D such that the four points A, B, C, D could be approximately the result of applying a projective transformation to the points $a = 0, b = 1, c = 2, d = 3$ on the x-axis? Try it, in Figure 2.64, then calculate cross-ratio(A, B, C, D). Draw your own conclusions.

The theorems of geometry

We have introduced projective transformations and Möbius transformations. We have mentioned some basic deep geometry results, such as Theorems 2.7.4, 2.7.10, and 2.7.30, the cross-ratio theorem, and illustrated to some extent what they mean for sets, measures and pictures. But there are many more results that we have not mentioned, such as Steiner's porism, Pappus' theorem, La Hire's theorem, the three tangents and three chords theorem and so on.

Figure 2.63 See also Figures 2.61 and 2.62. Here a picture of flowers and birds, represented both on $\mathbb{R}^2$ and on $\mathbb{D}_+$, is transformed by a projective transformation. Here and in Figures 2.61 and 2.62 some picture information has been lost owing to numerical effects arising because points have been mapped too close to L_∞ or to the boundary of $\mathbb{D}_+$. Can you find some examples of such information loss?

This introduction of ours may serve as an invitation to you to consider afresh the existing body of theorems. Begin by reading or rereading a good work on geometry, such as the very practical book [25], the more abstract books by Coxeter [26] and by Berger [23] or the good brief historical review of geometry in *Encylopaedia Brittanica* [94]. Take theorems from such sources as appeal to you and think about what they may say explicitly and specifically about transformations of pictures, over and above what they say about transformations of points, lines, planes and conics. What do they not tell you? What things that you might think are obvious from a visual colourful point of view, what visual intuitions that have not yet been captured by mathematics do these theorems suggest but not prove?

Figure 2.64 Using your intuition alone, can you sketch the locations of the missing windows? Do so, then calculate some cross-ratios. Are your answers close to 1.33?

2.8 Transformations on code spaces

Here we describe some simple transformations, on code spaces, that are relevant to fractals. These transformations are continuous and they can be interpreted geometrically in terms of affine transformations acting on trees in $\mathbb{R}^2$. Moreover, remarkably, they preserve a quantity – quite unlike angle, length, or cross-ratio – that is related to information-carrying capacity. This fact enriches our theme of the connections between code spaces and meristems.

Recall that we introduced the code spaces $\Omega_\mathcal{A}$ and $\Omega'_\mathcal{A}$ in Chapter 1. Also, recall that we can think of $\Omega_{\{0,1\}} \cup \Omega'_{\{0,1\}}$ as a tree-like structure, which in this section we call simply a tree, embedded in $\mathbb{R}^2$; finite strings of zeros and ones, points of $\Omega'_{\{0,1\}}$, are represented by the nodes of the tree while the points of $\Omega_{\{0,1\}}$, infinite strings of zeros and ones, correspond to the tips of the twigs, the canopy of the tree, as illustrated in Figure 2.65.

Let $\omega \in \Omega'_\mathcal{A}$. Then we define the **branch transformation** $f_\omega : \Omega_\mathcal{A} \cup \Omega'_\mathcal{A} \to \Omega_\mathcal{A} \cup \Omega'_\mathcal{A}$ by

$$f_\omega(\sigma) = \omega\sigma \quad \text{for all } \sigma \in \Omega_\mathcal{A} \cup \Omega'_\mathcal{A}.$$

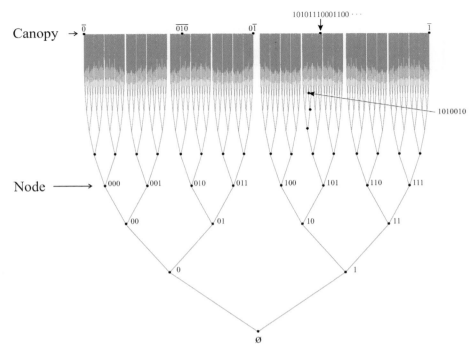

Figure 2.65 The code space $\Omega_{\{0,1\}} \cup \Omega'_{\{0,1\}}$ can be thought of as a tree whose nodes represent $\Omega'_{\{0,1\}}$ and whose canopy represents $\Omega_{\{0,1\}}$. The points with addresses $\bar{0}$, $\bar{1}$ and $\overline{010}$ are examples of periodic points for the shift transformation.

For example,

$$f_1(0100101 \cdots) = 10100101 \cdots$$

It is easy to see that f_ω is one-to-one and continuous with respect to the natural topology. Also $f_\omega(\Omega_A) \subset \Omega_A$ (and $f_\omega(\Omega'_A) \subset \Omega'_A$), so we can restrict f_ω to Ω_A. We denote this restricted transformation by $f_\omega : \Omega_A \to \Omega_A$. In terms of the tree representation, f_ω maps the whole tree onto the branch at the node ω, as illustrated in Figure 2.66(iii). In this representation the transformation is a similitude acting on $\mathbb{R}^2$.

EXERCISE 2.8.1   *Let $\omega, \nu \in \Omega'_A$. Prove that $f_\omega \circ f_\nu = f_{\omega\nu}$.*

EXERCISE 2.8.2   *Let I denote $[0, 1] \subset \mathbb{R}^2$ minus all the points that possess two binary addresses (cf. Exercise 1.4.3). What does f_ω look like geometrically when interpreted as acting on I? What happens if you try to include points with two binary addresses?*

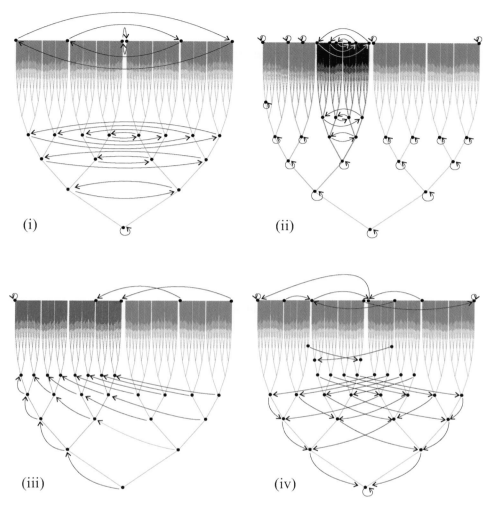

Figure 2.66 Examples of continuous transformations on the code spaces $\Omega_{\{0,1\}} \cup \Omega'_{\{0,1\}}$ represented as transformations acting on a tree in $\mathbb{R}^2$. In (iii) the branch transformation f_1 behaves like a similitude. The shift transformation S is illustrated in (iv). The flip transformations $flip_\varnothing$ and $flip_{01}$ are illustrated in (i) and (ii).

The **shift transformation** $S : \Omega_{\mathcal{A}} \cup \Omega'_{\mathcal{A}} \to \Omega_{\mathcal{A}} \cup \Omega'_{\mathcal{A}}$ is defined by

$$S(\sigma) = \begin{cases} \sigma_2\sigma_3 \cdots \in \Omega_{\mathcal{A}} & \text{when } \sigma = \sigma_1\sigma_2\sigma_3 \cdots \in \Omega_{\mathcal{A}}, \\ \sigma_2\sigma_3 \cdots \sigma_k \in \Omega'_{\mathcal{A}} & \text{when } \sigma = \sigma_1\sigma_2 \cdots \sigma_k \in \Omega'_{\mathcal{A}}, \end{cases}$$

and $S(\varnothing) = \varnothing$. S is continuous with respect to the natural topology. Also, when $|\mathcal{A}| > 1$ it is many-to-one. Indeed, when $\mathcal{A} = \{0, 1\}$ we have

$$S^{-1}(\sigma) = \{f_0(\sigma), f_1(\sigma)\}$$

for all $\sigma \in \Omega_A \cup \Omega'_A \setminus \{\varnothing\}$. We might say that f_0 and f_1 are *branches* of the inverse of S.

The shift transformation is illustrated in Figure 2.66(iv).

EXERCISE 2.8.3   *Show that, when* $\mathcal{A} = \{0, 1\}$,

$$(S \circ S)^{-1}(\sigma) = \{f_{00}(\sigma), f_{01}(\sigma), f_{10}(\sigma), f_{11}(\sigma)\}$$

for all $\sigma \in \Omega_A$.

EXERCISE 2.8.4   *Describe the shift transformation* $S : \Omega_{\{0,1\}} \to \Omega_{\{0,1\}}$ *in terms of arithmetical operations on the point* $0.\sigma_1\sigma_2\sigma_3 \cdots \in [0, 1] \subset \mathbb{R}^2$ *corresponding to* $\sigma \in \Omega_{\{0,1\}}$.

A subset $\Omega \subset \Omega_A$ is called a **shift-invariant subspace** of Ω_A if

$$S(\Omega) = \Omega.$$

Note that a shift-invariant subspace Ω of Ω_A need not be, according to our terminology, an invariant set for $S : \Omega_A \to \Omega_A$ because we have not required that $S^{-1}(\Omega) = \Omega$. However, it is an invariant set for $S|_\Omega : \Omega \to \Omega$.

There are various types of shift-invariant subspaces. They relate to geometrical aspects of fractals as well as to information theory. They include: (i) subsets of Ω_A defined in terms of the periodic points of S, see below; (ii) sets of points defined via topological entropy, see below; (iii) sets of points defined via fractal tops, discussed in Chapter 4; (iv) sets of points defined via stationary Markov processes; see for example [88].

We need the concept of periodic points elsewhere, so we define it generally here.

DEFINITION 2.8.5   Let $\mathbb{X}$ be a space and let $T : \mathbb{X} \to \mathbb{X}$ be a transformation. A point $p \in \mathbb{X}$ is called a **periodic point** of T of **period** $n \geq 1$, $n \in \mathbb{N}$, iff $T^{\circ n}(p) = p$ and $T^{\circ k}(p) \neq p$ for all $k = 1, 2, \ldots, n - 1$, where

$$T^{\circ k}(p) := \underbrace{T(T(\cdots(T(p))\cdots))}_{k \text{ times}}.$$

A periodic point of T of period 1 is simply a fixed point of T.

Examples of periodic points of the shift transformation $S : \Omega_{\{0,1\}} \to \Omega_{\{0,1\}}$ are

$$\overline{0} = 000 \cdots, \quad \overline{010} = 010010010 \cdots \text{ and } \overline{01 \cdots 11} = 01 \cdots 1101 \cdots 1101 \cdots.$$

Here $\overline{0}$ is of period 1, $\overline{010}$ is of period 3 and $\overline{01 \cdots 11}$ is of the same period as the number of symbols under the bar. Examples of shift-invariant subspaces of $S : \Omega_{\{0,1\}} \to \Omega_{\{0,1\}}$ are the set of all periodic points of S and the set of points $\{p, S(p), S^{\circ 2}(p), \ldots, S^{\circ n-1}(p)\}$, where $p \in \Omega_{\{0,1\}}$ is a periodic point of S of period $n \geq 1$. Figure 2.65 includes illustrations of some periodic points.

Other types of shift-invariant subspace can be defined using a quantity that is related to information theory.

DEFINITION 2.8.6   Let $\sigma \in \Omega_\mathcal{A}$ and $k \in \{1, 2, \ldots\}$. A **substring** of length k of σ is a point $\sigma' \in \Omega'_\mathcal{A}$ such that $|\sigma'| = k$ and there exists $n \in \mathbb{Z}^+$ for which $\sigma'_i = \sigma_{i+n}$ for all $i = 1, 2, \ldots, k$.

DEFINITION 2.8.7   The **topological entropy** of a point $\sigma \in \Omega_\mathcal{A}$ is defined to be

$$h(\sigma) = \lim_{k \to \infty} \frac{1}{k} \log_2 |U_k(\sigma)|,$$

where $U_k(\sigma)$ is the set of all distinct substrings of σ of length k.

The following theorem tells us that this definition works.

THEOREM 2.8.8   *The limit $h(\sigma)$ exists for all $\sigma \in \Omega_\mathcal{A}$.*

PROOF   This follows from the observation

$$|U_{mk}(\sigma)| \leq |U_m(\sigma)|\,|U_k(\sigma)|.$$

See [88], p. 132. $\square$

The topological entropy of a point $\sigma \in \Omega_\mathcal{A}$ is a measure of the **information-carrying capacity** of the string σ. It takes account of the diversity of substrings of σ but not of the relative frequencies of occurrence of the different substrings.

Let $\widehat{\sigma} \in \Omega_\mathcal{A}$ be such that all the elements of $\Omega'_\mathcal{A}$ are substrings of $\widehat{\sigma}$. Then clearly, on the one hand,

$$h(\sigma) \leq h(\widehat{\sigma}) = \lim_{k \to \infty} \frac{1}{k} \log_2 |\mathcal{A}|^k = \log_2 |\mathcal{A}| \quad \text{for all } \sigma \in \Omega_\mathcal{A}.$$

On the other hand, if $s \in \mathcal{A}$ then

$$h(\overline{s}) = 0$$

because $\overline{s}$ contains only one substring of length k for each $k \in \mathbb{Z}^+$. The following theorem tells us that the topological entropy is invariant both under shift transformations and branch transformations.

THEOREM 2.8.9   *Let $S : \Omega_\mathcal{A} \to \Omega_\mathcal{A}$ be a shift transformation. Then*

$$h(S(\sigma)) = h(\sigma) \quad \text{for all } \sigma \in \Omega_\mathcal{A}.$$

Let $\omega \in \Omega'_\mathcal{A}$ and let $f_\omega : \Omega_\mathcal{A} \to \Omega_\mathcal{A}$ be a branch transformation. Then

$$h(f_\omega(\sigma)) = h(\sigma) \quad \text{for all } \sigma \in \Omega_\mathcal{A}.$$

PROOF   Let $\sigma = \sigma_1 \sigma_2 \sigma_3 \cdots \in \Omega_\mathcal{A}$ and $k \in \{2, 3, \ldots\}$. Then

$$|U_k(\sigma_2 \sigma_3 \cdots)| \leq |U_k(\sigma_1 \sigma_2 \sigma_3 \cdots)| \leq 1 + |U_k(\sigma_2 \sigma_3 \cdots)|$$

so we have

$$\lim_{k\to\infty}\frac{1}{k}\log_2|U_k(\sigma_2\sigma_3\cdots)|\le h(\sigma)\le\lim_{k\to\infty}\frac{1}{k}\log_2(1+|U_k(\sigma_2\sigma_3\cdots)|).$$

We now use $x=|U_k(\sigma_2\sigma_3\ldots)|$ in the estimate

$$\log_2 x\le\log_2(1+x)<\log_2 x+\frac{1}{x\ln 2},$$

which is valid for $x\ge 1$. From this it follows that both sides converge to $h(\sigma_2\sigma_3\cdots)=h(S(\sigma))$. It follows that

$$h(f_\omega(\sigma))=h(\omega\sigma)=h\big(S^{\circ|\omega|}(\omega\sigma)\big)=h(\sigma).$$

$\square$

We remark that other transformations, which increase the topological entropy of the points upon which they act, are used in data compression. Such transformations are much harsher and are related to transformations that change fractal dimension. They are discussed in Section 4.15.

EXERCISE 2.8.10 *Show that the set of points $\sigma\in\Omega_A$ such that $h(\sigma)=1.3$ is a shift-invariant subset of Ω_A.*

EXERCISE 2.8.11 *Estimate the topological entropy of the point*

$$\sigma=0100100010101010000101000101010001010100010000000010\cdots\in\Omega_{\{0,1\}}$$

wherein the symbol 0 always follows the symbol 1.

The shift transformation admits diverse invariant measures. For example, let $p\in\Omega_A$ be a periodic point of S of period n. Let $\mu_p\in\mathbb{P}(\Omega_A)$ denote a measure that assigns mass $1/n$ to each point in the set $\{p,S(p),S^{\circ 2}(p),\ldots,S^{\circ(n-1)}(p)\}$, called the orbit of p, and zero mass to the complement of the orbit of p. Then μ_p is invariant under the shift transformation. The measure described in Example 2.3.13 is also invariant under the shift transformation.

We note that $S:\Omega_A\to\Omega_A$ is not invertible and so does not admit invariant 'pictures', that is, picture functions whose domains lie in Ω_A rather than in say $\mathbb{R}^2$. However, the closely related transformation $S:\Omega_A^2\to\Omega_A^2$, where $\Omega_A^2=\Omega_A\times\Omega_A$ is the code space of doubly infinite sequences of symbols from the alphabet A, defined by

$$S(\sigma,\omega)=(S\sigma,\sigma_1\omega)\quad\text{for all }(\sigma,\omega)\in\Omega_A^2,$$

is continuous and invertible. This transformation may be represented in $\mathbb{R}^2$ by means of a suitable embedding transformation $\xi:\Omega_A^2\to\mathbb{R}^2$ such that $\xi(\Omega_A^2)$ is of the form $C\times C$, where C is a Cantor set. That is, S may be represented by $\xi\circ S\circ\xi^{-1}:C\times C\to C\times C$. The images in Figure 2.7 may be viewed as invariant pictures for a transformation of this kind, in the case where the gaps in the Cantor set are infinitesimal.

The group of transformations generated by $S : \Omega^2_{\mathcal{A}} \to \Omega^2_{\mathcal{A}}$ conserves the topological entropy of both components of each point on which it acts, since

$$(h(\sigma), h(\omega)) = (h(S\sigma), h(\sigma_1\omega)) \quad \text{for all } (\sigma, \omega) \in \Omega^2_{\mathcal{A}}.$$

Finally we define **nodal flip** transformations, for which the alphabet is $\mathcal{A} = \{0, 1\}$. Let $\omega \in \Omega'_{\mathcal{A}}$. Define $flip_\omega : \Omega_{\mathcal{A}} \cup \Omega'_{\mathcal{A}} \to \Omega_{\mathcal{A}} \cup \Omega'_{\mathcal{A}}$ by

$$flip_\omega(\sigma) = \begin{cases} \sigma & \text{when } \sigma_1\sigma_2 \cdots \sigma_{|\omega|} \neq \omega, \\ \sigma_1\sigma_2 \cdots \sigma_{|\omega|}\sigma'_{|\omega|+1}\sigma'_{|\omega|+2}\sigma'_{|\omega|+3} \cdots & \text{when } \sigma_1\sigma_2 \cdots \sigma_{|\omega|} = \omega, \end{cases}$$

where $0' = 1$ and $1' = 0$; see Figure 2.66(i), (ii). Nodal flip transformations are continuous, one-to-one and onto, and pairs of such transformations commute. Furthermore we can compose infinite sequences of such transformations to obtain new continuous invertible transformations on code space; see Exercise 2.8.12 below.

In Chapter 4, in connection with fractal tops and colour-stealing, we show how to use continuous invertible transformations on code space to define transformations on subsets of $\mathbb{R}^2$ that are continuous almost everywhere. Also, we show how you can apply such transformations to some beautiful pictures to obtain other beautiful pictures.

EXERCISE 2.8.12   *Let $\{\omega^{(n)} \in \Omega'_{\mathcal{A}}\}^\infty_{n=1}$ be such that $|\omega^{(n)}| \leq |\omega^{(n+1)}|$ for all n. Define $F_n : \Omega_{\mathcal{A}} \cup \Omega'_{\mathcal{A}} \to \Omega_{\mathcal{A}} \cup \Omega'_{\mathcal{A}}$ by*

$$F_n = flip_{\omega^{(1)}} \circ flip_{\omega^{(2)}} \circ \cdots \circ flip_{\omega^{(n)}}$$

for each $n \in \{1, 2, 3, \dots\}$. Show that $\{F_n\}^\infty_{n=1}$ converges uniformly with respect to the metric d_Ω to a continuous invertible function $F : \Omega_{\mathcal{A}} \cup \Omega'_{\mathcal{A}} \to \Omega_{\mathcal{A}} \cup \Omega'_{\mathcal{A}}$. Show that $F^{-1} = F$.

In this section we have played with the fact that code space can be embedded in a tree in $\mathbb{R}^2$. This allows us to handle some transformations on code space by using classical geometrical transformations on $\mathbb{R}^2$. We can also embed code space in diverse other geometrical structures in $\mathbb{R}^2$, such as products of Cantor sets and not-quite-touching Sierpinski triangles. Then we may define transformations on $\mathbb{R}^2$ that map these new structures into themselves. By such means we may define and think about transformations on code space in terms of transformations of a more classical type.

CHAPTER 3

Semigroups on sets, measures and pictures

3.1 Introduction

In this chapter we introduce semigroups and groups and explain how certain of them act upon sets, measures and pictures. Groups of transformations play a definitive role in classical geometry. Semigroups of transformations play an essential role in fractal geometry. What properties of the objects upon which they act are preserved by all the elements of a semigroup or group?

You can find various projective transformations that map parts of the picture in Figure 3.1 into itself, and parts of Figure 3.2 into itself. You can also find various projective transformations that map a given conic section in $\mathbb{R}^2$ into itself. To what extent is a set, measure or picture defined by a collection of transformations that leave it invariant? Clearly, a wallpaper picture is not completely defined by the group of transformations under which it is invariant. But in later chapters we will prove that certain fractal sets, measures and pictures are completely defined by IFS semigroups that leave them invariant. IFS semigroups are sets of transformations that are generated by an IFS. We introduce IFS semigroups in this chapter. Given a picture, how do we look for semigroups of transformations that map the picture into itself? We need to develop some feel for such matters.

Figure 3.3 shows part of a spiral of flowers produced as follows: first the initial flower at the upper right is rotated about, and contracted towards, the centre of the spiral to produce the second flower; then the second flower is transformed by the same clockwise rotation and contraction to produce the third flower, and so on. That is, the multitude of flowers is produced by the repeated application of the same transformation to *different flowers*. This iterative action of a single transformation is equivalent to the action of an infinite sequence of transformations on a *single flower*. This set of transformations, one associated with each flower in the picture, is an example of a semigroup. Any pair of elements of the semigroup can be combined to make another element of the semigroup.

Figure 3.1 'A conspicuous system of veins branches into the leaf blade ... The veins form a structural framework for the blade ... Each vein contains xylem and phloem; and each is usually surrounded by a bundle sheath, composed of cells so tightly packed together that there are few spaces between them. In most cases the branching of the veins is such that no mesophyll cell is far removed from a veinlet; in one study the veins were found to attain a combined length of 102 cm per square centimeter of leaf blade.' [57], p. 207.

Figure 3.2 The picture comprising just the trees can be mapped into itself by a projective transformation. A different projective transformation is needed to map the picture of the road (approximately) into itself because the poles and the road lie in different planes, in three dimensions.

Figure 3.3 This shows part of a spiral of flowers produced by applying all elements belonging to a semigroup of transformations to the (mathematical) picture corresponding to a single morning-glory flower image, the one at the beginning of the spiral at the top right. In this case the semigroup is generated by a single transformation $f : \mathbb{R}^2 \to \mathbb{R}^2$. There is much more going on here than meets the eye. What?

In particular we notice that each member of the semigroup maps the spiral of flowers, that is *the whole picture*, into itself. This illustrates the general principle that appropriate semigroups of transformations may be used to construct pictures that are mapped into themselves by the transformations of the semigroup.

Early on in this chapter we will introduce the **tops union** of two pictures and show how it leads to interesting and enjoyable examples of semigroups of pictures, which we call **tops semigroups**. We use examples of tops semigroups to help build up familiarity with the tops union, to illustrate the idea that an 'attractor' is a set of pictures and to show how random iteration may be used to explore semigroups of pictures.

The main focus of this chapter is on the orbits of sets, measures and pictures under IFS semigroups of transformations. These orbits, in turn, are used to define uniquely certain sets, measures and pictures, which we call **orbital sets, orbital measures** and **orbital pictures** respectively. In a sense that we will make precise, some of these objects may be thought of as 'tilings'. The tiles are themselves sets, measures or pictures and are constructed from elements of the appropriate orbit. Figure 3.4 is a simple example of an orbital picture in which the tiles are segments of flower pictures.

Orbital sets, measures and pictures are ubiquitous in fractal geometry. The reason, as we shall see, is that they always obey a **self-referential equation** which

Figure 3.4 Part of a wallpaper picture. The whole picture has domain $\mathbb{R}^2$. What other information, apart from that which you can glean from the portion shown here, do you need in order to completely define the wallpaper picture?

expresses the orbital object in terms of transformations of the IFS applied to the object or, in the case of orbital pictures, parts of the object.

We pay particular attention to **orbital pictures**. Each of these remarkable pictures is constructed from an orbit of pictures under IFS semigroups, with the help of the tops union. They can possess fascinating code space structures, topological invariants and beautiful segments. By looking at orbital pictures we obtain insights into how to identify IFS semigroups associated with real-world pictures.

It is important not to confuse orbital sets, measures and pictures with fractal sets, fractal measures and fractal tops, which we will later associate with IFS semigroups. The latter objects, various kinds of 'attractor' of the IFS, are essentially limit sets of the former. In Chapter 4 we will explore limit sets associated with IFS semigroups as objects in their own right. A very simple example of a limit set of an orbital picture is the dot at the centre of the spiral of flowers in Figure 3.3; it is invariant under the transformations of the semigroup. The limit set of an IFS semigroup acting on a set may be a fractal set, called the set attractor of the IFS. The limit set of an IFS semigroup acting on a measure may be a fractal measure, called the measure attractor of the IFS. But the limit set of an IFS semigroup acting on a picture is much harder to pin down; what colour, for example, is the dot at the centre of the spiral of flowers? This realization motivated the discovery of **fractal tops**, which are described in Chapter 4.

Images constructed using IFS semigroups have their own invariance properties under the set of transformations that generate the semigroup. This echoes Felix Klein's elegant concept that geometrical properties are the invariants of the associated group of transformations. Klein (1849–1925) considered geometry to be concerned with very tangible objects – mathematically perfect spheres and cones that you could almost touch. According to him a geometry is a space together with a group of transformations that leave the space invariant. Here, in a similar way, we may define a semigroup geometry to be a space of mathematical objects such as sets, measures or pictures, together with a semigroup of transformations that unifies the space by leaving properties of the objects invariant. We find fractal geometry to be much concerned with semigroup geometries.

Many fractal geometrical objects, be they sets, measures or pictures, and their relationships to the semigroups that define them can be partly understood in terms of the shift transformation and its inverse branches, which generate a kind of semigroup geometry on code space. The code space $\Omega'_{\mathcal{A}}$ underlying an IFS semigroup tiling assigns addresses to the tiles. It allows us to manipulate the semigroup symbolically and to relate the limit sets of the tiles, be they sets of points or measures or fractal tops, to $\Omega_{\mathcal{A}}$. The different relationships between code spaces and the sets, measures or pictures with which they may be associated can provide invariants of the sets, measures or pictures. These invariants may possess some independence from the specific class of transformations used to define the sets, measures or pictures. As a very simple example, the picture in Figure 3.5 is associated with an IFS semigroup that is, from a code space point of view, entirely equivalent to the one used to generate Figure 3.3. But the pictures themselves, in terms of the deformations from one flower to the next flower to the next, are quite different. For example, in Figure 3.5, the green leaf is sometimes lanceolate and sometimes ovate, whereas in Figure 3.3 it has the same shape everywhere. Instead of saying that the two pictures are similar because they are related by an affine transformation, we may instead say that they are related because they have in common a certain code space structure.

We have seen already, in Chapter 2, how projective transformations and Möbius transformations acting on sets, measures and pictures produce images which, by and large, depend continuously on the coefficients that define the transformations. We will find, and not be surprised, that this continuous dependence can extend to orbital sets, measures and pictures constructed using projective or Möbius transformations. The flexibility and adjustability of such orbital objects means that they may be used in biological modelling, computer graphics and many other situations where one wants to construct and adjust fractal geometrical models in order to approximate given information.

Figure 3.5 The semigroup of transformations associated with this picture is algebraically equivalent to the one associated with Figure 3.3. But the changes from one flower to the next are different. Sometimes the green leaf is thin and sometimes it is foreshortened.

Furthermore, many projective transformations and Möbius transformations can be described efficiently with small arrays of discrete data, and, as pointed out in Chapter 2, projective transformations tend to occur in natural ways in connection with real-world images. These facts make the use of orbital pictures, constructed using IFS semigroups of projective transformations, appealing for potential applications in image compression, segmentation and representation.

Contents of this chapter

In Section 3.2 we define a semigroup and illustrate some 'visual' examples by showing how two subsets of $\mathbb{R}^2$ may be combined to define a new subset of $\mathbb{R}^2$, how, in the 'tops semigroup', two pictures may be combined to define a new picture and how two normalized measures may be combined to yield a normalized measure. As a means to build familiarity with ideas needed later on, we illustrate how random iteration may be used to explore tops semigroups.

In Section 3.3 we introduce semigroups of transformations and IFS semigroups. Semigroups of transformations arise in a deep and natural manner from models of the physical world; for example, simple autonomous equations that provide models for everyday physical phenomena are a source of semigroups of elementary transformations. We include examples of semigroups of transformations on $\mathbb{R}^2$, of rational transformations on the Riemann sphere, of dynamical systems and of transformations on code spaces. We introduce the idea of the orbit of a point under a semigroup of transformations and describe the relationship between such orbits and code spaces.

In Section 3.4 we consider IFS semigroups acting on sets, define the orbit of a set under an IFS semigroup and the associated orbital set and note the self-referential equation which the latter obeys. We define **semigroup tilings** of orbital sets and provide a necessary and sufficient condition for the orbit of a set under an IFS semigroup to yield a semigroup tiling. We observe that many colourful pictures related to Julia sets, generated for example using Fractint, see [92], represent semigroup tilings.

In Section 3.5 we consider IFS semigroups acting on pictures, define the orbit of a picture under an IFS semigroup and also define the associated orbital picture. We shall devote much space to orbital pictures since this material is new to fractal geometry and appears to have many exciting applications. We discuss the computation of some orbital pictures and show how they obey their own special type of self-referential equation. We define the panels and the code space of an orbital picture, relate these concepts to a symbolic dynamical system and use invariants of the latter to define topological invariants of orbital pictures. We introduce and illustrate the concepts of the diversity, the growth rate of the diversity and the 'space of limiting pictures' associated with an orbital picture. We also discuss two types of tiling of orbital pictures. We mention several other pictures that can be defined in terms of the orbit of a picture, including the **underneath picture** and pictures generated using an associated tops semigroup. We use the Henon transformation to illustrate an orbital picture associated with a geometrically intricate dynamical system. We conclude Section 3.5 with a discussion of the applications of orbital pictures.

In Section 3.6 we consider IFS semigroups acting on measures, define the orbit of a measure under an IFS semigroup and also define the associated orbital measure. We prove that an orbital measure is uniquely defined by a self-referential equation, which it obeys in very general circumstances.

In Section 3.7 we treat groups of transformations as examples of semigroups of transformations with the special property that inverses of all transformations are included. This allows us to apply the theory of Sections 3.4–3.6, which is fundamentally 'fractal', to orbital sets, orbital measures and orbital pictures associated with finitely generated groups of transformations. There exists a vast body of literature concerning the relationships between geometry, tilings and group theory; see for example, [23], [42], [73] and [89]. We shall not describe or review this area but simply connect it with some aspects of fractal geometry, particularly the new concept of orbital pictures.

Also in Section 3.7 we provide a brief survey of geometries associated with different families of transformations, including 'geometries on code space', with an emphasis on the properties of pictures. The associated groups are sources of transformations for the construction of fractal objects associated with different classical geometries and different types of invariance property.

3.2 Semigroups

Definition of a semigroup

A semigroup may be defined whenever there is a simple rule for combining pairs of mathematical objects to produce new mathematical objects of the same kind.

DEFINITION 3.2.1 A **semigroup** is a set $\mathcal{S}$ together with a function, called a **binary operation**, $b : \mathcal{S} \times \mathcal{S} \to \mathcal{S}$ that is **associative**, that is, $b(s, b(t, u)) = b(b(s, t), u)$, for all $s, t, u \in \mathcal{S}$. The binary operation may be denoted $b(s, t) = s \bigcirc t$ for all $s, t \in \mathcal{S}$, where $\bigcirc$ is called the **binary operator**. The semigroup may be denoted by $(\mathcal{S}, \bigcirc)$.

The order in which one evaluates binary operations makes no difference to the final result; for example,

$$(s_1 \bigcirc s_2) \bigcirc (s_3 \bigcirc s_4) = (s_1 \bigcirc (s_2 \bigcirc s_3)) \bigcirc s_4$$

for all $s_1, s_2, s_3, s_4 \in \mathcal{S}$, as is readily proved using the associativity of $\bigcirc$. It follows that an expression such as

$$s_1 \bigcirc s_2 \bigcirc \cdots \bigcirc s_n \tag{3.2.1}$$

defines a unique element of $\mathcal{S}$ for all $s_1, s_2, \ldots, s_n \in \mathcal{S}$ and for all $n = 1, 2, 3, \ldots$ Notice, though, that if one changes the order in which the elements $s_1, s_2, \ldots, s_n \in \mathcal{S}$ in the composition (3.2.1) appear then the result of the composition may change. For example, in general $s_1 \bigcirc s_2 \neq s_2 \bigcirc s_1$.

Two examples of semigroups are $(\mathbb{R}, \times)$ and $(\mathbb{R}, +)$, where $\times$ denotes the multiplication of numbers and $+$ denotes addition. Both $\times$ and $+$ are well known to be associative operations. Each of these semigroups contains many **sub-semigroups**. A sub-semigroup is a subset of a semigroup that is a semigroup in its own right, using the same binary operation.

EXERCISE 3.2.2 *Let* $\mathbb{N} = \{1, 2, 3, \ldots\}$, $\mathbb{Z}^+ = \{0, 1, 2, 3, \ldots\}$ *and* $\mathbb{Z} = \{\ldots, -2, -1, 0, 1, 2, 3, \ldots\}$. *Let* $\mathbb{C} = \{x + iy : x, y \in \mathbb{R}\}$ *denote the set of complex numbers, with* $i = \sqrt{-1}$. *Verify that each object in the following two chains of inclusions represents a semigroup:*

$$(\{1\}, \times) \subset (\mathbb{N}, \times) \subset (\mathbb{Z}^+, \times) \subset (\mathbb{Z}, \times) \subset (\mathbb{R}, \times) \subset (\mathbb{C}, \times);$$

$$(\{0\}, +) \subset (\mathbb{Z}^+, +) \subset (\mathbb{Z}, +) \subset (\mathbb{R}, +) \subset (\mathbb{C}, +).$$

An important type of semigroup is $(M_2(\mathbb{R}), \cdot)$ where $M_2(\mathbb{R})$ denotes the set of 2×2 matrices with real coefficients and the binary operation indicated by the raised point represents matrix multiplication. More generally, $(M_N(\mathbb{F}), \cdot)$ is a semigroup for all $N = 1, 2, 3, \ldots$, where $M_N(\mathbb{F})$ is the set of $N \times N$ matrices all of whose coefficients lie in $\mathbb{F}$ and $\mathbb{F}$ may be for example $\mathbb{N}$, $\mathbb{Z}^+$, $\mathbb{Z}$, $\mathbb{R}$ or $\mathbb{C}$.

EXERCISE 3.2.3   *Let $A \in M_2(\mathbb{F})$, where $\mathbb{F} \in \{\mathbb{N}, \mathbb{Z}^+, \mathbb{Z}, \mathbb{R}, \mathbb{C}\}$. Show that $(\{A^n : n \in \mathbb{N}\}, \cdot)$ is a semigroup, where $A^1 = A$ and $A^{m+1} = A \cdot A^m$ for $m = 1, 2, \ldots$*

EXERCISE 3.2.4   *Let $A, B \in M_2(\mathbb{R})$, where the matrix B is invertible. Show that $(\{B \cdot A^n \cdot B^{-1} : n \in \mathbb{N}\}, \cdot)$ is a semigroup.*

A different type of semigroup is $(\Omega'_A, \bigcirc)$, where Ω'_A is the code space of finite strings of symbols from the finite alphabet $\mathcal{A}$ and the binary operation is $\sigma \bigcirc \upsilon = \sigma \upsilon$ for all $\sigma, \upsilon \in \Omega'_A$. For example, if $\sigma = 111$ and $\upsilon = 000$ then $\sigma \bigcirc \upsilon = 111000$.

DEFINITION 3.2.5   Let $\mathcal{S}$ be a semigroup. A **sub-semigroup** of $\mathcal{S}$ is a semigroup which is contained in $\mathcal{S}$ and which has the same binary operation as $\mathcal{S}$. Let $\widetilde{\mathcal{S}} \subset \mathcal{S}$. The **semigroup generated by** $\widetilde{\mathcal{S}}$ is defined to be the smallest sub-semigroup of $\mathcal{S}$ that contains $\widetilde{\mathcal{S}}$.

EXERCISE 3.2.6   *Verify that the semigroup generated by $\widetilde{\mathcal{S}}$ is well defined. To do this, demonstrate that (i) there exists a semigroup $\mathcal{T}$ that consists of all possible finite compositions, under the operation of the semigroup, of the elements of $\widetilde{\mathcal{S}}$ and (ii) any semigroup that contains $\widetilde{\mathcal{S}}$ must also contain $\mathcal{T}$.*

The above examples of semigroups are 'symbolic' or 'algebraic' because the elements of the semigroups are themselves collections of symbols or formulas. But in the following subsections we illustrate semigroups whose elements are sets, pictures or measures.

Semigroups of sets

The union of two subsets of a space is a new subset of the space. So a simple example of a semigroup operation is $\cup$, the union operation. Let $\mathbb{S}(\mathbb{R}^2)$ denote the space of all subsets of $\mathbb{R}^2$. Then $(\mathbb{S}(\mathbb{R}^2), \cup)$ is a semigroup. It possesses many fascinating sub-semigroups, for example those illustrated in Figure 3.6.

EXERCISE 3.2.7   *Construct and illustrate your own example of a sub-semigroup of $(\mathbb{S}(\mathbb{R}^2), \cup)$.*

Semigroups of picture segments: tops semigroups

It is often convenient to think of a picture as being a combination of other pictures. This leads us to the following example of a semigroup, which we call a **tops semigroup**. Tops semigroups are related to fractal tops, to be discussed in Chapter 4. They enable us to illustrate some basic ideas that occur in more

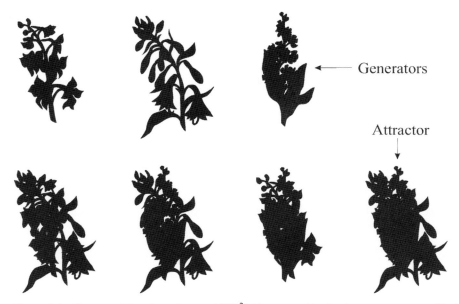

Figure 3.6 Elements of the sub-semigroup of $(\mathbb{S}(\mathbb{R}^2), \cup)$ generated by the three sets represented in the top row. What is special about the element labelled 'attractor'? See also Section 3.2.

technically complicated situations, including those involving superfractals. But tops semigroups are interesting in their own right because they can be used to describe large families of related pictures, which can be sampled by random iteration.

DEFINITION 3.2.8   A **picture segment** is a picture. A picture $\mathfrak{P}_1$ is said to be **segment of a picture** $\mathfrak{P}_2$ if the domain of $\mathfrak{P}_1$ is contained in the domain of $\mathfrak{P}_2$. When $\mathfrak{P}_1$ and $\mathfrak{P}_2$ are pictures the notation

$$\mathfrak{P}_1 \subset \mathfrak{P}_2$$

means that $\mathfrak{P}_1$ is a segment of $\mathfrak{P}_2$.

When it is clear from the context that we are talking about a picture segment, we may refer to it simply as a **segment**. Watch out for pictures of worms . . .

DEFINITION 3.2.9   Let $\Pi = \Pi_{\mathfrak{C}}(\mathbb{X})$ denote the space of pictures with colour space $\mathfrak{C}$. The **tops union** $\mathfrak{P}_1 \uplus \mathfrak{P}_2$ of $\mathfrak{P}_1, \mathfrak{P}_2 \in \Pi$ is the picture $\mathfrak{P}_1 \uplus \mathfrak{P}_2 \in \Pi$ defined by

$$\mathfrak{P}_1 \uplus \mathfrak{P}_2 : D_{\mathfrak{P}_1 \uplus \mathfrak{P}_2} \subset \mathbb{X} \to \mathfrak{C},$$

where

$$D_{\mathfrak{P}_1 \uplus \mathfrak{P}_2} = D_{\mathfrak{P}_1} \cup D_{\mathfrak{P}_2}$$

and

$$\mathfrak{P}_1 \uplus \mathfrak{P}_2(x) = \begin{cases} \mathfrak{P}_1(x) & \text{if } x \in D_{\mathfrak{P}_1}, \\ \mathfrak{P}_2(x) & \text{if } x \in D_{\mathfrak{P}_2} \backslash D_{\mathfrak{P}_1}, \end{cases}$$

for all $x \in D_{\mathfrak{P}_1 \uplus \mathfrak{P}_2}$ and for all $\mathfrak{P}_1, \mathfrak{P}_2 \in \Pi$.

We will use the tops union repeatedly in later sections, as well as here. We say that two segments or pictures are disjoint if their domains are disjoint. Given $\mathfrak{P}_1, \mathfrak{P}_2 \in \Pi$ we define $\mathfrak{P}_1 \backslash \mathfrak{P}_2$ to be the picture whose domain is $D_{\mathfrak{P}_1} \backslash D_{\mathfrak{P}_2}$ and whose values are given by

$$(\mathfrak{P}_1 \backslash \mathfrak{P}_2)(x) = \mathfrak{P}_1(x) \quad \text{for all } x \in D_{\mathfrak{P}_1 \backslash \mathfrak{P}_2}.$$

EXERCISE 3.2.10 *Verify that the binary operation $\uplus$ is associative but not commutative.*

EXERCISE 3.2.11 *Let $f : \mathbb{X} \to \mathbb{X}$ be one-to-one. Prove that*

$$f(\mathfrak{P}_1 \uplus \mathfrak{P}_2) = f(\mathfrak{P}_1) \uplus f(\mathfrak{P}_2) \quad \text{for all } \mathfrak{P}_1, \mathfrak{P}_2 \in \Pi.$$

EXERCISE 3.2.12 *Let $f : \mathbb{X} \to \mathbb{X}$ be one-to-one. Prove that*

$$f(\mathfrak{P}_1 \backslash \mathfrak{P}_2) = f(\mathfrak{P}_1) \backslash f(\mathfrak{P}_2) \quad \text{for all } \mathfrak{P}_1, \mathfrak{P}_2 \in \Pi.$$

EXERCISE 3.2.13 *Show that*

$$\mathfrak{P}_1 \uplus \mathfrak{P}_2 = \mathfrak{P}_1 \uplus (\mathfrak{P}_2 \backslash \mathfrak{P}_1) \quad \text{for all } \mathfrak{P}_1, \mathfrak{P}_2 \in \Pi.$$

Notice that we can decompose a picture $\mathfrak{P}$ into two segments $\mathfrak{P}_1$ and $\mathfrak{P}_2$ by choosing two domains $D_{\mathfrak{P}_1}$ and $D_{\mathfrak{P}_2}$ such that $D_{\mathfrak{P}} = D_{\mathfrak{P}_1} \cup D_{\mathfrak{P}_2}$. We have not required that the segments have disjoint domains. Now we define two pictures $\mathfrak{P}_1 : D_{\mathfrak{P}_1} \to \mathfrak{C}$ and $\mathfrak{P}_2 : D_{\mathfrak{P}_2} \to \mathfrak{C}$ by

$$\mathfrak{P}_k(x) = \mathfrak{P}(x) \quad \text{for all } x \in D_{\mathfrak{P}_k} \text{ and for } k = 1, 2. \tag{3.2.2}$$

It follows that these two pictures agree for all $x \in D_{\mathfrak{P}_1} \cap D_{\mathfrak{P}_2}$, and consequently that

$$\mathfrak{P} = \mathfrak{P}_1 \uplus \mathfrak{P}_2 = \mathfrak{P}_2 \uplus \mathfrak{P}_1.$$

More generally, if two pictures $\mathfrak{P}_3$ and $\mathfrak{P}_4$ are such that they disagree at some point belonging to the intersection of their domains, then

$$\mathfrak{P}_3 \uplus \mathfrak{P}_4 \neq \mathfrak{P}_4 \uplus \mathfrak{P}_3.$$

DEFINITION 3.2.14 The semigroup $(\Pi, \uplus)$ is called the **tops semigroup**. Given $\Gamma \subset \Pi$, the smallest sub-semigroup of $(\Pi, \uplus)$ that contains Γ is called the **tops semigroup generated by** Γ.

Figure 3.7 illustrates the pictures in the tops semigroup generated by three segments.

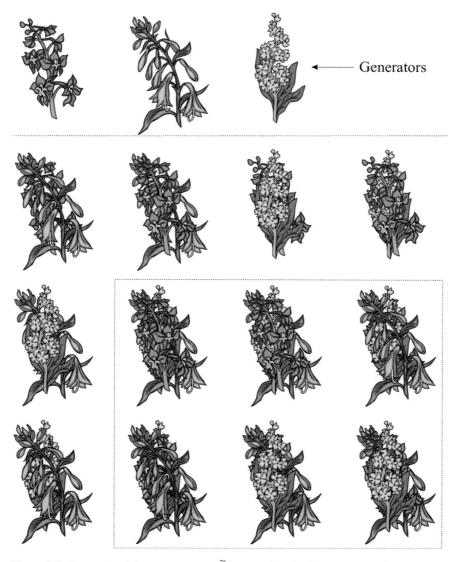

Figure 3.7 Illustration of the tops semigroup $\widetilde{\Pi}$ generated by the three pictures in the upper row. The attractor A of this semigroup is represented by the six pictures at the lower right. Verify that if $\mathfrak{P} \in \widetilde{\Pi}$ and $\mathfrak{Q} \in A$ then $\mathfrak{P} \uplus \mathfrak{Q} \in A$.

THEOREM 3.2.15 *Let $\widetilde{\Pi}$ denote the tops semigroup generated by the finite set of pictures $\{\mathfrak{P}_1, \mathfrak{P}_2, \ldots, \mathfrak{P}_N\} \subset \Pi$. Then $\widetilde{\Pi}$ is a finite set. Define $f_i : \widetilde{\Pi} \to \widetilde{\Pi}$ by $f_i(\mathfrak{P}) = \mathfrak{P}_i \uplus \mathfrak{P}$ for all $\mathfrak{P} \in \widetilde{\Pi}$ and define $F : \mathbb{S}(\widetilde{\Pi}) \to \mathbb{S}(\widetilde{\Pi})$ by*

$$F(B) = f_1(B) \cup f_2(B) \cup \cdots \cup f_N(B) \quad \textit{for all } B \in \mathbb{S}(\widetilde{\Pi}),$$

where the points of $\mathbb{S}(\widetilde{\Pi})$, *the space of subsets of* $\widetilde{\Pi}$, *each consist of a set of pictures. Then there exists a unique point* $A \in \mathbb{S}(\widetilde{\Pi})$, *i.e. a set of pictures, such that*

$$A = F(A)$$

and moreover

$$\lim_{k \to \infty} F^{\circ k}(B) = A$$

for all $B \in \mathbb{S}(\widetilde{\Pi})$.

PROOF   Let $\mathcal{A} = \{1, 2, \dots, N\}$ and define $f_\sigma = f_{\sigma_1} \circ f_{\sigma_2} \circ \cdots \circ f_{\sigma_{|\sigma|}}$ for all $\sigma = \sigma_1 \sigma_2 \cdots \sigma_{|\sigma|} \in \Omega'_{\mathcal{A}}$. Then

$$f_\sigma(\mathfrak{P}) = \mathfrak{P}_{\sigma_1} \uplus \mathfrak{P}_{\sigma_2} \uplus \cdots \uplus \mathfrak{P}_{\sigma_{|\sigma|}} \uplus \mathfrak{P} \quad \text{for all } \sigma \in \Omega'_{\mathcal{A}}, \mathfrak{P} \in \widetilde{\Pi}.$$

It is readily verified that

$$f_\sigma = f_{\widetilde{\sigma}} \quad \text{for all } \sigma \in \Omega'_{\mathcal{A}},$$

where $\widetilde{\sigma}$ is obtained from σ by deleting all but the leftmost occurrence of each symbol in $\mathcal{A}$, so that for example

$$f_{132142} = f_{1324}, \quad f_{1222222} = f_{12} \quad \text{and } f_{111121} = f_{12}.$$

From Exercise 3.2.6 we know that every element of $\widetilde{\Pi}$ can be written in the form

$$\mathfrak{P}_{\sigma_1} \uplus \mathfrak{P}_{\sigma_2} \uplus \cdots \uplus \mathfrak{P}_{\sigma_{|\sigma|}}, \tag{3.2.3}$$

for some $\sigma \in \Omega'_{\mathcal{A}}$. So it follows that every element of $\widetilde{\Pi}$ can be written as in Equation (3.2.3) with $|\sigma| \leq N$, which tells us that $\widetilde{\Pi}$ is a finite set. It also follows that

$$F^{\circ(N+l)}(B) = \bigcup_{\sigma \in \Omega'_{\mathcal{A}}, |\sigma| = N+l} \left\{ f_{\sigma_1} \uplus f_{\sigma_2} \uplus \cdots \uplus f_{\sigma_{|\sigma|}}(B) \right\}$$

$$= \bigcup_{\sigma \in Perm(1,2,\dots,N)} \left\{ \mathfrak{P}_{\sigma_1} \uplus \mathfrak{P}_{\sigma_2} \uplus \cdots \uplus \mathfrak{P}_{\sigma_N} \right\}$$

for all $l = 0, 1, 2, \dots$, where $Perm(1, 2, \dots, N)$ denotes the set of strings σ, of length N, all of whose components are distinct. $\qquad\square$

DEFINITION 3.2.16   The set of pictures A defined in Theorem 3.2.15 is called the **attractor** of the tops semigroup generated by the finite set of picture segments $\{\mathfrak{P}_1, \mathfrak{P}_2, \dots, \mathfrak{P}_N\}$.

The attractor A of the tops semigroup illustrated in Figure 3.7 is represented by the set of six pictures at the lower right.

Notice the following **random iteration algorithm**, which may be used to sample A. This description is informal. Let $\{\mathfrak{F}_1, \mathfrak{F}_2, \mathfrak{F}_3\}$ denote the three pictures in the top row of Figure 3.7, which generate the semigroup. Define a sequence of pictures $\mathfrak{P}_1, \mathfrak{P}_2, \mathfrak{P}_3, \ldots$ by choosing $\mathfrak{P}_1 = \mathfrak{F}_{\sigma_1}$ and

$$\mathfrak{P}_{n+1} = f_{\sigma_n}(\mathfrak{P}_n) = \mathfrak{F}_{\sigma_n} \uplus \mathfrak{P}_n \quad \text{for } n = 1, 2, \ldots,$$

where, for each n, independently of all other choices, $\sigma_n = 1$ with probability $\frac{1}{6}$, $\sigma_n = 2$ with probability $\frac{1}{2}$ and $\sigma_n = 3$ with probability $\frac{1}{3}$. Look at the sequence $\mathfrak{P}_1, \mathfrak{P}_2, \mathfrak{P}_3, \ldots$ What will we see? The theory of **Markov processes**, see for example [37], Chapter XV, tells us it is almost certain, after some finite number of iterations N, that we will see $\mathfrak{P}_n \in A$ for all $n \geq N$. That is, the random sequence is 'attracted' to A. Moreover, with very high probability, the sequence of pictures will then behave 'ergodically', jumping around from picture to picture of the attractor, spending on average a certain fixed fraction of the 'time' on each element of the attractor. This highly probable eventual behaviour of the sequence of pictures is referred to as a **stationary state** of the Markov process.

More precisely, the possible pictures on the attractor are $\mathfrak{F}_1 \uplus \mathfrak{F}_2 \uplus \mathfrak{F}_3, \mathfrak{F}_2 \uplus \mathfrak{F}_1 \uplus \mathfrak{F}_3, \mathfrak{F}_3 \uplus \mathfrak{F}_2 \uplus \mathfrak{F}_1, \mathfrak{F}_1 \uplus \mathfrak{F}_3 \uplus \mathfrak{F}_2, \mathfrak{F}_3 \uplus \mathfrak{F}_1 \uplus \mathfrak{F}_2$ and $\mathfrak{F}_2 \uplus \mathfrak{F}_3 \uplus \mathfrak{F}_1$, which may be labelled 1, 2, 3, 4, 5 and 6, respectively. Then the probability of transition from picture i to picture j on the attractor is $p_{i,j}$, where $(p_{i,j})$ is the **stochastic matrix**

$$P = \begin{pmatrix} \frac{1}{6} & \frac{1}{2} & 0 & 0 & \frac{1}{3} & 0 \\ \frac{1}{6} & \frac{1}{2} & \frac{1}{3} & 0 & 0 & 0 \\ 0 & 0 & \frac{1}{3} & \frac{1}{6} & 0 & \frac{1}{2} \\ 0 & \frac{1}{2} & 0 & \frac{1}{6} & \frac{1}{3} & 0 \\ 0 & 0 & 0 & \frac{1}{6} & \frac{1}{3} & \frac{1}{2} \\ \frac{1}{6} & 0 & \frac{1}{3} & 0 & 0 & \frac{1}{2} \end{pmatrix}.$$

The stationary state is described by the unique vector of probabilities

$$p = (p_1, p_2, p_3, p_4, p_5, p_6)$$

such that

$$pP = P, \quad p_i > 0 \text{ for } i = 1, 2, \ldots, 6, \quad \sum_{i=1}^{6} p_i = 1. \tag{3.2.4}$$

The number p_i gives the average fraction of the pictures in the random sequence $\mathfrak{P}_1, \mathfrak{P}_2, \mathfrak{P}_3, \ldots$ equal to the ith picture on the attractor; that is, almost always,

$$p_i = \lim_{K \to \infty} K^{-1}\{\text{number of times picture } i \text{ occurs in } \mathfrak{P}_1, \mathfrak{P}_2, \mathfrak{P}_3, \ldots, \mathfrak{P}_K\}.$$

On solving Equation (3.2.4), using the Maple engine in [87], we find that

$$p = \left(\tfrac{1}{10}, \tfrac{1}{6}, \tfrac{1}{4}, \tfrac{1}{15}, \tfrac{1}{12}, \tfrac{1}{3}\right).$$

We may think of this stationary state as being described by a probability measure μ on the field generated by the pictures, with $\mu(\mathfrak{F}_1 \uplus \mathfrak{F}_2 \uplus \mathfrak{F}_3) = \frac{1}{10}$, $\mu(\mathfrak{F}_2 \uplus \mathfrak{F}_1 \uplus \mathfrak{F}_3) = \frac{1}{6}$, $\mu(\mathfrak{F}_3 \uplus \mathfrak{F}_2 \uplus \mathfrak{F}_1 \cup \mathfrak{F}_1 \uplus \mathfrak{F}_3 \uplus \mathfrak{F}_2) = \frac{1}{4} + \frac{1}{15}$ and so on.

Thus we see how one may sample the elements of a semigroup by means of random iteration, actually a Markov process, thereby learning something about the semigroup. In fact, in this case, what we 'see' are elements of the attractor of the semigroup, sampled according to a certain probability distribution on the attractor.

EXERCISE 3.2.17 *Choose the probabilities in the above discussion to be $\sigma_n = 1$ with probability $\frac{1}{10}$, $\sigma_n = 2$ with probability $\frac{1}{5}$ and $\sigma_n = 3$ with probability $\frac{7}{10}$. Estimate the probability that $\mathfrak{P}_{100} = \mathfrak{F}_1 \uplus \mathfrak{F}_2 \uplus \mathfrak{F}_3$.*

Another example of a tops semigroup is given in Figure 3.8. Here, the semigroup is generated by pictures of the playing cards A♢, Q♢, Q♠, K♡, J♢ and A ♣, each positioned at a fixed angle. Again we may assign probabilities to the pictures and then sample the semigroup by means of the random iteration algorithm. This example provides a visual note of the connection between semigroups of pictures and probability theory.

Figure 3.9 shows members of the attractor of a tops semigroup generated by pictures of fallen leaves. Following the above discussion, we see how it is possible to generate probability measures on spaces of pictures, and how we may sample such spaces, even when they are vast, by means of random iteration.

EXERCISE 3.2.18 *Let $\mathfrak{P}_i \in \Pi$ for $i = 1, 2, 3, 4$. Verify that*

$$(\mathfrak{P}_1 \uplus \mathfrak{P}_2 \uplus \mathfrak{P}_3 \uplus \mathfrak{P}_4)(x) = \begin{cases} \mathfrak{P}_1(x) & \text{if } x \in D_1 := D_{\mathfrak{P}_1}, \\ \mathfrak{P}_2(x) & \text{if } x \in D_2 := D_{\mathfrak{P}_2} \backslash D_1, \\ \mathfrak{P}_3(x) & \text{if } x \in D_3 := D_{\mathfrak{P}_3} \backslash D_2, \\ \mathfrak{P}_4(x) & \text{if } x \in D_4 := D_{\mathfrak{P}_4} \backslash D_3. \end{cases}$$

The following exercise gives an example of how to embed the semigroup $(\mathbb{S}(\mathbb{R}^2), \cup)$ in the tops semigroup $(\Pi_{\mathfrak{C}}(\mathbb{R}^2), \uplus)$ in such a way that the operation of $\cup$ on $\mathbb{S}(\mathbb{R}^2)$ is equivalent to the operation of $\uplus$ on the embedded elements in $\Pi_{\mathfrak{C}}(\mathbb{R}^2)$.

EXERCISE 3.2.19 *Let the colour space $\mathfrak{C}$ be such that $0 \in \mathfrak{C}$ and $1 \in \mathfrak{C}$. Let $\mathfrak{P}_0 : \mathbb{R}^2 \to \mathfrak{C}$ denote an endless 'blank' picture, that is, $\mathfrak{P}_0(x) = 0$ for all $x \in \mathbb{R}^2$. Let $S_1, S_2 \in \mathbb{S}(\mathbb{R}^2)$ and let $\mathfrak{P}_{S_i} : S_i \to \mathfrak{C}$ be defined by $\mathfrak{P}_{S_i}(x) = 1$ for all $x \in S_i$. Let χ_S denote the characteristic function of $S \subset \mathbb{R}^2$. Show that*

  (i) *$\mathfrak{P}_{S_i} \uplus \mathfrak{P}_0 = \chi_{S_i}$ for $i = 1, 2$;*
  (ii) *$\mathfrak{P}_{S_1} \uplus \mathfrak{P}_{S_2} = \mathfrak{P}_{S_2} \uplus \mathfrak{P}_{S_1} = \mathfrak{P}_{S_1 \cup S_2}$;*
  (iii) *if $\xi : \mathbb{S}(\mathbb{R}^2) \to \Pi_{\mathfrak{C}}(\mathbb{R}^2)$ is defined by $\xi(S) = \mathfrak{P}_S$ for all $S \in \mathbb{S}(\mathbb{R}^2)$ then ξ is one-to-one and hence an embedding, and moreover*

$$\xi(S_1 \cup S_2) = \mathfrak{P}_{S_1} \uplus \mathfrak{P}_{S_2} \quad \text{for all } S_1, S_2 \in \mathbb{S}(\mathbb{R}^2).$$

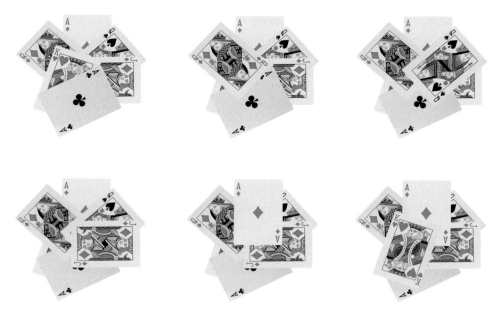

Figure 3.8 Pictures belonging to the attractor of the tops semigroup generated by pictures of A◇, Q◇, Q♠, K♡, J◇ and A♣.

Figure 3.9 Pictures of the 'forest floor' belonging to a tops semigroup generated by pictures of individual leaves.

The surfaces of some moons are pockmarked with disk-shaped craters. Model pictures of these surfaces may be generated by pretending that meteors of randomly different sizes hit the moon at randomly different places, overlaying craters on craters. Such pictures may be treated as random fractal pictures generated by *tremas*, a word apparently coined by Mandelbrot; see [64], pp. 305–8. Vast collections of such pictures may be explored by random iteration.

EXERCISE 3.2.20   *See if you can find, on the internet, simulations of pictures of falling dead leaves; use a search engine such as Google. What is the difference between the behaviour, over time, of pictures of fallen leaves on a glass table, on which they steadily accumulate starting from a clean surface, viewed (a) from above and (b) from below?*

Semigroups of measures

The sum of two Borel measures is a Borel measure. The weighted average of two probability measures on $\square \subset \mathbb{R}^2$ is a probability measure on $\square$. Thus $(\mathbb{P}(\square), \heartsuit)$ is a semigroup of probability measures, where we define $\mu \heartsuit \nu = \frac{1}{3}\mu + \frac{2}{3}\nu$. If we think of μ and ν as greyscale pictures, then $\mu \heartsuit \nu$ is a weighted average of the two pictures. Clearly $\mu \heartsuit \nu \neq \nu \heartsuit \mu$ in general.

EXERCISE 3.2.21   *Let S denote the sub-semigroup of $(\mathbb{P}(\square), \heartsuit)$ generated by two distinct measures $\mu_0, \mu_1 \in \mathbb{P}(\square)$. Describe S. For example, think of μ_0 and μ_1 as greyscale pictures and then describe the set of pictures in S. Can you set up an addressing function $f : \Omega'_{\{0,1\}} \to S$? Better still, can you describe an addressing function $f : \Omega_{\{0,1\}} \cup \Omega'_{\{0,1\}} \to \overline{S}$, where $\overline{S}$ denotes the closure of S?*

3.3   Semigroups of transformations

Semigroups of transformations are central to this book because we use them to define and manipulate sets, pictures and measures. They play a key role in fractal geometry.

DEFINITION 3.3.1   A **semigroup of transformations** on a space $\mathbb{X}$ is a semigroup $(\mathcal{S}(\mathbb{X}), \circ)$, where $\mathcal{S}(\mathbb{X})$ consists of transformations from $\mathbb{X}$ into $\mathbb{X}$ and where the binary operation is composition. That is, $f \circ g$ is the transformation defined by

$$f \circ g(x) = f(g(x)) \quad \text{for all } x \in \mathbb{X}.$$

The composition of functions is an associative operation because

$$f_1 \circ (f_2 \circ f_3)(x) = f_1(f_2(f_3(x))) = (f_1 \circ f_2)(f_3(x)) = (f_1 \circ f_2) \circ f_3(x)$$

whenever $f_1, f_2, f_3 : \mathbb{X} \to \mathbb{X}$. We will tend to drop the explicit reference to the binary operation for semigroups when the operation is obvious, for example, the composition of functions. So we may say 'S is a semigroup' or '$\mathcal{S}(\mathbb{X})$ is a semigroup of transformations (on the space $\mathbb{X}$)'. We will look mainly at semigroups of transformations on spaces, such as $\mathbb{R}^2$, that are related to pictures.

Examples of semigroups of transformations

Here we introduce the main semigroups of transformations that we need for fractal geometry and superfractals. The most important of these for the purposes of this book are IFS semigroups, in particular those built from projective and Möbius transformations. Elsewhere we use these semigroups to form semigroup tilings, fractal sets and measures, fractal tops and superfractals.

Semigroups of linear transformations

The set of linear transformations that map a linear space such as $\mathbb{R}^2$ into itself forms a semigroup, because if both f and g are linear transformations then so is $f \circ g$. If two linear transformations f_1 and f_2 are represented by matrices A_1 and A_2 respectively then it is readily verified that the linear transformation $f_1 \circ f_2$ is represented by the matrix $A_1 \cdot A_2$. So we may use the semigroups of matrices to study semigroups of linear transformations, and vice versa.

Recall that the domain of a transformation is an important part of its definition. So, for example, let $\mathcal{T}$ denote the set of linear transformations that map a certain set $D \subset \mathbb{R}^2$ into itself. Then $\mathcal{S} := \{ f|_D : f \in \mathcal{T} \}$ is a semigroup. Notice too that there is no requirement of invertibility on the transformations in a semigroup. Let $f \in \mathcal{T}$. Then the transformation $f|_D : D \to D$ may not be one-to-one for one of the following reasons: (i) there are points outside D that are mapped by f into D; (ii) the determinant of the matrix that represents f may be zero.

Semigroups of Möbius transformations

The composition of two Möbius transformations acting on the space $\mathbb{R}^2 \cup \{\infty\}$, or equivalently $\widehat{\mathbb{C}}$, is a new Möbius transformation. So the set of Möbius transformations on $\mathbb{R}^2 \cup \{\infty\}$ is an example of a semigroup. Interesting sub-semigroups of Möbius transformations are generated by small sets of Möbius transformations with integer coefficients.

Suppose that M is a set of Möbius transformations that map a domain $D \subset \mathbb{R}^2 \cup \{\infty\}$ into itself. Then the set of transformations obtained by restricting the transformations of M to D is a semigroup. Although a Möbius transformation is always invertible, the corresponding transformation restricted to D may not be one-to-one.

Semigroups of projective transformations

The set of projective transformations acting on $\mathbb{R}^2 \cup L_\infty$ or $\mathbb{RP}^2$ forms a semigroup of transformations. Sets of projective transformations with a common restriction, for example those that share a fixed point or map a particular subset such as a conic section into itself, also form semigroups. Semigroups of projective transformations, restricted to a domain that they map into itself, can also be constructed.

Again, although a projective transformation is always invertible, such restricted transformations may not be.

Semigroups of transformations on code spaces

It is possible to form diverse semigroups of transformations on a code space. We note in particular the semigroup of transformations generated by the shift transformation, which is not invertible when $|\mathcal{A}| > 1$.

EXERCISE 3.3.2 *For each $\sigma \in \Omega'_A$ the corresponding branch transformation $f_\sigma : \Omega_A \to \Omega_A$ is given by $f_\sigma(\omega) = \sigma\omega = \sigma_1\sigma_2 \cdots \sigma_{|\sigma|}\omega_1\omega_2 \cdots$ for all $\omega \in \Omega_A$. Show that $\{f_\sigma : \sigma \in \Omega'_A\}$ is a semigroup of transformations.*

IFS semigroups

An **iterated function system**, or **IFS**, consists of a finite sequence of transformations that map from a space to itself. An IFS may be denoted by

$$\{\mathbb{X}; f_1, f_2, \ldots, f_N\},$$

where $f_i : \mathbb{X} \to \mathbb{X}$ for $i = 1, 2, \ldots, N$ and $N \geq 1$ is an integer. Thus we may refer to 'the IFS $\{\mathbb{X}; f_1, f_2, \ldots, f_N\}$'. Please look back at Chapter 2, around Theorem 2.4.15, where we first introduced IFSs. Typically we consider IFSs in which the space $\mathbb{X}$ is a metric space, the transformations are Lipschitz or strictly contractive, i.e. $L < 1$, and there is more than one transformation. When the transformations are contractions and the space $\mathbb{X}$ is complete the IFS is called a contractive IFS. A contractive IFS is referred to as a 'hyperbolic' IFS in [9] and possesses a unique attractor, or fractal set, the fixed point of the associated contraction mapping on $\mathbb{H}(\mathbb{X})$. We will often denote the attractor set of a contractive IFS by the symbol A.

DEFINITION 3.3.3 An **IFS semigroup** is a semigroup of transformations generated by an IFS.

We will use the notation $\mathcal{S}_{\{\mathbb{X}; f_1, f_2, \ldots, f_N\}}$, or $\mathcal{S}_{\{f_1, f_2, \ldots, f_N\}}(\mathbb{X})$ or more briefly $\mathcal{S}_{\{f_1, f_2, \ldots, f_N\}}$, to denote the IFS semigroup generated by the IFS $\{\mathbb{X}; f_1, f_2, \ldots, f_N\}$. In this chapter we are interested in the orbits of sets, measures and pictures under IFS semigroups and in the sets, measures and pictures that can be constructed from these orbits.

EXERCISE 3.3.4 *Let $(\mathbb{X}, d)$ be a metric space. Show that the set of Lipschitz transformations on $\mathbb{X}$ forms a semigroup.*

EXERCISE 3.3.5 *Let $(\mathbb{X}, d)$ be a metric space. Show that the set of Lipschitz transformations on $\mathbb{X}$ with Lipschitz constant $L < 1$ forms a semigroup.*

EXERCISE 3.3.6 *Let $(\mathbb{X}, d)$ be a metric space. Construct an example to show that the set of Lipschitz functions with a fixed Lipschitz constant $L > 1$ does not in general form a semigroup.*

From this point, if this is new material for you, you might like to omit the final three examples of semigroups and skip ahead to the next subsection.

Semigroups of rational transformations on the Riemann sphere

There are many types of semigroups of tranformations that act on 'flat' spaces such as $\mathbb{R}^2$. We note in particular that the set of rational functions of a complex variable, that is, ratios of complex polynomials in $z \in \widehat{\mathbb{C}}$, forms a semigroup of transformations on the Riemann sphere. Such semigroups are related to complex analytic dynamical systems and to graceful families of fractals such as Julia sets. The set of complex polynomials and the set of rational functions of degree 1, namely the Möbius transformations, are each sub-semigroups. Again, new semigroups may be obtained by restricting the domains of the transformations.

Semigroups associated with dynamical systems

There is a close relationship between dynamical systems and fractals, and techniques used in dynamical systems theory are useful in connection with IFSs and IFS semigroups. Conversely, fractal geometry informs dynamical systems theory.

The study of the semigroup generated by a single transformation $f : \mathbb{X} \to \mathbb{X}$ is essentially the study of the corresponding **dynamical system**, denoted by $\{\mathbb{X}; f\}$. Studies of dynamical systems tend to focus on the case where f is invertible – see for example [56]. The **orbit** of a point $x_0 \in \mathbb{X}$ under the dynamical system $\{\mathbb{X}; f\}$ is the sequence of points $\{x_n = f^{\circ n}(x_0) : n = 0, 1, 2, \ldots\}$; note that the orbit includes the initial point x_0. Studies of dynamical systems are primarily concerned with the structure of their orbits, the limiting behaviour of their orbits, ergodic properties, recurrence properties (dealing with questions such as 'When does an orbit return arbitrarily close to its starting point?') and properties which are invariant under changes of coordinates.

Topological dynamics, for example, is concerned with groups of homeomorphisms and semigroups of continuous transformations on compact metric spaces. Dynamical systems theory uses in particular the study of dynamical systems on code space Ω_A, called **symbolic dynamical systems**, together with mappings between code space and other spaces, for example $\mathbb{R}^2$, to explain aspects of the behaviour of dynamical systems acting on the latter spaces.

Semigroups associated with autonomous systems

One notable circumstance where semigroups of transformations arise is in connection with any model physical system whose state $x(t)$ at time $t \geq 0$ can be determined fully from a knowledge of both its state at any earlier time $s \geq 0$ and the time elapsed, $t - s$. We call such systems **autonomous**.

An autonomous system always behaves in the same way when it is started off in the same way; it runs to its own clock, not an external one. Indeed a perfect wind-up clock is an example of an autonomous system. Autonomous systems occur frequently in the physical sciences; any experiment in physics which can be repeated over and over again to produce the same behaviour, regardless of the date and time, and which may be initialized at any of its states may be represented by such a model. Often the model involves an array of integro-differential equations that incorporate the model assumptions, physical laws, etc. which govern its behaviour. Some of these systems model physical processes that influence the shape and look of the world around us.

Autonomous systems may be associated with conservation laws and invariance properties in fluid dynamics, classical mechanics, electrostatics and so on. We are interested in them because the colour and intensity of the light emitted or reflected by real-world objects moving in an approximately autonomous system, such as waves on the sea or clouds in the sky or the rings of Saturn, finds its way into real-world pictures; we expect to find some sort of trace or record of these systems in invariance properties of parts of pictures under appropriate semigroups of transformations.

Let $\mathbb{X}$ denote the set of possible states of an autonomous system. It could describe, for example, the height of a plant, the coordinates and momentum of a particle, the number of sharks and fishermen in a model for interacting species, the positions of the hands on a clockface or possible combinations of colours and forms in a picture that changes with time according to certain rules.

We define $F_t : \mathbb{X} \to \mathbb{X}$ to be the transformation that maps the state of an autonomous system at time $t = 0$ to its state at time $t \geq 0$. The transformation F_t is sometimes called an **evolution operator**. Since $F_t(F_s(x)) = F_{t+s}(x)$ for all $x \in \mathbb{X}$, it follows that

$$F_t \circ F_s = F_{t+s} \quad \text{for all } s, t \geq 0.$$

This implies that

$$(\{F_t : t \geq 0\}, \circ) \tag{3.3.1}$$

is a semigroup of transformations. Since this semigroup depends upon a single parameter, t, it is called a **one-parameter semigroup**.

Let $F_t : \mathbb{X} \to \mathbb{X}$ with $t \geq 0$ be an evolution operator, let $x \in \mathbb{X}$ and let $\mathcal{O}(x)$ be the orbit of x; see Definition 3.3.8 below. When $\mathbb{X} \subset \mathbb{R}^2$ the set of orbits of an autonomous system may provide what is called a **phase portrait** of the system. We think of a phase portrait as being a picture, maybe in black and white, showing the orbits of many different points simultaneously. We notice that

$$F_t(\mathcal{O}(x)) = F_t(\{F_{\widetilde{t}}(x) : \widetilde{t} \geq 0\}) = \{F_{\widetilde{t}}(x) : \widetilde{t} \geq t\} \subset \mathcal{O}(x) \quad \text{for all } t \geq 0.$$

Now let $\mathbb{X}_0 \subset \mathbb{X}$ and define $T(\mathbb{X}_0) = \cup \{\mathcal{O}(x) : x \in \mathbb{X}_0\}$. Then

$$F_t(T(\mathbb{X}_0)) \subset T(\mathbb{X}_0). \tag{3.3.2}$$

So an autonomous system yields a semigroup of transformations, Equation (3.3.1), and a collection of sets, Equation (3.3.2), each of which is mapped into itself by every transformation of the semigroup. We can think of some phase portraits as being pictures that are invariant under semigroups of transformations.

EXERCISE 3.3.7   *Let* $\mathbb{X} = \mathbb{R}^2$ *and* $(x, y) = (x(t), y(t)) \in \mathbb{R}^2$ *evolve according to the pair of differential equations*

$$\frac{dx}{dt} = -\alpha y, \quad \frac{dy}{dt} = \beta x \quad \text{for all } t \geq 0, \tag{3.3.3}$$

subject to the initial condition $(x(0), y(0)) = (x_0, y_0)$, *where* $\alpha, \beta > 0$ *and* (x_0, y_0) *is any point in* $\mathbb{R}^2$. *Show that the corresponding evolution operator* $F_t : \mathbb{R}^2 \to \mathbb{R}^2$ *is defined by the* 2×2 *matrix*

$$F_t = \begin{pmatrix} \cos \sqrt{\alpha\beta}t & -\sqrt{\dfrac{\alpha}{\beta}} \sin \sqrt{\alpha\beta}t \\ \sqrt{\dfrac{\beta}{\alpha}} \sin \sqrt{\alpha\beta}t & \cos \sqrt{\alpha\beta}t \end{pmatrix}. \tag{3.3.4}$$

Verify that $F_t \cdot F_s = F_{t+s}$. *Show that, for all points* (x, y) *on any orbit of the system,*

$$\beta x^2 + \alpha y^2 = constant. \tag{3.3.5}$$

Describe subsets of $\mathbb{R}^2$ *that are mapped into themselves by all transformations of the semigroup.*

Orbits of semigroups

DEFINITION 3.3.8   An **orbit** of a semigroup $\mathcal{S}(\mathbb{X})$ is a subset of $\mathbb{X}$ of the form

$$\mathcal{O}(x) = \{x\} \cup \{f(x) : f \in \mathcal{S}(\mathbb{X})\}$$

for some $x \in \mathbb{X}$. $\mathcal{O}(x)$ is called the **orbit of the point** x. A semigroup is said to be **discrete** iff the orbit $\mathcal{O}(x) \subset \mathbb{X}$ consists of isolated points for all $x \in \mathbb{X}$. A semigroup is said to be **continuous** iff, for any given $x \in \mathbb{X}$, there is a continuous function $f : [0, \infty) \to \mathbb{X}$ such that the orbit $\mathcal{O}(x)$ can be written in the form $\mathcal{O}(x) = \{f(t) : t \in [0, \infty)\}$.

An example of an orbit $\mathcal{O}(x)$ of a point x under a continuous semigroup is illustrated in Figure 3.10. The semigroup of transformations is $\{f_\theta : \theta \in [0, \infty)\}$, where $f_\theta : \mathbb{R}^2 \to \mathbb{R}^2$ is defined by

$$f_\theta(x, y) = \left(xr^{2\theta} \cos \theta - yr^{2\theta} \sin \theta, \ xr^{2\theta} \sin \theta + yr^{2\theta} \cos \theta\right). \tag{3.3.6}$$

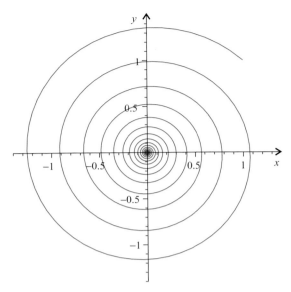

Figure 3.10 Orbit of the point $(1, 1)$ under the continuous semigroup of transformations defined in Equation (3.3.6).

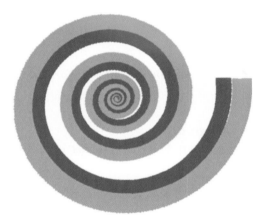

Figure 3.11 Example of a picture that is invariant under a continuous semigroup of transformations.

In the figure, $r = 0.975$ and $x = (1, 1)$. In this case the orbit is actually invariant under the semigroup, that is,

$$\mathcal{O}(x) = \bigcup \{f_\theta(\mathcal{O}(x)) : \theta \in [0, \infty)\}.$$

See also Figure 3.11.

A visual example of an orbit of a semigroup is the set of points defined by the tips of the green leaves in Figure 3.3. Clearly in this case we are dealing with a discrete semigroup. See also Figure 3.12.

EXERCISE 3.3.9   *Let $\alpha \in \mathbb{R}$ and let $f_\alpha : \mathbb{R} \to \mathbb{R}$ be defined by $f_\alpha(x) = \alpha \cdot x$. Show that $\{f_\alpha : \alpha \in \mathbb{R}\}$ is a continuous semigroup and that $\{f_\alpha : \alpha \in \mathbb{Z}\}$ is a discrete semigroup.*

EXERCISE 3.3.10   *Let $r = 0.99$. For $\theta \in [0, \infty)$ define $f_\theta : \mathbb{R}^2 \to \mathbb{R}^2$ by*

$$(f_\theta(x, y))^T = \begin{pmatrix} r^\theta \cos \theta & -r^\theta \sin \theta \\ r^\theta \sin \theta & r^\theta \cos \theta \end{pmatrix} \begin{pmatrix} x \\ y \end{pmatrix}.$$

Sketch the orbit of the point $(1, 1)$ under the semigroup of transformations $\{f_\theta : \theta \in [0, \infty)\}$.

Consider the IFS semigroup $\mathcal{S}_{\{f\}}(\mathbb{X})$ generated by a single transformation $f : \mathbb{X} \to \mathbb{X}$. We define $f^{\circ 0}(x) = x$ for all $x \in \mathbb{X}$, $f^{\circ 1} = f$ and $f^{\circ(n+1)} = f \circ f^{\circ n}$ for $n = 1, 2, 3, \ldots$ Then

$$\mathcal{S}_{\{f\}}(\mathbb{X}) := \{f^{\circ n} : n = 0, 1, 2, \ldots\}.$$

Clearly

$$f^{\circ n} \circ f^{\circ m} = f^{\circ(n+m)} \quad \text{for all } m, n = 0, 1, 2, \ldots$$

In this case the orbit $\mathcal{O}(x)$ of the point $x \in \mathbb{X}$ under the semigroup $\mathcal{S}_{\{f\}}(\mathbb{X})$ is

$$\mathcal{O}(x) = \{f^{\circ n}(x) : n = 0, 1, 2, \ldots\}.$$

Notice that, knowing the IFS, we can treat this orbit as a sequence.

Suppose that the semigroup $S_{\{M\}}(\mathbb{R}^2)$ is generated by a linear transformation $M : \mathbb{R}^2 \to \mathbb{R}^2$ represented by the matrix M. Then, since

$$(f^{\circ n}(x, y))^T = M^n \begin{pmatrix} x \\ y \end{pmatrix},$$

it follows that

$$\mathcal{S}_{\{M\}}(\mathbb{R}^2) = \{M^n : n = 0, 1, 2, \ldots\}$$

where $M^0 := I$, the identity matrix.

EXERCISE 3.3.11   *Plot the orbit of the point $(1, 1) \in \mathbb{R}^2$ under the semigroup of transformations $\mathcal{S}_{\{M\}}(\mathbb{R}^2)$, where M is the linear transformation represented by*

$$M = \begin{pmatrix} 0.9 & 0.1 \\ 0.2 & 0.3 \end{pmatrix}.$$

Is this semigroup discrete?

Now consider the orbit of the point $x \in \mathbb{X}$ under the IFS semigroup $\mathcal{S}_{\{f_1, f_2, \ldots, f_N\}}(\mathbb{X})$. We notice that

$$\mathcal{S}_{\{f_1, f_2, \ldots, f_N\}}(\mathbb{X}) = \left\{ f_\sigma : \sigma \in \Omega'_{\{1,2,\ldots,N\}} \right\}$$

where $f_\sigma := f_{\sigma_1} \circ f_{\sigma_2} \circ \cdots \circ f_{\sigma_{|\sigma|}}$ for all $\sigma \in \Omega'_{\{1,2,\dots,N\}}$, and we define $f_\varnothing : \mathbb{X} \to \mathbb{X}$ by $f_\varnothing(x) = x$ for all $x \in \mathbb{X}$. It follows that

$$\mathcal{O}(x) = \left\{ f_\sigma(x) : \sigma \in \Omega'_{\{1,2,\dots,N\}} \right\}.$$

This provides us with an addressing function $\phi : \Omega'_{\{1,2,\dots,N\}} \to \mathcal{O}(x)$ defined by

$$\phi(\sigma) = f_\sigma(x) \quad \text{for all } \sigma \in \Omega'_{\{1,2,\dots,N\}},$$

for each $x \in \mathbb{X}$. When the IFS is contractive this addressing function can be extended continuously to $\Omega'_{\{1,2,\dots,N\}} \cup \Omega_{\{1,2,\dots,N\}}$, as described in the following theorem.

THEOREM 3.3.12 *Let $\mathbb{X}$ be a complete metric space. Let the transformations $f_n : \mathbb{X} \to \mathbb{X}$ be contractions, that is, strictly contractive functions, for $n = 1, 2, \dots, N$ where $N \geq 1$ is an integer. Let A denote the attractor of the IFS $\{\mathbb{X}; f_1, f_2, \dots, f_N\}$. That is, A is the unique compact nonempty set that obeys*

$$A = f_1(A) \cup f_2(A) \cup \cdots \cup f_N(A).$$

Let $x \in \mathbb{X}$ and let $\mathcal{O}(x)$ denote the orbit of x under the IFS semigroup. Then there is a continuous transformation

$$\phi : \Omega'_{\{1,2,\dots,N\}} \cup \Omega_{\{1,2,\dots,N\}} \to \mathcal{O}(x) \cup A,$$

defined by

$$\phi(\sigma) = \begin{cases} f_\sigma(x) & \text{when } \sigma \in \Omega'_{\{1,2,\dots,N\}}, \\ \lim_{n\to\infty} f_{\sigma_1 \sigma_2 \cdots \sigma_n}(x) & \text{when } \sigma = \sigma_1 \sigma_2 \cdots \in \Omega_{\{1,2,\dots,N\}}. \end{cases}$$

PROOF The underlying topology is the natural topology on the code space $\Omega'_{\{1,2,\dots,N\}} \cup \Omega_{\{1,2,\dots,N\}}$, which we discussed at length in Chapter 1. The main points to demonstrate are that $\lim_{n\to\infty} f_{\sigma_1 \sigma_2 \cdots \sigma_n}(x)$ exists and that the resulting function ϕ is continuous. Both follow from the contractivity of the functions $f_1, f_2, \dots, f_N$ and the completeness of the space $\mathbb{X}$. See [48], Theorem 3.1(3). $\qquad\square$

EXERCISE 3.3.13 *Prove Theorem 3.3.12.*

3.4 Orbits of sets under IFS semigroups

DEFINITION 3.4.1 Let $\mathcal{S}(\mathbb{X})$ be a semigroup and let $C \subset \mathbb{X}$ with $C \neq \varnothing$. Then the **orbit of the set C** under the semigroup $S(\mathbb{X})$ is the set of subsets of $\mathbb{X}$ defined by

$$\mathcal{O}(C) = \{C\} \cup \{f(C) : f \in \mathcal{S}(\mathbb{X})\}.$$

Figure 3.12   What is the orbit of the top left corner of the second largest frame in this picture, under the semigroup of transformations implied by this picture?

Notice that $\mathcal{O}(C)$ is a set of sets of points. We will write

$$P = \bigcup \mathcal{O}(C)$$

to denote the union of all the sets in the orbit $\mathcal{O}(C)$. We call $P = \bigcup \mathcal{O}(C)$ the **orbital set** associated with the semigroup $S(\mathbb{X})$ acting on the set C.

Orbits of sets under semigroups of transformations are illustrated in Figures 3.13–3.16 and 3.18. Notice that the sets in the orbits may have nonempty intersections or they may all be separated from one another. Notice too that there is no requirement that the set C be connected; for example, C could be the union of the fish in the four corners of Figure 3.13.

When $\mathbb{X}$ is a 'flat' space such as $\mathbb{R}^2$ we think of P as a black-and-white picture. The following theorem says that this picture is the union of transformed copies of itself together with the set C. In order to describe this picture, we need to know only C and the set of transformations that generate the IFS. In this context we sometimes call C a **condensation set**; see for example [9] or [46]. We will also, later, refer to **condensation pictures** and **condensation measures**. Be careful not to confuse 'condensation set' with 'the set of condensation points *of* a set'. The latter refers, in other texts, to an unrelated concept.

THEOREM 3.4.2   *Let $\mathcal{O}(C)$ denote the orbit of a nonempty subset C of $\mathbb{X}$ under the IFS semigroup $\mathcal{S}_{\{f_1, f_2, \dots, f_N\}}(\mathbb{X})$. Let $P = \bigcup \mathcal{O}(C)$. Then P obeys the*

Figure 3.13 Some sets in the orbits of each of the four sets represented by the fish at the four corners, under an IFS semigroup generated by a single projective transformation. One orbit is marked in red. Can you identify, in blue, sets in the orbit of the fish in the bottom left corner?

*following equality, known as a **self-referential equation***:

$$P = C \cup f_1(P) \cup f_2(P) \cup \cdots \cup f_N(P). \tag{3.4.1}$$

PROOF   We have

$$C \cup f_1(P) \cup f_2(P) \cup \cdots \cup f_N(P)$$

$$= C \cup \left\{ \bigcup_n f_n(\cup \mathcal{O}(C)) \right\}$$

$$= C \cup \left\{ \bigcup_n f_n \left(\cup \{ f_\sigma(C) : \sigma \in \Omega_{\{1,2,\ldots,N\}} \} \right) \right\}$$

$$= C \cup \left\{ \bigcup \{ f_\sigma(C) : \sigma \in \Omega_{\{1,2,\ldots,N\}}, |\sigma| \geq 1 \} \right\} = P.$$

$\square$

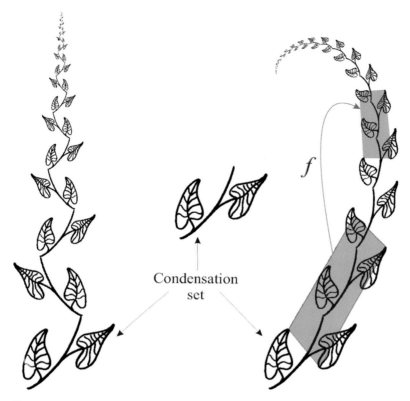

Figure 3.14   Two different semigroup tilings of orbital sets generated by a single projective transformation f acting on a condensation set C, the leafy spring at the centre of the figure. The intersection of any filled rectangle that does not meet C with the orbital picture is the image, under this transformation, of the intersection of a filled quadrilateral with the orbital picture.

Theorem 3.4.2 says that the orbital set P is a fixed point of the transformation $F_C : \mathbb{S}(\mathbb{X}) \to \mathbb{S}(\mathbb{X})$ defined by

$$F_C(B) = C \cup f_1(B) \cup f_2(B) \cup \cdots \cup f_N(B)$$

for all $B \in \mathbb{S}(\mathbb{X})$, and that

$$\bigcup \mathcal{O}(C) = \bigcup \mathcal{O}(P).$$

When the underlying space $\mathbb{X}$ is a compact metric space, C is compact and the IFS consists of strictly contractive transformations, we have $F_C(\mathbb{H}(\mathbb{X})) \subset \mathbb{H}(\mathbb{X})$, where $\mathbb{H}(\mathbb{X})$ is the space of nonempty compact subsets of $\mathbb{X}$. In this case, if we restrict F_C to $\mathbb{H}(\mathbb{X})$ then it becomes a strict contraction with respect to the Hausdorff metric. In this case, as in Theorem 2.4.15, the orbital set P is unique in $\mathbb{H}(\mathbb{X})$. Moreover, this unique fixed point depends continuously on the transformations in the IFS and on the condensation set C. In other words, in this strictly contractive case, if you change the IFS slightly and the set C by a small amount, as measured by the Hausdorff metric, then the orbital picture P will change only a little.

This continuity property is useful in image-modelling applications, where one seeks a set C and an IFS such that the set of points in the orbit of the IFS semigroup acting on the set C is an approximation, perhaps an elegant one, to a given set. For more on this, see Chapter 4.

DEFINITION 3.4.3   Let $\mathcal{S}_{\{f_1, f_2, \ldots, f_N\}}(\mathbb{X})$ be an IFS semigroup. If $C \subset \mathbb{X}$ is nonempty and such that $f_\sigma(C) \cap f_\upsilon(C) = \varnothing$ whenever $\sigma, \upsilon \in \Omega'_{\{1,2,\ldots N\}}$ with $\sigma \neq \upsilon$ then the orbit of C is called an IFS **semigroup tiling** of the set $\cup \mathcal{O}(C)$, and each set $f_\sigma(C)$, for $\sigma \in \Omega'_{\{1,2,\ldots N\}}$, is called a **semigroup tile**. In this case σ is called the **address** of the tile $f_\sigma(C)$. We say that the semigroup, acting on C, **generates** the semigroup tiling.

Examples of semigroup tilings of sets are illustrated in Figures 3.13–3.15, 3.17, and 3.18. Notice that the object tiled need not be two dimensional – it is a fractal in Figure 3.16. The transformations may not be one-to-one. The tiles may be of diverse sizes and shapes. Many pictures in this book contain IFS semigroup tilings. Polygon tilings of some attractors of IFSs for contractive affine maps have been documented by Fathauer [36]. He refers to these tilings as fractal tilings.

The following property is quite a natural one: at least, I have often encountered situations where it applies when considering IFS semigroups of contractive transformations.

DEFINITION 3.4.4   Let $\mathcal{S}_{\{f_1, f_2, \ldots, f_N\}}(\mathbb{X})$ be an IFS semigroup and $C \subset \mathbb{X}$ be a nonempty set. Then the orbit $\mathcal{O}(C)$ is said to be **layered** iff

$$\bigcap_{n=1}^{\infty} F^{\circ n}(P) = \varnothing,$$

where $P = \bigcup \mathcal{O}(C)$ is the associated orbital set and $F : \mathbb{S}(\mathbb{X}) \to \mathbb{S}(\mathbb{X})$ is defined by

$$F(B) = f_1(B) \cup f_2(B) \cup \cdots \cup f_N(B) \quad \text{for all } B \text{ in } \mathbb{S}(\mathbb{X}).$$

An orbit is layered, roughly speaking, if the 'limit set' of the sequence of sets $\{F^{\circ k}(P) : k = 1, 2, 3, \ldots\}$ does not intersect P.

EXERCISE 3.4.5   *Let $C \subset \mathbb{R}^2$ denote the circle of radius 1 centred at the origin. Let $f_1 : \mathbb{R}^2 \to \mathbb{R}^2$ be the similitude $f_1(x, y) = (\frac{1}{2}x, \frac{1}{2}y + 1)$ and let $f_2 : \mathbb{R}^2 \to \mathbb{R}^2$ be the similitude $f_2(x, y) = (\frac{1}{2}x + 1, \frac{1}{2}y)$. Show that the orbit of C under the IFS semigroup $\{\mathbb{R}^2; f_1, f_2\}$ is layered.*

Figure 3.15   Two semigroup tilings generated by an IFS semigroup with $N = 2$. The regular tiling on the right continues onwards to the right and downwards without limit, but the tiling on the left approaches the green canopy, the attractor of the IFS.

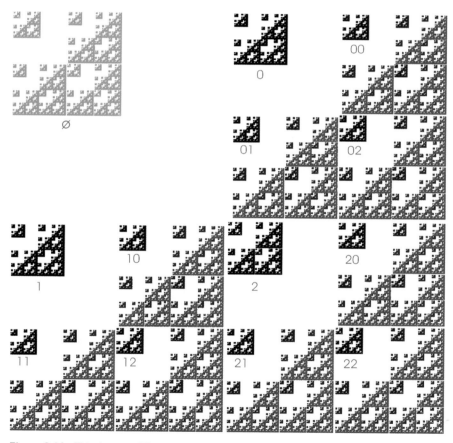

Figure 3.16   This shows an IFS semigroup tiling and addresses for some of the tiles. The condensation set, with address $\varnothing$, in green at the top left, is itself a fractal set. Successive generations of tiles are smaller and smaller. The semigroup orbit in this case is actually *layered*, see Definition 3.4.4. The attractor of the IFS, the limit set of the tiling, shown in blue, is not part of the tiling but the tiles approach it arbitrarily closely.

The following theorem provides a convenient necessary and sufficient condition for a layered orbit $\mathcal{O}(C)$ to yield a semigroup tiling of $P = \bigcup \mathcal{O}(C)$. It is relevant to image compression. Given a set of the form P, we would like to represent it as efficiently as possible; therefore we may wish to choose the condensation set C to be as small as possible, without changing the orbital set P. This theorem fills me with wonder because it seems almost magical that all the tiles fit together neatly once the initial ones do.

THEOREM 3.4.6 *Let $\mathcal{S}_{\{f_1, f_2, \ldots, f_N\}}(\mathbb{X})$ be an IFS semigroup of one-to-one transformations. Let $\mathcal{O}(C)$ be a layered orbit of a nonempty set $C \subset \mathbb{X}$. Let*

$$C_0 = C \backslash (f_1(C) \cup f_2(C) \cup \cdots \cup f_N(C)).$$

Then $\bigcup \mathcal{O}(C) = \bigcup \mathcal{O}(C_0)$. Also, $\mathcal{O}(C)$ is a semigroup tiling of $P = \bigcup \mathcal{O}(C)$ if and only if

$$C \cap f_n(P) = \varnothing \text{ and } f_n(P) \cap f_m(P) = \varnothing \quad \text{for } n \neq m, \tag{3.4.2}$$

for all $n, m \in \{1, 2, \ldots, N\}$.

PROOF From Theorem 3.4.2 we have $P = C \cup f_1(P) \cup f_2(P) \cup \cdots \cup f_N(P) = C \cup F(P)$. It follows that

$$P = C_0 \cup F(P) \tag{3.4.3}$$

because $f_n(P)$ contains $f_n(C)$ for all $n \in \{1, 2, \ldots, N\}$. We substitute from Equation (3.4.3) into itself to obtain

$$P = \left\{ \bigcup_{\sigma \in \Omega'_{\{1,2,\ldots,N\}}, |\sigma| \leq 1} f_\sigma(C_0) \right\} \cup F^{\circ 2}(P).$$

By induction, we have

$$P = \left\{ \bigcup_{\sigma \in \Omega'_{\{1,2,\ldots,N\}}, |\sigma| \leq k-1} f_\sigma(C_0) \right\} \cup F^{\circ k}(P)$$

for all $k \in \{1, 2, 3, \ldots\}$. It now follows that if $x \in P$ then x belongs to the right-hand side of the latter equation; since the orbit is layered there exists $k \in \{1, 2, 3, \ldots\}$ such that $x \notin F^{\circ k}(P)$ and therefore x belongs to the expression within braces for some k, which in turn implies that $x \in f_\sigma(C_0)$ for some $\sigma \in \Omega'_{\{1,2,\ldots,N\}}$. Hence $P \subset \mathcal{O}(C_0)$. But also, since $C_0 \subset C$, we must have $\mathcal{O}(C_0) \subset P$. So $P = \mathcal{O}(C_0)$ as desired.

Now assume that Equation (3.4.2) is true. It follows that

$$f_\sigma(C) \cap f_\omega(C) = \varnothing \quad \text{for all } \sigma, \omega \in \Omega'_{\{1,2,\ldots,N\}}, \quad |\sigma| \leq 1, |\omega| \leq 1, \sigma \neq \omega,$$

We proceed by induction. Let us assume that, for some integer $K \geq 1$, we have

$$f_\sigma(C) \cap f_\omega(C) = \varnothing \quad \text{for all } \sigma, \omega \in \Omega'_{\{1,2,\ldots,N\}}, \quad |\sigma| \leq K, |\omega| \leq K, \sigma \neq \omega. \tag{3.4.4}$$

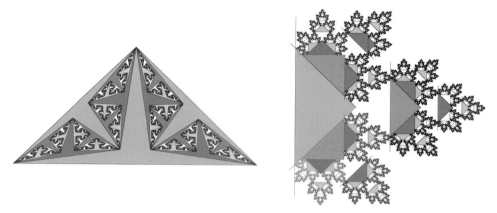

Figure 3.17   Two examples of IFS semigroup tilings. The triangle on the left is tiled with the orbit of a six-sided figure, under an IFS of two affine transformations. The limit set of the set of triangular tiles, on the right, is the attractor of an IFS of three transformations.

Figure 3.18   The left-hand picture illustrates the points in the orbit of a set, the flower picture at centre left, under a semigroup generated by a single Möbius transformation. This orbital set is in fact a semigroup tiling, as illustrated by the image at centre right. The image on the far right is the initial tile.

Using the assumed one-to-oneness of the transformations and the fact that $f_n(P) \cap f_m(P) = \varnothing$ for $n \neq m$, when $n, m \in \{1, 2, \dots, N\}$, it follows that

$$f_\sigma(C) \cap f_\omega(C) = \varnothing \quad \text{for all } \sigma, \omega \in \Omega'_{\{1,2,\dots,N\}},$$
$$2 \leq |\sigma| \leq K + 1, 2 \leq |\omega| \leq K + 1, \sigma \neq \omega.$$

The assumption $C \cap f_n(P) = \varnothing$ implies that $C \cap f_\sigma(C) = \varnothing$ for $|\sigma| = K$, and this in turn implies that $f_n(C) \cap f_\sigma(C) = \varnothing$ for $|\sigma| = K + 1$ and $n \in \{1, 2, \dots, N\}$, again because the maps are one-to-one. It follows that Equation (3.4.4) holds with K replaced by $K + 1$. This completes the inductive step and implies that $f_\sigma(C) \cap f_\omega(C) = \varnothing$ for all $\sigma, \omega \in \Omega'_{\{1,2,\ldots,N\}}$, $\sigma \neq \omega$. Hence $\mathcal{O}(C)$ forms a semigroup tiling of P. To prove the converse, we simply note that if any statement in Equation (3.4.2) is false then the corresponding pair of putative tiles 'overlap' one another, which implies that $\mathcal{O}(C)$ is not a semigroup tiling of P. $\square$

If the ranges of the transformations $f_n(\mathbb{X})$ are disjoint then the second set of conditions in Equation (3.4.2) is satisfied. Such IFSs, whose transformations have disjoint ranges, occur in connection with set attractors that are 'just touching' or totally disconnected, as will be discussed in Chapter 4. Also, semigroup tilings associated with a dynamical system $f : \mathbb{X} \to \mathbb{X}$ may occur when there is an IFS associated with the inverse of f. For example, in Section 3.5 we show that the Henon transformation generates fascinating tilings because it is one-to-one and onto.

Also, in complex analytic dynamics, when f is a rational function on the Riemann sphere the ranges of the branches of f^{-1} intersect only at certain isolated points. These inverse branch transformations generate astonishing tilings. For example, let J be the Julia set of the dynamical system $\{\widehat{\mathbb{C}}; f(z) = (z - \lambda)^2\}$, where $\lambda \in \mathbb{C}$ is a parameter. Then J is essentially the fractal set, the attractor, of the IFS

$$\{\mathbb{C}; f_1(z) = \lambda + \sqrt{z}, \ f_2(z) = \lambda - \sqrt{z}\};$$

see for example [9]. In this case $f_1(P) \cap f_1(P) = \varnothing$ whenever $P \subset \mathbb{C}$ and $0 \notin P$. It follows that if one chooses $C \subset \mathbb{C}\backslash\{0\}$ such that $C \cap J = \varnothing$, which ensures that the orbit of C is layered, then it follows from Theorem 3.4.6 that the set C_0 generates a semigroup tiling under the IFS semigroup. In fact, pictures of J are often produced by means of the 'escape time' algorithm; see for example [78] or [9]. Points $z \in \mathbb{C}$ are assigned colours according to the least integer n such that $f^{\circ n}(z) \in C_0$, where $C_0 \subset \mathbb{C}$ has been chosen so that $C_0 \cap J = \varnothing$. The resulting beautiful pictures, artistic and harmonious, illustrate the Julia set by colouring points that do not belong to the Julia set; see Figure 3.19, for example. For us, however, such pictures comprise two different, more substantial, mathematical entities: semigroup tilings, and pictures that are invariant, or mapped into themselves, under certain transformations.

EXERCISE 3.4.7 *Identify the addresses of the tiles in the IFS semigroup tilings illustrated in Figure 3.15. State your assumptions.*

EXERCISE 3.4.8 *Let $\mathcal{O}(C)$ denote the orbit of a set $C \subset \mathbb{X}$ under the semigroup $\mathcal{S}_{\{f_1, f_2, \ldots, f_N\}}(\mathbb{X})$. Assume that the orbit is layered and that the transformations are one-to-one. Define*

$$C' := \bigcup_{n \in \{1,2,\ldots,N\}} f_n^{-1}(f_n(C) \cap C).$$

Figure 3.19   Semigroup tiling associated with a Julia set. The condensation set, the first tile, is the outer necklace. Each of the two 'second-generation' tiles, whose addresses are 0 and 1, is formed from a copy of the original necklace that has been cut so that it forms a strand instead of a loop. These two cut necklaces meet each other to make the second necklace, which has twice as many beads as the first one. The different colours indicate different tiles, for the first four generations.

Show that

$$\left(\bigcup \mathcal{O}(C')\right) \cup C = \bigcup \mathcal{O}(C)$$

and illustrate this result using an elegant overlapping orbit of a set C under a nontrivial semigroup of transformations on $\mathbb{R}^2$.

3.5   Orbits of pictures under IFS semigroups

The underlying ideas in this section are not the same as those in Section 3.4 for this reason: whereas the union of two sets is a new set, the union of two pictures is not defined. This leads us to use the tops union $\mathbb{U}$ to define orbital pictures. An orbital picture is a picture that is specified, uniquely and naturally, in terms of an orbit of pictures under an IFS semigroup, using the tops union. But the tops union is not commutative. This has the consequence that an orbital picture may be more intricate mathematically than a corresponding orbital set.

In this section we define, establish properties of and illustrate orbital pictures; we often find illustrations of them to be visually exciting and beautiful. Orbital pictures have applications to fractal image compression and computer graphics. We believe that they also have applications in image recognition, cryptography, number theory and bioinformatics. We will mention these applications in later sections.

Although the theory of orbital pictures has been inspired by the idea of IFSs acting on pictures, it is important to remember, as you read the mathematics, that this theory is simply mathematics and may be read as such, without regard for physical pictures.

After defining orbital pictures we establish the theoretical basis for the deterministic computation of them. We show how orbital pictures obey their own special type of self-referential equation and define the panels and the code space of an orbital picture. We prove that the shift transformation maps this space into itself, and this provides us with a symbolic dynamical system and various topological invariants of orbital pictures, including the orbital growth rate and a certain symbolic entropy. We illustrate the corresponding dynamics on the panels of an orbital picture and mention the critical relationship between an orbital picture and the attractor of an IFS, when both are defined. We also illustrate and discuss a family of examples, related to pictures of flowers in a field that stretches to the horizon, for which we can say something about the symbolic entropy and the diversity. The diversity of an orbital picture is another invariant and, together with the orbital growth rate, provides quantitative information, which is invariant under homeomorphism, about the way that orbital pictures look. When the diversity equals infinity we define another quantity, the growth rate of diversity, which is bounded above by the growth rate of periodic cycles. We introduce the space of limiting pictures associated with an orbital picture, which is used to define the diversity and the growth rate of diversity. Finally we introduce the concept of orbital tiling (in contrast with semigroup tiling by pictures, which is essentially the same as tiling a set by images of the set under a semigroup of transformations) and underneath pictures. Applications of orbital pictures are mentioned. A transformation called the Henon mapping is used to illustrate that the orbital picture generated by a single transformation can be quite complicated.

We restrict our attention to semigroups of one-to-one transformations, because only invertible transformations can be applied to pictures. We will tend to think of $\mathbb{X}$ as being a 'flat' space such as $\mathbb{R}^2$, so that pictures whose domains lie in $\mathbb{X}$ may be illustrated. The symbol $\Pi = \Pi_{\mathfrak{C}}(\mathbb{X})$ denotes the space of pictures whose domains lie in $\mathbb{X}$ and whose ranges lie in a fixed colour space $\mathfrak{C}$.

DEFINITION 3.5.1   Let $\mathfrak{P}_0 \in \Pi$ denote a picture with its domain in $\mathbb{X}$. Then the **orbit of the picture** $\mathfrak{P}_0$ under the semigroup $S(\mathbb{X})$ is defined to be the set of pictures

$$\mathcal{O}(\mathfrak{P}_0) = \{ f(\mathfrak{P}_0) : f \in \mathcal{S}(\mathbb{X}) \}.$$

Notice that the set of domains of the pictures in an orbit of a picture $\mathfrak{P}_0$ is the orbit of a set, the domain of $\mathfrak{P}_0$. In general these domains will overlap, so we are led to use the tops union to define a picture of the orbit of a picture. In turn this means that we need the set of pictures in the orbit to be countable, so we restrict our attention to orbits of pictures generated by IFS semigroups.

The orbital picture $\mathfrak{P} = \uplus \mathcal{O}(\mathfrak{P}_0)$

A picture of the orbit of a picture $\mathfrak{P}_0$ under an IFS semigroup $\mathcal{S}_{\{f_1, f_2, \ldots, f_N\}}(\mathbb{R}^2)$ is obtained as follows. (i) Arrange the pictures in the orbit as a sequence, $\{\mathfrak{P}_n\}_{n=0}^{\infty}$, starting with $\mathfrak{P}_0$. (ii) Let $\mathfrak{P}_1'$ be the picture whose domain is the union of the domains of $\mathfrak{P}_0$ and $\mathfrak{P}_1$ and whose colour values agree with $\mathfrak{P}_0$ on $D_{\mathfrak{P}_0}$, the domain of $\mathfrak{P}_0$, and agree with $\mathfrak{P}_1$ on the rest of its domain. (When the domains of $\mathfrak{P}_0$ and $\mathfrak{P}_1$ overlap, the resulting picture $\mathfrak{P}_1'$ looks like $\mathfrak{P}_0$ with the picture $\mathfrak{P}_1$ sticking out from underneath it.) (iii) Next, similarly combine $\mathfrak{P}_1'$ with $\mathfrak{P}_2$ to produce $\mathfrak{P}_2'$, which may look like $\mathfrak{P}_1'$ with $\mathfrak{P}_2$ sticking out from underneath it. (iv) Continue in this manner to make a sequence of pictures $\{\mathfrak{P}_n'\}_{n=1}^{\infty}$, which, in turn, defines a limiting picture that we denote by $\mathfrak{P} = \uplus \mathcal{O}(\mathfrak{P}_0)$.

Now we will describe this construction more specifically.

DEFINITION 3.5.2   Let $\mathcal{S}_{\{f_1, f_2, \ldots, f_N\}}(\mathbb{X})$ be an IFS semigroup and let $\mathfrak{P}_0 \in \Pi$ be the space of pictures with domains in $\mathbb{X}$. The **canonical sequence** of pictures $\{\mathfrak{P}_n\}_{n=0}^{\infty}$ in the orbit of $\mathfrak{P}_0$ is defined by

$$\mathfrak{P}_n = f_{\sigma(n)}(\mathfrak{P}_0) \quad \text{for all } n = 0, 1, 2, \ldots \tag{3.5.1}$$

where $\sigma(n) = c^{-1}(n)$, $f_{\varnothing} = I$, the identity map, $c : \Omega'_{\{1,2,\ldots N\}} \to \{0, 1, 2, \ldots\}$ is defined by $c(\varnothing) = 0$ and

$$c(\sigma) = \sum_{n=1}^{|\sigma|} \sigma_n N^{|\sigma|-n} \quad \text{for all } \sigma \in \Omega'_{\{1,2,\ldots N\}} \text{ with } \sigma \neq \varnothing. \tag{3.5.2}$$

The function $c : \Omega'_{\{1,2,\ldots N\}} \to \{0, 1, 2, \ldots\}$ assigns a unique index in $\{0, 1, 2, \ldots\}$ to each element of the code space, and it is invertible. So, for example, when $N = 2$ we have

$$c^{-1}(0) = \varnothing, \quad c^{-1}(1) = 1, \quad c^{-1}(2) = 2, \quad c^{-1}(3) = 11,$$
$$c^{-1}(4) = 12 \quad \text{and} \quad c^{-1}(83) = 121211.$$

THEOREM 3.5.3   *Let $\{\mathfrak{P}_n\}_{n=0}^{\infty}$ denote the canonical sequence of pictures in the orbit of $\mathfrak{P}_0 \in \Pi$ under the IFS semigroup $S_{\{f_1, f_2, \ldots, f_N\}}(\mathbb{X})$. Let*

$$\mathfrak{P}_n' = \mathfrak{P}_0 \uplus \mathfrak{P}_1 \uplus \mathfrak{P}_2 \uplus \cdots \uplus \mathfrak{P}_n \tag{3.5.3}$$

for $n = 0, 1, 2, \ldots$ Then there exists a unique picture $\mathfrak{P} = \mathfrak{P}(\mathfrak{P}_0)$ such that $D_{\mathfrak{P}} = \bigcup \mathcal{O}(D_{\mathfrak{P}_0})$ and such that, given $x \in D_{\mathfrak{P}}$, $\mathfrak{P}(x) = \mathfrak{P}_n'(x)$ for some $n \in \{0, 1, 2, \ldots\}$.

PROOF   Notice that, by construction,

$$D_{\mathfrak{P}} = \bigcup_{n=0}^{\infty} D_{\mathfrak{P}_n} = \bigcup_{n=0}^{\infty} D_{\mathfrak{P}_n'}.$$

Therefore given any $x \in D_{\mathfrak{P}}$ we can find $n \in \{0, 1, 2, \ldots\}$ such that $x \in D_{\mathfrak{P}'_n}$. So we need only to prove the uniqueness of the value of $\mathfrak{P}(x) = \mathfrak{P}'_n(x)$. Suppose that $n' \in \{0, 1, 2, \ldots\}$ is such that $x \in D_{\mathfrak{P}'_{n'}}$ and, without loss of generality, $n' \geq n$. Then $x \in D_{\mathfrak{P}'_n} \subset D_{\mathfrak{P}'_{n'}}$ and $\mathfrak{P}_{n'} = \mathfrak{P}_n \Cup \mathfrak{P}_{n'}$, so $\mathfrak{P}_{n'}(x) = \mathfrak{P}_n(x)$. $\qquad\square$

DEFINITION 3.5.4 The unique picture $\mathfrak{P}$ in Theorem 3.5.3 is called the **picture of the orbit of the picture** $\mathfrak{P}_0$ under the IFS semigroup $S_{\{f_1, f_2, \ldots, f_N\}}(\mathbb{X})$. It is denoted by

$$\mathfrak{P} := \Cup \mathcal{O}(\mathfrak{P}_0)$$

and we can write

$$\mathfrak{P} = \mathfrak{P}_0 \Cup \mathfrak{P}_1 \Cup \mathfrak{P}_2 \Cup \cdots$$

We may refer to $\mathfrak{P} = \Cup \mathcal{O}(\mathfrak{P}_0)$ as the **orbital picture** of $\mathfrak{P}_0$ (under the IFS semigroup $S_{\{f_1, f_2, \ldots, f_N\}}(\mathbb{X})$.)

An example of a picture of an orbit of a picture $\mathfrak{P}_0$ under an IFS semigroup is shown in Figure 3.20. The IFS consists of three transformations and its associated fractal set, the set attractor of the IFS, is a Sierpinski triangle. Notice how the domain of $\mathfrak{P}_0$, the part corresponding to the red flower with stamens, *overlaps the attractor of the IFS*. As a result, all the pictures in the orbit of $\mathfrak{P}_0$ overlap the Sierpinski triangle.

Other simple examples of pictures of orbits of pictures under IFS semigroups are shown in Figures 3.21–3.25 and 3.55. The manner in which these images were computed is explained below. The two affine transformations for the orbital picture of buttercups in Figures 3.21, and the close-up in Figure 3.22, are illustrated in Figure 3.36; they are given by

$$f_1(x, y) = (0.7x, 0.7y + 0.3), \quad f_2(x, y) = (0.7x + 0.3, 0.7y + 0.3). \quad (3.5.4)$$

The set attractor of this IFS is the line segment ab in Figure 3.21.

The Möbius transformations used in Figure 3.24 are

$$f_1(z) = \frac{4z + 1 - i}{(1 + i)z + 4} \quad \text{and} \quad f_2(z) = \frac{4z - 1 - i}{(-1 + i)z + 4}, \quad (3.5.5)$$

together with their inverses, and the viewing window is specified by $-1 \leq x \leq 1$ and $-1 \leq y \leq 1$. An invariant set for the IFS in this case is a circle; $N = 4$ and the picture on the left is actually $\mathfrak{P}'_{\frac{4(4^4-1)}{3}}$. The three Möbius transformations

$$f_1(z) = \frac{z}{2iz - 1}, \quad f_2(z) = \frac{z + 1}{2} \quad \text{and} \quad f_3(z) = \frac{(1 - i)z + 1}{z + 1 + i} \quad (3.5.6)$$

are used in Figure 3.25.

Notice that there was potential arbitrariness in the choice of the integers that we assigned to the elements of the IFS semigroup. We made a convenient choice, once

Figure 3.20   Picture of an orbit of a condensation picture, of flowers and a ribbon, under an IFS semigroup with $N = 3$. This is not a picture tiling, according to our definition, because bits of the condensation picture are missing from the 'tiles'. Can you see how the original picture, lower left, overlaps the attractor of the IFS?

and for all, and we will keep to it because it has some wonderful consequences. One consequence is that $\mathfrak{P}$ obeys a self-referential equation; see Theorem 3.5.8. Another is that we can construct a code space and a symbolic dynamical system for the orbital picture, as described in a later subsection. Yet another is that we can compute, efficiently, approximations to $\mathfrak{P} = \Cup \mathcal{O}(\mathfrak{P}_0)$, as we describe next.

Computation of orbital pictures

Notice that $\mathfrak{P}'_n$ is a function of $\mathfrak{P}_0$. Specifically $\mathfrak{P}'_n : \Pi \rightarrow \Pi$ is given by

$$\mathfrak{P}'_n(\mathfrak{P}_0) = f_{\sigma(0)}(\mathfrak{P}_0) \Cup f_{\sigma(1)}(\mathfrak{P}_0) \Cup f_{\sigma(2)}(\mathfrak{P}_0) \cdots \Cup f_{\sigma(n)}(\mathfrak{P}_0)$$

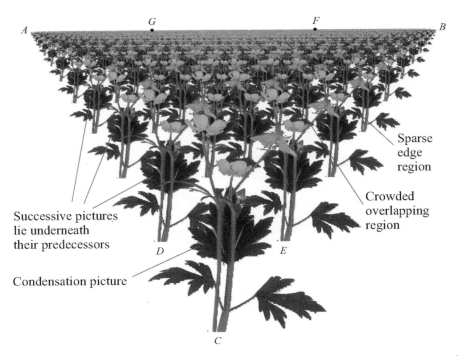

Figure 3.21 Picture $\mathfrak{P}$ of the orbit of a picture of a buttercup $\mathfrak{P}_0$ under an IFS semigroup $S_{\{f_1, f_2\}}(\mathbb{R}^2)$. See also Figures 3.22 and 3.36. Can you find the two transformations for which $\mathfrak{P} = \mathfrak{P}_0 \uplus f_1(\mathfrak{P}) \uplus f_2(\mathfrak{P})$? How would the picture of the orbit of $\mathfrak{P}_0$ under the IFS semigroup $S_{\{f_2, f_1\}}(\mathbb{R}^2)$ differ from $\mathfrak{P}$?

for all $\mathfrak{P}_0 \in \Pi$ and for all $n \in \{0, 1, 2, \ldots\}$. Recall that $f_\varnothing$ denotes the identity transformation.

THEOREM 3.5.5   *Let $\mathfrak{P} = \uplus \mathcal{O}(\mathfrak{P}_0)$ denote the picture of the orbit of $\mathfrak{P}_0 \in \Pi$ under the IFS semigroup $S_{\{f_1, f_2, \ldots, f_N\}}(\mathbb{X})$. Let $\{\mathfrak{P}_n\}_{n=0}^{\infty}$ denote the canonical sequence of pictures in $\mathcal{O}(\mathfrak{P}_0)$. Then, remarkably,*

$$\mathfrak{P}'_{\frac{N(N^k-1)}{N-1}}(\mathfrak{P}_0) = (\mathfrak{P}'_N)^{\circ k}(\mathfrak{P}_0) \quad \text{for all } k = 0, 1, 2, \ldots \text{ and all } \mathfrak{P}_0 \in \Pi, \quad (3.5.7)$$

where $\mathfrak{P}'_n = \mathfrak{P}_0 \uplus \mathfrak{P}_1 \uplus \mathfrak{P}_2 \uplus \cdots \uplus \mathfrak{P}_n$, as in Equation (3.5.3).

PROOF   We prove this result for the case $N = 2$. We will use induction on k in Equation (3.5.7). To keep the notation clean let us write $F = \mathfrak{P}'_N$. Then we are trying to show that, for all $k = 0, 1, 2, \ldots$,

$$\mathfrak{P}'_{\frac{N(N^k-1)}{N-1}}(\mathfrak{P}_0) = F^{\circ k}(\mathfrak{P}_0),$$

Endlessly
changing
patterns
of flowers

Regions
with
diverse
shapes

Figure 3.22   Deep in the snowy field of buttercups. This is a close-up of part of Figure 3.21; it contains a wealth of *different* segments made of overlapping buttercups. Notice how the yellow flowers dominate as we approach the horizon. Remember, we are in flatland here! This is not a projection of a three-dimensional scene, as normally used in computer graphics.

Figure 3.23   The left-hand image illustrates the condensation picture $\mathfrak{P}_0$. The right-hand image represents part of the orbital picture of $\mathfrak{P}_0$ under the IFS semigroup consisting of the two Möbius transformations in Equation (3.5.5). Can you find the picture on the left in the picture on the right?

Figure 3.24   The left-hand image illustrates part of the orbital picture of the picture in the square frame in the middle, under the IFS semigroup consisting of the two Möbius transformations in Equation (3.5.5) *and their inverses*. See the main text. The right-hand picture shows the corresponding underneath picture. Why isn't more of the central tile missing in the right-hand image?

Figure 3.25   Picture of an orbit of a picture under a semigroup of Möbius transformations generated by those in Equation (3.5.6). This is actually a picture tiling because the pictures in the orbit do not overlap.

which is clearly true when $k = 0$. We suppose that it is true up to k, and consider $F^{\circ(k+1)}(\mathfrak{P}_0)$. This inductive hypothesis implies that

$$\mathfrak{P}'_{\frac{2(2^k-1)}{2-1}}(\mathfrak{P}_0)$$

$$= F^{\circ k}(\mathfrak{P}_0)$$

$$= \mathfrak{P}_0 \uplus \big(f_1(\mathfrak{P}_0) \uplus f_2(\mathfrak{P}_0)\big) \uplus \big(f_{11}(\mathfrak{P}_0) \uplus f_{12}(\mathfrak{P}_0) \uplus f_{21}(\mathfrak{P}_0) \uplus f_{22}(\mathfrak{P}_0)\big)$$

$$\uplus \big(f_{111}(\mathfrak{P}_0) \uplus \cdots \uplus f_{222}(\mathfrak{P}_0)\big) \uplus \cdots \uplus \Big(\underbrace{f_{1\ldots11}}_{k \text{ times}}(\mathfrak{P}_0) \uplus \cdots \uplus \underbrace{f_{2\ldots22}}_{k \text{ times}}(\mathfrak{P}_0)\Big)$$

$$(3.5.8)$$

for all $\mathfrak{P}_0 \in \Pi$. So we consider

$$F^{\circ(k+1)}(\mathfrak{P}_0) = F^{\circ k}(F(\mathfrak{P}_0)) = F^{\circ k}\big(\mathfrak{P}_0 \uplus f_1(\mathfrak{P}_0) \uplus f_2(\mathfrak{P}_0)\big).$$

Replacing $\mathfrak{P}_0$ by $\mathfrak{P}_0 \uplus f_1(\mathfrak{P}_0) \uplus f_2(\mathfrak{P}_0)$ in Equation (3.5.8) we now find that

$$F^{\circ(k+1)}(\mathfrak{P}_0)$$

$$= \big(\mathfrak{P}_0 \uplus f_1(\mathfrak{P}_0) \uplus f_2(\mathfrak{P}_0)\big)$$

$$\uplus \Big(f_1\big(\mathfrak{P}_0 \uplus f_1(\mathfrak{P}_0) \uplus f_2(\mathfrak{P}_0)\big) \uplus f_2\big(\mathfrak{P}_0 \uplus f_1(\mathfrak{P}_0) \uplus f_2(\mathfrak{P}_0)\big)\Big)$$

$$\uplus \Big(f_{11}\big(\mathfrak{P}_0 \uplus f_1(\mathfrak{P}_0) \uplus f_2(\mathfrak{P}_0)\big) \uplus f_{12}\big(\mathfrak{P}_0 \uplus f_1(\mathfrak{P}_0) \uplus f_2(\mathfrak{P}_0)\big)\Big)$$

$$\uplus \Big(f_{21}\big(\mathfrak{P}_0 \uplus f_1(\mathfrak{P}_0) \uplus f_2(\mathfrak{P}_0)\big) \uplus f_{22}\big(\mathfrak{P}_0 \uplus f_1(\mathfrak{P}_0) \uplus f_2(\mathfrak{P}_0)\big)\Big)$$

$$\uplus \Big(f_{111}\big(\mathfrak{P}_0 \uplus f_1(\mathfrak{P}_0) \uplus f_2(\mathfrak{P}_0)\big) \uplus \cdots \uplus f_{222}\big(\mathfrak{P}_0 \uplus f_1(\mathfrak{P}_0) \uplus f_2(\mathfrak{P}_0)\big)\Big)$$

$$\uplus \cdots \uplus \Big(\underbrace{f_{1\ldots11}}_{k \text{ times}}\big(\mathfrak{P}_0 \uplus f_1(\mathfrak{P}_0) \uplus f_2(\mathfrak{P}_0)\big) \uplus \cdots \uplus \underbrace{f_{2\ldots22}}_{k \text{ times}}\big(\mathfrak{P}_0 \uplus f_1(\mathfrak{P}_0) \uplus f_2(\mathfrak{P}_0)\big)\Big)$$

This simplifies to

$$F^{\circ(k+1)}(\mathfrak{P}_0)$$

$$= \mathfrak{P}_0 \uplus f_1(\mathfrak{P}_0) \uplus f_2(\mathfrak{P}_0)$$

$$\uplus f_1(\mathfrak{P}_0) \uplus f_{11}(\mathfrak{P}_0) \uplus f_{12}(\mathfrak{P}_0) \uplus f_2(\mathfrak{P}_0) \uplus f_{21}(\mathfrak{P}_0) \uplus f_{22}(\mathfrak{P}_0)$$

$$\uplus f_{11}(\mathfrak{P}_0) \uplus f_{111}(\mathfrak{P}_0) \uplus f_{112}(\mathfrak{P}_0) \uplus f_{12}(\mathfrak{P}_0) \uplus f_{121}(\mathfrak{P}_0) \uplus f_{122}(\mathfrak{P}_0)$$

$$\uplus f_{21}(\mathfrak{P}_0) \uplus f_{211}(\mathfrak{P}_0) \uplus f_{212}(\mathfrak{P}_0) \uplus f_{22}(\mathfrak{P}_0) \uplus f_{221}(\mathfrak{P}_0) \uplus f_{222}(\mathfrak{P}_0)$$

$$\uplus f_{111}(\mathfrak{P}_0) \uplus f_{1111}(\mathfrak{P}_0) \uplus f_{1112}(\mathfrak{P}_0) \uplus \cdots \uplus f_{222}(\mathfrak{P}_0) \uplus f_{2221}(\mathfrak{P}_0)$$

$$\uplus f_{2222}(\mathfrak{P}_0) \uplus \cdots \uplus \underbrace{f_{1\ldots11}}_{k \text{ times}}(\mathfrak{P}_0) \uplus \underbrace{f_{1\ldots11}1}_{k \text{ times}}(\mathfrak{P}_0) \uplus \underbrace{f_{1\ldots11}2}_{k \text{ times}}(\mathfrak{P}_0)$$

$$\uplus \cdots \uplus \underbrace{f_{2\ldots22}}_{k \text{ times}}(\mathfrak{P}_0) \uplus \underbrace{f_{2\ldots22}1}_{k \text{ times}}(\mathfrak{P}_0) \uplus \underbrace{f_{2\ldots22}2}_{k \text{ times}}(\mathfrak{P}_0).$$

In turn this simplifies to

$$F^{\circ(k+1)}(\mathfrak{P}_0) = \mathfrak{P}_0 \uplus f_1(\mathfrak{P}_0) \uplus f_2(\mathfrak{P}_0)$$
$$\uplus f_{11}(\mathfrak{P}_0) \uplus f_{12}(\mathfrak{P}_0) \uplus f_{21}(\mathfrak{P}_0) \uplus f_{22}(\mathfrak{P}_0)$$
$$\uplus f_{111}(\mathfrak{P}_0) \uplus f_{112}(\mathfrak{P}_0) \uplus f_{121}(\mathfrak{P}_0) \uplus f_{122}(\mathfrak{P}_0)$$
$$\uplus f_{211}(\mathfrak{P}_0) \uplus f_{212}(\mathfrak{P}_0) \uplus f_{221}(\mathfrak{P}_0) \uplus f_{222}(\mathfrak{P}_0)$$
$$\uplus f_{1111}(\mathfrak{P}_0) \uplus f_{1112}(\mathfrak{P}_0) \uplus \cdots \uplus f_{2221}(\mathfrak{P}_0) \uplus f_{2222}(\mathfrak{P}_0)$$
$$\uplus \cdots f_{\underbrace{1\cdots11}_{k+1 \text{ times}}}(\mathfrak{P}_0) \uplus f_{\underbrace{1\cdots12}_{k+1 \text{ times}}}(\mathfrak{P}_0)$$
$$\uplus \cdots f_{\underbrace{2\cdots21}_{k+1 \text{ times}}}(\mathfrak{P}_0) \uplus f_{\underbrace{2\cdots22}_{k+1 \text{ times}}}(\mathfrak{P}_0)$$
$$= \mathfrak{P}'_{\frac{2(2^{k+1}-1)}{2-1}}.$$

This almost completes the proof.

We need also to show that the result is remarkable! Equation (3.5.7) implies that

$$\mathfrak{P}'_{\frac{N(N^l-1)}{N-1}}\left(\mathfrak{P}'_{\frac{N(N^m-1)}{N-1}}(\mathfrak{P}_0)\right) = \mathfrak{P}'_{\frac{N(N^{l+m}-1)}{N-1}}(\mathfrak{P}_0)$$

for all $l, m \in \{0, 1, 2, \dots\}$. But in general

$$\mathfrak{P}'_L(\mathfrak{P}'_M(\mathfrak{P}_0)) \notin \{\mathfrak{P}'_n\}_{n=0}^{\infty},$$

as you may readily verify by choosing $L = M = N$. $\qquad\square$

Theorem 3.5.5 tells us that we can compute approximations to $\mathfrak{P} = \uplus \mathcal{O}(\mathfrak{P}_0)$ by recursion. For example, we can compute an approximation to, say, the sequence of functions in the mapping $(\mathfrak{P}'_N)^{\circ 4}$, use it to apply this mapping to $\mathfrak{P}_0$ to obtain $\mathfrak{P}'_{\frac{N(N^4-1)}{N-1}}(\mathfrak{P}_0)$, then apply it again to yield $(\mathfrak{P}'_N)^{\circ 4}$ applied to $\mathfrak{P}'_{\frac{N(N^4-1)}{N-1}}(\mathfrak{P}_0)$, to obtain $\mathfrak{P}'_{\frac{N(N^8-1)}{N-1}}(\mathfrak{P}_0)$, and so on. This type of recursion may be used quite efficiently, as only a few iterates are needed to produce a 'high-order' approximation to the orbital picture. In some cases, for example when all the transformations are strict contractions, this allows us to minimise the growth rate of the cumulative error due to successive rounding errors by keeping low the required number of iterates both of functions and of pictures.

In the top row of Figure 3.26 we illustrate four approximants to an orbital picture. The approximants are

$$\mathfrak{P}_0, \quad \mathfrak{P}'_{340}(\mathfrak{P}_0), \quad \mathfrak{P}'_{87380}(\mathfrak{P}_0) \quad \text{and} \quad \mathfrak{P}'_{22369620}(\mathfrak{P}_0).$$

They were computed in three steps, according to

$$\mathfrak{P}'_{340}(\mathfrak{P}_0) = (\mathfrak{P}'_4)^{\circ 4}(\mathfrak{P}_0), \quad \mathfrak{P}'_{87380}(\mathfrak{P}_0) = (\mathfrak{P}'_4)^{\circ 4}(\mathfrak{P}'_{340}(\mathfrak{P}_0))$$

Figure 3.26   The top row shows four approximants, from left to right, to the orbital picture of the buttercup $\mathfrak{P}_0$. The bottom row shows four underneath pictures. In this case the sequence does not converge to some final picture; instead, a restless sequence of textures is produced. See the main text.

and

$$\mathfrak{P}'_{22369620}(\mathfrak{P}_0) = (\mathfrak{P}'_4)^{\circ 4}(\mathfrak{P}'_{87380}(\mathfrak{P}_0)).$$

In this case the IFS semigroup was generated by the four projective transformations

$$f_n(x, y) = \left(\frac{a_n x + b_n y + c_n}{g_n x + h_n y + j_n}, \frac{d_n x + e_n y + f_n}{g_n x + h_n y + j_n}\right), \quad n = 1, 2, 3, 4, \quad (3.5.9)$$

where the coefficients are given in Table 3.1.

The set attractor of this IFS is the domain of the textured green and yellow leaf-shaped segment that is the bottom right element of Figure 3.26; this was discussed briefly in the Introduction. What you can see from the top row in Figure 3.26 is that the sequence of approximants converges efficiently to an approximation to the orbital picture, which ceases to change, at viewing resolution, if further iterations are effected.

Table 3.1   *Coefficients for the IFS used in Figure 3.26*

n	a_n	b_n	c_n	d_n	e_n	f_n	g_n	h_n	j_n
1	19.05	0.72	1.86	−0.15	16.9	−0.28	5.63	2.01	20.0
2	0.2	4.4	7.5	−0.3	−4.4	−10.4	0.2	8.8	15.4
3	96.5	35.2	5.8	−131.4	−6.5	19.1	134.8	30.7	7.5
4	−32.5	5.81	−2.9	122.9	−0.1	−19.9	−128.1	−24.3	−5.8

We may refer to algorithms for the computation of approximants to orbital pictures based on Theorem 3.5.5, as above, as being **deterministic**. This is in contrast to **random** iteration algorithms, such as the chaos game algorithm, which are discussed in Chapter 4.

We note that Theorem 3.5.5 implies that the orbital picture of $\mathfrak{P}_0$ is the same as the orbital picture of $\mathfrak{P}_0 \uplus f_1(\mathfrak{P}_0) \uplus f_2(\mathfrak{P}_0) \uplus \cdots \uplus f_N(\mathfrak{P}_0)$. A little algebra then provides us with the following result.

COROLLARY 3.5.6   *Let* $\mathfrak{P}(\mathfrak{P}_0) = \uplus \mathcal{O}(\mathfrak{P}_0)$ *denote the orbital picture of* $\mathfrak{P}_0 \in \Pi$ *under the IFS semigroup* $S_{\{f_1, f_2, \ldots, f_N\}}(\mathbb{X})$. *Let*

$$\widetilde{\mathfrak{P}}_0 = \left(f_1(\mathfrak{P}_0) \uplus f_2(\mathfrak{P}_0) \uplus \cdots \uplus f_N(\mathfrak{P}_0) \right) \backslash \mathfrak{P}_0.$$

Then

$$\mathfrak{P}(\mathfrak{P}_0) = \mathfrak{P}_0 \uplus \mathfrak{P}(\widetilde{\mathfrak{P}}_0).$$

EXERCISE 3.5.7   *Prove Corollary 3.5.6. Look at some orbital pictures and identify* $\widetilde{\mathfrak{P}}_0$ *and* $\mathfrak{P}(\widetilde{\mathfrak{P}}_0)$.

The self-referential equation obeyed by some orbital pictures

The definition of an orbital picture may be expressed as

$$\mathfrak{P} = \uplus \, \mathcal{O}(\mathfrak{P}_0)$$
$$= \mathfrak{P}_0 \uplus f_1(\mathfrak{P}_0) \uplus f_2(\mathfrak{P}_0) \uplus \cdots \uplus f_N(\mathfrak{P}_0)$$
$$\uplus f_{11}(\mathfrak{P}_0) \uplus f_{12}(\mathfrak{P}_0) \uplus \cdots \uplus f_{1N}(\mathfrak{P})$$
$$\uplus f_{21}(\mathfrak{P}_0) \uplus f_{22}(\mathfrak{P}_0) \uplus \cdots \uplus f_{2N}(\mathfrak{P}_0) \uplus \cdots.$$

Thus we can always write an orbital picture as a union of disjoint segments, which we call **global segments**, of the form $f_n(\mathfrak{R}_n) \subset \mathfrak{P}$ for $n = 1, 2, \ldots, N$,

$$\mathfrak{P} = \mathfrak{P}_0 \uplus f_1(\mathfrak{R}_1) \uplus f_2(\mathfrak{R}_2) \uplus \cdots \uplus f_N(\mathfrak{R}_N), \tag{3.5.10}$$

where $\mathfrak{R}_n \subset \mathfrak{P}$ for $n = 1, 2, \ldots, N$. Typically each global segment contains multiple 'tiles'.

We refer to Equation (3.5.10) as a self-referential equation because it says that the orbital picture $\mathfrak{P}$ is the disjoint union of $\mathfrak{P}_0$ with at most N transformations of segments of itself. It is this self-referencing property that makes many orbital pictures, including wallpaper patterns, beautiful and mysterious. The orbital pictures illustrated in Figures 3.20, 3.21, 3.23 and 3.24 involve overlapping 'tiles'. Look at each of these pictures, to visualize how it obeys a self-referential equation like (3.5.10).

Under the condition (*) in the following theorem, the segments $\mathfrak{R}_1, \mathfrak{R}_2, \ldots, \mathfrak{R}_N$ can be chosen to be the whole orbital picture. These conditions might at first sight look difficult to check. But they apply in quite simple situations, for example if the $f_n(\mathfrak{P})\backslash\mathfrak{P}_0$ for $n = 1, 2, \ldots, N$ are disjoint, or if the $f_n(\mathfrak{P})$ are disjoint, or if the sets $f_n(\mathbb{X})$ are disjoint or if $N = 1$.

THEOREM 3.5.8   *Let* $\mathfrak{P} = \uplus\mathcal{O}(\mathfrak{P}_0)$ *denote the orbital picture of* $\mathfrak{P}_0 \in \Pi$ *under the IFS semigroup* $\mathsf{S}_{\{f_1, f_2, \ldots, f_N\}}(\mathbb{X})$, *and suppose that (*) for each* $n = 1, 2, \ldots, N - 1$ *the following set of pictures is disjoint:*

$$f_n(\mathfrak{P})\backslash\big(\mathfrak{P}_0 \uplus f_1(\mathfrak{P}_0) \uplus f_2(\mathfrak{P}_0) \uplus \cdots \uplus f_n(\mathfrak{P}_0)\big)$$

and

$$f_m(\mathfrak{P}_0)\backslash\big(\mathfrak{P}_0 \uplus f_1(\mathfrak{P}_0) \uplus f_2(\mathfrak{P}_0) \uplus \cdots \uplus f_{m-1}(\mathfrak{P}_0)\big)$$

for $m = n + 1, \ldots, N$. *Then the orbital picture obeys the self-referential equation*

$$\mathfrak{P} = \mathfrak{P}_0 \uplus f_1(\mathfrak{P}) \uplus f_2(\mathfrak{P}) \uplus \cdots \uplus f_N(\mathfrak{P}). \qquad (3.5.11)$$

PROOF   As in the proof of Theorem 3.5.5 we write

$$F^{\circ k}(\mathfrak{P}_0) := \mathfrak{P}'_{\frac{N(N^k-1)}{N-1}}(\mathfrak{P}_0).$$

Then we start by proving that under the condition (*) we have, for all $k = 0, 1, 2, \ldots$,

$$F^{\circ(k+1)}(\mathfrak{P}_0) = \mathfrak{P}_0 \uplus f_1(F^{\circ k}(\mathfrak{P}_0)) \uplus f_2(F^{\circ k}(\mathfrak{P}_0)) \uplus \cdots \uplus f_N(F^{\circ k}(\mathfrak{P}_0)),$$

$$(3.5.12)$$

for all $\mathfrak{P}_0 \in \Pi$. We will demonstrate this result for the case $N = 2$. The general case is a straightforward generalization of the same ideas. We proceed by induction. When $k = 0$ and $N = 2$, Equation (3.5.12) reads

$$F^{\circ 1}(\mathfrak{P}_0) = \mathfrak{P}_0 \uplus f_1(\mathfrak{P}_0) \uplus f_2(\mathfrak{P}_0),$$

which is true. Suppose that Equation (3.5.12) is true for $k = 0, 1, \ldots, K$. Then, choosing $k = K$, $N = 2$ and $\mathfrak{P}_0$ to be $F^{\circ 1}(\mathfrak{P}_0) = \mathfrak{P}_0 \uplus f_1(\mathfrak{P}_0) \uplus f_2(\mathfrak{P}_0)$ in

Equation (3.5.12), we have

$$F^{\circ(K+1)}(F^{\circ 1}(\mathfrak{P}_0)) = \mathfrak{P}_0 \uplus f_1(\mathfrak{P}_0) \uplus f_2(\mathfrak{P}_0) \uplus f_1\big(F^{\circ K}(F^{\circ 1}(\mathfrak{P}_0))\big)$$
$$\uplus f_2\big(F^{\circ K}(F^{\circ 1}(\mathfrak{P}_0))\big)$$

for all $\mathfrak{P}_0 \in \Pi$. By Theorem 3.5.5 it follows that

$$F^{\circ(K+2)}(\mathfrak{P}_0) = \mathfrak{P}_0 \uplus f_1(\mathfrak{P}_0) \uplus f_2(\mathfrak{P}_0) \uplus f_1\big(F^{\circ(K+1)}(\mathfrak{P}_0)\big) \uplus f_2\big(F^{\circ(K+1)}(\mathfrak{P}_0)\big)$$

for all $\mathfrak{P}_0 \in \Pi$. The key idea now comes. We can rewrite the last equation as

$$F^{\circ(K+2)}(\mathfrak{P}_0) = \mathfrak{P}_0 \uplus f_1(\mathfrak{P}_0) \uplus \big(f_2(\mathfrak{P}_0)\backslash(\mathfrak{P}_0 \uplus f_1(\mathfrak{P}_0))\big)$$
$$\uplus \big(f_1(F^{\circ(K+1)}(\mathfrak{P}_0))\backslash(\mathfrak{P}_0 \uplus f_1(\mathfrak{P}_0))\big) \uplus f_2\big(F^{\circ(K+1)}(\mathfrak{P}_0)\big).$$

We can commute the terms $f_2(\mathfrak{P}_0)\backslash(\mathfrak{P}_0 \uplus f_1(\mathfrak{P}_0))$ and $f_1\big(F^{\circ(K+1)}(\mathfrak{P}_0)\backslash(\mathfrak{P}_0 \uplus f_1(\mathfrak{P}_0))\big)$, because $F^{\circ(K+1)}(\mathfrak{P}_0) \subset \mathfrak{P}(\mathfrak{P}_0)$ implies that

$$f_1\big(F^{\circ(K+1)}(\mathfrak{P}_0)\big)\backslash\big(\mathfrak{P}_0 \uplus f_1(\mathfrak{P}_0)\big) \subset f_1(\mathfrak{P}(\mathfrak{P}_0))\backslash\big(\mathfrak{P}_0 \uplus f_1(\mathfrak{P}_0)\big),$$

and the latter picture is disjoint from $f_2(\mathfrak{P}_0)\backslash(\mathfrak{P}_0 \uplus f_1(\mathfrak{P}_0))$ by condition (*).

It now follows that

$$F^{\circ(K+2)}(\mathfrak{P}_0) = \mathfrak{P}_0 \uplus f_1(\mathfrak{P}_0) \uplus \big(f_1(F^{\circ(K+1)}(\mathfrak{P}_0))\backslash(\mathfrak{P}_0 \uplus f_1(\mathfrak{P}_0))\big)$$
$$\uplus \big(f_2(\mathfrak{P}_0)\backslash(\mathfrak{P}_0 \uplus f_1(\mathfrak{P}_0))\big) \uplus f_2\big(F^{\circ(K+1)}(\mathfrak{P}_0)\big),$$

which is the same as

$$F^{\circ(K+2)}(\mathfrak{P}_0) = \mathfrak{P}_0 \uplus f_1(\mathfrak{P}_0) \uplus f_1\big(F^{\circ(K+1)}(\mathfrak{P}_0)\big) \uplus f_2(\mathfrak{P}_0) \uplus f_2\big(F^{\circ(K+1)}(\mathfrak{P}_0)\big)$$
$$= \mathfrak{P}_0 \uplus f_1\big(F^{\circ(K+1)}(\mathfrak{P}_0)\big) \uplus f_2\big(F^{\circ(K+1)}(\mathfrak{P}_0)\big),$$

where, in the last step, we have used $f_1(\mathfrak{P}_0) \subset f_1(F^{\circ(K+1)}(\mathfrak{P}_0))$ and $f_2(\mathfrak{P}_0) \subset f_2(F^{\circ(K+1)}(\mathfrak{P}_0))$.

Hence Equation (3.5.12) is true when $k = K + 1$, which implies completion of the induction. Hence Equation (3.5.12) is true for $K = 0, 1, 2, \ldots$ By letting K tend to infinity, we obtain Equation (3.5.11). $\qquad\square$

It is tempting to think that $\mathfrak{P} = \uplus \, \mathcal{O}(\mathfrak{P}_0)$ is the unique solution of the self-referential equation (3.5.11). This is not the case, as the following example shows. Let $\mathfrak{P}_0$ have domain $\{(x, y) \in \mathbb{R}^2 : 0 \leq x \leq 1, 0 \leq y \leq 1\}$, let $f_1(x, y) = (\frac{1}{2}x + 2, \frac{1}{2}y)$ and $f_2(x, y) = (\frac{1}{2}x, \frac{1}{2}y + 2)$. Let A denote the closed line segment that joins the pair of points $(0, 4)$ and $(4, 0)$. Then A is the attractor of the IFS $\{\mathbb{R}^2; f_1, f_2\}$ and obeys $A = f_1(A) \cup f_2(A)$, and it is disjoint from the domain of $\mathfrak{P}$. Let $\mathfrak{P}_A$ denote a picture of constant colour, with domain A. Then

$$\mathfrak{P}_A = f_1(\mathfrak{P}_A) \uplus f_2(\mathfrak{P}_A) = f_2(\mathfrak{P}_A) \uplus f_1(\mathfrak{P}_A).$$

Figure 3.27   An example of a picture $\widetilde{\mathfrak{P}}$ which obeys the self-referential equation $\widetilde{\mathfrak{P}} = \mathfrak{P}_0 \uplus f_1(\widetilde{\mathfrak{P}}) \uplus$ $f_2(\widetilde{\mathfrak{P}})$ but which is not the orbital picture $\mathfrak{P}$ of the buttercup $\mathfrak{P}_0$. The difference between $\widetilde{\mathfrak{P}}$ and $\mathfrak{P}$ is the red segment, whose domain is the fractal set, the attractor of the IFS.

Now let $\widetilde{\mathfrak{P}} = \mathfrak{P} \uplus \mathfrak{P}_A = \mathfrak{P}_A \uplus \mathfrak{P}$. Then it is readily verified that

$$\widetilde{\mathfrak{P}} \neq \mathfrak{P}, \quad \mathfrak{P} = \mathfrak{P}_0 \uplus f_1(\mathfrak{P}) \uplus f_2(\mathfrak{P}) \quad \text{and} \quad \widetilde{\mathfrak{P}} = \mathfrak{P}_0 \uplus f_1(\widetilde{\mathfrak{P}}) \uplus f_2(\widetilde{\mathfrak{P}}).$$

See for example Figure 3.27.

A commonly used technique in the fractal compression of a given picture $\mathfrak{P}$ involves seeking a set of segments S of $\mathfrak{P}$ each of which can be transformed, under one of a given family of transformations T, into a segment belonging to a *given* set of segments S' of $\mathfrak{P}$; see for example [12], [53] or [38]. Typically the given segments S' are obtained by chopping the domain of $\mathfrak{P}$ into square blocks, with little regard for the geometry of $\mathfrak{P}$. Figure 3.28 illustrates that domains of the segments $f_1(\mathfrak{R}_1), f_2(\mathfrak{R}_2), \ldots, f_N(\mathfrak{R}_N)$ occurring in Equation (3.5.10) may be very complicated even when the domain of $\mathfrak{P}_0$ is rectangular. *This suggests that, in the future development of fractal image compression technology, more*

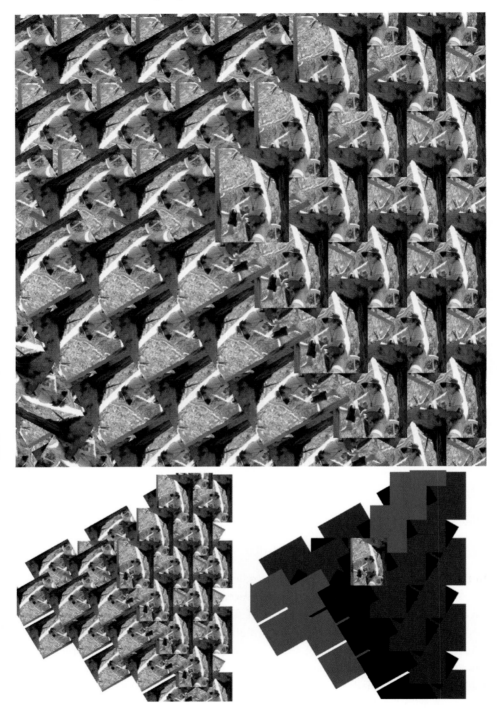

Figure 3.28 The main image here represents part of an orbital picture $\mathfrak{P}$ associated with the IFS semi-group in Equation (3.5.13). The bottom right image shows the condensation picture and, in three shades of blue, the domains of the segments $f_1(\mathfrak{R}_1)$, $f_2(\mathfrak{R}_2)$ and $f_3(\mathfrak{R}_3)$ (see Equation (3.5.10)), each intersected with the domain of the segment of $\mathfrak{P}$ shown on the left.

Figure 3.29 The photographs on the left and right show two quite distinct leaves taken off the same plant, like the one in the middle, which was growing near Lake Padden in northern Washington State, U.S.A. in June 2003. It seems as though the branching veins crowd together, in the leaf on the right, and either stop growing, or go 'underneath'. Can orbits of pictures be used to model the geometry of leaf veins? Can an underlying code space be identified, yielding biologically meaningful topological invariants?

attention should be given to the geometry of the segments into which pictures are partitioned. Without such attention, the compression would be inefficient for many orbital pictures; given the 'fractal' and self-referential character of the latter, it would seem to be a minimum requirement for fractal compression to work well, at least for orbital pictures where N is small.

The IFS used in Figure 3.28 is

$$\{\mathbb{R}^2; f_1(x, y) = \left(\tfrac{1}{2}(-x + \sqrt{3}y), \tfrac{1}{2}(\sqrt{3}x - y)\right),$$
$$f_2(x, y) = \left(x + \tfrac{1}{2}, y - \tfrac{1}{2}\sqrt{3}\right), f_3(x, y) = \left(x + \tfrac{1}{2}, y + \tfrac{1}{2}\sqrt{3}\right)\}, \quad (3.5.13)$$

and the visible part of the orbital picture corresponds to the window $-3 \leq x \leq 3$ and $-3 \leq y \leq 3$.

EXERCISE 3.5.9 *Identify the segments $f_1(\mathfrak{P})$, $f_2(\mathfrak{P})$ and $f_3(\mathfrak{P})$ in the picture $\mathfrak{P}$ in Figure 3.55. Also, humour your author: draw a complicated domain D within the domain of one of these segments, say $f_1(\mathfrak{P})$, and identify a larger domain $\widetilde{D}$, within the domain of the whole picture, such that $f_1(\mathfrak{P}|_{\widetilde{D}}) = \mathfrak{P}|_D$. Notice how your domain may contain parts of the boundaries of many picture tiles.*

The code space of the orbital picture

In this subsection we define and investigate code spaces of orbital pictures. This relates to our meristem theme, as we discuss below; see also the caption of Figure 3.29.

Code spaces of orbital pictures, with a few side conditions, enable us to: (i) establish the existence of invariant quantities associated with orbital pictures, including the growth rate of periodic cycles and the topological entropy; (ii) establish a dynamical system on panels, i.e. certain segments of orbital pictures, see below; (iii) construct a certain 'space of limiting pictures', $L_{\mathfrak{P}_0}$, from the set of panels; (iv) relate some of the limiting pictures, elements of $L_{\mathfrak{P}_0}$, to the periodic cycles of the dynamical system. These constructions (i)–(iv) provide us with insight into the observed visual richness of some orbital pictures and help us to imagine, for example, the ever changing diversity of the flowers in Figure 3.21 as the horizon is approached.

Later, in Chapter 4, when we are discussing the theory and applications of fractal tops, we will attach great significance to the code spaces of orbital pictures; we will find that they correspond to special subsets of fractal sets and that they have applications in computer graphics.

The addresses and panels of an orbital picture

The orbital picture $\mathfrak{P} = \uplus \, \mathcal{O}(\mathfrak{P}_0)$ possesses a very interesting and useful code space. To reveal this structure, we decompose $\mathfrak{P}$ into special 'tiles', which we call panels, defined in the following manner. We define a sequence of subsets $\{D_n\}_{n=0}^{\infty}$ inductively in terms of the canonical sequence $\{\mathfrak{P}_n = f_{\sigma(n)}(\mathfrak{P}_0)\}_{n=0}^{\infty}$ by

$$D_0 = D_{\mathfrak{P}_0}, \quad \mathfrak{P}'_n = \mathfrak{P}_0 \uplus \mathfrak{P}_1 \uplus \cdots \uplus \mathfrak{P}_n \quad \text{and} \quad D_n = D_{\mathfrak{P}'_n} \setminus D_{\mathfrak{P}'_{n-1}}$$

$$\text{for } n = 0, 1, 2, \ldots$$

Notice that D_0 is nonempty. Let $\{D_{n_k}\}_{k=0}^{K}$ denote the subsequence of $\{D_n\}_{n=0}^{\infty}$ which consists of those D_n that are nonempty, where $K \geq 0$ is either a finite integer or else ∞. We will write $\{0, 1, 2, \ldots, K\}$ when $K = \infty$ to denote the set $\{0, 1, 2, \ldots\}$.

DEFINITION 3.5.10   Let $\{D_{n_k}\}_{k=0}^{K}$ be defined as above. A **panel** of the orbital picture $\mathfrak{P} = \uplus \, \mathcal{O}(\mathfrak{P}_0)$ is defined to be a picture of the form

$$\mathfrak{Q}_{n_k} = \mathfrak{P}'_{n_k}|_{D_{n_k}} \quad \text{for some } k \in \{0, 1, 2, \ldots, K\}.$$

The code $\sigma(n_k) \in \Omega'_{\{1,2,\ldots,N\}}$, as in Definition 3.5.2, is defined to be **the address of the panel**

$$\mathfrak{Q}_{\sigma(n_k)} := \mathfrak{Q}_{n_k}.$$

The set of panels of the orbital picture is the set denoted by

$$P_{panels}(\mathfrak{P}_0) = \{\mathfrak{P}' \in \Pi : \mathfrak{P}' = \mathfrak{Q}_{n_k} \quad \text{for some } k \in \{0, 1, 2, \ldots, K\}\}.$$

When $k \geq 1$ and $\mathfrak{P}_{n_k}$ overlaps $\mathfrak{P}'_{n_k-1}$, the panel $\mathfrak{Q}_{n_k}$ is the segment of the picture $\mathfrak{P}_{n_k}$ that 'sticks out from underneath' the picture $\mathfrak{P}'_{n_k-1} = \mathfrak{P}_0 \uplus f_1(\mathfrak{P}_0) \uplus f_2(\mathfrak{P}_0) \uplus \cdots \uplus f_{\sigma(n_k-1)}(\mathfrak{P}_0)$. Clearly $\mathfrak{P}$ is the disjoint union of the pictures in the sequence $\{\mathfrak{Q}_{n_k}\}_{k=0}^{K}$, and we have

$$\mathfrak{P} = \uplus_{k=0}^{K} \mathfrak{Q}_{n_k},$$

where the order in which the panels are combined, in the tops union $\uplus_{k=0}^{K} \mathfrak{Q}_{n_k}$, makes no difference.

Let Ω' denote $\Omega'_{\{1,2,\ldots,N\}}$ and let Ω denote $\Omega_{\{1,2,\ldots,N\}}$. Then let us define

$$\Omega'_{\mathfrak{P}_0} = \{\sigma(n_k) : k \in \{0, 1, 2, \ldots, K\}\}$$

to be the **set of addresses of the orbital picture** $\mathfrak{P} = \uplus \, \mathcal{O}(\mathfrak{P}_0)$, so that

$$P_{panels} = \left\{\mathfrak{Q}_\sigma : \sigma \in \Omega'_{\mathfrak{P}_0}\right\}.$$

Let $\overline{\Omega'_{\mathfrak{P}_0}} \subset \Omega' \cup \Omega$ denote the closure of $\Omega'_{\mathfrak{P}_0}$ in the natural topology on $\Omega' \cup \Omega$; see Chapter 1. Let

$$\Omega_{\mathfrak{P}_0} = \overline{\Omega'_{\mathfrak{P}_0}} \cap \Omega.$$

Then $\Omega_{\mathfrak{P}_0}$ is the set of points in the closure of $\Omega'_{\mathfrak{P}_0}$ that are not in $\Omega'_{\mathfrak{P}_0}$. We call

$$\overline{\Omega'_{\mathfrak{P}_0}} = \Omega_{\mathfrak{P}_0} \cup \Omega'_{\mathfrak{P}_0}$$

the **code space of the orbital picture** $\mathfrak{P} = \uplus \, \mathcal{O}(\mathfrak{P}_0)$.

In the above definition we assume, as elsewhere, that when we are given an orbital picture we know the condensation set and semigroup by which it is generated.

EXERCISE 3.5.11   *Prove that $\Omega_{\mathfrak{P}_0}$ is a closed subset of Ω.*

EXERCISE 3.5.12   *Consider the code space $\Omega_{\mathfrak{P}_0}$ of the orbital picture of $\mathfrak{P}_0$ when the IFS is $\{\mathbb{R}^2; f_1(x, y) = \left(\frac{1}{2}x, \frac{1}{2}y\right), f_2(x, y) = \left(\frac{1}{2}(x + 1), \frac{1}{2}y\right)\}$. Show that the set attractor A of the IFS is the closed line segment that connects the points $x = 0$ and $x = 1$ on the x-axis. Let $C \subset \mathbb{R}^2$ be a nonempty closed set such that $A \cap C = \varnothing$. Show that $\Omega_{A \cup C} = \Omega$ and that $\Omega_A = \varnothing$.*

Shift transformation on the code space of an orbital picture

The following theorem says that the space $\Omega_{\mathfrak{P}_0} \cup \Omega'_{\mathfrak{P}_0}$ is mapped into itself by the shift transformation. This enables us to define certain topological invariants of orbital pictures.

THEOREM 3.5.13   *Let $\mathfrak{P}$ denote the orbital picture of $\mathfrak{P}_0 \in \Pi$ under the IFS semigroup $\mathcal{S}_{\{f_1, f_2, \ldots, f_N\}}(\mathbb{X})$. Let $\Omega_{\mathfrak{P}_0} \cup \Omega'_{\mathfrak{P}_0}$ denote the code space of $\mathfrak{P}$. Let*

$$S : \Omega' \cup \Omega \to \Omega' \cup \Omega$$

denote the shift operator on code space discussed at the end of Chapter 2. Then

$$S\left(\Omega_{\mathfrak{P}_0} \cup \Omega'_{\mathfrak{P}_0}\right) \subset \Omega_{\mathfrak{P}_0} \cup \Omega'_{\mathfrak{P}_0};$$

in particular,

$$S\left(\Omega'_{\mathfrak{P}_0}\right) \subset \Omega'_{\mathfrak{P}_0} \quad and \quad S\left(\Omega_{\mathfrak{P}_0}\right) \subset \Omega_{\mathfrak{P}_0}.$$

PROOF   Let $\sigma \in \Omega'_{\mathfrak{P}_0}$. If $\sigma = \varnothing$ then, by the definition of the action of the shift transformation, $S(\sigma) = \sigma \in \Omega'_{\mathfrak{P}_0}$. If $|\sigma| = 1$ then $S(\sigma) = \varnothing \in \Omega'_{\mathfrak{P}_0}$. If $|\sigma| = n > 1$, let us write $\sigma = \sigma_1 \sigma_2 \sigma_3 \cdots \sigma_n \in \Omega'_{\mathfrak{P}_0}$. Then $D_{c(\sigma_1 \sigma_2 \sigma_3 \cdots \sigma_n)} \neq \varnothing$. Recall that the sequence $\{D_l\}_{l=0}^{\infty}$ is defined, as above, by $D_l = D_{\mathfrak{P}'_l} \setminus D_{\mathfrak{P}'_{l-1}}$. We need to prove that the address $S(\sigma) = \sigma_2 \sigma_3 \cdots \sigma_n$ corresponds to a panel, that is, $D_{c(\sigma_2 \sigma_3 \cdots \sigma_n)} \neq \varnothing$.

So, suppose that $D_{c(\sigma_2 \sigma_3 \cdots \sigma_n)} = \varnothing$. Since $D_{c(\sigma_1 \sigma_2 \sigma_3 \cdots \sigma_n)} \neq \varnothing$ it follows that we can find $x \in D_{c(\sigma_1 \sigma_2 \sigma_3 \cdots \sigma_n)}$. Thus there exists $x_0 \in D_{\mathfrak{P}_0} = D_0$ such that $x = f_{\sigma_1}(f_{\sigma_2 \sigma_3 \cdots \sigma_n}(x_0))$. Since $D_{c(\sigma_2 \sigma_3 \cdots \sigma_n)} = \varnothing$ it follows that $f_{\sigma_2 \sigma_3 \cdots \sigma_n}(x_0) \notin D_{c(\sigma_2 \sigma_3 \cdots \sigma_n)}$. But $f_{\sigma_2 \sigma_3 \cdots \sigma_n}(x_0) \in D_{\mathfrak{P}_{c(\sigma_2 \sigma_3 \cdots \sigma_n)}}$, which implies that $f_{\sigma_2 \sigma_3 \cdots \sigma_n}(x_0) \in D_{\mathfrak{P}'_{c(\sigma_2 \sigma_3 \cdots \sigma_n)}}$. Noting that $D_{c(\sigma_2 \sigma_3 \cdots \sigma_n)} = D_{\mathfrak{P}'_{c(\sigma_2 \sigma_3 \cdots \sigma_n)}} \setminus D_{\mathfrak{P}'_{c(\sigma_2 \sigma_3 \cdots \sigma_n)-1}}$, we have $f_{\sigma_2 \sigma_3 \cdots \sigma_n}(x_0) \in D_{\mathfrak{P}'_{c(\sigma_2 \sigma_3 \cdots \sigma_n)-1}}$. Therefore $f_{\sigma_2 \sigma_3 \cdots \sigma_n}(x_0) = f_{\widetilde{\sigma}}(x_0)$ for some $\widetilde{\sigma} \in \Omega'_{\mathfrak{P}_0}$ with $c(\widetilde{\sigma}) < c(\sigma_2 \sigma_3 \cdots \sigma_n) - 1$. It follows that $x = f_{\sigma_1}(f_{\widetilde{\sigma}}(x_0)) = f_{\sigma_1 \widetilde{\sigma}}(x_0)$ where $c(\sigma_1 \widetilde{\sigma}) < c(\sigma_1 \sigma_2 \sigma_3 \cdots \sigma_n)$, which implies that $D_{c(\sigma_1 \sigma_2 \sigma_3 \cdots \sigma_n)} = \varnothing$, which is a contradiction. We conclude that $D_{c(\sigma_2 \sigma_3 \cdots \sigma_n)} \neq \varnothing$, that $\mathfrak{Q}_{c(S(\sigma))}$ is indeed a panel and hence that $S(\sigma) \in \Omega'_{\mathfrak{P}_0}$. This proves that $S(\Omega'_{\mathfrak{P}_0}) \subset \Omega'_{\mathfrak{P}_0}$.

Finally, using the continuity of $S : \Omega' \cup \Omega \to \Omega' \cup \Omega$, we have $S(\overline{\Omega'_{\mathfrak{P}_0}}) \subset \overline{\Omega'_{\mathfrak{P}_0}}$.
□

It is appropriate here to mention the **transformation of the colours of a picture** by means of a mapping $C : \mathfrak{C} \to \mathfrak{C}$, where $\mathfrak{C}$ is the colour space. Let $\mathfrak{P} \in \Pi = \Pi_{\mathfrak{C}}$. Then $C(\mathfrak{P})$ is the picture whose domain is $D_{\mathfrak{P}}$ and whose colour at the point $x \in D_{\mathfrak{P}}$ is $C(\mathfrak{P}(x))$. The key distinction between $C(\mathfrak{P})$ where $C : \mathfrak{C} \to \mathfrak{C}$ and $H(\mathfrak{P})$ where $H : \mathbb{X} \to \mathbb{X}$ lies with the domains of H and C.

Typically the colour space $\mathfrak{C}$ is discrete and so we can endow it with the discrete topology. But it may be for example $\{(R, G, B) \in \mathbb{R}^3 : 0 \leq R, G, B \leq 255\}$, in which case we can give it the natural topology of $\mathbb{R}^3$. In any case, it makes sense for us to refer to a homeomorphism $C : \mathfrak{C} \to \mathfrak{C}$.

DEFINITION 3.5.14   Two pictures $\mathfrak{P}, \widetilde{\mathfrak{P}} \in \Pi$ are said to be **topologically equivalent** iff there is a homeomorphism $H : \mathbb{X} \to \mathbb{X}$ and a homeomorphism $C : \mathfrak{C} \to \mathfrak{C}$ such that

$$\widetilde{\mathfrak{P}} = C(H(\mathfrak{P})).$$

EXERCISE 3.5.15   *Show that* $C(H(\mathfrak{P})) = H(C(\mathfrak{P}))$.

Figure 3.30 provides an illustration of two different-looking pictures that are topologically equivalent. The following theorem tells us that if two pictures are topologically equivalent and one of them is an orbital picture then the other is also an orbital picture, with the same code space structure.

THEOREM 3.5.16   *Let $\mathfrak{P}$ and $\widetilde{\mathfrak{P}}$ be topologically equivalent pictures. Let $\mathfrak{P}$ be the orbital picture of $\mathfrak{P}_0 \in \Pi_{\mathfrak{C}}$ under the IFS semigroup $\mathcal{S}_{\{f_1, f_2, \ldots, f_N\}}$. Then $\widetilde{\mathfrak{P}}$*

Figure 3.30   These two leaves are related by a homeomorphism. If one is an orbital picture then the other can also be represented as an orbital picture, with the same code space structure.

is the orbital picture of $\widetilde{\mathfrak{P}}_0 = C(H(\mathfrak{P}_0))$ *under the IFS semigroup*

$$\mathcal{S}_{\{Hf_1H^{-1}, Hf_2H^{-1}, \dots, Hf_NH^{-1}\}}$$

where $H \times C : \mathbb{X} \times \mathfrak{C} \to \mathbb{X} \times \mathfrak{C}$ *is the homeomorphism that provides the equivalence between* $\mathfrak{P}$ *and* $\widetilde{\mathfrak{P}}$. *In particular, both orbital pictures have the same code space structure; that is,*

$$\Omega'_{\mathfrak{P}_0} = \Omega'_{\widetilde{\mathfrak{P}}_0} \quad and \quad \Omega_{\mathfrak{P}_0} = \Omega_{\widetilde{\mathfrak{P}}_0}.$$

PROOF   This result follows immediately once it is shown that $\widetilde{\mathfrak{P}}$ is the orbital picture of $\widetilde{\mathfrak{P}}_0 = C(H(\mathfrak{P}_0))$ under the IFS semigroup $\mathcal{S}_{\{Hf_1H^{-1}, Hf_2H^{-1}, \dots, Hf_NH^{-1}\}}$. But this is a direct consequence of the fact that a homeomorphism is one-to-one and invertible, which in turn implies that $\mathfrak{Q}$ is a panel of the orbital picture of $\mathfrak{P}_0$ iff $\widetilde{\mathfrak{Q}} = C(H(\mathfrak{Q}))$ is a panel of the orbital picture of $\widetilde{\mathfrak{P}}_0 = C(H(\mathfrak{P}_0))$ under the IFS semigroup $\mathcal{S}_{\{Hf_1H^{-1}, Hf_2H^{-1}, \dots, Hf_NH^{-1}\}}$ and that the addresses of these two panels are the same. Notice that the continuity of H ensures that the functions in the IFS $\{\mathbb{X}; \widetilde{f}_1, \widetilde{f}_2, \dots, \widetilde{f}_N\}$, where $\widetilde{f}_n = Hf_nH^{-1}$ for $n = 1, 2, \dots, N$, are continuous when those of $\{\mathbb{X}; f_1, f_2, \dots, f_N\}$ are continuous.   □

Two orbital pictures equivalent to each other are illustrated in Figure 3.31.

This invariance of the code space of an orbital picture under homeomorphism, as expressed in Theorem 3.5.16, together with Theorem 3.5.13 enables us to define certain real numbers which are unchanged by homeomorphisms and which may capture the visual richness of the orbital picture. These topological invariants arise from deep within dynamical systems theory; see for example [56] and references therein. We summarize them, for shift transformations, in what follows.

Symbolic invariants of orbital pictures

The mapping

$$S : \Omega'_{\mathfrak{P}_0} \cup \Omega_{\mathfrak{P}_0} \to \Omega'_{\mathfrak{P}_0} \cup \Omega_{\mathfrak{P}_0}$$

is a continuous mapping from a compact metric space into itself. Hence, from [56], pp. 105–9, we discover that it possesses:

(i) a well-defined **number of periodic points**, of period n,

$$C_n(\mathfrak{P}_0) := \left| \left\{ \sigma \in \Omega'_{\mathfrak{P}_0} \cup \Omega_{\mathfrak{P}_0} : S^{\circ n}(\sigma) = \sigma \right\} \right| \quad \text{for each } n = 1, 2, 3, \dots;$$

(ii) a well-defined **growth rate for periodic cycles**,

$$C_{\mathfrak{P}_0} := \limsup_{n \to \infty} \frac{\log_2 C_n(\mathfrak{P}_0)}{n};$$

(iii) a well-defined ζ-function (the **zeta-function**),

$$\zeta_{\mathfrak{P}_0}(z) = \exp \sum_{n=1}^{\infty} \frac{C_n(\mathfrak{P}_0)}{n} z^n,$$

where $z \in \mathbb{C}$ and the series converges for $|z| < (\log_e 2) C_{\mathfrak{P}_0}$;

(iv) a well-defined **topological entropy**

$$h_{\mathfrak{P}_0} := h_{top}(S : \Omega_{\mathfrak{P}_0} \to \Omega_{\mathfrak{P}_0}).$$

The topological entropy h_{top} of a dynamical system $f : \mathbb{X} \to \mathbb{X}$ is defined formally in the next part of the subsection. We use $\log_2$ rather than $\log_e$ in our definitions because we are interested in questions relating to information theory.

We may refer to the quantities $C_n(\mathfrak{P}_0)$, $C_{\mathfrak{P}_0}$, $\zeta_{\mathfrak{P}_0}(z)$ and $h_{\mathfrak{P}_0}$ as being associated with the orbital picture from which the dynamical system arises; so for example we will say in full that $h_{\mathfrak{P}_0}$ is the **symbolic entropy of the orbital picture** of $\mathfrak{P}_0 \in \Pi$ under the IFS semigroup $S_{\{f_1, f_2, \dots, f_N\}}$. More briefly, we may say that $h_{\mathfrak{P}_0}$ is the **entropy of $\mathfrak{P}$ modulo $\mathfrak{P}_0$**.

Notice that $C_n(\mathfrak{P}_0) \geq 1$ because $S(\varnothing) = \varnothing$. Also, $C_n(\mathfrak{P}_0) \leq N^n + 1$ because the number of periodic points of period n for $S : \Omega_{\{1,2,\dots,N\}} \to \Omega_{\{1,2,\dots,N\}}$ is N^n. It follows that

$$0 \leq C_{\mathfrak{P}_0} \leq \log_2 N.$$

The zeta-function of an orbital picture always possesses singularities on the circle $|z| = 2^{-C_{\mathfrak{P}_0}}$ and may be a meromorphic function on the whole complex plane. For example, when $\Omega_{\mathfrak{P}_0} = \Omega_{\{1,2,\dots,N\}}$ we have $C_{\mathfrak{P}_0} = \log_2 N$ and

$$\zeta_{\mathfrak{P}_0}(z) = \exp \left(z + \sum_{n=1}^{\infty} \frac{N^n}{n} z^n \right) = \frac{\exp z}{1 - Nz}.$$

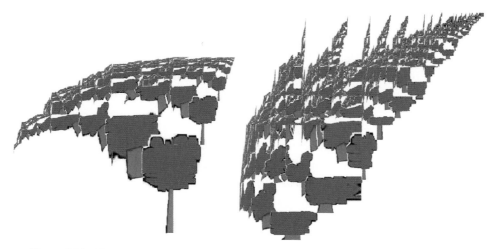

Figure 3.31   These two images correspond to orbital pictures which are homeomorphic to the orbital picture represented at the bottom left of Figure 3.33, so the topological entropies are the same. The fractal dimensions of their limit sets are quite different, however.

Zeta-functions associated with dynamical systems have been much studied; see for example [83]. I have mentioned the zeta-function of an orbital picture because it seems to me such an extraordinary thing that we can assign, in a meaningful manner based on the analysis of patterns, an analytic function to a class of *pictures,* albeit pictures of a quite special type.

We discuss formally the topological entropy in the next part of the subsection. In some cases, it is equal to the growth rate of periodic orbits

$$h_{\mathfrak{P}_0} = C_{\mathfrak{P}_0}$$

and can be estimated accurately; see Figure 3.32. This is true, for example, when $S : \Omega_{\mathfrak{P}_0} \to \Omega_{\mathfrak{P}_0}$ is related, in an appropriate way, to a transitive topological Markov shift; see [56], p. 176, and [77].

A simple example of a code space associated with an orbital picture is provided by the orbital picture in Figure 3.3, for which

$$\Omega'_{\mathfrak{P}_0} = \{\varnothing, 1, 11, 111, 1111, \dots\} \quad \text{and} \quad \Omega_{\mathfrak{P}_0} = \{11111\cdots\}.$$

In this case the growth rate of periodic cycles and the symbolic entropy are zero.

For the orbital picture in Figure 3.21 the symbolic entropy and the growth rate of periodic orbits is $-\log_2 0.7 = 0.5145\cdots$ We were able to compute this entropy, and the entropies of the orbital pictures in Figure 3.32, because in each case the mapping $S : \Omega_{\mathfrak{P}_0} \to \Omega_{\mathfrak{P}_0}$ is related to the piecewise linear mapping $R_\beta : [0, 1] \to [0, 1]$ defined by

$$R_\beta(x) = (\beta x)\,\mathrm{mod}\,1,$$

where $\beta \in (1, \infty)$ is a parameter. The topological entropy of $R_\beta(x)$ is $\log_2 \beta$. See for example [77] and [82] and also [88].

EXERCISE 3.5.17 *On the basis of your own guesses, arrange, in order of increasing growth rate of periodic cycles, the orbital pictures illustrated in Figure 3.33. Some close-ups are shown in Figure 3.34. Provide a rationale for your guesses.*

Aside: Topological entropy of a dynamical system

You could skip this section on a first reading, but you should come back to it later.

Here we follow [56], p. 108. Topological entropy is the most important numerical invariant related to the diversity or 'growth' of orbits of points. It represents the number of orbits of points, under the dynamical system, that are distinguishable with arbitrarily fine but finite precision. Let $f : \mathbb{X} \to \mathbb{X}$ be a continuous mapping from a compact metric space $(\mathbb{X}, d)$ to itself. Define an increasing sequence of metrics $\{d_n : \mathbb{X} \times \mathbb{X} \to [0, \infty)\}_{n=1}^\infty$ by

$$d_n(x, y) = \max_{0 \leq k \leq n-1} d\left(F^{\circ k}(x), F^{\circ k}(y)\right).$$

You should verify that this equation does indeed define a metric for each $n \in \{1, 2, 3, \dots\}$. Let

$$\mathcal{B}_n(x, \epsilon) = \{y \in \mathbb{X} : d_n(x, y) < \epsilon\}$$

denote the open ball of centre x and radius $\epsilon > 0$ in the metric d_n. Let $\mathcal{N}_n(\epsilon)$ denote the minimum number of such balls needed to cover $\mathbb{X}$. This number is finite because $\mathbb{X}$ is compact. Let

$$h_d(f, \epsilon) = \limsup_{n \to \infty} \frac{1}{n} \log_2 \mathcal{N}_n(\epsilon).$$

This is a monotone decreasing function of $\epsilon \in (0, 1)$ and hence has a finite or infinite limit as ϵ approaches zero through positive values. The topological entropy of the dynamical system $f : \mathbb{X} \to \mathbb{X}$ is defined to be

$$h_{top}(f) := \lim_{\epsilon \to 0, \epsilon > 0} h_d(f, \epsilon).$$

The remarkable fact is that this quantity is independent of the metric, so long as the metric defines the same topology. See [56], p. 109.

The following theorem gives some properties of the topological entropy that may be useful towards its calculation in specific examples.

THEOREM 3.5.18 *Let $f : \mathbb{X} \to \mathbb{X}$ be a continuous mapping from a compact metric space $(\mathbb{X}, d)$ to itself. Let $m \geq 1$ be an integer.*
 (i) If $\mathbb{Y} \subset \mathbb{X}$ is closed and such that $f(\mathbb{Y}) = \mathbb{Y}$ then $h_{top}(f|_\mathbb{Y}) \leq h_{top}(f)$.
 (ii) If $\mathbb{X} = \bigcup_{i=1}^m \mathbb{Y}_i$, where $\mathbb{Y}_i$ is closed and $f(\mathbb{Y}_i) = \mathbb{Y}_i$ for $i = 1, 2, \dots, m$, then
  $h_{top}(f) = \max\{h_{top}(f|_{\mathbb{Y}_i}) : i = 1, 2, \dots, m\}.$

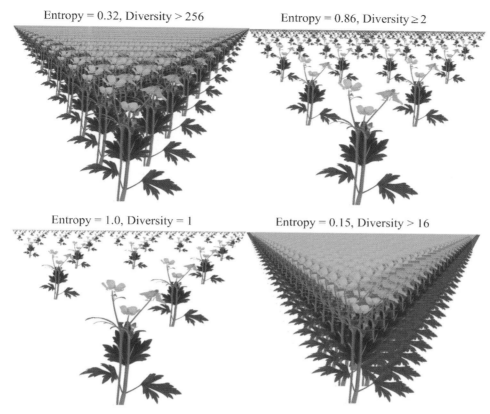

Entropy = 0.32, Diversity > 256

Entropy = 0.86, Diversity ≥ 2

Entropy = 1.0, Diversity = 1

Entropy = 0.15, Diversity > 16

Figure 3.32 Four 'buttercup field' orbital pictures, with estimates of their entropies and diversities, modulo the buttercup. In these cases the entropy tells the growth rate of the number of periodic cycles of little pictures, i.e. panels, as you approach the horizon. The diversity tells how many non-homeomorphic segments of the buttercup are contained in the set of panels. See Definition 3.5.29.

(iii) $h_{top}(f^{\circ m}) = m h_{top}(f)$.

(iv) If $\widetilde{f} : \widetilde{\mathbb{X}} \to \widetilde{\mathbb{X}}$ is a continuous mapping from a compact metric space $(\widetilde{\mathbb{X}}, \widetilde{d})$ to itself then $f : h_{top}(f \times \widetilde{f}) = h_{top}(f) + h_{top}(\widetilde{f})$.

PROOF   See [56], p. 111.   □

When the domain of a picture $\mathfrak{P} \in \Pi_{\mathfrak{C}}$ is compact, we may define the topological entropy of $\mathfrak{P}$ to be the greatest lower bound for the set of entropies of all the homeomorphisms $H \times C : \mathbb{X} \times \mathfrak{C} \to \mathbb{X} \times \mathfrak{C}$ such that

$$H(C(\mathfrak{P})) = \mathfrak{P}.$$

That is,

$$h(\mathfrak{P}) := \inf\{h_{top}(H \times C : \mathbb{X} \times \mathfrak{C} \to \mathbb{X} \times \mathfrak{C}) : H(C(\mathfrak{P})) = \mathfrak{P}\}. \quad (3.5.14)$$

This quantity is a topological invariant and may correlate with the amount of information needed to describe the picture in terms of a dynamical system. However, $h(\mathfrak{P})$ takes no account of the amount of information needed to describe transformations $H \times C$ whose entropies are close to $h(\mathfrak{P})$.

Dynamics on panels and orbital pictures

Theorem 3.5.13 allows us to construct a dynamical system on the panels of an orbital picture. We simply define

$$T : P_{panels} \rightarrow P_{panels}$$

by

$$T(\mathfrak{Q}_\sigma) = \mathfrak{Q}_{S(\sigma)} \quad \text{for all } \mathfrak{Q}_\sigma \in P_{panels},$$

where we recall that $\mathfrak{Q}_\sigma := \mathfrak{Q}_{c(\sigma)}$, where c is the counting function in Equation (3.5.2) in Definition 3.5.2, giving the canonical ordering. We now define an addressing function

$$\phi : \Omega'_{\mathfrak{P}_0} \rightarrow P_{panels}$$

by

$$\phi(\sigma) = \mathfrak{Q}_\sigma \quad \text{for all } \sigma \in \Omega'_{\mathfrak{P}_0}.$$

Then ϕ is one-to-one and onto; hence ϕ is invertible.

The relationship between the action of T on P_{panels} and S on $\Omega'_{\mathfrak{P}_0}$ may be represented by the diagram

$$
\begin{array}{ccc}
\Omega'_{\mathfrak{P}_0} & \overset{\phi}{\leftrightarrow} & P_{panels} \\
S \downarrow & & \downarrow \ T \\
\Omega'_{\mathfrak{P}_0} & \underset{\phi}{\leftrightarrow} & P_{panels}
\end{array}.
$$

Here we have used double-headed arrows to emphasize that $\phi : \Omega'_{\mathfrak{P}_0} \rightarrow P_{panels}$ is invertible.

The action of T on the space of panels may be extended to yield a mapping on the orbital picture itself, $T : \mathfrak{P} \rightarrow \mathfrak{P}$, by defining $T : D_\mathfrak{P} \rightarrow D_\mathfrak{P}$ as

$$T(x) = f_{\sigma_1}^{-1}(x) \quad \text{when } x \in \mathfrak{D}_{\mathfrak{Q}_\sigma}, \quad \text{for all } \mathfrak{Q}_\sigma \in P_{panels}.$$

Then the orbital picture $\mathfrak{P}$ is mapped into itself by $T : D_\mathfrak{P} \rightarrow D_\mathfrak{P}$, that is,

$$T(\mathfrak{P}) \subset \mathfrak{P}.$$

(In particular, when $T : D_\mathfrak{P} \rightarrow D_\mathfrak{P}$ is continuous, we have

$$h(\mathfrak{P}) \leq h_{top}(T : \mathfrak{P} \rightarrow \mathfrak{P}) \leq h_{\mathfrak{P}_0},$$

Figure 3.33 Orbital pictures with various different symbolic entropies. The condensation picture is the tree. See also Figure 3.34. Which orbital picture seems to have the highest growth rate of periodic cycles? Which one seems to contain the greatest diversity of panels?

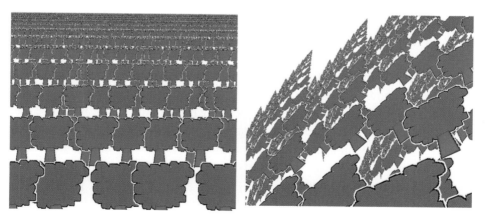

Figure 3.34 Close-ups of two orbital pictures in Figure 3.33. Notice the diversity of shapes, caused by the many ways in which the trees overlap.

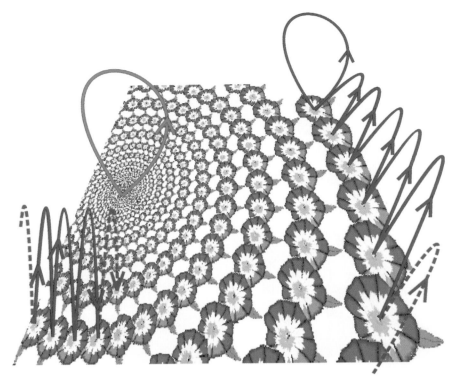

Figure 3.35 Illustration of the dynamical system $T : P_{panels} \to P_{panels}$, showing with red arrows the action on some of the panels. The panels are parts of flowers with a small part missing; each flower is mapped to the next flower out around the spiral, while the last flower is mapped to itself. The blue arrow represents a related transformation on the attractor of the IFS.

where $h(\mathfrak{P})$ is defined in Equation (3.5.14). We expect that the same relationship will hold when $T : D_{\mathfrak{P}} \to D_{\mathfrak{P}}$ is not continuous and the set of points where T is discontinuous makes no contribution to h_{top}.)

Two examples of the dynamics of $T : P_{panels} \to P_{panels}$ are given in Figures 3.35 and 3.37; see Figure 3.36 for an illustration of the mappings used in connection with Figure 3.37.

In Figure 3.35 the panels are flowers; each flower is mapped to the next flower out along the spiral, while the last flower is mapped to itself. So $T : D_{\mathfrak{P}} \to D_{\mathfrak{P}}$ maps each point in a flower to the corresponding point in the next flower. The blue arrow represents an extension of the dynamics of $T : D_{\mathfrak{P}} \to D_{\mathfrak{P}}$ to the attractor of the IFS. Similarly, in Figure 3.37, the action of T on the domain of the visible parts of buttercups has been extended to define an action, again indicated by blue arrows, on a limit set, the horizon.

You can get an intuitive feel for this extension of $T : D_{\mathfrak{P}} \to D_{\mathfrak{P}}$ to the limit set by looking at Figure 3.37 and analysing how the dynamics of $T : P_{panels} \to P_{panels}$ acts on panels close to the horizon. In Chapter 4 we will show that this intuitively

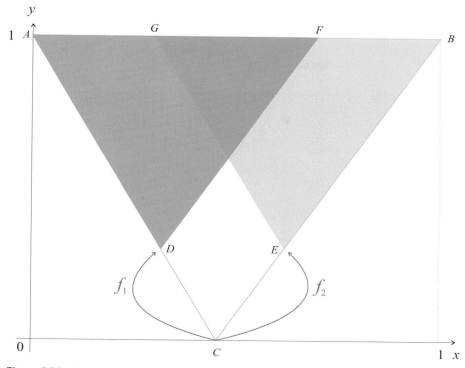

Figure 3.36 Illustration of the action of the transformations, in Equation (3.5.4), used in Figures 3.21, 3.22 and 3.37. We have $f_1(ABC) = AFD$ and $f_2(ABC) = GBE$. It is helpful to think of the triangle AFD as lying on top of triangle GBE.

glimpsed dynamical system is actually the 'tops' dynamical system restricted to the subset $A_{\mathfrak{P}_0}$ of the attractor A of the IFS, which 'peeks out from underneath the orbital picture', as illustrated in Figure 3.38. $A_{\mathfrak{P}_0}$ is defined with the continuous addressing function $\phi : \Omega \to A$ from Theorem 3.3.12 by

$$A_{\mathfrak{P}_0} := \phi\left(\Omega_{\mathfrak{P}_0}\right).$$

EXERCISE 3.5.19 *Figure 3.38 illustrates orbital pictures for the IFS semigroup generated by the three transformations*

$$f_1(x, y) = (0.5x + 0.25, \ 0.5y + 0.4),$$
$$f_2(x, y) = (0.355x - 0.355y + 0.266, \ 0.355x + 0.355y + 0.078), \qquad (3.5.15)$$
$$f_3(x, y) = (0.355x + 0.355y + 0.378, \ -0.355x + 0.355y + 0.434)$$

Mark some arrows between panels in the orbital picture at the top right of Figure 3.38 to illustrate the action of the dynamical system $\{P_{panels}, T\}$. Deduce a

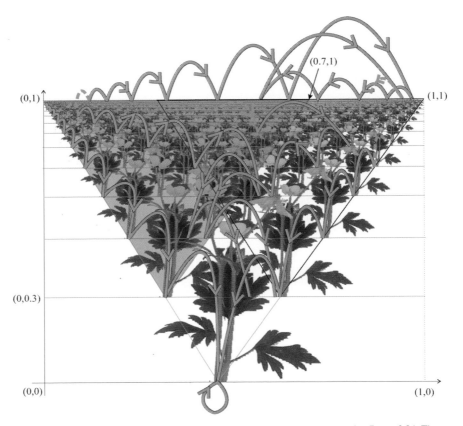

Figure 3.37 Illustration of the dynamical system $T : P_{panels} \to P_{panels}$; see also Figure 3.36. The action of T on visible parts of the buttercups may be extended to define an action, represented by the blue arrows, on the limit set, the horizon. The topological entropy of this limiting system, $-\log_2 0.7$, is a measure of the complexity of the orbital picture. See the main text.

consistent action for T on the 'limit set' of the orbital picture, illustrated in various colours in the top left panel of Figure 3.38.

EXERCISE 3.5.20   *Prove that $A_{\mathfrak{P}_0}$, as defined above, is a closed set.*

EXERCISE 3.5.21   *Consider the set-up in Exercise 3.5.12. Show that $A \subset \mathfrak{P}(A \cup C)$ and $A_{A\cup C} = A$. So in this case we have $A_{A\cup C} \subset \mathfrak{P}(A \cup C)$, and none of $A_{A\cup C}$ would be seen 'peeking out from underneath the orbital picture'. Show also that $\mathfrak{P}(A) = A$ and $A_A = \varnothing$.*

Further examples of panels and the associated dynamical systems are illustrated in Figures 3.39–3.45.

Figures 3.39–3.41 illustrate the panels $\{\mathfrak{Q}_\sigma : \sigma \in \Omega'_{\mathfrak{P}_0}\}$ in Figures 3.21 and 3.22. The colours of the panels have been modified to produce a new set of panels

Figure 3.38 (Top right) The orbital picture of the condensation picture $\mathfrak{P}_0$, as in Figure 3.1; (bottom left) the underneath picture and, in colours different from green, the attractor of the IFS; (top left) the first few generations of the orbital picture, with the attractor 'peeking out from underneath'; (bottom right) the orbital picture when a smaller condensation set $\mathfrak{P}'_0$ is used. Do the visible parts of the leaves in the bottom left image represent a picture tiling?

$\{\widetilde{\mathfrak{Q}}_\sigma : \sigma \in \Omega'_{\mathfrak{P}_0}\}$ with the aid of a semigroup of homeomorphisms $\{C_\sigma : \mathfrak{C} \to \mathfrak{C} : \sigma \in \Omega'_{\mathfrak{P}_0}\}$, according to

$$\widetilde{\mathfrak{Q}}_\sigma = C_\sigma(\mathfrak{Q}_\sigma) \quad \text{for all } \sigma \in \Omega'_{\mathfrak{P}_0}.$$

We should notice the diversity of the shapes and forms of the panels, and the emergence of new patterns, as we zoom in deeper and deeper towards the distant horizon. We will formalize this intuition in the next part of the subsection. This sequence of figures illustrates how orbital pictures may be used in graphics for video games to produce, in a simple way, scenery which possesses rich patterns that change as the user 'travels towards the horizon'.

Figure 3.39   This illustrates the panels of the orbital picture in Figure 3.21. The colours of the segments are modifed from one panel to the next by means of an invertible mapping on the colour space. See the main text. Two successive zooms towards the horizon are shown in Figures 3.40 and 3.41.

Figure 3.40   A zoom towards the horizon in Figure 3.39. See also Figure 3.41.

Figure 3.41   A deeper zoom towards the horizon in Figure 3.39. What shapes are visible at this resolution but not clearly visible in Figure 3.39?

In Figure 3.42 we have illustrated the panels of the orbital picture of a bright-green leaf silhouette, $\mathfrak{P}_0$, situated inside the attractor set $\square$, a filled square, under an IFS of four similitudes, each of which maps $\square$ onto one of its four quarters. Different colours are used to illustrate the panels (otherwise the orbital picture would look like a green $\square$.) Let us say that a panel is larger or smaller than a second panel if it is a segment of a leaf that is respectively larger or smaller than the leaf of which the second panel is a segment. Then the transformation $T : P_{panels} \to P_{panels}$ maps the largest segment, $\mathfrak{P}_0$, to itself and every other panel to one of the next larger panels. Notice that there are various different-shaped panels of the same size. In this case the limit set $A_{\mathfrak{P}_0}$ includes the boundary of $\square$ together with various fractal crosses that project into the interior of $\square$. Clearly there is a great diversity of panels in any neighborhood of $A_{\mathfrak{P}_0}$.

It is interesting to compare Figure 3.42 with Figure 3.43. In the latter the attractor is again $\square$ but this time the four maps in the IFS are the similitudes $f_i : \mathbb{C} \to \mathbb{C}$ defined by

$$f_1(z) = 0.7z, \qquad f_2(z) = 0.6z + 0.4,$$
$$f_3(z) = 0.66z + 0.34i, \quad f_4(z) = 0.5z + 0.5(1+i). \tag{3.5.16}$$

These similitudes are such that $f_i(\square) \cap f_j(\square)$ has a nonempty interior for each $i, j \in \{1, 2, 3, 4\}$. A close-up of Figure 3.43 is shown in Figure 3.44. In this case the limit set $A_{\mathfrak{P}_0}$ is simply the boundary of $\square$ and the growth rate of periodic cycles is lower than for the situation in Figure 3.42. But Figure 3.43 *seems* more complicated than Figure 3.42. Is it? In the next part of the subsection, which now follows, we will show a way in which such pictures may be compared.

The space of limiting pictures and the diversity of segments in the orbital picture
The code space $\Omega'_{\mathfrak{P}_0}$ provides an addressing scheme for the panels of the orbital picture. But what is the significance of $\Omega_{\mathfrak{P}_0}$? Can we find pictures, some sort of magnified limiting panels, that correspond to sequences of points in $\Omega'_{\mathfrak{P}_0}$? Can we find such pictures that also correspond to periodic cycles of the dynamical system $\{S, \Omega_{\mathfrak{P}_0}\}$? And can we find a way to discuss the number of fundamentally 'different' panels that occur in an orbital picture?

To answer these questions we construct a wonderful new metric space whose elements are, essentially, segments of $\mathfrak{P}_0$ that are homeomorphic either to panels of the orbital picture or to certain limiting pictures. We will restrict our attention to the case where $(\mathbb{X}, d)$ is a compact metric space. But the main ideas are much more generally applicable.

We need a few definitions and concepts first. Let $\mathfrak{P}_0 \in \Pi = \Pi_{\mathfrak{C}}(\mathbb{X})$ have compact domain $D_{\mathfrak{P}_0} \subset \mathbb{X}$. Then we define $S_{segments}(\mathfrak{P}_0)$ to be the **space of segments of $\mathfrak{P}_0$** whose domains are compact and nonempty. Given any segment $\mathfrak{R}$ of $\mathfrak{P}_0$ we can form a corresponding segment $\overline{\mathfrak{R}} \in S_{segments}(\mathfrak{P}_0)$, which we will call the

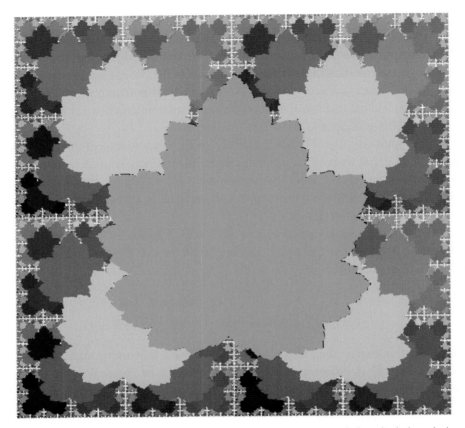

Figure 3.42 Orbital picture of a leaf silhouette $\mathfrak{P}_0$, taken from a photo, with the individual panels shown in different colours. Notice the diversity of visible coloured shapes. In this case the attractor of the IFS is 'just touching', in contrast with that used in Figure 3.43, which is 'overlapping'.

closure of the segment $\mathfrak{R}$, by taking the domain of $\overline{\mathfrak{R}}$, $D_{\overline{\mathfrak{R}}}$, to be the closure of the domain of $\mathfrak{R}$. We define

$$\overline{\mathfrak{R}}(x) = \mathfrak{P}_0(x) \quad \text{for all } x \in D_{\overline{\mathfrak{R}}} = \overline{D_{\mathfrak{R}}}.$$

Then it is easy to see that $(S_{segments}(\mathfrak{P}_0), d)$ is a compact metric space, where

$$d(\mathfrak{R}_1, \mathfrak{R}_2) = d_{\mathbb{H}(\mathbb{X})}(D_{\mathfrak{R}_1}, D_{\mathfrak{R}_2}) \quad \text{for all } \mathfrak{R}_1, \mathfrak{R}_2 \in S_{segments}(\mathfrak{P}_0)$$

and where $d_{\mathbb{H}(\mathbb{X})}$ denotes the Hausdorff distance function defined in Chapter 1.

Let us say that $\{\mathfrak{R}_n \in S_{segments}(\mathfrak{P}_0)\}_{n=1}^{\infty}$ is a **nested sequence of segments** iff

$$\mathfrak{R}_1 \supset \mathfrak{R}_2 \supset \mathfrak{R}_3 \supset \cdots$$

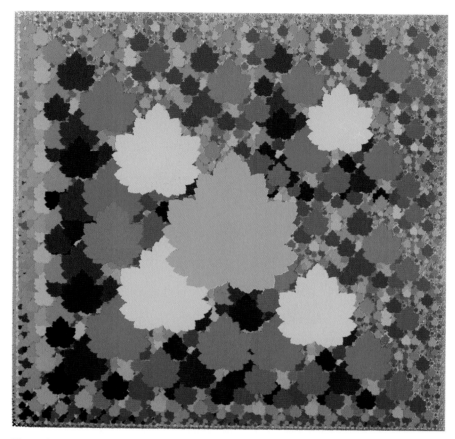

Figure 3.43 The panels have been assigned various colours. The IFS is given by Equation (3.5.16) and is 'overlapping' in contrast to that used in Figure 3.42. A close-up of this picture is shown in Figure 3.44. See also Figure 3.45. In the limit of infinite magnification, what shapes might you see?

Then *any nested sequence of segments of $\mathfrak{P}_0$ converges to a unique element of* $S_{segments}(\mathfrak{P}_0)$, because the corresponding sequence of domains forms a decreasing (nested) sequence of compact sets.

Now let $\mathfrak{P}$ denote the orbital picture of $\mathfrak{P}_0$ under the IFS semigroup $\mathcal{S}_{\{f_1, f_2, ..., f_N\}}$. Then we define a mapping

$$\Lambda : P_{panels} \rightarrow S_{segments}(\mathfrak{P}_0)$$

by

$$\Lambda(\mathfrak{Q}_\sigma) = \overline{f_\sigma^{-1}(\mathfrak{Q}_\sigma)} = \overline{f_{\sigma_{|\sigma|}}^{-1} \circ f_{\sigma_{|\sigma|-1}}^{-1} \circ \cdots \circ f_{\sigma_1}^{-1}(\mathfrak{Q}_\sigma)}.$$

In other words, $\Lambda(\mathfrak{Q}_\sigma)$ is the closure of the unique segment of $\mathfrak{P}_0$ which is transformed to the panel $\mathfrak{Q}_\sigma$ under a transformation that belongs to the set of transformations $\{f_\sigma : \sigma \in \Omega'_{\mathfrak{P}_0}\}$.

Figure 3.44 Close-up of part of Figure 3.43. Again, notice the emergence of new shapes and forms as the resolution is increased! In this case, is the space of limiting pictures finite or infinite?

Next we show that this definition can be extended to a subset $\overleftarrow{\Omega}_{\mathfrak{P}_0} \subset \Omega$, which we define as follows.

DEFINITION 3.5.22   Let $\Omega'_{\mathfrak{P}_0}$ denote the set of addresses of the orbital picture $\mathfrak{P} = \uplus \mathcal{O}(\mathfrak{P}_0)$. Then

$$\overleftarrow{\Omega}_{\mathfrak{P}_0} := \left\{ \omega = \omega_1 \omega_2 \cdots \in \Omega : \omega_n \omega_{n-1} \cdots \omega_1 \in \Omega'_{\mathfrak{P}_0} \text{ for each } n \in \{1, 2, \dots\} \right\}.$$

THEOREM 3.5.23   *Let $(\mathbb{X}, d)$ be a compact metric space. Let the domain of $\mathfrak{P}_0 \in \Pi = \Pi_{\mathfrak{C}}(\mathbb{X})$ be compact. Let $\{\mathfrak{Q}_\sigma : \sigma \in \Omega'_{\mathfrak{P}_0}\}$ denote the set of panels of the orbital picture of $\mathfrak{P}_0$ under the IFS semigroup $\mathcal{S}_{\{\mathbb{X}; f_1, f_2, \dots, f_N\}}$. Then there exists a well-defined mapping $\Phi : \Omega'_{\mathfrak{P}_0} \cup \overleftarrow{\Omega}_{\mathfrak{P}_0} \to S_{segments}(\mathfrak{P}_0)$ specified by*

$$\Phi(\sigma) = \begin{cases} \Lambda(\mathfrak{Q}_\sigma) & \text{when } \sigma \in \Omega'_{\mathfrak{P}_0}, \\ \lim_{n \to \infty} \Lambda(\mathfrak{Q}_{\sigma_n \sigma_{n-1} \cdots \sigma_1}) & \text{when } \sigma_1 \sigma_2 \cdots \in \overleftarrow{\Omega}_{\mathfrak{P}_0}. \end{cases}$$

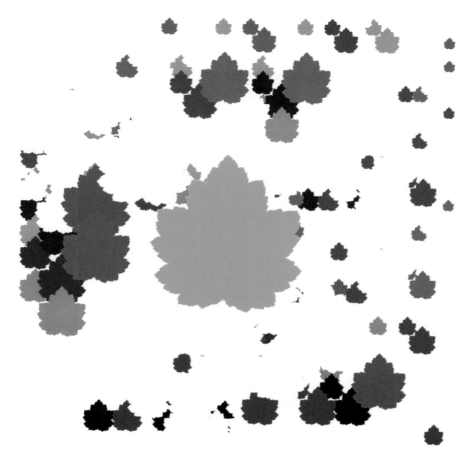

Figure 3.45 Some of the panels in Figure 3.43. Can you identify a panel here whose domain is disconnected? Roughly, how many distinct shapes are shown here?

Furthermore, we have

$$\Phi(S(\sigma)) \subset \Phi(\sigma) \quad \textit{for all } \sigma \in \Omega'_{\mathfrak{P}_0},$$

where $S : \Omega'_{\mathfrak{P}_0} \to \Omega'_{\mathfrak{P}_0}$ *denotes the shift transformation.*

PROOF   We notice that when $\sigma \in \Omega'_{\mathfrak{P}_0}$ the definition of Λ is straightforward.
It follows at once that

$$f_{\sigma_1}^{-1}\left(\mathfrak{Q}_{\sigma_1 \sigma_2 \cdots \sigma_{|\sigma|}}\right) \subset \mathfrak{Q}_{\sigma_2 \cdots \sigma_{|\sigma|}}$$

for each given $\sigma \in \Omega'_{\mathfrak{P}_0}$. Hence, on applying the transformation

$$f_{\sigma_2 \cdots \sigma_{|\sigma|}}^{-1} = f_{\sigma_{|\sigma|}}^{-1} \circ f_{\sigma_{|\sigma|-1}}^{-1} \circ \cdots \circ f_{\sigma_2}^{-1}$$

to both sides and taking the closure, we obtain

$$\Lambda(\mathfrak{Q}_\sigma) \subset \Lambda(\mathfrak{Q}_{S(\sigma)}) \quad \text{for all } \sigma \in \Omega'_{\mathfrak{P}_0},$$

which proves the last statement in the theorem.

Now let $\sigma_1\sigma_2\cdots \in \overleftarrow{\Omega}_{\mathfrak{P}_0}$ and choose $\sigma = \sigma_n\sigma_{n-1}\cdots\sigma_1$ for $n \in \{1, 2, \ldots\}$. Then $\sigma \in \Omega'_{\mathfrak{P}_0}$ and consequently

$$\Lambda(\mathfrak{Q}_{\sigma_n\sigma_{n-1}\cdots\sigma_1}) \subset \Lambda(\mathfrak{Q}_{\sigma_{n-1}\cdots\sigma_1}) \quad \text{for each } n \in \{1, 2, \ldots\}.$$

It follows that

$$\{\Lambda(\mathfrak{Q}_{\sigma_n\sigma_{n-1}\cdots\sigma_1}) \in S_{segments}(\mathfrak{P}_0)\}_{n=1}^\infty$$

is a nested sequence of segments of $\mathfrak{P}_0$ and so converges to a unique element of $S_{segments}(\mathfrak{P}_0)$. □

DEFINITION 3.5.24   The space defined using the transformation Φ in Theorem 3.5.23,

$$L_{\mathfrak{P}_0} := \Phi\big(\Omega'_{\mathfrak{P}_0} \cup \overleftarrow{\Omega}_{\mathfrak{P}_0}\big),$$

is called the **space of limiting pictures** associated with the orbital picture of $\mathfrak{P}_0$ under the IFS semigroup $S_{\{f_1, f_2, \ldots, f_N\}}$.

Let us define the closure $\overline{\mathfrak{Q}_\sigma}$ of a panel $\mathfrak{Q}_\sigma \in P_{panels}$ by

$$\overline{\mathfrak{Q}_\sigma} = f_\sigma\big(\overline{f_\sigma^{-1}(\mathfrak{Q}_\sigma)}\big).$$

Then each point in $\Phi(\Omega'_{\mathfrak{P}_0})$ corresponds to a set of panels in P_{panels} whose closures are homeomorphic. Indeed if $\sigma, \omega \in \Omega'_{\mathfrak{P}_0}$ with $\Phi(\sigma) = \Phi(\omega)$ then

$$\overline{\mathfrak{Q}_\omega} = f_\omega\big(f_\sigma^{-1}(\overline{\mathfrak{Q}_\sigma})\big).$$

The following theorem tells us that corresponding to each periodic orbit of the shift transformation acting on the space $\Omega_{\mathfrak{P}_0}$ there is at least one point in $\overleftarrow{\Omega}_{\mathfrak{P}_0}$.

THEOREM 3.5.25   *Let $(\mathbb{X}, d)$ be a compact metric space. Let the domain of $\mathfrak{P}_0 \in \Pi = \Pi_{\mathfrak{C}}(\mathbb{X})$ be compact. Let $\{\mathfrak{Q}_\sigma : \sigma \in \Omega'_{\mathfrak{P}_0}\}$ denote the set of panels of the orbital picture of $\mathfrak{P}_0$ under the IFS semigroup $S_{\{\mathbb{X}; f_1, f_2, \ldots, f_N\}}$. Let $\rho \in \Omega_{\mathfrak{P}_0}$ be a periodic point for the shift transformation $S : \Omega_{\mathfrak{P}_0} \to \Omega_{\mathfrak{P}_0}$ of period $k \in \{1, 2, \ldots\}$. That is,*

$$\rho = \overline{\rho_1\rho_2\cdots\rho_k}.$$

Then at least one of the points

$$\overline{\rho_k\rho_{k-1}\cdots\rho_1}, \quad \overline{\rho_{k-1}\rho_{k-2}\cdots\rho_1\rho_k}, \quad \ldots, \quad \overline{\rho_1\rho_k\cdots\rho_3\rho_2}$$

belongs to $\overleftarrow{\Omega}_{\mathfrak{P}_0}$. When $N = 2$ there exist examples where

$$\overline{\rho_2\rho_1} \in \overleftarrow{\Omega}_{\mathfrak{P}_0} \quad \text{but} \quad \overline{\rho_1\rho_2} \notin \overleftarrow{\Omega}_{\mathfrak{P}_0}.$$

PROOF   Since $\rho \in \Omega_{\mathfrak{P}_0}$ it follows that there exist two sequences of integers $\{m_l\}_{l=1}^{\infty}$ and $\{n_l\}_{l=1}^{\infty}$ such that

$$0 < m_1 < m_2 < \cdots$$

and

$$n_l \in \{1, 2, \ldots, k\} \quad \text{for each } l = 1, 2, \ldots,$$

with

$$\underbrace{(\rho_1 \rho_2 \cdots \rho_k)}_{m_l \text{ times}} \rho_1 \rho_2 \cdots \rho_{n_l} \in \Omega'_{\mathfrak{P}_0} \quad \text{for all } l = 1, 2, 3, \ldots$$

One value of the index n_l must be repeated infinitely many times; let us denote such a value by $s \in \{1, 2, \ldots, k\}$. It follows that there exists a sequence of integers $\{q_l\}_{l=1}^{\infty}$,

$$0 < q_1 < q_2 < \cdots,$$

such that

$$\underbrace{(\rho_1 \rho_2 \cdots \rho_k)}_{q_l \text{ times}} \rho_1 \rho_2 \cdots \rho_s \in \Omega'_{\mathfrak{P}_0} \quad \text{for all } l = 1, 2, 3, \ldots$$

With the help of applications of $S : \Omega'_{\mathfrak{P}_0} \to \Omega'_{\mathfrak{P}_0}$, it now follows that

$$\rho_t \rho_{t+1} \cdots \rho_k \underbrace{(\rho_1 \rho_2 \cdots \rho_k)}_{r \text{ times}} \rho_1 \rho_2 \cdots \rho_s \in \Omega'_{\mathfrak{P}_0}$$

for any integer $r \geq 0$ and any $t \in \{1, 2, \ldots, k\}$. It also follows similarly that $\rho_s \in \Omega'_{\mathfrak{P}_0}$, $\rho_{s-1}\rho_s \in \Omega'_{\mathfrak{P}_0}$, $\ldots$ and $\rho_1 \rho_2 \cdots \rho_s \in \Omega'_{\mathfrak{P}_0}$. Hence

$$\rho_s \rho_{s-1} \cdots \rho_1 \overline{\rho_k \rho_{k-1} \cdots \rho_1} \in \overleftarrow{\Omega}_{\mathfrak{P}_0},$$

which implies that

$$\overline{\rho_s \rho_{s-1} \cdots \rho_1 \rho_k \rho_{k-1} \ldots \rho_{s+1}} \in \overleftarrow{\Omega}_{\mathfrak{P}_0}.$$

This proves the first part of the theorem.

To prove the second part we consider the IFS $\{\mathbb{R}^2; f_1, f_2\}$, where

$$f_1(x, y) = (0.5x, \ 0.5y + 1), \quad f_2(x, y) = (1 - x, \ y).$$

Let us choose the domain of $\mathfrak{P}_0$ to be the filled unit square $\square$. Then since $f_2(\square) = \square$ it follows that $2 \notin \Omega'_{\mathfrak{P}_0}$. Therefore $\Omega'_{\mathfrak{P}_0}$ contains no address that terminates in the symbol 2. Remember that if $\sigma \in \Omega'_{\mathfrak{P}_0}$ then $S(\sigma) \in \Omega'_{\mathfrak{P}_0}$. Hence $\overleftarrow{\Omega}_{\mathfrak{P}_0}$ contains no address that commences with the symbol 2.

But it is readily verified, by induction, that $\overline{12}$ belongs to $\Omega_{\mathfrak{P}_0}$. Hence at least one of $\overline{12}$ and $\overline{21}$ belongs to $\overleftarrow{\Omega}_{\mathfrak{P}_0}$. We conclude that $\overline{12} \in \overleftarrow{\Omega}_{\mathfrak{P}_0}$ and $\overline{21} \notin \overleftarrow{\Omega}_{\mathfrak{P}_0}$. A related, but different example is illustrated in Figure 3.46. $\square$

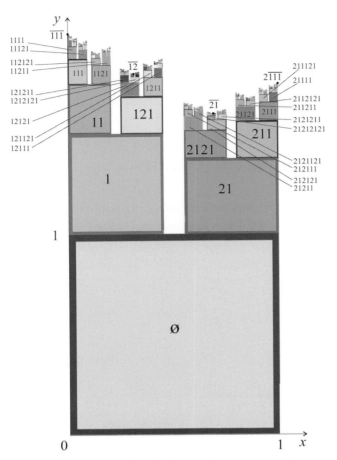

Figure 3.46   The panels of an orbital picture, illustrated using various colours to distinguish them, together with some addresses. The condensation picture $\mathfrak{P}_0$ corresponds to the largest square region, with address $\varnothing$. In this example $\overline{12} \in \Omega_{\mathfrak{P}_0}$, $\overline{21} \in \Omega_{\mathfrak{P}_0}$ and $\overline{12} \in \overleftarrow{\Omega}_{\mathfrak{P}_0}$ but $\overline{21} \notin \overleftarrow{\Omega}_{\mathfrak{P}_0}$. Why is there no panel with address 2?

EXERCISE 3.5.26   *Find the IFS used to generate the orbital picture whose panels are illustrated in Figure 3.46.*

EXERCISE 3.5.27   *Define an IFS semigroup $\mathcal{S}_{\{f_1, f_2, f_3\}}(\mathbb{R}^2)$ and condensation picture $\mathfrak{P}_0$ such that $\overline{123} \in \overleftarrow{\Omega}_{\mathfrak{P}_0}$ but $\overline{231} \notin \overleftarrow{\Omega}_{\mathfrak{P}_0}$ and $\overline{321} \notin \overleftarrow{\Omega}_{\mathfrak{P}_0}$.*

EXERCISE 3.5.28   *Show that the code space $\Omega'_{\mathfrak{P}_0} \cup \Omega_{\mathfrak{P}_0}$ for the example used at the end of the proof of Theorem 3.5.25 can be obtained from the code space $\Omega'_{\{1,X\}} \cup \Omega_{\{1,X\}}$ by replacing the symbol X, wherever it occurs, by the string 12. The symbol X has been used here, rather than the symbol 2, to help you to avoid confusions when making the replacements.*

In Figure 3.47 we have illustrated some parts of some boundaries of segments belonging to the space of limiting pictures in the case of the orbital picture

Figure 3.47   The internal boundaries within this picture demarcate parts of boundaries of segments in the space of limiting pictures, in relation to the orbital picture in Figure 3.42.

illustrated in Figure 3.42. In Figure 3.48 we have illustrated some of the segments belonging to the space of limiting pictures corresponding to buttercup-field orbital pictures like those illustrated in Figure 3.32. Let us denote these limiting pictures by $Q_\omega(\lambda)$, where $\omega \in \Omega'_{\mathfrak{P}_0}$ is the address and $\lambda \in \{0.7, 0.8, 0.9\}$ is a parameter that specifies the IFS,

$$\{\mathbb{R}^2; f_1(x, y) = (\lambda x, \lambda y + 1 - \lambda), f_2(x, y) = (\lambda x + 1 - \lambda, \lambda y + 1 - \lambda)\}.$$

Look at the top left and bottom right pictures in Figure 3.32. You will notice that the panels on the left-hand side and right-hand side of each picture, which look something like half buttercup-plants, seem to have converged after few iterations, so that

$$Q_{1111}(\lambda) = Q_{111\cdots 1}(\lambda) \quad \text{and} \quad Q_{2222}(\lambda) = Q_{222\cdots 2}(\lambda).$$

In Figure 3.48 the limiting pictures $Q_{1212}(\lambda)$ and $Q_{2121}(\lambda)$ become more fragmented, into torn-up fragments of yellow petals, as λ increases. We note from Figure 3.48 that the domain of a panel of an orbital picture may be disconnected even though the domain of $\mathfrak{P}_0$ is connected.

Figures 3.43 and 3.44 provide further illustrations of the wide variety of pictures that we can expect to find in the space of limiting pictures. In this case, is the space of limiting pictures finite or infinite?

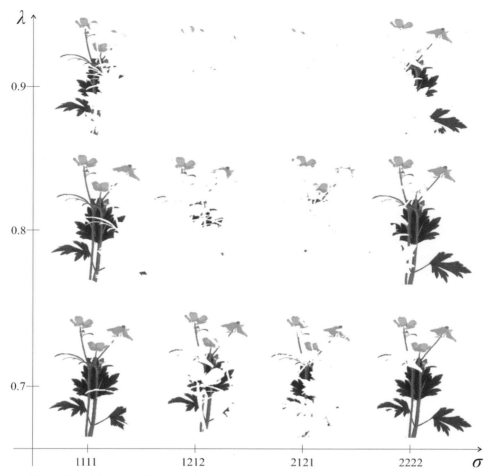

Figure 3.48 Elements of the space of limiting pictures associated with some buttercup fields. The parameter values are 0.7, 0.8 and 0.9 and the corresponding addresses are 0000, 0101, 1010 and 1111. The domain of which of these segments possesses the greatest number of connected components?

It seems clear that the size of the space of limiting pictures, $|L_{\mathfrak{P}_0}|$, is an interesting parameter both mathematically and descriptively, as a means to capture the visual complexity of some orbital pictures. But when $|L_{\mathfrak{P}_0}| = \infty$ we need a finer parameter, so we make the following definition.

DEFINITION 3.5.29　Let $(\mathbb{X}, d)$ be a compact metric space, let $\mathfrak{P}(\mathfrak{P}_0)$ be the orbital picture of $\mathfrak{P}_0 \in \Pi_{\mathfrak{C}}(\mathbb{X})$ and let $L_{\mathfrak{P}_0}$ denote the associated space of limiting pictures. The **diversity** of the orbital picture is $|L_{\mathfrak{P}_0}| \in \{1, 2, \dots\} \cup \{\infty\}$. When $|L_{\mathfrak{P}_0}| = \infty$ the (exponential) **rate of growth of diversity** (in the orbital picture) is defined to be

$$\limsup_{n \to \infty} \frac{1}{n} \log_2 \left| \Phi\big(\{\sigma \in \Omega'_{\mathfrak{P}_0} : |\sigma| = n\}\big) \right|.$$

The latter limit exists because $1 \leq |\Phi(\{\sigma \in \Omega'_{\mathfrak{P}_0} : |\sigma| = n\})| \leq N^n$ for $n = 0, 1, \ldots$

Let us look at some examples. In the case of Figure 3.3 we have $|L_{\mathfrak{P}_0}| = 1$. For the orbital picture in Figure 3.32 with the highest symbolic entropy we again have $|L_{\mathfrak{P}_0}| = 1$ while the orbital pictures with entropies 0.15 and 0.32 clearly have $|L_{\mathfrak{P}_0}| > 1$. Indeed, for the family of IFSs considered in connection with Figure 3.48 it appears that for some values of the parameter $\lambda \in (0.5, 1)$ the value of $|L_{\mathfrak{P}_0}|$ is finite while for others, related to 'β-numbers', which have certain number-theoretic properties, $|L_{\mathfrak{P}_0}|$ is infinite; the growth rate of diversity may be the same as the growth rate of periodic cycles, namely the symbolic entropy, in these cases. See for example [17]. The growth rate of diversity seems to provide an independent measure of the visual complexity of some orbital pictures.

Code spaces of orbital pictures, tree-like or not tree-like

We digress briefly here to illustrate how the code space of an orbital picture may have the structure of a 'pruned tree' and how in other cases it may not be tree-like. This digression serves to increase our familiarity with orbital pictures.

In some cases the structure of $\Omega_{\mathfrak{P}_0} \cup \Omega'_{\mathfrak{P}_0}$ is tree-like, in the sense that $\Omega_A \cup \Omega'_A$ is tree-like, as seen in Figure 1.15. Consider the following examples, associated with the family of IFSs

$$\left\{ \square; \ f_1(x, y) = \left(\lambda x, \tfrac{1}{3}(y + 2)\right), \ f_2(x, y) = \left(\lambda x + 1 - \lambda, \tfrac{1}{3}(y + 2)\right) \right\} \quad (3.5.17)$$

where $0 < \lambda < 1$. The fractal set, the attractor of this IFS, is the line segment A that connects the pair of points $(0, 1)$ and $(1, 1)$ in $\mathbb{R}^2$. We choose $\mathfrak{P}_0$ to be a picture of a block, with domain

$$D_{\mathfrak{P}_0} = \left\{ (x, y) \in \mathbb{R}^2 : 0 \leq x \leq 0.97, 0 \leq y \leq \tfrac{1}{3} \right\}.$$

The resulting patterns of blocks, the orbital pictures, for $\lambda = 0.6, 0.66, 0.7$ and 0.8, are illustrated in Figure 3.49.

In Figure 3.50 we have labelled the visible blocks, the panels, by their addresses. Each 'tree of gaps between blocks' converges to the line segment A and provides a different coding or addressing system for the unit interval. These codings all have the following property: if $\sigma \in \Omega'_{\mathfrak{P}_0} \cup \Omega_{\mathfrak{P}_0}$ then $1\sigma \in \Omega'_{\mathfrak{P}_0} \cup \Omega_{\mathfrak{P}_0}$ and, if also $\sigma_1 = 2$, then $2\sigma \in \Omega'_{\mathfrak{P}_0} \cup \Omega_{\mathfrak{P}_0}$; it follows that in these cases the code spaces of the orbital pictures are 'pruned trees', the trees of gaps between the blocks.

In Figures 3.51 and 3.52 we show examples which are not tree-like. The IFS used in these figures belongs to the family of examples

$$f_1(x, y) = (-\lambda y, -\lambda x + 1), \quad f_2(x, y) = (\lambda x + 1, \lambda y + 1 - \lambda) \quad (3.5.18)$$

with $\lambda = 0.6$. It is quite easy to see that the code space includes the codes $\{\varnothing, 1, 2, 11, 12, 22, 111, 112, 122, 211, 221, 222\}$ but not the code 21 because

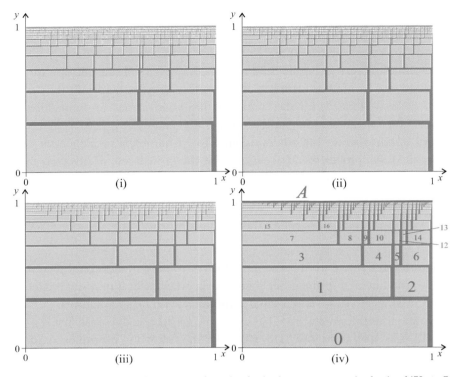

Figure 3.49 Examples of different sets of panels of orbital pictures, using the family of IFSs in Equation (3.5.17), for (i) $\lambda = 0.6$, (ii) $\lambda = 0.66$, (iii) $\lambda = 0.7$ and (iv) $\lambda = 0.8$. The domain of $\mathfrak{P}_0$ is the rectangle at the bottom of each picture. In the limit, each different orbital picture is associated with a different addressing scheme for the points in the interval $[0, 1]$. See also Figure 3.50.

the corresponding picture in the orbit of $\mathfrak{P}_0$ is hidden underneath $\mathfrak{P}_0$. Hence the code space $\Omega'_{\mathfrak{P}_0} \cup \Omega_{\mathfrak{P}_0}$ is not tree-like in this case; see Figure 3.53.

Figure 3.52 illustrates the relationship between the orbital picture, the underneath picture and the attractor of the associated IFS.

EXERCISE 3.5.30 *Write down the addresses of the larger panels in Figure 3.51. Identify some addresses in $\Omega'_{\{1,2\}}$ that do not correspond to panels in this orbital picture. Show that the set of panel addresses $\Omega'_{\mathfrak{P}_0}$ in this case is not tree-like.*

Picture tilings and panellings

We now distinguish between picture tilings and panellings. The idea of a IFS semigroup *picture* tiling is the same as that of an IFS semigroup tiling: non-overlapping picture tiles are obtained by applying *all* the elements of the semigroup to the condensation picture. Illustrations of IFS semigroup picture tilings are provided by Figures 3.3, 3.25, 3.54 and 3.55. In Figure 3.54 the picture tiles are leafy annuli.

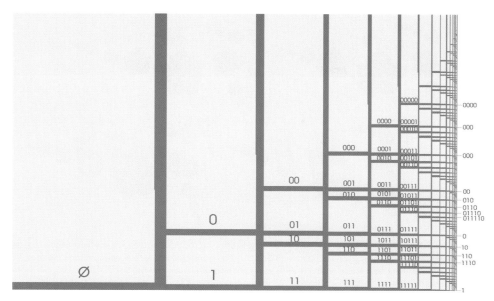

Figure 3.50   Addresses for some of the panels in the orbital picture (iv) in Figure 3.49, corresponding to $\lambda = 0.8$. The addresses are written in the alphabet $\{0, 1\}$ rather than $\{1, 2\}$. The figure has been rotated clockwise through $90°$. The addresses cascade into an addressing scheme for a line interval and are related to fractal tops, discussed in Chapter 4.

Figure 3.51   See Exercise 3.5.30. Choose the square leaf tile to be the condensation picture $\mathfrak{P}_0$. Find an IFS of two affine transformations such that this figure represents the orbital picture of $\mathfrak{P}_0$ under the IFS semigroup. Write down the addresses of some pictures in the orbit of $\mathfrak{P}_0$ that are *not* part of this orbital picture.

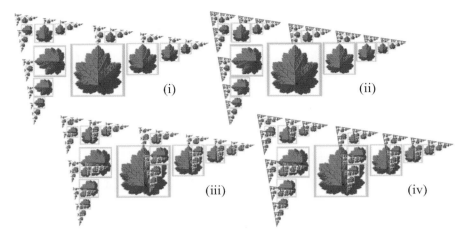

Figure 3.52 The code space structure is not tree-like for this example or the example illustrated in Figure 3.46. Here the IFS is that given in Equation (3.5.18). (i) The orbital picture for the condensation picture $\mathfrak{P}_0$, which looks like a square tile with a leaf on it; (ii) the orbital picture and the set $A_{\mathfrak{P}_0}$ 'peeking out from underneath'; (iii) the underneath picture; (iv) the underneath picture plus the attractor A of the IFS.

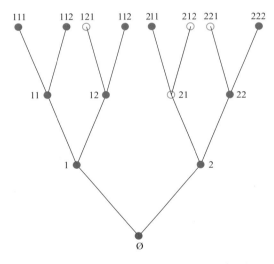

Figure 3.53 Points $\sigma \in \Omega'_{\mathfrak{P}_0}$ with $|\sigma| \leq 3$ associated with Figure 3.52 are here represented as some of the nodes on a tree-like structure, as defined in graph theory. The presence of the nodes with addresses 2 and 211 and the absence of the nodes corresponding to the address 21 means that $\Omega'_{\mathfrak{P}_0}$ is not tree-like.

DEFINITION 3.5.31   Let $\mathcal{S}_{\{f_1, f_2, \ldots, f_N\}}(\mathbb{R}^2)$ be an IFS semigroup and let $\mathfrak{P}_0$ be a picture with domain $D_{\mathfrak{P}_0} \subset \mathbb{R}^2$. Let the orbit $\mathcal{O}(D_{\mathfrak{P}_0})$ be a semigroup tiling of the set $\bigcup \mathcal{O}(D_{\mathfrak{P}_0})$. Then the orbit $\mathcal{O}(\mathfrak{P}_0)$ of $\mathfrak{P}_0$ is called a semigroup tiling of the picture $\uplus \mathcal{O}(\mathfrak{P}_0)$ or a **picture tiling**. Each picture $f_\sigma(\mathfrak{P}_0)$, for $\sigma \in \Omega'_{\{1,2,\ldots N\}}$, is called a *semigroup* **picture tile** and σ is called the **address** of the picture tile $f_\sigma(\mathfrak{P}_0)$. We say that the semigroup, acting on the picture $\mathfrak{P}_0$, **generates** the picture tiling $\mathcal{O}(\mathfrak{P}_0)$.

Figure 3.54   Here an orbital picture is tiled by leafy annuli. Notice how this picture also looks like an underneath picture. Underneath pictures can be used to help find tilings!

In Figure 3.56 we show three examples of IFS semigroup picture tilings. These are especially interesting. In each case, let the IFS that generates the semigroup be called *IFS*#1. Then, in each case, the domain of the orbital picture is the attractor of a just-touching IFS (see Chapter 4) *IFS*#2, such that *IFS*#1 ⊂ *IFS*#2. Let *IFS*#3 = *IFS*#2*IFS*#1, meaning the IFS whose transformations consist of those in *IFS*#2 that are not in *IFS*#1. Then the domain of the condensation picture consists of the union of the sets obtained by applying the transformations in *IFS*#3 to the attractor of *IFS*#2. For example, in the case of the fern picture in Figure 3.56, *IFS*#2 is given by the four projective transformations represented, as in Equation (3.5.9), by the data in the following table:

n	a_n	b_n	c_n	d_n	e_n	f_n	g_n	h_n	j_n
1	0.85	0.04	0.0	−0.04	0.85	1.60	0.0	0.0	1.0
2	0.0	0.0	0.0	0.16	0.0	0.0	0.0	0.0	1.0
3	0.200	−0.26	0.0	0.23	0.22	0.8	0.0	0.0	1.0
4	−0.15	0.28	0.0	0.26	0.24	0.4	0.0	0.0	1.0

Figure 3.55   Example of an IFS semigroup tiling of a picture. The domain of the picture is the complement of a Sierpinski triangle in the space $\mathbb{X} = \square$.

while *IFS*#1 consists of the first and third transformations and *IFS*#3 consists of the second and fourth transformations.

EXERCISE 3.5.32   *Find the IFSs used to make the middle and bottom picture tilings in Figure 3.56. The domains of these two orbital pictures are examples of* **reptiles***, namely attractors of IFSs that can be used to tile $\mathbb{R}^2$; see for example [40] or simply type 'fractal reptiles' into your favourite internet search engine.*

It is useful, for applications such as image compression, to think of an orbital picture of finite diversity as a kind of tiling that we call a **panelling**. In a panelling the 'tiles' are panels as illustrated in Figure 3.57, where we contrast picture tilings with panellings. An orbital picture of finite diversity is always a panelling, but may be a tiling only if the diversity is 1. When an orbital picture is a panelling, it can

Figure 3.56 In each of these IFS semigroup picture tilings the domain of the condensation picture, shown at the right, is a subset of the set attractor of another related IFS. Can you describe the IFSs that generate the orbital pictures on the left?

(i)   (ii)   (iii)   (iv)

Figure 3.57 Examples of panellings of orbital pictures. (i) A panelling of diversity 2. (ii), (iii) Panellings of diversity 1 that are also IFS semigroup tilings. (iv) A panelling of diversity 7. Can you think of a panelling of diversity 1 that is not an IFS semigroup tiling?

be constructed from the set of tiles obtained by applying the IFS semigroup to a finite set of condensation pictures, the elements of the space of limiting pictures. When $|\Omega'_{\mathfrak{P}_0}| < \infty$, the orbital picture is a panelling that consists of finitely many tiles, as illustrated in Figure 3.57(iii).

Figure 3.51 illustrates what appears to be an example of a panelling where $|\Omega'_{\mathfrak{P}_0}| = \infty$ and the diversity equals 1, yet this is not a picture tiling since some transforms of the square leaf-tile $\mathfrak{P}_0$ overlap one another. We say 'appears' because we have not ruled out that some other IFS semigroup of transformations applied to $\mathfrak{P}_0$ could achieve the same orbital picture with no overlaps.

EXERCISE 3.5.33 *Write down the addresses for some of the picture tiles in Figure 3.55. Assume that the IFS is* $\{\Box; f_1, f_2, f_3\}$ *where* $\Box = \{(x, y) \in \mathbb{R}^2 : 0 \le x \le 1, 0 \le y \le 1\}$, $f_1(x, y) = (\frac{1}{2}x, \frac{1}{2}(y + 1))$, $f_2(x, y) = (\frac{1}{2}(x + 1), \frac{1}{2}(y + 1))$ *and* $f_3(x, y) = (\frac{1}{2}(x + 1), \frac{1}{2}y)$.

As in the case of IFS semigroup tilings of sets, pictures that are picture tilings can be represented with some efficiency. So how may we find $\mathfrak{P}_0$ such that $\mathcal{O}(\mathfrak{P}_0)$ is a picture tiling? One approach is to look for a set C such that $\mathcal{O}(C)$ is a tiling, as in Theorem 3.4.6, then choose $\mathfrak{P}_0$ so that $D_{\mathfrak{P}_0} = C$, again as in Theorem 3.4.6. Another approach is to look underneath $\uplus \mathcal{O}(\mathfrak{P}_0)$, as will be discussed in the following subsection.

Underneath the orbital picture

Given an IFS semigroup $\mathcal{S}_{\{f_1, f_2, \dots, f_N\}}$ and a condensation picture $\mathfrak{P}_0$ we define an **underneath picture** to be a picture belonging to the sequence $\{\mathfrak{P}''_n\}_{n=1}^\infty$, where

$$\mathfrak{P}''_n = f_{\sigma(n)}(\mathfrak{P}_0) \uplus f_{\sigma(n-1)}(\mathfrak{P}_0) \uplus \cdots \uplus f_1(\mathfrak{P}_0) \uplus \mathfrak{P}_0 = f_{\sigma(n)}(\mathfrak{P}_0) \uplus \mathfrak{P}''_{n-1}$$

and $\mathfrak{P}''_0 := \mathfrak{P}_0$. Some examples of underneath pictures are shown in Figure 3.58.

Figure 3.58 The underneath pictures of the four orbital pictures in Figure 3.33. We have flipped the figure horizontally, so you can more easily imagine that you have turned Figure 3.33 upside down.

Notice that the underneath picture $\mathfrak{P}_n''$ is obtained from $\mathfrak{P}_{n-1}''$ by putting $f_{\sigma(n)}(\mathfrak{P}_0)$ on top of $\mathfrak{P}_{n-1}''$. As a consequence the sequence of underneath pictures does not converge in general. Consider the case of a contractive IFS whose set attractor possesses a nonempty interior and let x be a point in this interior. Assume too that $D_{\mathfrak{P}_0}$ possesses a nonempty interior. Then for infinitely many values of n, say $n = n_k$ for $k = 1, 2, \ldots$, it will occur that x belongs to the domain of $f_{\sigma(n_k)}(\mathfrak{P}_0)$. Furthermore, it can clearly occur that the sequence of colours $\{f_{\sigma(n_k)}(\mathfrak{P}_0)(x)\}_{k=1}^{\infty}$ does not converge. This is illustrated in the bottom row of underneath pictures in Figure 3.26. In this case the set attractor of the IFS is the leaf-shaped region, with a grainy yellow and green pattern, in the bottom right image. We actually computed many more pictures in this sequence of underneath pictures, and found that the parts of the pictures whose domains intersect the attractor of the IFS seemed never to settle down; a restless sequence of beautiful textures was observed. We will show, in Section 4.8, that a model explanation for this effect, which we call the **texture effect**, lies with the ergodic theorem. It thus appears that a novel application of underneath pictures, and in particular of the ergodic theorem, is to the production of rich textures for computer graphics applications.

Although the sequence $\{\mathfrak{P}_n''\}_{n=1}^{\infty}$ does not generally define a limiting picture it sometimes does. For example, the picture tiling in Figure 3.3 is of this type. In this case the domain of the condensation picture and the images of this domain under the IFS semigroup do not intersect the attractor of the IFS. The same situation occurs in the top right and bottom left images in Figure 3.58.

E X E R C I S E 3.5.34 *Suppose that you are given a picture $\mathfrak{P}$ and two transformations f_1, $f_2 : \mathbb{R}^2 \to \mathbb{R}^2$, and you know that $\mathfrak{P}$ represents the orbit of a picture $\mathfrak{P}_0$ under the IFS semigroup $\mathcal{S}_{\{f_1, f_2\}}(\mathbb{R}^2)$. How would you find $\mathfrak{P}_0$? Now suppose that you do not know f_1 or f_2. What can you say now? Suppose for example that you know that f_1 and f_2 are similitudes, but that is all. Can you design an algorithm, some sort of iterative procedure, to find f_1 and f_2?*

There are clearly many pictures that we can associate with an orbit of pictures, in addition to the orbital picture and the underneath pictures. An interesting family of such pictures is provided by the tops semigroup generated by the infinite set of pictures $\{\mathfrak{P}_n = f_{\sigma(n)}(\mathfrak{P}_0) : n = 0, 1, 2, \ldots \}$; see Section 3.2. It may be explored by a random iteration similar to that in Section 3.2 but using infinitely many pictures instead of finitely many. One may, for example, associate the probability $p_{\sigma_1} p_{\sigma_2} \cdots p_{\sigma_{|\sigma|}}$ with the picture $f_{\sigma}(\mathfrak{P}_0)$, for all $\sigma \in \Omega'_{\{1,2,\cdots,N\}}$, where $p_{\sigma_1} p_{\sigma_2} \cdots p_{\sigma_0}$ means p_0 and where the p_n are non-negative numbers such that $p_0 + p_1 + \cdots + p_N = 1$.

E X E R C I S E 3.5.35 *Show that*

$$\sum_{\sigma \in \Omega'_{\{1,2,\cdots,N\}}} p_{\sigma_1} p_{\sigma_2} \cdots p_{\sigma_{|\sigma|}} = 1,$$

where the p_{σ_n} are non-negative numbers such that $p_0 + p_1 + \cdots + p_N = 1$.

E X E R C I S E 3.5.36 *As a special project choose a simple IFS semigroup and an interesting condensation picture $\mathfrak{P}_0$ and explore the associated tops semigroup mentioned in the last paragraph above, using random iteration. How does the look and 'feel' of the pictures that you obtain change when the probabilities are altered?*

The Henon transformation

Up to this point we have illustrated orbital pictures generated by IFS semi-groups made of quite simple transformations, such as projective and Möbius

Figure 3.59 Some elements of an orbit of a picture $\mathfrak{P}_0$ of a 'Morning Glory' flower, shown in the leftmost panel, under the IFS semigroup generated by the Henon transformation, Equation (3.5.19), with $a = 1.0001$ and $c = 0.45$. The lower left corner has coordinates $(-1.5, -2.0)$ while the upper right corner has coordinates $(1.7, 0.8)$. Where are the flowers going? See Figure 3.60.

transformations. Although most of the theory of orbital pictures is more generally applicable, we have emphasized pictures associated with contractive IFSs. So it is worthwhile here to consider briefly an example that involves a much more complicated mapping, namely the Henon transformation, which has been much studied from a dynamical systems point of view, [45]. Our goal is to emphasize the generality of the theory of orbital pictures and to illustrate that even with only one transformation there may be very complicated structure and hugely deformed picture tiles. In so doing we contact standard dynamical systems theory from the novel point of view of orbital pictures.

Orbits of dynamical systems, that is, of IFS semigroups generated by a single transformation, may lie on or be attracted to geometrically complicated structures called **strange attractors**, often by dint of a certain level of *complication* in the single underlying transformation. For example, consider the semigroup generated by the Henon transformation $f_{Henon} : \mathbb{R}^2 \to \mathbb{R}^2$, defined by

$$f_{Henon}(x, y) = (y + 1 - ax^2, \; cx), \qquad (3.5.19)$$

where a and c are real numbers; for example $a = 1.4$ and $c = 0.3$. This transformation stretches and bends pictures upon which it acts, as illustrated in Figure 3.59. Figure 3.59 illustrates from left to right the pictures

$$\mathfrak{P}_0, \quad \mathfrak{P}_0 \uplus f_{Henon}(\mathfrak{P}_0) \quad \text{and} \quad \mathfrak{P}_0 \uplus f_{Henon}(\mathfrak{P}_0) \uplus f^{\circ 2}_{Henon}(\mathfrak{P}_0),$$

where $\mathfrak{P}_0$ is a picture of a 'Morning Glory' flower. It is seen that f_{Henon} moves some pairs of points further apart while moving other pairs closer together. This behaviour contrasts with that in Figure 3.3, where the underlying transformation moves all pairs of points closer together. All orbits of points under the latter transformation converge to a single point. But some orbits of the Henon transformation are much more complicated: Figure 3.60 shows a plot of one million points of the orbit of the point $(0.5, 0.5)$, outlining the structure of an associated attractor, defined below.

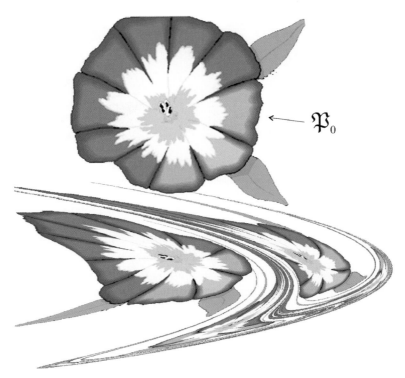

Figure 3.60 Where have all the flowers gone? See also Figure 3.59. This orbital picture, generated by the Henon transformation, represents a picture tiling. Also shown, in red, is a set of points obtained by computing one million points of the orbit of the point $(0.5, 0.5)$, and plotting all except the first thousand. This set represents an attractor of the Henon transformation. The flowers have gone towards a strange attractor, getting thoroughly bent out of shape in the process.

EXERCISE 3.5.37   *Verify that the Henon transformation is invertible. Figure 3.61 shows several elements of an orbit of pictures generated by f_{Henon}^{-1} when $a = 1.4$ and $c = 0.3$ in Equation (3.5.19).*

EXERCISE 3.5.38   *Plot the orbits of various points in $\square = \{(x, y) \in \mathbb{R}^2 : -1.5 \le x, y \le 1.5\}$ under the semigroup of transformations generated by f_{Henon} defined in Equation (3.5.19). Which points in $\square$, according to your computations, have orbits that remain in $\square$? Which ones escape?*

DEFINITION 3.5.39   Let $(\mathbb{X}, d)$ be a metric space. A compact set $A \subset \mathbb{X}$ is called an **attractor** of a dynamical system $f : \mathbb{X} \to \mathbb{X}$ if there exists a neighbourhood V of A such that $f(V) \subset V$ and

$$A = \bigcap_{n=1}^{\infty} f^{\circ n}(V).$$

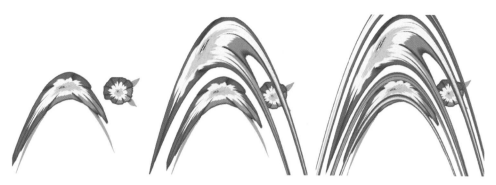

Figure 3.61   Three underneath pictures associated with the IFS semigroup generated by f_{Henon}^{-1} when $a = 1.4$ and $c = 0.3$ in Equation (3.5.19). The successive flowers in the orbit of the condensation picture are being drawn towards a **repeller** of f_{Henon}.

THEOREM 3.5.40   *Let A be an attractor of an invertible dynamical system $f : \mathbb{X} \to \mathbb{X}$ and let V be a neighbourhood of A such that $f(V) \subset V$. Let $\mathfrak{P}_0$ be a condensation picture, with compact domain $D_{\mathfrak{P}_0}$ such that $D_{\mathfrak{P}_0} \subset V \backslash A$. Let $\mathfrak{P}$ be the orbital picture generated by the IFS semigroup $\mathcal{S}_{\{f\}}(\mathbb{X})$ acting on $\mathfrak{P}_0$. Then $\mathfrak{P}$ is either a picture tiling or has diversity 2.*

PROOF   Let $C \subset V \backslash A$ be nonempty. Then we claim that $\mathcal{O}(C)$ is layered; see Definition 3.4.4. Let $x \in \bigcap_{n=1}^{\infty} f^{\circ n}(C)$. Then $x \in \bigcap_{n=1}^{\infty} f^{\circ n}(V) = A$. It follows that $f^{-1}(x) \in f^{-1}(\bigcap_{n=1}^{\infty} f^{\circ n}(V))$, which says that $f^{-1}(x)$ belongs, in particular, to $f^{\circ(n-1)}(V)$ for $n = 2, 3, \ldots$, which in turn means that $f^{-1}(x) \in A$. But x does not belong to A. So we conclude that $\bigcap_{n=1}^{\infty} f^{\circ n}(C) = \varnothing$, that is, $\mathcal{O}(C)$ is layered.

Now we use the fact that $D_{\mathfrak{P}_0}$ is contained in $V \backslash A$ to deduce that $f^{\circ n}(D_{\mathfrak{P}_0})$ is contained in $V \backslash A$ for all $n = 0, 1, 2, \ldots$ and hence that $\bigcup \mathcal{O}(D_{\mathfrak{P}_0}) = \bigcup_{n=0}^{\infty} f^{\circ n}(D_{\mathfrak{P}_0}) \subset V \backslash A$. Thus the orbit of $D_{\mathfrak{P}_0}$ under the IFS semigroup $\mathcal{S}_{\{f\}}(\mathbb{X})$ is layered.

It now follows from Theorem 3.4.6, wherein we take $C = D_{\mathfrak{P}_0}$, that $D_{\mathfrak{P}_0}$ is a semigroup tiling of $P = \bigcup \mathcal{O}(D_{\mathfrak{P}_0})$ iff $D_{\mathfrak{P}_0} \cap f(D_{\mathfrak{P}_0}) = \varnothing$. Hence, by Definition 3.5.31, $\uplus \mathcal{O}(\mathfrak{P}_0)$ is a picture tiling iff $D_{\mathfrak{P}_0} \cap f(D_{\mathfrak{P}_0}) = \varnothing$.

So, suppose that $\uplus \mathcal{O}(\mathfrak{P}_0)$ is not a picture tiling; then $D_{\mathfrak{P}_0} \cap f(D_{\mathfrak{P}_0}) \neq \varnothing$. Consequently, $C_0 = D_{f(\mathfrak{P}_0)} \backslash D_{\mathfrak{P}_0}$ must be nonempty and hence $\mathfrak{Q}_1$ is a panel distinct from $\mathfrak{P}_0$. The orbital picture generated by $\mathfrak{Q}_1$ is a tiling since it is layered and, as can be readily checked, $D_{\mathfrak{Q}_1} \cap f(D_{\mathfrak{Q}_1}) = \varnothing$. Moreover, $\uplus \mathcal{O}(\mathfrak{Q}_1)$ and $\mathfrak{P}_0$ are disjoint pictures and $\uplus \mathcal{O}(\mathfrak{P}_0) = \mathfrak{P}_0 \uplus \mathcal{O}(\mathfrak{Q}_1)$. Hence, the space of limiting pictures contains exactly two distinct elements, $\mathfrak{P}_0$ and $f^{-1}(\mathfrak{Q}_1)$.   □

Applications of orbital pictures

Here we speculate briefly on possible applications of orbital pictures.

Orbital pictures have obvious applications to computer graphics. Indeed, many standard pictures of fractals, such as Julia sets surrounded by ribbons of colour, and tree-like sets where each branch looks like a small copy of the trunk, may be interpreted as orbital pictures. From computational experiments it appears that some orbital pictures vary smoothly in appearance when the condensation picture and the IFS are varied, so it is clearly possible to design attractive-looking pictures using orbital pictures. To make a fresh-looking advertisement on the internet one might use an IFS whose set attractor is in the shape of a corporate logo with a condensation set that is a picture of a brand of the company and then adjust parameters to animate the resulting orbital picture. Orbital pictures may also be used to generate intricate textures and patterns that may be wrapped around wire-frame models to fill in backgrounds in synthetic imagery. Some related ideas are explored in [95].

In some cases it may be possible to decompose a picture approximately into a tops union of orbital pictures. In turn the condensation pictures may themselves be approximated by orbital pictures in the same manner. If such a recursive decomposition is possible, with some stability, then a new type of method for image approximation and compression would result, distinct from block-based fractal image compression, as described in [53], [12] and [38] for example.

The code space of an orbital picture and the diversity or growth rate of diversity of an orbital picture are parameters that may be applied to the problem of classifying real-world pictures and textures. Quantities which one might associate with real-world pictures such as photographs and which are based on these types of ideas would be invariant under homeomorphism. Such quantities would be of a character altogether different from those based on fractal dimension, which are invariant under transformations that provide equivalent metrics but are not robust against more ferocious transformations.

For example, one may wish to compare pictures of leaves of different plants. The boundaries of the leaves could have different experimental fractal dimensions yet the pictures might be well described by equivalent orbital pictures. In such a case one might define an empirical diversity and use it to classify and compare the leaves.

The applications of orbital pictures to biological modelling may be considered as refinements of approaches, already used with some success, based on the ideas of Lindemeyer; see for example [54] and [80]. It is in the non-commutative inter-action, via the tops union, of the images of the condensation picture under the semigroup that enriches the approach via orbital pictures; this interaction reminds me of the way in which, in the expression of a genetic code, some genes become active only in certain circumstances.

For biological modelling applications it is interesting to apply what we call **orbit-stealing**. Let two related IFS semigroups $\mathcal{S}_{\{f_1, f_2, \dots, f_N\}}(\mathbb{X})$ and $\mathcal{S}_{\{\tilde{f}_1, \tilde{f}_2, \dots, \tilde{f}_N\}}(\mathbb{X})$ and two pictures $\mathfrak{P}_0$ and $\mathfrak{Q}_0 \in \Pi(\mathbb{X})$ be given. Use $\mathfrak{P}_0$ to construct $\mathfrak{P}(\mathfrak{P}_0)$ and, in particular, the code space $\Omega'_{\mathfrak{P}_0}$. Let the addresses in $\Omega'_{\mathfrak{P}_0}$, in order, be given by the sequence $\{\sigma_{\mathfrak{P}_0}(n) : n = 0, 1, 2, \dots\}$. Then we define a stolen orbital picture by

$$\widetilde{\mathfrak{P}}(\mathfrak{P}_0 \to \mathfrak{Q}_0) := \mathfrak{Q}_0 \uplus \widetilde{f}_{\sigma_{\mathfrak{P}_0}(1)}(\mathfrak{Q}_0) \uplus \widetilde{f}_{\sigma_{\mathfrak{P}_0}(2)}(\mathfrak{Q}_0) \cdots,$$

in an obvious manner. $\mathfrak{P}_0$ and $\mathfrak{Q}_0$ may represent leaves of the same species of plant, at different ages. Suppose that we have successfully modelled an aspect of the geometry of the first plant by means of the orbital picture $\mathfrak{P}(\mathfrak{P}_0)$. Then $\widetilde{\mathfrak{P}}(\mathfrak{P}_0 \to \mathfrak{Q}_0)$ is a possible model for the corresponding aspect of the second plant, which may be younger or older.

The same idea can be applied in computer graphics: once a particular code space structure $\Omega'_{\mathfrak{P}_0}$ has been found to produce a beautiful and harmonious picture, it may be applied over and over again to other condensation pictures $\mathfrak{Q}_0$ to obtain different, potentially lovely, synthetic content.

Other questions that may lead to applications for orbital pictures are as follows. Suppose that you fix an IFS semigroup $\mathcal{S}_{\{f_1, f_2, \dots f_N\}}(\mathbb{X})$. Then how does the diversity $|L_{\mathfrak{P}_0}|$ of the orbital picture $\mathfrak{P}(\mathfrak{P}_0) = \uplus \mathcal{O}(\mathfrak{P}_0)$ depend upon $\mathfrak{P}_0$? What is the relationship between $|L_{\mathfrak{P}_0}|$ and the numbers $|L_{f_n^{-1}(\mathfrak{P}_0)}|$ for $n \in \{1, 2, \dots, N\}$? Are there number-theoretic relationships that may be established in special cases? Given a set $\Sigma \subset \Omega'_{\{1,2,\dots,N\}} \cup \Omega_{\{1,2,\dots,N\}}$ such that $S(\Sigma) \subset \Sigma$, when can an IFS and a condensation picture $\mathfrak{P}_0$ be found such that $\Sigma = \Omega'_{\mathfrak{P}_0} \cup \Omega_{\mathfrak{P}_0}$? Such questions lead naturally to the speculation that orbital pictures may be used in cryptography.

We note the following construction. Let an IFS of contractive transformations $\{\mathbb{R}^2; f_1(x, y), \dots, f_N(x, y)\}$, with set attractor $A \subset \mathbb{R}^2$, be given. Then construct the IFS

$$\{\mathbb{R}^3; g_1(x, y, z), \dots, g_N(x, y, z)\},$$

where $g_n(x, y, z) = (f_n(x, y), \frac{1}{2}z)$. The attractor of the latter IFS is the set A in the plane $z = 0$. Now let $\mathfrak{P}_0$ denote a three-dimensional picture whose domain does not intersect the plane $z = 0$. Then the corresponding code space $\Omega_{\mathfrak{P}_0}$ provides a symbolic representation of the attractor set A that is quite distinct in general from the usual code space representation. We discuss this representation further in Section 4.13.

An orbital picture in three dimensions does not model a physical picture, of course. It may instead be thought of as an accretion of solid multicoloured chunks of material. Such chunky structures might be used to model complicated objects made of many types of material or geometrical aspects of the physiology of a plant.

3.6   Orbits of measures under IFS semigroups

DEFINITION 3.6.1   Let $\mathbb{X}$ be a topological space, let $S(\mathbb{X})$ be a semigroup of continuous transformations and let υ be a Borel measure on $\mathbb{X}$. Then the **orbit of**

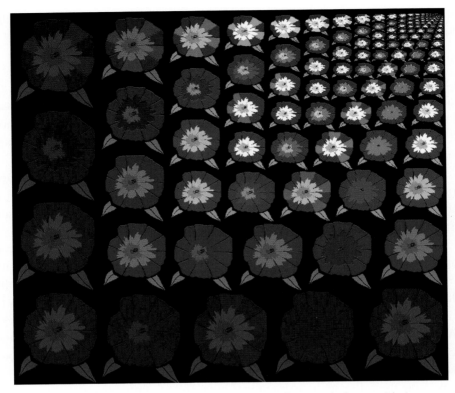

Figure 3.62 The orbit of the measure represented by the flower at the bottom right is represented by the sequence of successively brighter flowers going up on the right. The successive flowers cease to become brighter after approximately six iterations because of saturation effects, which also cause changes in colour. Does the corresponding sequence of measures converge to a limiting measure?

the measure υ under the semigroup $S(\mathbb{X})$ is the set of Borel measures

$$\mathcal{O}(\upsilon) = \{f(\upsilon) : f \in \mathcal{S}(\mathbb{X})\}.$$

Some pictures of measures belonging to orbits of measures under a semigroup generated by an affine transformation in $\mathbb{R}^2$ are illustrated in Figure 3.62. Notice that this image contains pictures of the measures in the orbit of the measure represented by the flowers in the bottom row and the left-hand column.

How can we make a single measure out of an orbit of measures? The natural and simple thing to do is to 'add them all up' with appropriate weights. To be able to do this easily we restrict our attention to orbits of measures generated by IFS semigroups.

DEFINITION 3.6.2   An **IFS with probabilities** is an IFS $\{\mathbb{X}; f_1, f_2, \ldots, f_N\}$ together with a set of probabilities, non-negative real numbers $p_1, p_2, \ldots, p_N$ such that $p_1 + p_2 + \cdots + p_N = 1$. The probability p_n is associated with the function f_n for $n = 1, 2, \ldots, N$. An IFS with probabilities may be denoted

$$\{\mathbb{X}; f_1, f_2, \ldots, f_N; p_1, p_2, \ldots, p_N\}.$$

The following theorem is notable because it applies in very general circumstances. It is not required that the space $\mathbb{X}$ is compact or even complete; nor is it required that the transformations in the IFS be contractive, or even contractive on average.

THEOREM 3.6.3   *Let $\mathbb{X}$ be a topological space and let $\{\mathbb{X}; f_1, f_2, \ldots, f_N; p_1, p_2, \ldots, p_N\}$ be an IFS with probabilities, where $f_n : \mathbb{X} \to \mathbb{X}$ is continuous for each $n \in \{1, 2, \ldots, N\}$. Let $0 < p_0 \leq 1$ and let $\upsilon_0 \in \mathbb{P}(\mathbb{X})$, the space of normalized Borel measures on $\mathbb{X}$. Then the Borel measure $\upsilon \in \mathbb{P}(\mathbb{X})$ defined by*

$$\upsilon = p_0 \upsilon_0 + \sum_{\sigma \in \Omega'_{\{1,2,\ldots,N\}}, |\sigma| \geq 1} p_0 (1 - p_0)^{|\sigma|} p_{\sigma_1} p_{\sigma_2} \cdots p_{\sigma_{|\sigma|}} f_\sigma(\upsilon_0) \qquad (3.6.1)$$

is the unique solution of the self-referential equation

$$\upsilon = p_0 \upsilon_0 + (1 - p_0)\big(p_1 f_1(\upsilon) + p_2 f_2(\upsilon) + \cdots + p_N f_N(\upsilon)\big). \qquad (3.6.2)$$

PROOF   Let $B \in \mathcal{B}(\mathbb{X})$ be a Borel subset of $\mathbb{X}$. Then the value $\upsilon(B)$ is well defined, because the series

$$p_0 \upsilon_0(B) + \sum_{\sigma \in \Omega'_{\{1,2,\ldots,N\}}, |\sigma| \geq 1} p_0 (1 - p_0)^{|\sigma|} p_{\sigma_1} p_{\sigma_2} \cdots p_{\sigma_{|\sigma|}} f_\sigma(\upsilon_0)(B)$$

consists of non-negative terms and is bounded above, term by term, by the absolutely convergent series

$$p_0 + \sum_{\sigma \in \Omega'_{\{1,2,\ldots,N\}}, |\sigma| \geq 1} p_0 (1 - p_0)^{|\sigma|} p_{\sigma_1} p_{\sigma_2} \cdots p_{\sigma_{|\sigma|}} = 1.$$

Hence $\upsilon : \mathcal{B}(\mathbb{X}) \to [0, 1]$. Notice that $\upsilon(\mathbb{X}) = 1$. Let us define

$$\rho_0 = \upsilon_0 \quad \text{and} \quad \rho_n = \sum_{\sigma \in \Omega'_{\{1,2,\cdots,N\}}, |\sigma| = n} p_{\sigma_1} p_{\sigma_2} \cdots p_{\sigma_{|\sigma|}} f_\sigma(\upsilon_0) \quad \text{for } n = 1, 2, \ldots$$

Then it is readily verified that $\rho_n \in \mathbb{P}(\mathbb{X})$, and we can rewrite Equation (3.6.1) as

$$\upsilon = \sum_{n=0}^{\infty} p_0 (1 - p_0)^n \rho_n.$$

Now, referring back to Definition 2.3.9, let $\{\mathcal{O}_m \in \mathcal{B}(\mathbb{X}) : m = 1, 2, \ldots\}$ be a sequence such that $\bigcup_{m=1}^{\infty} \mathcal{O}_m \in \mathcal{B}(\mathbb{X})$ and

$$\mathcal{O}_{m_1} \cap \mathcal{O}_{m_2} = \varnothing$$

for all $m_1, m_2 \in \mathbb{N}$ with $m_1 \neq m_2$. Then

$$\sum_{m=0}^{\infty} \upsilon(\mathcal{O}_m) = \sum_{m=1}^{\infty} \sum_{n=0}^{\infty} p_0 (1 - p_0)^n \rho_n(\mathcal{O}_m).$$

Since the series on the right is absolutely convergent, we can interchange the order in which the two summations are evaluated, which yields

$$\sum_{m=1}^{\infty} \upsilon(\mathcal{O}_m) = \sum_{n=0}^{\infty} p_0(1-p_0)^n \sum_{m=1}^{\infty} \rho_n(\mathcal{O}_m)$$

$$= \sum_{n=0}^{\infty} p_0(1-p_0)^n \rho_n\left(\bigcup_{m=1}^{\infty} \mathcal{O}_n\right) = \upsilon\left(\bigcup_{m=1}^{\infty} \mathcal{O}_n\right).$$

It follows that υ is indeed a measure on $\mathcal{B}(\mathbb{X})$ and, since $\upsilon(\mathbb{X}) = 1$, that $\upsilon \in \mathbb{P}(\mathbb{X})$.

To prove that this measure obeys Equation (3.6.2), we note that since all the series involved are absolutely convergent, it suffices to show that the algebra works out correctly, term by term. Substituting from Equation (3.6.1) into the right-hand side of Equation (3.6.2) we find that

r.h.s. of Equation (3.6.2)

$$= p_0\upsilon_0 + (1-p_0)\sum_{n=1}^{N} p_n f_n\Bigg(p_0\upsilon_0$$

$$+ \sum_{\sigma \in \Omega'_{\{1,2,\dots,N\}}, |\sigma|\geq 1} p_0(1-p_0)^{|\sigma|} p_{\sigma_1} p_{\sigma_2} \cdots p_{\sigma_{|\sigma|}} f_\sigma(\upsilon_0)\Bigg)$$

$$= p_0\upsilon_0 + (1-p_0)\sum_{n=1}^{N} p_n f_n\Bigg(p_0\upsilon_0 + \sum_{m=1}^{N}(1-p_0)p_0 p_m f_m(\upsilon_0)$$

$$+ \sum_{\sigma \in \Omega'_{\{1,2,\dots,N\}}, |\sigma|\geq 2} p_0(1-p_0)^{|\sigma|} p_{\sigma_1} p_{\sigma_2} \cdots p_{\sigma_{|\sigma|}} f_\sigma(\upsilon_0)\Bigg)$$

$$= p_0\upsilon_0 + \sum_{n=1}^{N}(1-p_0)p_0 p_n f_n(\upsilon_0) + \sum_{n=1}^{N}\sum_{m=1}^{N} p_0(1-p_0)^2 p_n p_m f_n(f_m(\upsilon_0))$$

$$+ \sum_{n=1}^{N}\sum_{\sigma \in \Omega'_{\{1,2,\dots,N\}}, |\sigma|\geq 2} (1-p_0)^{|\sigma|+1} p_0 p_n p_{\sigma_1} p_{\sigma_2} \cdots p_{\sigma_{|\sigma|}} f_n(f_\sigma(\upsilon_0))$$

$$= p_0\upsilon_0 + \sum_{\sigma \in \Omega'_{\{1,2,\dots,N\}}, |\sigma|\geq 1} p_0(1-p_0)^{|\sigma|} p_{\sigma_1} p_{\sigma_2} \cdots p_{\sigma_{|\sigma|}} f_\sigma(\upsilon_0)$$

$$= \text{l.h.s. of Equation (3.6.2)}.$$

In order to prove uniqueness, suppose that $\widetilde{\upsilon} \in \mathbb{P}(\mathbb{X})$ obeys

$$\widetilde{\upsilon} = p_0\upsilon_0 + (1-p_0)\big(p_1 f_1(\widetilde{\upsilon}) + p_2 f_2(\widetilde{\upsilon}) + \cdots + p_N f_N(\widetilde{\upsilon})\big).$$

Then, by repeatedly substituting from the left-hand side into the right-hand side, we find that $\widetilde{\upsilon}$ can be represented by the same absolutely convergent series as υ, whence $\widetilde{\upsilon} = \upsilon$. $\square$

DEFINITION 3.6.4   The measure $\upsilon = \upsilon(\upsilon_0) \in \mathbb{P}(\mathbb{X})$ in Theorem 3.6.3 is called the **orbital measure** associated with the IFS semigroup $\mathcal{S}_{\{f_1, f_2, \dots, f_N\}}(\mathbb{X})$ and with the numbers $p_0, p_1, \dots, p_N$ acting on the measure $\upsilon_0 \in \mathbb{P}(\mathbb{X})$.

Notice that the expressions above could have been written down and handled more succinctly by introducing the linear operator $L : \mathbb{P}(\mathbb{X}) \to \mathbb{P}(\mathbb{X})$ defined by

$$ L\mu = \sum_{n=1}^{N} p_n f_n(\mu) \quad \text{for all } \mu \in \mathbb{P}(\mathbb{X}). $$

L acts linearly on the space of all possible linear combinations of Borel measures on $\mathbb{X}$. We call L the **Markov operator** associated with the IFS. Using this notation, the self-referential equation (3.6.2) reads

$$ \mu = p_0 \upsilon_0 + (1 - p_0) L\mu, $$

and the series expansion in Equation (3.6.1) can be written as

$$ \mu = p_0 (1 - (1 - p_0)L)^{-1} \upsilon_0 $$
$$ = p_0 \sum_{m=0}^{\infty} (1 - p_0)^m L^m \upsilon_0. $$

We did not introduce L earlier because we wanted to display and manipulate the full series expansions, show the parallels and distinctions between orbital pictures and orbital measures and specifically illustrate how the probability $p_0(1 - p_0)^{|\sigma|}$ is associated with the measure $f_\sigma(\upsilon_0)$. When represented as a picture, each term in the series corresponds to a contribution or component of the picture; for example, each term in the series may correspond to a distinct 'semigroup measure tile', as in Figure 3.63. This suggests how one might define an IFS semigroup measure tiling.

Pictures of orbital measures corresponding to various simple IFS semigroups acting on $\mathbb{R}^2$ are illustrated in Figures 3.63–3.67. The manner in which these pictures were computed is described below.

Figures 3.64 and 3.65 relate to condensation measures that are drawn by the IFS towards the 'horizon', namely a line segment in $\mathbb{R}^2$, the set attractor of the IFS. Figure 3.65 is particularly interesting because it illustrates not only how elementary orbital measures can be used to produce synthetic, real-looking, pictures but also how subtle changes in these pictures can be produced by making small changes in the probabilities. On the right $p_1 = p_2 = 0.5$, on the left p_1 is approximately 0.4 and p_2 is approximately 0.6, and in both cases p_0 is very close to zero, see below. The horizon on the left in Figure 3.65 looks threatening in contrast with the bright distant sky on the right.

It is worth comparing Figure 3.66 with Figure 2.23. The latter illustrates the convergence of the sequence of measures $\{L^n \mu_0\}_{n=1}^{\infty}$ to the measure attractor of the same IFS with slightly different probabilities, where $\mu_0 \in \mathbb{P}(\mathbb{R}^2)$ is similar to the condensation measure used in Figure 3.66.

Figure 3.63 Picture of an orbital measure of an IFS semigroup generated by two contractive similitudes. The condensation measure is represented by the bottom shield-shaped tile. The probabilities on the maps are such that successive shields on the left are darker and darker, while those on the right are successively lighter.

The IFS with probabilities used in Figure 3.67 is

$$\left\{ \mathbb{R}^2; \left(\frac{x}{2}, \frac{y}{2}\right), \left(\frac{x+1}{2}, \frac{y}{2}\right), \left(\frac{x}{2}, \frac{y+1}{2}\right), \left(\frac{x+1}{2}, \frac{y+1}{2}\right); \right.$$

$$\left. p_1 = \frac{4}{25}, p_2 = \frac{5}{25}, p_3 = \frac{7}{25}, p_3 = \frac{9}{25} \right\}.$$

The set attractor is the filled unit square $\square$ with lower left corner at the origin. The support of the orbital measure represented in Figure 3.67 is contained in $\square$. A comparison of Figure 3.67 and Figure 3.42 provides a striking contrast between an orbital measure and a closely related orbital picture.

EXERCISE 3.6.5 *Let $\mathbb{X} = [0, 1) \subset \mathbb{R}$ with the usual topology. Let $\mathcal{S}_{\{f\}}(\mathbb{X})$ be the semigroup generated by the function $f : [0, 1) \to [0, 1)$ defined by $f(x) = \frac{1}{2} + \frac{1}{2}x$. Let $\upsilon_0 \in \mathbb{P}([0, 1))$ denote a normalized Borel measure all of whose mass is contained in $[0, \frac{1}{2})$. That is, $\upsilon_0([0, \frac{1}{2})) = 1$ and $\upsilon_0(([\frac{1}{2}, 1)) = 0$. Then the associated*

Figure 3.64 Two pictures of orbital measures generated by IFS semigroups. Each IFS consists of two similitudes and has as its limit set a horizontal line segment, located near the top of each picture. The condensation measure is represented by the flower picture in the bottom left corner of each picture. In the orbital measure pictured on the left the probabilities and contractivity factors for the two maps are equal; on the right the probabilities and contractivity factors are different. Saturation effects cause parts of the picture with intense measure to be represented by maximum white, namely $R = G = B = 255$.

Figure 3.65 Each picture illustrates an orbital measure generated by an IFS semigroup. The same two transformations and the same condensation measure are used in each case. Can you spot them? The difference is in the probabilities.

orbital measure $\upsilon \in \mathbb{P}(\mathbb{X})$ uniquely satisfies

$$\upsilon = p_0 \upsilon_0 + (1 - p_0) f(\upsilon) = \sum_{n=0}^{\infty} p_0 (1 - p_0)^n f^{\circ n}(\upsilon_0).$$

What happens as $p_0 \to 0$? Do we get a solution to $\upsilon = f(\upsilon)$ with $\upsilon \in \mathbb{P}(\mathbb{X})$? Show that for each $x \in [0, 1)$ we have

$$\lim_{p_0 \to 0} \upsilon([0, x]) = 0.$$

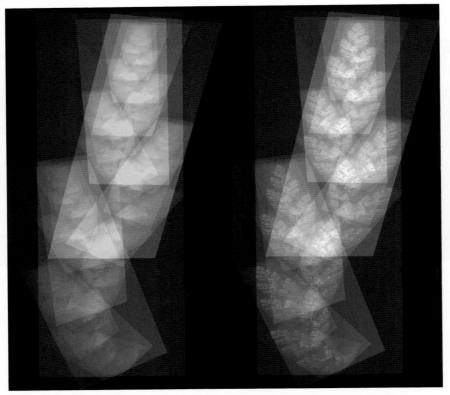

Figure 3.66 Pictures of the orbital measure of an IFS semigroup. The IFS consists of two projective transformations in $\mathbb{R}^2$; its measure attractor is pictured in shades of blue in the right-hand image, superimposed on the orbital picture. The condensation measure is uniform over a rectangular region that contains the set attractor of the IFS.

Figure 3.67 On the left is a picture of the measure attractor of the IFS in Equation (3.6.3). On the right is shown the orbital measure generated by the corresponding IFS semigroup, applied to a condensation measure that is uniformly distributed on a leaf-shaped region, similar to the main leaf in Figure 3.42. Notice the luminous shades of green and the way the shape of the measure attractor influences the orbital measure.

Conclude that we do not obtain, in the limit, a solution to $\upsilon = f(\upsilon)$ with $\upsilon \in \mathbb{P}(\mathbb{X})$. What happens if the interval $[0, 1)$ is replaced by $[0, 1]$?

Next we describe the type of method that we used to compute the approximate pictures of orbital measures shown in Figures 3.63–3.67. Let $\upsilon_0 \in \mathbb{P}(\mathbb{X})$, $0 < p_0 < 1$, and an IFS $\{\mathbb{X}; f_1, f_2, \ldots, f_N; p_1, p_2, \ldots, p_N\}$ be given, where $\mathbb{X} = \square \subset \mathbb{R}^2$. Let $\mathcal{F} : \mathbb{P}(\square) \to \mathbb{P}(\square)$ be defined by

$$\mathcal{F}(\upsilon) = p_0 \upsilon_0 + (1 - p_0) L\upsilon \quad \text{for all } \upsilon \in \mathbb{P}(\square).$$

Then, by what we have been saying above, the sequence $\{\mathcal{F}^{\circ k}(\upsilon_0) \in \mathbb{P}(\square)\}_{k=1}^{\infty}$ converges to the orbital measure υ; namely, given any $\epsilon > 0$ there exists an integer l such that $|\mathcal{F}^{\circ k}(\upsilon_0)(B) - \upsilon(B)| < \epsilon$ for all $k > l$, uniformly for all Borel subsets $B \in \mathcal{B}(\square)$.

It follows that we can compute a sequence of approximations to the value of υ for any array of pixels, successively, one step at a time. Specifically, let a resolution $W \times H$ be selected and construct the discretization $\{\square_{w,h} : w = 1, 2, \ldots, W, h = 1, 2, \ldots, H\}$ of $\square$, as discussed in Section 2.2. Then observe that the sequence of digital pictures $\{\mathfrak{P}^{(k)} : \square \to [0, \infty)\}_{k=0}^{\infty}$, whose pixels are $\mathfrak{P}_{w,h}^{(k)} = \mathcal{F}^{\circ k}(\upsilon_0)(\square_{w,h})$ for $k = 0, 1, 2, \ldots$, satisfies

$$\mathfrak{P}_{w,h}^{(k+1)} = \mathfrak{P}^{(k+1)}(\square_{w,h}) = \mathcal{F}(\mathcal{F}^{\circ k}(\upsilon_0))(\square_{w,h})$$

$$= p_0 \upsilon_0(\square_{w,h}) + (1 - p_0) \sum_{n=1}^{N} p_n f_n(\mathcal{F}^{\circ k}(\upsilon_0))(\square_{w,h}).$$

Notice that $\mathfrak{P}_{w,h}^{(0)} = \upsilon_0(\square_{w,h})$. Given $\mathfrak{P}^{(k)}$, we can form approximations to each term inside the last summation and thus produce an approximation to $\mathfrak{P}^{(k+1)}$. Suppose that we have already computed an approximation $\widetilde{\mathfrak{P}}^{(k)}$ to $\mathfrak{P}^{(k)}$. Then for example we may approximate $\mathfrak{P}^{(k+1)}$ by

$$\widetilde{\mathfrak{P}}_{w,h}^{(k+1)} := p_0 \mathfrak{P}_{w,h}^{(0)} + (1 - p_0) \sum_{n=1}^{N} p_n \sum_{(w',h') \in Q(n,w,h)} \widetilde{\mathfrak{P}}_{w',h'}^{(k)},$$

where $Q(n, w, h)$ is the set of indices (w', h') corresponding to pixel domains $\square_{w',h'}$ whose centre points, say, are mapped into $\square_{w,h}$, that is,

$$Q(n, w, h) = \{(w', h') \in \{1, 2, \ldots, W\} \times \{1, 2, \ldots, H\} : f_n(c_{w',h'}) \in \square_{w,h}\},$$

where $c_{w,h}$ denotes a selected representative point in $\square_{w,h}$. This type of approximation produces pictures that are accurate to viewing resolution when the transformations are sufficiently contractive. In other cases we use the inverse of the maps f_n to provide approximations for the contribution $f_n(\mathcal{F}^{\circ k}(\upsilon_0))(\square_{w,h})$ in terms of $\mathfrak{P}^{(k)}$; for example, in some cases we use the approximation

$$f_n(\mathcal{F}^{\circ k}(\upsilon_0))(\square_{w,h}) \simeq \frac{\text{area of } f_n^{-1}(\square_{w,h})}{\text{area of } \square_{w,h}} \widetilde{\mathfrak{P}}^{(k)}(\square_{w'(n,w,h),h'(n,w,h)}),$$

where $w'(n, w, h), h'(n, w, h)$ is the index of the pixel domain in which lies the point $f_n^{-1}(c_{w,h})$. Here we may approximate the ratio of areas using the Wronskian of the transformation f_n, as described in Section 2.7 for the case of projective transformations. In general, a good understanding of the specific way in which the transformations of the IFS deform the space, as described in Chapter 2, is very helpful in the construction of good approximations to pictures of orbital measures. Some problems in the discretization of IFSs have been analyzed in [79].

In working with sequences of approximate digital pictures of orbital measures, we also run into effects caused by the finite range of values in the colour space $\mathfrak{C}$. The expressions above assume that the colour space is of the form $[0, \infty) \subset \mathbb{R}$. In practice $\mathfrak{C}$ may be $\{0, 1, 2, \dots, 255\}$. To deal with this, we not only discretize the values of $\widetilde{\mathfrak{P}}^{(k)}$ but also replace those that exceed 255 by 255, which leads to colour saturation effects such as those mentioned in the captions of some of the figures.

If we divide Equation (3.6.1) by p_0 we obtain

$$\widehat{\upsilon}(p_0) := \frac{1}{p_0}\upsilon = \upsilon_0 + p_1 f_1(\upsilon) + \dots + p_N f_N(\upsilon).$$

Namely, we get a picture of $\upsilon_0 + \sum(1 - p_0)^{|\sigma|} p_\sigma f_\sigma(\upsilon_0)$, which, when p_0 approaches zero, approaches the expression

$$\upsilon_0 + \sum_{|\sigma| \geq 1} p_\sigma f_\sigma(\upsilon_0).$$

This expression represents an 'unbounded measure' because

$$\upsilon_0(\mathbb{X}) + \sum_{|\sigma| \geq 1} p_\sigma f_\sigma(\upsilon_0)(\mathbb{X}) = \infty.$$

Nonetheless, it is straightforward to make approximate pictures of this 'unbounded measure' using the same techniques as above, because saturation effects stop the divergence. This allows us to make approximate pictures of orbital measures when p_0 is very small. The two pictures in Figure 3.65 are of this kind; the bright horizon on the right would be utterly dazzling if not for saturation. Imagine it.

3.7   Groups of transformations

A group of transformations is a special type of semigroup – every transformation possesses an inverse that is also in the group. A group of transformations acting upon a picture of a seahorse is illustrated in Figure 3.68. An important difference between Figure 3.68 and Figure 3.3 is that each seahorse is the image of another seahorse under some transformation in the group. In Figure 3.3, however, one flower has no pre-image. Another example of a group of transformations, this time acting on subsets of $\mathbb{R}^2$, is illustrated in Figure 3.69.

We have chosen to introduce groups of transformations with the complicated and initially slightly confusing image in Figure 3.68 in order to emphasize the

Figure 3.68 A group of Möbius transformations acts on a leafy seahorse on the Riemann sphere $\widehat{\mathbb{C}}$. Think of the picture as a map of most of the surface of the sphere. Then you may imagine that the source of the seahorses is the centre of a two-dimensional reverse whirlpool. Seemingly, they grow as they swirl outwards from the source, and some are hidden from view, on the other side of the sphere. Eventually they appear to be caught by a second whirlpool. But which is the source and which is the sink?

richness and visual complexity that may be associated with the underlying simple idea of a group – a parade of identical horses prancing round a carousel, say, hardly has the same intricacy. In our example, not only is each seahorse a different size, it is also a different shape.

DEFINITION 3.7.1 A group $(\mathcal{G}, \bigcirc)$ is a semigroup with the following properties:

(i) there is a **unit element** $I \in \mathcal{G}$ with the property

$$I \bigcirc g = g \bigcirc I = g \quad \text{for all } g \in \mathcal{G};$$

(ii) given any $g \in \mathcal{G}$ there is an element $g^{-1} \in \mathcal{G}$, called the inverse of g, with the property

$$g^{-1} \bigcirc g = g \bigcirc g^{-1} = I.$$

A **subgroup** of $(\mathcal{G}, \bigcirc)$ is a group of the form $(\widetilde{\mathcal{G}}, \bigcirc)$, where $\widetilde{\mathcal{G}} \subset \mathcal{G}$.

Figure 3.69 The orbit of a single set, which looks like a fish, under a group of transformations. Properties that all the fish have in common are geometrical properties of this 'fish geometry'.

Examples of groups are: the positive rational numbers with $\times$ as the binary operation, the unit element being the number 1; the set of invertible $n \times n$ matrices for some $n \in \mathbb{N}$, the unit element being the identity matrix; the set of permutations $\mathcal{G}_A$ of the alphabet $\mathcal{A}$, in which case the group consists of the set of one-to-one invertible transformations from $\mathcal{A}$ into itself and the unit element is $I : \mathcal{A} \to \mathcal{A}$ where $I(x) = x$ for all $x \in \mathbb{X}$. $\mathcal{G}_A$ is called the **permutation group**.

DEFINITION 3.7.2   A **group of transformations** on a space $\mathbb{X}$ is a group $(\mathcal{G}(\mathbb{X}), \circ)$, where $\mathcal{G}(\mathbb{X})$ consists of one-to-one invertible transformations from $\mathbb{X}$ onto $\mathbb{X}$, where the binary operation is composition and where:
 (i) the unit element is the **identity transformation** $I : \mathbb{X} \to \mathbb{X}$, with $I(x) = x$ for all $x \in \mathbb{X}$;
 (ii) whenever $f \in \mathcal{G}$ we have $f^{-1} \in \mathcal{G}$, where f^{-1} is the inverse of f.

Two important examples of groups of transformations are the group of projective transformations $\mathcal{P} : \mathbb{R}^2 \cup L_\infty \to \mathbb{R}^2 \cup L_\infty$ and the group of Möbius transformations $\mathcal{M} : \widehat{\mathbb{R}^2} \to \widehat{\mathbb{R}^2}$, which we discussed in detail in Chapter 2.

EXERCISE 3.7.3 *Let* $(\mathcal{G}(\mathbb{X}), \circ)$ *be a group of transformations on* $\mathbb{X}$, *and let* $T : \mathbb{X} \to \mathbb{X}$ *be an invertible transformation. Let*

$$\widetilde{\mathcal{G}}(\mathbb{X}) = \{T \circ g \circ T^{-1} : g \in \mathcal{G}(\mathbb{X})\}.$$

Prove that $(\widetilde{\mathcal{G}}(\mathbb{X}), \circ)$ *is a group of transformations on* $\mathbb{X}$. *We say that two IFS semigroups* $\overline{\mathcal{S}}(\mathbb{X})$ *and* $\widetilde{\mathcal{S}}(\mathbb{X})$ *are* **conjugate** *iff there exists an invertible transformation* $T : \mathbb{X} \to \mathbb{X}$ *such that* $\widetilde{\mathcal{S}}(\mathbb{X}) = \{T \circ f \circ T^{-1} : f \in \mathcal{S}(\mathbb{X})\}$. *So, for example, two IFS semigroups* $\mathcal{S}_{\{f_1, f_2, \ldots, f_N\}}(\mathbb{X})$ *and* $\mathcal{S}_{\{\widetilde{f}_1, \widetilde{f}_2, \ldots, \widetilde{f}_N\}}(\mathbb{X})$ *are conjugate when* $\widetilde{f}_n = T \circ f_n \circ T^{-1}$ *for* $n = 1, 2, \ldots, N$.

We are interested in groups of transformations when they are IFS semigroups. Accordingly, we will use the notation

$$\mathcal{G}_{\{f_1, f_2, \ldots, f_N\}}(\mathbb{X})$$

to denote the IFS semigroup $\mathcal{S}_{\{f_1, f_2, \ldots, f_N\}}(\mathbb{X})$ only when $\mathcal{S}_{\{f_1, f_2, \ldots, f_N\}}(\mathbb{X})$ is, in fact, a group of transformations. In this case we will call the IFS semigroup an **IFS group**.

EXERCISE 3.7.4 *Show, by means of an example, that an IFS semigroup of invertible transformations is not necessarily an IFS group.*

EXERCISE 3.7.5 *Let* $\mathcal{G}_{\{f_1, f_2, \ldots, f_N, f_1^{-1}, f_2^{-1}, \ldots, f_N^{-1}\}}(\mathbb{X})$ *be an IFS group, and let* $\widetilde{\mathbb{X}} \subset \mathbb{X}$ *have the property that* $f_n(\widetilde{\mathbb{X}}) \subset f_n(\mathbb{X})$. *Show, by means of an example, that it does not follow that the set of functions* $\{f_1, f_2, \ldots, f_N, f_1^{-1}, f_2^{-1}, \ldots, f_N^{-1}\}$ *generates an IFS group on* $\widetilde{\mathbb{X}}$.

An IFS group is normally called a **finitely generated group** of transformations. By referring to a finitely generated group of transformation as an IFS group, however, we signal that we are treating it as an IFS semigroup rather than from the point of view of group theory.

We tend to think of IFS semigroups as being associated with IFSs of contractive, or on average contractive, transformations. Similarly we tend to think of an IFS group as being generated by a set of contractive transformations *and their inverses*. But we do not include these prejudices in the definitions of IFS semigroups and IFS groups because this would be overly restrictive. For example, our broader definition allows us to transpose the theory of orbital sets, measures and pictures, discussed in Sections 3.4–3.6, from IFS semigroups to IFS groups.

In Figure 3.69 we give an example of an orbital set generated by an IFS group of Möbius transformations and in Figures 3.24, 3.68 and 3.79 examples of orbital pictures generated by IFS groups of Möbius transformations. Two examples of

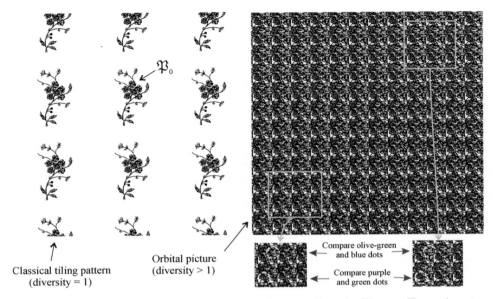

↑
Classical tiling pattern
(diversity = 1)

Orbital picture
(diversity > 1)

Compare olive-green
← and blue dots →

Compare purple
← and green dots →

Figure 3.70 Parts of two orbital pictures generated by a crystallographic IFS group. The condensation picture used on the right is a larger version of the one used on the left. The panelling on the right has diversity greater than 4.

orbital pictures generated by IFS groups of projective transformations are shown in Figure 3.75 and examples of orbital pictures generated by IFS groups of euclidean transformations are shown in Figures 3.70 and 3.72–3.74. We discuss some of these examples in the geometry subsections below.

Notice that for IFS groups many different addresses in code space may correspond to the same sequence of transformations. For example, suppose that the IFS group is $\{\mathbb{X}; f_1, f_2, f_3, f_4\}$ where $f_3 = f_1^{-1}$ and $f_4 = f_2^{-1}$. Then

$$f_{1132314}(\mathfrak{P}_0) = f_{12314}(\mathfrak{P}_0) = f_{124}(\mathfrak{P}_0) = f_1(\mathfrak{P}_0),$$

for all $\mathfrak{P}_0 \in \Pi$. This has obvious consequences for the computation of orbital sets, measures and pictures associated with IFS groups. To generate addresses without this redundancy, in this case, notice that 1 must be followed by 1, 2 or 4, 2 must be followed by 1, 2 or 3 and so on. Thus the set of all addresses in $\Omega'_{\{1,2,3,4\}}$ of length n, which contains 4^n distinct strings, can be reduced, by cancellation of adjacent inverse transformations, to a set containing $4 \times 3^{n-1}$ addresses. To compute an orbital picture associated with this IFS group, we need only consider the reduced set of addresses.

The structure of code spaces associated with IFS groups in the case of four maps, as above, is described very fully, in the context of Möbius transformations, in the book *Indra's Pearls*; see [73].

Figure 3.71 Pictures of five different projective **IFS objects** associated with the same IFS; see Equation (3.5.9). The objects are (i) a set attractor, (ii) an orbital picture, (iii) a colour-rendered fractal top, (iv) a colour-rendered measure attractor and (v) an orbital set. The geometrical property of being a projective IFS object belongs to projective geometry.

The general theory of groups of transformations has been widely studied and there exists a vast body of literature concerning the relationships between geometry, tilings and group theory; see for example [23], [42], [73], [89] and references therein. We shall not describe or review this area, which is essentially classical geometry.

Here we want to connect IFS theory and the associated fractals, orbital sets, orbital measures and pictures, IFS semigroup picture tilings and panellings and so on to classical geometry. We do this, in part, by informally allowing semigroups as well as groups to define geometries.

Figure 3.72 Portions of orbital pictures made using the third crystallographic group. Compare with Figures 3.70 and 3.73. The figure on the right does not represent part of a standard wallpaper pattern. Why?

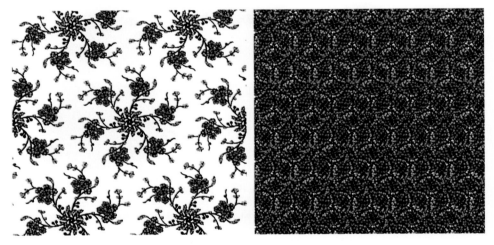

Figure 3.73 Orbital pictures generated by the fifth cystallographic group. On the left the tiles are non-overlapping, and a classical wallpaper pattern is the result. But the pattern on the right is not a semigroup tiling because there are different panels, and the pattern varies subtly across the picture. Can you see some of these variations?

EXERCISE 3.7.6   *Let $\{\mathbb{X}; f_1, f_2, \ldots, f_N\}$ denote an IFS of invertible transformations. Prove that $\mathcal{S}_{\{f_1, f_2, \ldots, f_N, f_1^{-1}, f_2^{-1}, \ldots, f_N^{-1}\}}(\mathbb{X})$ is an IFS group.*

EXERCISE 3.7.7   *Let*

$$R_\theta = \begin{pmatrix} \cos 2\pi\theta & \sin 2\pi\theta \\ -\sin 2\pi\theta & \cos 2\pi\theta \end{pmatrix}.$$

Show that $\mathcal{S}_{\{R_\theta\}}(\mathbb{R}^2)$ is a group if and only if θ is a rational number.

EXERCISE 3.7.8   *An invertible affine transformation* $f : \mathbb{R}^2 \to \mathbb{R}^2$ *may be defined by*

$$(f(x, y))^T = \begin{pmatrix} a & b \\ c & d \end{pmatrix} \begin{pmatrix} x \\ y \end{pmatrix} + \begin{pmatrix} e \\ f \end{pmatrix},$$

where $a, b, c, d, e, f \in \mathbb{R}$ *with* $ad - bc \neq 0$. *Show that the set of all such transformations forms a group on* $\mathbb{R}^2$. *Show that the set of affine transformations of the form* R_θ, *see Exercise 3.7.7, with* θ *rational forms a subgroup of the group of affine transformations on* $\mathbb{R}^2$.

Geometries and IFS objects

Klein's elegant idea was that a group of transformations acting on a space defines a geometry.

DEFINITION 3.7.9   Let $\mathcal{G}(\mathbb{X})$ be a group of transformations. Let the transformations of $\mathcal{G}(\mathbb{X})$ act upon the space of subsets, $\mathbb{S}(\mathbb{X})$, according to

$$T(S) = \{T(x) : x \in S\} \quad \text{for all } S \in \mathbb{S}(\mathbb{X}),$$

as in Chapter 1. Then $\mathcal{G}(\mathbb{X})$ is called a **geometry**, and properties of members of $\mathbb{S}(\mathbb{X})$ that are invariant under all the transformations of the group are called **geometrical properties** (of the geometry).

Let us also refer informally, from time to time, to properties that are invariant under the transformations of a semigroup as geometrical properties (of the semigroup).

Given a geometry $\mathcal{G} = \mathcal{G}(\mathbb{X})$, we extend it to allow the transformations to act on the space of pictures $\Pi_{\mathfrak{C}}(\mathbb{X})$ and the space of normalized Borel measures $\mathbb{P}(\mathbb{X})$, when these spaces are well defined. So, for example, when $\mathbb{X} = \mathbb{R}^2$ we can talk about an invariance property of an orbital picture under the transformations belonging to $\mathcal{G}$ as being a geometrical property of $\mathcal{G}$.

Note that the geometry of an IFS group whose transformations are contained in a geometry $\mathcal{G}$ may have more properties than $\mathcal{G}$ because the smaller the set of transformations, the more invariants they are likely to share. See Exercise 3.7.14 below, for example.

There are many different objects that may be associated with an IFS. They include: orbital sets, measures and pictures; set attractors; measure attractors and fractal tops; panels and tiles. We call them **IFS objects**. An IFS object is defined by an IFS and possible ancillary information such as a condensation set, measure or picture. Typically it is constructed by repeated applications of the transformations of the IFS. So, when the IFS belongs to a geometry $\mathcal{G}$ its IFS objects tend to have properties related to $\mathcal{G}$. In Figure 3.71 we show five different IFS objects, all associated with the same IFS of projective transformations.

When the transformations in an IFS belong to an overarching distinctive group or semigroup $\mathcal{G}$, such as the group of projective transformations, the group of Möbius transformations or the semigroup generated by the inverse branches of a rational function on the Riemann sphere, the corresponding IFS objects tend to have their own distinctive 'look and feel', which depends upon the geometrical properties of $\mathcal{G}$. For example, pictures of set attractors of IFSs of similitudes tend to contain angular features, and distinctive patterns of features, that are repeated at all scales of observation. Attractors of Möbius transformations tend to contain angular features that are repeated at all scales of magnification and patterns of features that are seen to recur in a distorted form, owing to the changing ratios between distances. Attractors of projective transformations tend to contain diverse angles and distorted shapes, yet similar incidences and cross-ratios will be repeated over and over again at different scales. If you are shown a picture of an attractor of one of these types, you will rarely mistake it for being of another type; see for example Figure 4.5.

Such a distinctive 'look and feel' derives at least in part from the following two factors. (i) The IFS objects have properties that are inherited from $\mathcal{G}$. (ii) The IFS objects themselves provide new properties of $\mathcal{G}$.

Here we elaborate in a general way on these two points. Then, in the following subsections, we discuss properties of specific important geometries and relate them to (i) and (ii).

(i) *The IFS objects have properties that are inherited from $\mathcal{G}$.* If a condensation set, a condensation measure or a condensation picture possesses a certain geometrical property then the elements of the corresponding orbits under the IFS semigroup will share that property. In turn, these shared properties will be echoed within the corresponding orbital set, measure or picture.

For example, if a panel $\mathfrak{Q}_\sigma$ of an orbital picture possesses a certain geometrical property then the panels $\mathfrak{Q}_{S^{\circ n}(\sigma)}$, $n = 1, 2, \ldots, |\sigma|$, will also have that property. If the IFS consists of Möbius transformations and if the condensation picture possesses a circular boundary then the corresponding panels will possess boundaries that are finite unions of arcs of circles. If a semigroup tiling is generated by an IFS of invertible affine transformations applied to a triangular condensation set then the tiles will be triangular.

Quite generally, it follows from the self-referential equations obeyed by some IFS objects, such as Equations (3.4.1), (3.5.10), (3.5.11) and (3.6.2), as well as those obeyed by set attractors, measure attractors and fractal tops, that an IFS object typically possesses global features (that is, relating to many tiles, segments or panels or to the whole of itself) that are repeated in the object via transformations belonging to the IFS. Since the transformations belong to $\mathcal{G}$, these repeated global features share properties of $\mathcal{G}$. For example Equation (3.5.10) tells us that

Figure 3.74 Orbital pictures generated by the fifth crystallographic group applied to a buttercup picture. Neither is a wallpaper pattern – subtle differences occur in some of the patterns.

any orbital picture $\mathfrak{P}$, with a sufficiently rich code space, contains global segments $f_n(\mathfrak{R}_n)$ that are the images of global segments $\mathfrak{R}_n \subset \mathfrak{P}$. The geometrical properties of $\mathfrak{R}_n$ are shared by $f_n(\mathfrak{R}_n)$: an orbital picture generated by a semigroup of euclidean transformations is the union of a finite set of rigid transformations of segments of itself, for instance. You can readily identify parts of global repeated patterns, indicated by distinctive angles and distances, in the orbital pictures illustrated in Figures 3.70 and 3.72–3.74.

(ii) *The IFS objects themselves provide new properties of $\mathcal{G}$.* Let $\mathfrak{P}$ denote an orbital picture of an IFS semigroup contained in $\mathcal{G}$. Then if $\mathcal{G}$ is the group of affine transformations we may say that $\mathfrak{P}$ is an **affine orbital picture**. We will use similar terminology to describe other IFS objects. So for example we may refer to a projective set attractor, a Möbius orbital measure or an affine fractal top.

Let $\mathfrak{P}$ denote an orbital picture generated by an IFS semigroup contained in a group $\mathcal{G}$. Then $g(\mathfrak{P})$ is an orbital picture generated by an IFS semigroup contained in $\mathcal{G}$, for all $g \in \mathcal{G}$. So, for example, the property of being an affine orbital picture is a geometrical property of affine geometry, $\mathcal{G}_{affine}$. This is analogous to saying that the property of being a polygon is a property of affine geometry.

Let $\mathfrak{P}$ be an orbital picture whose code space is given as $\Omega'_{\mathfrak{P}_0} \subset \Omega'_{\{1,2,...,N\}}$. Then we say that $\mathfrak{P}$ is an **orbital picture with code space structure** $\Omega'_{\mathfrak{P}_0}$.

Now let $\mathfrak{P}$ denote an affine orbital picture with code space structure $\Omega'_{\mathfrak{P}_0}$. Let $g \in \mathcal{G}_{affine}$. Then it is readily proved that $g(\mathfrak{P})$ is an affine orbital picture with the same code space structure. Thus, the property of being an affine orbital picture with a certain code space structure is a property of affine geometry. This is analogous to saying that the property of being a polygon with a certain number of vertices is a property of affine geometry or that the property of being a triangle with certain angles at the vertices is a property of euclidean geometry.

We can also use invariants associated with dynamical systems, such as the growth rate of periodic orbits, entropy or zeta-functions or related quantities such as the diversity of an orbital picture, to define properties of geometries. For example, let $\mathfrak{P}$ denote an orbital picture with symbolic entropy 0.8. Then it is readily proved that $g(\mathfrak{P})$ is also an orbital picture with symbolic entropy 0.8, for all $g \in \mathcal{G}$. That is, the property of being an orbital picture with a certain symbolic entropy is a geometrical property of any geometry to which the IFS semigroup belongs.

In Chapter 4 we will extend the notion of code space structure to attractors and fractal tops of contractive IFSs. Then you will see that the following general principle applies: *code space structure is a geometrical property.* That is, let $\mathcal{F}$ be an IFS contained in a group $\mathcal{G}$ and let O be an IFS object generated by $\mathcal{F}$; then $g(O)$ has the same code space structure as O for all $g \in \mathcal{G}$. So for example the property of being a projective set attractor of an IFS, with a certain code space structure, is a property of projective geometry; and the property of being a Möbius fractal top with a certain code space structure is a property of Möbius geometry.

The idea of code space structure as a geometrical invariant becomes particularly exciting when we discover the fractal homeomorphism theorem in Chapter 4: this theorem says that set attractors of IFSs have the same code space structure if and only if they are homeomorphic.

EXERCISE 3.7.10 *Let $\mathfrak{P}$ denote an orbital picture generated by an IFS semigroup contained in a group $\mathcal{G}$. Prove that $g(\mathfrak{P})$ is an orbital picture generated by an IFS semigroup contained in $\mathcal{G}$, for all $g \in \mathcal{G}$.*

EXERCISE 3.7.11 *Let $\mathfrak{P}$ denote an affine orbital picture with code space structure $\Omega'_{\mathfrak{P}_0}$. Let $g \in \mathcal{G}_{affine}$. Prove that $g(\mathfrak{P})$ is an affine orbital picture with code space structure $\Omega'_{\mathfrak{P}_0}$.*

Euclidean geometry

Euclidean geometry in two dimensions involves two concepts: (i) a plane and (ii) the transformations that rigidly move the plane upon itself. By (i) we mean the euclidean plane, which we represent by $\mathbb{R}^2$, as well as subsets of it: lines, circles, triangles, fractals and so on. In this plane we can measure angles between lines and distances between points. By (ii) we mean the euclidean transformations, the set of all mappings that take the plane to itself while preserving angles and distances. euclidean geometry comes into being as the interplay between the plane and the euclidean transformations; this interplay reveals most of what we know about both these entities.

Euclidean geometry is represented by the set of all transformations $E : \mathbb{R}^2 \to \mathbb{R}^2$ defined by

$$E(x, y) = (s(x \cos \theta - y \sin \theta) + e,\ x \sin \theta + y \cos \theta + f) \quad \text{for all } x, y \in \mathbb{R}^2$$

for some set of parameters $e, f, \theta \in \mathbb{R}$; $s \in \{-1, +1\}$.

In addition to preserving distances and angles, the group of euclidean transformations acting on $\mathbb{R}^2$ and its subsets has the remarkable property that it admits only seventeen fundamentally different classical euclidean tilings; see [23], vol. 1, Section 1.7, pp. 11–22.

We now explain more carefully what this last statement means. We first note that any given tile, in standard terminology, may correspond to many different sequences of transformations from the group applied to the fundamental tile, namely the condensation picture. But in the theory of IFS semigroups we distinguish between IFS tilings, where each tile has exactly one address in code space, and panellings, where the 'tiles' are panels and possess unique addresses in the space $\Omega'_{\mathfrak{P}_0}$. Thus a tiling under a group of transformations, in standard nomenclature, corresponds to what we call a panelling of diversity 1.

We say that two panellings are **conjugate** iff the associated IFSs are conjugate under a transformation T, see Exercise 3.7.3, and the associated orbital pictures are related by $\widetilde{\mathfrak{P}} = T(\mathfrak{P})$. We define a **classical euclidean tiling** to be a panelling of diversity 1, of an orbital picture whose domain is $\mathbb{R}^2$, associated with an IFS group of euclidean transformations, for which the domain of the condensation picture is compact and connected.

Then, by our statement above that 'the group of euclidean transformations acting on $\mathbb{R}^2$ admits only seventeen fundamentally different classical euclidean tilings' we mean more precisely that the picture of any classical euclidean tiling is an orbital picture of an IFS group that is conjugate under an affine transformation to an element of a set of seventeen distinct IFS groups of euclidean transformations. Of course, any element of the set may be replaced by any IFS group that is conjugate to it under an affine transformation. Five of these IFS groups, called the **crystallographic groups**, may be generated by the following IFSs:

$$\{\mathbb{R}^2; (x + 1, y), (x, y - 1), (x - 1, y), (x, y + 1)\},$$
$$\{\mathbb{R}^2; (-x, -y), (x, y - 1), (x, y + 1)\},$$
$$\left\{\mathbb{R}^2; \left(-\tfrac{1}{2}x - \tfrac{\sqrt{3}}{2}y, \tfrac{\sqrt{3}}{2}x - \tfrac{1}{2}y\right), \left(x + \tfrac{1}{2}, y - \tfrac{\sqrt{3}}{2}\right), \left(x + \tfrac{1}{2}, y + \tfrac{\sqrt{3}}{2}\right)\right\},$$
$$\{\mathbb{R}^2; (-y - 1, x - 1), (-x, -y), (y + 1, -x - 1)\},$$
$$\left\{\mathbb{R}^2; (x + 1, y), \left(\tfrac{1}{2}x - \tfrac{\sqrt{3}}{2}y, \tfrac{\sqrt{3}}{2}x + \tfrac{1}{2}y\right), (x - 1, y)\right\}.$$

Here each transformation is denoted by the result of applying it to the point $(x, y) \in \mathbb{R}^2$. The remaining twelve IFS groups may be obtained by composing some transformations in the above IFSs with an improper rotation such as $(-x, y)$; see [23], vol. 1, p. 19.

In Figures 3.70, 3.72, and 3.73 we showed pairs of orbital pictures corresponding to three different crystallographic groups. In each case, basically the same condensation picture, $\mathfrak{P}_0$, illustrated in Figure 3.70, is used. The units of the viewing windows on the right are larger, with the consequence that the condensation sets on the right are in effect larger than on the left. On the left the transformed copies of the condensation picture are non-overlapping and the result is a classical wallpaper pattern. On the right, however, some transformed copies of $\mathfrak{P}_0$ overlap and the resulting pattern, almost a wallpaper pattern, varies subtly across the picture. By inspection, one finds that the pictures on the right are panellings of diversity greater than 4.

In Figure 3.74 the condensation picture represents our friend the buttercup. We have not shown here systematically the many types of wonderful orbital pictures that may be generated by the tiling groups. Great diversity, a wealth of different types of harmonious pictures, may be produced, for example merely by changing the ordering of the maps and the position and scaling of the condensation picture. Are modern wallpaper printing machinery and paper-hangers up to the task of decorating your dining room with orbital pictures?

Since any rigid transformation is an invertible affine transformation, euclidean geometry also displays all the properties of affine geometry, including fractal dimension.

Affine geometry

Two-dimensional affine geometry is defined by the group $\mathcal{G}_{affine}$, which consists of all invertible affine transformations acting on the space $\mathbb{R}^2$. Angles and distances are not preserved but triangles are mapped onto triangles, ellipses onto ellipses, hyperbolas onto hyperbolas, parabolas onto parabolas and parallel lines to parallel lines. The properties of being triangular, elliptical or parabolic etc. all belong to affine geometry. Since the transformations are also homeomorphisms, topological properties such as openness, compactness, connectedness, perfection etc. belong to affine geometry too. Moreover, since an invertible affine transformation is a metric transformation, fractal dimension is a property of affine geometry; see Section 1.14.

Let us say whimsically that a picture has the 'modernist property' iff it contains a domain whose boundary is a parallelogram, an elliptical feature, an open set coloured a certain shade of red ($R = 242$, $G = 160$, $B = 148$) and a subset, in brightest blue, whose boundary has fractal dimension 1.79. Then the 'modernist property' belongs to affine geometry.

IFS objects associated with affine IFSs inherit properties from affine geometry. For example, an affine orbital picture $\mathfrak{P}$ may contain a global segment, made of multiple panels and possessing distinctive features of parallel lines, cross-ratios and triangular structures, that is mapped by a transformation of the IFS onto a different segment of $\mathfrak{P}$ with the same distinctive features. Such patterns may be repeated many times. Also, if the domain of the condensation picture is triangular then the boundaries of tiles and panels will be piecewise linear; and if the domain of the condensation set is constructed from finitely many pieces of hyperbolas then the domains of all the panels will be constructed from finitely many pieces of hyperbolas.

Also, affine IFS objects provide properties of affine geometry. In the same whimsical vein as above, let us say that a measure has the 'affine orbital measure property' iff it is an orbital measure generated by an IFS semigroup of affine transformations. Then the 'affine orbital measure property' belongs to affine geometry. You get the idea?

Some properties of affine geometry follow from the fact that it is a subset of projective geometry.

EXERCISE 3.7.12 *Show that a geometry is defined by the set of affine transformations whose linear parts have determinants equal to $+1$. Show that area is a property of this geometry.*

EXERCISE 3.7.13 *Show that the set of similitudes, that is, affine transformations that preserve angles, yields a geometry. This geometry is called **similitude geometry**.*

EXERCISE 3.7.14 *Let A' denote the set of affine transformations, on $\mathbb{R}^2$, of the special form*

$$\begin{pmatrix} a & 0 \\ c & d \end{pmatrix} \begin{pmatrix} x \\ y \end{pmatrix} + \begin{pmatrix} e \\ f \end{pmatrix}.$$

Let $\mathcal{G}'$ denote an IFS group whose transformations all belong to A'. Clearly, because $\mathcal{G}' \subset \mathcal{G}_{affine}$ the geometry $\mathcal{G}'$ has all the properties of affine geometry. Show that the geometry $\mathcal{G}'$ has the property of 'being a straight line parallel to the y-axis', and that $\mathcal{G}_{affine}$ does not have this property.

Projective geometry

Projective geometry, as discussed here, is defined by the group $\mathcal{G}_{projective}$, the set of all projective transformations acting on $\mathbb{R}^2 \cup L_\infty$, as discussed in Chapter 2. It contains euclidean and affine geometry. While angles and distances are not preserved, a rich structure of conserved properties remains; straight lines, sets of straight lines that have a point in common, sets of tangent lines to conic sections, conic sections, cross-ratios and so on are all preserved.

It is important to notice that *fractal dimension, defined using the euclidean metric, is not a property of projective geometry on* $\mathbb{R}^2 \cup L_\infty$. By this we mean the following. Let $S \subset \mathbb{R}^2$, and let $\mathcal{P} \in \mathcal{G}_{projective}$. Then $\mathcal{P}(S) \cap \mathbb{R}^2$ may have a fractal dimension different from that of $S \backslash L_D$, where $L_D = \mathcal{P}^{-1}(L_\infty)$, because $\mathcal{P}$ restricted to $\mathbb{R}^2 \backslash \mathcal{P}^{-1}(L_\infty)$ is generally not a metric transformation with respect to the euclidean metric. Typically $\mathcal{P}$ stretches euclidean distances by arbitrarily large factors.

For example, consider the orbit S of the point $(0, 0)$ under the semigroup generated by the IFS

$$\left\{ \mathbb{R}^2 : f_1(x, y) = \left(\frac{x}{2}, \frac{y+1}{2} \right), f_2(x, y) = \left(\frac{x+1}{2}, \frac{y}{2} \right) \right\};$$

S is a cloud of isolated points whose limit set, which is not included in S, is the line segment

$$A = \left\{ (x, y) \in \mathbb{R}^2 : x \geq 0; y \geq 0; x + y = 1 \right\}.$$

The fractal dimension of A is 1. Let $\mathcal{P}$ be a projective transformation that maps the line $x + y = 1$ to L_∞, such as that defined by

$$\mathcal{P}(x, y) = \left(\frac{x}{1 - x - y}, \frac{y}{1 - x - y} \right).$$

Then any bounded subset of $\mathcal{P}(A)$ consists of finitely many points and consequently has fractal dimension equal to 0.

This means that, in practice, two real pictures, one of which is, say, a perspective transformation of the other, may not have the same experimental fractal dimensions. While in practice the stretching may not be *arbitrarily* large, it may well be extreme compared with the ranges of scales over which the fractal dimension is supposed to provide a valid estimate.

Since the domains of IFS pictures associated with projective IFS groups may include points in L_∞, it is helpful to illustrate them on the unit disk $\mathbb{D}_+$ described in Section 2.7. The left-hand image in Figure 3.75 illustrates the orbital picture generated by the IFS group $\mathcal{G}_{\{\mathcal{P}, \mathcal{P}^{-1}\}}(\mathbb{D}_+)$, where $\mathcal{P}$ is the projective transformation associated with the matrix

$$\begin{pmatrix} 0.833 & 0.455 & 0.000 \\ -0.455 & 0.833 & 0.000 \\ 0.000 & 0.000 & 1.000 \end{pmatrix}.$$

Notice that this is an affine transformation that maps the line at infinity, L_∞, to itself. It causes orbits of points to spiral in towards the origin, away from L_∞. Its inverse causes orbits to spiral out towards the circular boundary of $\mathbb{D}_+$, which represents two copies of L_∞.

Figure 3.75 Two orbital pictures generated by projective IFS groups acting on $\mathbb{D}_+$. Both represent IFS picture tilings. On the left the boundary of $\mathbb{D}_+$ is mapped to itself. On the right the picture tiles cross the boundary of $\mathbb{D}_+$ and reappear. In each case infinitely many tiles crowd up against the invariant line. Notice the distortions of the tiles in the spiral on the right.

The right-hand image corresponds to the IFS group $\mathcal{G}_{\{\mathcal{P},\mathcal{P}^{-1}\}}(\mathbb{D}_+)$, where $\mathcal{P}$ is represented by the matrix

$$\begin{pmatrix} 0.76 & 0.415 & 0.0 \\ -0.415 & 0.76 & 0.0 \\ 0.68 & 0.0 & 1.0 \end{pmatrix}.$$

This is conjugate to an affine transformation because it maps a straight line in $\mathbb{R}^2 \cup L_\infty$ into itself. This straight line is half an ellipse on $\mathbb{D}_+$ and corresponds to the runkled part of the right-hand picture.

EXERCISE 3.7.15 *Calculate the formula for the conic section corresponding to L_∞ in the right-hand picture in Figure 3.75. To help do this, look back at Exercise 2.7.27.*

A vast range of tilings and orbital pictures is possible within projective geometry. This is demonstrated in tiny measure by the projective IFS objects illustrated in this book. An orbital picture that is clearly projective is shown in Figure 3.76.

Möbius geometry

The Möbius geometry, $\mathcal{G}_{M\ddot{o}bius}$, is defined by the group of Möbius transformations acting on the extended complex plane. These transformations are discussed in Chapter 2. They take the form

$$\mathcal{M}(z) = \frac{az + b}{cz + d} \quad \text{for } z \in \widehat{\mathbb{C}},$$

Figure 3.76 Example of a projective orbital picture. Compare it with the Möbius orbital picture in Figure 3.82 and the affine orbital picture in Figure 3.42.

where $a, b, c, d \in \mathbb{C}$ and $ad - bc \neq 0$. In this geometry generalized circles are mapped to generalized circles. Angles between intersecting circular arcs are preserved both in magnitude and orientation.

Inversive geometry, $\mathcal{G}_{inversive}$, is defined by the smallest group of transformations on $\widehat{\mathbb{C}}$ that includes the reflection $\mathcal{R}(z) = \overline{z}$; that is,

$$\mathcal{G}_{inversive} = \mathcal{G}_{M\ddot{o}bius} \cup \{\mathcal{M} \circ \mathcal{R} : \mathcal{M} \in \mathcal{G}_{M\ddot{o}bius}\}.$$

Inversive geometry does not have the property of oriented angles but does admit generalized circles and the magnitude of angles. Euclidean distance is not preserved.

Two-dimensional **hyperbolic geometry** may be represented by the subgroup of inversive geometry that maps the unit disk $\mathcal{D} \subset \widehat{\mathbb{C}}$, centred at the origin, onto itself. The corresponding Möbius transformations $\mathcal{M} : \mathcal{D} \to \mathcal{D}$ are defined by

$$\mathcal{M}(z) = \frac{az + b}{\bar{b}z + \bar{a}} \quad \text{for all } z \in \mathcal{D},$$

where $a, b \in \mathbb{C}$ with $|b| < a$.

Hyperbolic geometry was one of the most momentous mathematical discoveries of the nineteenth century; see [25], p. 261. It provided a two-dimensional geometry in which, given any line L and any point P not on L, there exist infinitely many lines through P that do not meet L. For more than two thousand years, since Euclid wrote his famous geometry books, generation after generation of mathematical thinkers asserted that this could not be true in the real physical world. They thought that the only possible geometry for physical space was euclidean geometry. Now hyperbolic geometry is considered as one of various possible models for the space in which the universe is located.

Tilings of the unit disk $\mathcal{D}$ associated with hyperbolic geometry, generated by various IFS groups, were popularized by the artist M. C. Escher; see for example [86]. Escher was fascinated by the different ways in which space could be cut up, methodically, into related shapes, reminiscent of animals, people and plants; his paintings suggest that there is something mysterious in geometrical transformations of shape and form. Escher was an artistic explorer, seeking visual geometrical properties of euclidean, Möbius and other geometries.

In effect, some of Escher's works exploit the fact that there are infinitely many fundamentally different tilings of the unit disk by generalized triangles. A generalized triangle is a three-sided figure whose sides are arcs of generalized circles. This is in striking contrast to the mere seventeen fundamentally different tilings allowed by euclidean geometry.

EXERCISE 3.7.16 *Type the phrase 'hyperbolic tilings' into Google or another internet search utility. Print out some pictures of hyperbolic tilings. Find the corresponding IFS groups.*

EXERCISE 3.7.17 *Show that if*

$$\mathcal{M}(z) = \frac{az + b}{cz + d} \quad then \quad \mathcal{M}^{-1}(z) = \frac{dz - b}{-cz + a}.$$

An IFS group is called discrete iff it is a discrete IFS semigroup.

Two important, interesting and closely related discrete groups of Möbius transformations are the **Sierpinski group** $\mathcal{G}_{Sierpinski}(\widehat{\mathbb{C}})$, which is associated with the IFS

$$\left\{ \widehat{\mathbb{C}}; \mathcal{M}_1(z) = \frac{z}{-2iz + 1}, \quad \mathcal{M}_2(z) = \frac{(1 - i)z - 1}{-z + (1 + i)}, \right.$$

$$\left. \mathcal{M}_3(z) = \mathcal{M}_1^{-1}(z), \quad \mathcal{M}_4(z) = \mathcal{M}_2^{-1}(z) \right\} \tag{3.7.1}$$

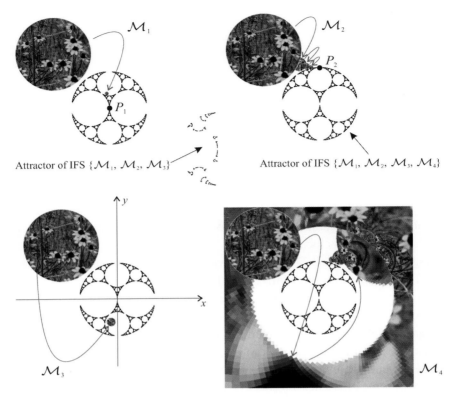

Figure 3.77 This illustrates the action of each of the four Möbius transformations in the IFS in Equation (3.7.1). It shows the result of applying each of these parabolic transformations to the condensation picture used in Figure 3.80, and the relationship to the limit set. The points P_1 and P_2 denote the fixed points of $\mathcal{M}_1$ and $\mathcal{M}_2$ respectively.

and the **modular group** $\mathcal{G}_{modular}(\widehat{\mathbb{C}})$, which is associated with the IFS

$$\left\{ \widehat{\mathbb{C}};\ \mathcal{M}_1(z) = \frac{(2-i)z - i}{-2z + i}, \quad \mathcal{M}_2(z) = \frac{-iz - i}{2z + (2+i)}, \right.$$

$$\left. \mathcal{M}_3(z) = \mathcal{M}_1^{-1}(z), \quad \mathcal{M}_4(z) = \mathcal{M}_2^{-1}(z) \right\}. \tag{3.7.2}$$

The four transformations $\mathcal{M}_1(z), \mathcal{M}_2(z), \mathcal{M}_3(z), \mathcal{M}_4(z) \in \mathcal{G}_{Sierpinski}(\widehat{\mathbb{C}})$ are parabolic; it may be helpful here to look back at Figure 2.34. Their actions are illustrated in Figure 3.77. For $\mathcal{M}_1(z) \in \mathcal{G}_{Sierpinski}(\widehat{\mathbb{C}})$, the fixed point is $z = 0$ and the fixed line is the imaginary axis. This transformation sweeps points lying in the left half-plane in a clockwise direction. The inverse, $\mathcal{M}_3(z) = \mathcal{M}_1^{-1}(z) \in \mathcal{G}_{Sierpinski}(\widehat{\mathbb{C}})$, has the same fixed point and fixed line as $\mathcal{M}_1(z)$ but the orientation of the sweeping motion is opposite. For the parabolic transformation $\mathcal{M}_2(z) \in \mathcal{G}_{Sierpinski}(\widehat{\mathbb{C}})$, the fixed point is $z = i$ and the fixed line is $\{z \in \widehat{\mathbb{C}} : z = x + i, x \in \mathbb{R} \cup \{\infty\}\}$. Points lying above the fixed line are swept in an anticlockwise direction.

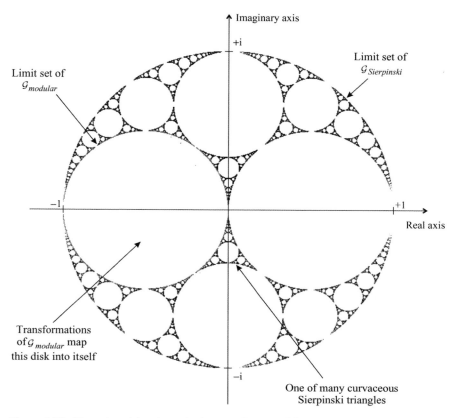

Figure 3.78 Illustration of the relationship between $\mathcal{G}_{modular}$ and $\mathcal{G}_{Sierpisnki}$, defined by the IFSs in Equations (3.7.2) and (3.7.1). The limit set of $\mathcal{G}_{Sierpisnki}$ is shown in black while the limit set of $\mathcal{G}_{modular}$ is the red circle. $\mathcal{G}_{modular}$ is a subgroup of $\mathcal{G}_{Sierpisnki}$.

The relationship between the limit sets of $\mathcal{G}_{Sierpinski}(\widehat{\mathbb{C}})$ and $\mathcal{G}_{modular}(\widehat{\mathbb{C}})$ is illustrated in Figure 3.78. These limit sets were computed using random iteration. We chose to represent the modular group using transformations that map the circle centred at $-\frac{1}{2}i$, of radius $\frac{1}{2}$, onto itself. The standard representation is obtained by conjugating the transformations here by a Möbius transformation that takes this circle to the upper half-plane. The modular group and its subgroups play an important role in the theory of continued fractions and number theory; see for example [73].

In the sequence of pictures (i)–(vi) in Figure 3.79 we illustrate the panels of a one-parameter family of orbital pictures associated with $\mathcal{G}_{Sierpinski}(\widehat{\mathbb{C}})$. In each picture, the domain of the condensation picture, shown in black, is the exterior of a circle centred at the origin; the radius R of this circle is decreased successively from $R = 2$ to $R = 1$, so that in effect we are zooming in on the circular region

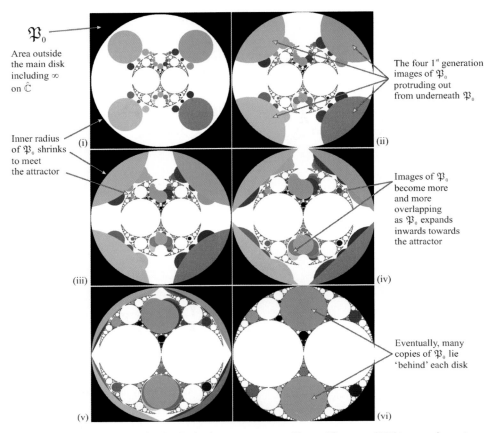

$\mathfrak{P}_0$

Area outside
the main disk
including ∞
on $\hat{\mathbb{C}}$

Inner radius
of $\mathfrak{P}_0$ shrinks
to meet
the attractor

The four 1ˢᵗ generation
images of $\mathfrak{P}_0$
protruding out
from underneath $\mathfrak{P}_0$

Images of $\mathfrak{P}_0$
become more
and more
overlapping
as $\mathfrak{P}_0$ expands
inwards towards
the attractor

Eventually, many
copies of $\mathfrak{P}_0$ lie
'behind' each disk

(i) (ii) (iii) (iv) (v) (vi)

Figure 3.79 Panels of a family of orbital pictures generated by an IFS group of Möbius transformations. In each case the condensation picture, shown in black, is the exterior of a circle centred at the origin, of decreasing radius, although this is masked by the continual zoom in towards the centre. The panels have been given different colours to distinguish them. In the last image, (vi), many panels have merged.

while its radius decreases. In (vi) $R = 1$ and the circle coincides with the outer boundary of the limit set of the group. In each picture the viewing window is $\{z = x + iy \in \mathbb{C} : -R \leq x \leq R, -R \leq y \leq R\}$. The panels are rendered in various colours. Inside each bubble of the limit set in which there is a panel, there is one disk-shaped panel and many crescent-shaped 'children'. As R decreases towards 1, the disk-shaped bubble approaches filling up the whole bubble and the children become like waning crescent moons; at $R = 1$ a quite famous type of picture, associated with the modular group, appears.

An example of an orbital picture associated with $\mathcal{G}_{Sierpinski}(\hat{\mathbb{C}})$ is illustrated in Figures 3.80 and 3.81. Figure 3.81 shows a magnification of part of Figure 3.80 to reveal some structures associated with limiting pictures. The sequences of panels labelled a, b and c correspond to distinctly different limiting pictures.

Examples of panels of orbital pictures associated with $\mathcal{G}_{modular}(\hat{\mathbb{C}})$ are shown in the right-hand images in Figures 3.82 and 3.83. In each figure the left-hand image

Condensation picture

Panel with piecewise circular boundary

Artifacts

Attractor

Figure 3.80 Orbital picture associated with the IFS group $\mathcal{G}_{Sierpinski}$. The boundaries of the domains of the panels are all finite unions of arcs of circles. The limit set of the IFS is labelled 'attractor'. The panels of the orbital picture crowd towards the attractor. The effects of digitization of the condensation picture mean that much of the picture is strewn with computational artifacts. The region inside the white rectangle is shown enlarged in Figure 3.81.

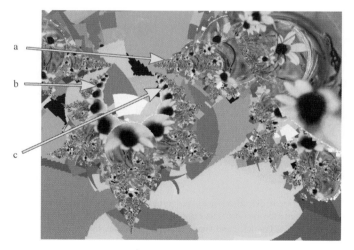

a

b

c

Figure 3.81 Zoom on part of Figure 3.80 revealing structures associated with limiting pictures. The sequences of panels labelled a, b and c correspond to distinctly different limiting pictures.

Figure 3.82   The panels of orbital pictures of two different IFS groups, $\mathcal{G}_{modular}(\widehat{\mathbb{C}})$ and $\mathcal{G}_{hyperbolic}(\widehat{\mathbb{C}})$, acting on the same condensation picture. The right-hand image shows, in various colours, some of the panels of an orbital picture generated by the modular group $\mathcal{G}_{modular}(\widehat{\mathbb{C}})$. The left-hand image is similar, but uses the IFS group $\mathcal{G}_{hyperbolic}(\widehat{\mathbb{C}})$ defined in Equation (3.7.3). Two different addressing schemes for the circle are implied.

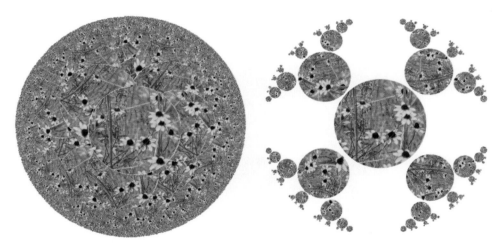

Figure 3.83   Two different orbital pictures, generated by IFS groups of Möbius transformations acting on the same condensation picture. The right-hand image shows an orbital picture generated by the modular group $\mathcal{G}_{modular}(\widehat{\mathbb{C}})$. The left-hand image is similar, but uses the IFS group $\mathcal{G}_{hyperbolic}(\widehat{\mathbb{C}})$ defined in Equation (3.7.3).

Figure 3.84 This floral pattern is an orbital picture in which the panels have different colour tones. It was generated by the Möbius IFS in Equation (3.7.3).

shows the orbital picture generated by an IFS group $\mathcal{G}_{hyperbolic}(\widehat{\mathbb{C}})$, using the same condensation picture as in the right-hand image. $\mathcal{G}_{hyperbolic}(\widehat{\mathbb{C}})$ corresponds to the IFS

$$\{\widehat{\mathbb{C}};\; \mathcal{M}_1(z) = \mathcal{M}_{0.3+0.3i,\pi/4}(z),\; \mathcal{M}_2(z) = \mathcal{M}_{0.35+0.35i,43\pi/36}(z),$$
$$\mathcal{M}_3(z) = \mathcal{M}_1^{-1}(z),\quad \mathcal{M}_4(z) = \mathcal{M}_2^{-1}(z)\}. \tag{3.7.3}$$

where $\mathcal{M}_{a,\theta}(z)$ denotes a member of the family of transformations defined in Equation (2.6.10). The transformations in $\mathcal{G}_{hyperbolic}(\widehat{\mathbb{C}})$ are hyperbolic and map the unit disk onto itself; each has two fixed points, one repulsive and one attractive, located on the boundary of the disk.

Another more artistic picture generated using the Möbius IFS in Equation (3.7.3) is shown in Figure 3.84. We have illustrated only a very few orbital pictures associated with Möbius IFS semigroups and groups, however. A wealth of others can be imagined. To obtain families of IFS objects associated with Möbius geometry, consider IFS groups and semigroups of transformations that share fixed points, or map from a fixed point of one to a fixed point of another, or share an invariant circle, or have invariant circles that are tangent to one another. See [73] for inspiration.

Code space geometries

Klein certainly had in mind that the underlying space for a geometry should be something like a surface, say of a sphere, or $\mathbb{R}^3$, and that the transformations should be quite 'geometrical' too. We can invent many other geometries, however; they may not really be quite so geometrical as the ones we have described and that were in Klein's mind. For example, we might work on $\mathbb{R}^2$ but take the group of transformations to be the set of homeomorphisms of $\mathbb{R}^2$ into itself. This geometry is relevant to fractal geometry, as we will see in Section 4.14.

It is useful to think about code space in geometrical terms. We introduced various families of transformations on code spaces in Chapter 2. Most of these, such as the shift transformation, are not invertible and do not give rise to geometries. But any homeomorphism $f : \Omega_{\mathcal{A}} \to \Omega_{\mathcal{A}}$ generates a group of transformations that conserve topological properties such as compactness, connectivity, boundaries, and so on. One example of a group of homeomorphisms is the group of permutations. This group is relevant to orbital pictures and fractal tops, both of which depend on the ordering of the functions in the IFS that produces them.

Let $\mathcal{G}_{\mathcal{A}}$ denote the permutation group for the alphabet $\mathcal{A}$. For each $p \in \mathcal{G}_{\mathcal{A}}$ define $f_p : \Omega'_{\mathcal{A}} \cup \Omega_{\mathcal{A}} \to \Omega'_{\mathcal{A}} \cup \Omega_{\mathcal{A}}$ by

$$f_p(\sigma) = p(\sigma_1)p(\sigma_2)p(\sigma_3)\cdots$$

Then $\mathcal{G}_{\Omega'_{\mathcal{A}} \cup \Omega_{\mathcal{A}}} = \{f_p : \Omega'_{\mathcal{A}} \cup \Omega_{\mathcal{A}} \to \Omega'_{\mathcal{A}} \cup \Omega_{\mathcal{A}} : p \in \mathcal{G}_{\mathcal{A}}\}$ is called the **permutation group on code space**. It is easy to see that each permutation f_p is a homeomorphism, that the topological entropy of a point in code space is invariant under each permutation and that shift-invariant subspaces are mapped into shift-invariant subspaces by each permutation.

EXERCISE 3.7.18   *Let $p : \{1, 2\} \to \{1, 2\}$ obey $p(1) = 2$ and $p(2) = 1$. Let Ω and $\widetilde{\Omega}$ denote the code spaces for orbital pictures generated by the IFS semigroups $\mathcal{S}_{\{f_1, f_2\}}(\mathbb{R}^2)$ and $\mathcal{S}_{\{f_2, f_1\}}(\mathbb{R}^2)$ respectively, acting on the same condensation picture. Suppose that $f_2(x, y) := -f_1(-x, -y)$ and that the condensation picture is invariant under the transformation $(x, y) \to (-x, -y)$. Show that $f_p(\Omega) = \widetilde{\Omega}$.*

CHAPTER 4

Hyperbolic IFSs, attractors and fractal tops

4.1 Introduction

In this chapter we introduce the newly discovered and very exciting subject of **fractal tops**. Fractal tops are simple to understand yet profound and lead at once to many potential applications. What is a fractal top? It is an addressing function for the set attractor of an IFS such that each point on the attractor has a *unique address, even in the overlapping case*! Fractal tops can be used to do the following things: (i) define pictures that are invariant under IFSs, in much the same way that the measure attractor and the set attractor are invariant; (ii) define transformations between different fractal sets; (iii) set up a uniquely defined dynamical system associated with any IFS and use the invariants of this dynamical system to define invariants for pictures; (iv) establish, in if-and-only-if fashion, when pairs of fractal sets are homeomorphic, see Figures 4.1 and 4.2; (v) produce beautiful special effects on still and video images, with diverse potential applications in image science; (vi) lead to an easily used wide-ranging definition of what a deterministic fractal is; (vii) handle topologically fractal sets in a manner that has serious analogies with the way in which cartesian coordinates can be used to handle classical geometry. A fractal top is illustrated in Figures 4.16 and 4.17, for example.

We begin by defining a hyperbolic IFS, its set attractor and its measure attractor. We then provide a simple way of writing down IFSs of projective and Möbius transformations, just to make it easy to tell one another which IFS we are talking about. We then discuss the chaos game algorithm and deterministic algorithms for computing set attractors and measure attractors. We also explain and illustrate the collage theorem, which is a useful tool for geometrical modelling using IFSs. At this stage we can contain ourselves no longer: we introduce fractal tops and explain how they can be used to colour-render fantastic pictures, which we say are produced by **tops plus colour-stealing**. We show how you can easily produce these pictures yourself, using a simple variant of the chaos game. Then we do some

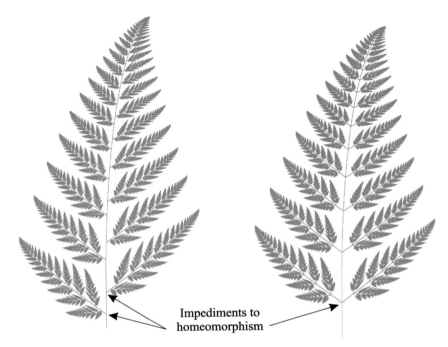

Impediments to
homeomorphism

Figure 4.1   The two mathematical ferns represented here are not topologically conjugate because their branching structures are different. Hence by the fractal homeomorphism theorem (see Section 4.14) **their code space structures are different**. But there exist transformations from one to the other that are 'nearly continuous'. See also Figure 4.2.

serious analysis. We define the tops dynamical system and an associated symbolic dynamical system; we show how pictures produced by tops plus colour-stealing are analogous to set attractors and measure attractors because they are fixed points of a contractive transformation defined using the IFS; and we establish a relationship with orbital pictures and other material in this chapter. Finally, inspired by what we have learnt, we introduce directed IFSs, which generalize IFSs in a very natural way.

In the back of your mind, as you read or scan this chapter, keep alive the theme of bioinformatics. What does this new material suggest in the way of new models in the biological science? Does it just look biological but really is not? Or is there something very deep here in the idea of treating protoplasmic things in the language of topology and sets of sets in code space?

Read on, and enjoy.

4.2   Hyperbolic IFSs

An **iterated function system** or **IFS**, as explained earlier, consists of a finite sequence of transformations $f_i : \mathbb{X} \to \mathbb{X}$ for $i = 1, 2, \ldots, N$ where $N \geq 1$ is an

Figure 4.2 There exist transformations that map from one mathematical fern to the other which are 'nearly' homeomorphisms. These transformations are easy to implement and have diverse applications. Carefully study these images to see how the form and colour are shifted.

integer and $\mathbb{X}$ is a space. It may be denoted by

$$\{\mathbb{X}; f_1, f_2, \ldots, f_N\} \quad \text{or} \quad \{\mathbb{X}; f_n, n = 1, 2, \ldots, N\}.$$

We use such terminology as 'the IFS $\{\mathbb{X}; f_1, f_2, \ldots, f_N\}$' and 'Let $\mathcal{F}$ denote an IFS'. We first introduced IFSs in the Introduction and Chapter 2. Typically, the space $\mathbb{X}$ is a metric space, the transformations are Lipschitz and there is more than one transformation.

An **IFS with probabilities** consists of an IFS together with a sequence of probabilities $p_1, p_2, \ldots, p_N$, positive real numbers such that $p_1 + p_2 + \cdots + p_N = 1$. An IFS with probabilities may be denoted

$$\{\mathbb{X}; f_1, f_2, \ldots, f_N; p_1, p_2, \ldots, p_N\}.$$

The probability p_n is associated with the transformation function f_n for each $n \in \{1, 2, \ldots, N\}$.

DEFINITION 4.2.1 Let $(\mathbb{X}, d)$ be a complete metric space. Let $\{f_1, f_2, \ldots, f_N\}$ be a finite sequence of strictly contractive transformations, $f_n : \mathbb{X} \to \mathbb{X}$, for $n = 1, 2, \ldots, N$. Then $\{\mathbb{X}; f_1, f_2, \ldots, f_N\}$ is called a strictly contractive IFS or a **hyperbolic IFS**.

Recall that a transformation $f_n : \mathbb{X} \to \mathbb{X}$ is strictly contractive iff there exists a number $l_n \in [0, 1)$ such that $d(f_n(x), f_n(y)) \leq l_n d(x, y)$ for all $x, y \in \mathbb{X}$. The number l_n is called a contractivity factor for f_n and the number

$$l = \max\{l_1, l_2, \ldots, l_N\}$$

is called a contractivity factor for the IFS.

We use such terminology as 'Let $\mathcal{F}$ denote a hyperbolic IFS with probabilities'. Although we often deal with hyperbolic IFSs, we tend to drop the adjective 'hyperbolic'. We may use an adjective such as affine, projective or Möbius when we want to describe the geometry to which the transformations of the IFS belong.

EXERCISE 4.2.2   *Let $f : \mathbb{R} \to \mathbb{R}$ be defined by $f(x) = \frac{1}{3}x + \frac{2}{3}$ for all $x \in \mathbb{R}$. Show that f is a contraction mapping with respect to the euclidean metric.*

EXERCISE 4.2.3   *Find the smallest square region $\square \subset \mathbb{R}^2$ such that $\{\square; f_1, f_2\}$ is a hyperbolic iterated function system, where*

$$f_1(x) = \tfrac{1}{3} R_\theta x + \left(\tfrac{1}{2}, 0\right) \quad and \quad f_1(x) = \tfrac{2}{3} R_\theta x \quad for\ all\ x \in \square \subset \mathbb{R}^2;$$

R_θ denotes an anticlockwise rotation through angle θ about the origin.

4.3   The set attractor and the measure attractor

Recall, from Theorems 2.4.6 and 2.4.8, that a hyperbolic IFS $\mathcal{F}$ possesses a unique **set attractor**, $A \in \mathbb{H}(\mathbb{X})$. The space $\mathbb{H}(\mathbb{X})$ is the set of nonempty compact subsets of $\mathbb{X}$.

The set attractor A is the unique fixed point of the strictly contractive transformation $\mathcal{F} : \mathbb{H}(\mathbb{X}) \to \mathbb{H}(\mathbb{X})$ defined by

$$\mathcal{F}(B) = f_1(B) \cup f_2(B) \cup \cdots \cup f_N(B). \tag{4.3.1}$$

The transformation $\mathcal{F} : \mathbb{H}(\mathbb{X}) \to \mathbb{H}(\mathbb{X})$ is strictly contractive with respect to the Hausdorff metric, with contractivity factor l. Note that we use the same symbol $\mathcal{F}$ for the IFS and for the transformation $\mathcal{F} : \mathbb{H}(\mathbb{X}) \to \mathbb{H}(\mathbb{X})$.

The set attractor A obeys the **self-referential equation**

$$A = f_1(A) \cup f_2(A) \cup \cdots \cup f_N(A).$$

An example of a set attractor, when the transformations are two similitudes on $\mathbb{R}^2$, is illustrated in Figure 4.3. The transformations are given in Table 4.1. This

Table 4.1   *Möbius IFS code for Figure 4.3. The attractor of this IFS is pictured below, as in the tables that follow. These transformations are actually similitudes, in contrast with those in Table 4.6*

n	a_R	a_I	b_R	b_I	c_R	c_I	d_R	d_I	p
1	1	1	0	0	0	0	2	0	0.47
2	−1	1	2	0	0	0	2	0	0.53

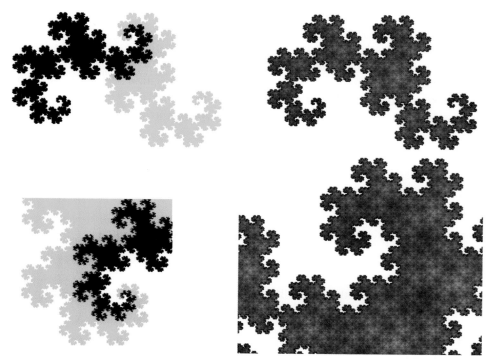

Figure 4.3   The top two images are pictures of the set attractor of an IFS and a measure attractor of the same IFS, given in Table 4.1. You can see on the left how the set attractor can be regarded as the union of two scaled copies of itself. The measure illustrated on the right is a superposition of two measures, each rescaled. Zooms are shown at the bottom of the figure.

attractor is known as the Heighway dragon. You can see quite clearly how it is the union of two scaled copies of itself. As described in the Introduction and justified in Section 4.5, we can use algorithms based on the chaos game to compute such pictures.

Closely related to the set attractor is the measure attractor. In dealing with measure attractors we restrict our attention to the case where $(\mathbb{X}, d)$ is a compact metric space, because this implies that $(\mathbb{P}(\mathbb{X}), d_{\mathbb{P}})$ is also a compact metric space. We do this purely for simplicity. There are many cases where this restriction is not needed. For example, given a hyperbolic IFS for which the metric space $\mathbb{X}$ is locally compact, that is, closed balls of finite radius are compact, we can redefine the IFS to act on a new space $\widetilde{\mathbb{X}} \subset \mathbb{X}$ that is compact; see Exercise 4.4.1. The space $\mathbb{R}^2$ is locally compact. So we will sometimes treat a hyperbolic IFS as though the underlying space were compact although in fact the specified underlying space $\mathbb{X}$ is not compact.

From Theorems 2.4.19 and 2.4.21, there exists a unique normalized measure $\mu \in \mathbb{P}(\mathbb{X})$, which is the fixed point of the transformation $\mathcal{F} : \mathbb{P}(\mathbb{X}) \to \mathbb{P}(\mathbb{X})$ defined by

$$\mathcal{F}(\xi) = \sum_{n=1}^{N} p_n f_n(\xi) \tag{4.3.2}$$

for all $\xi \in \mathbb{P}(\mathbb{X})$. Notice that we use the same symbol $\mathcal{F}$ for the IFS, for the transformation $\mathcal{F} : \mathbb{H}(\mathbb{X}) \to \mathbb{H}(\mathbb{X})$ and for the transformation $\mathcal{F} : \mathbb{P}(\mathbb{X}) \to \mathbb{P}(\mathbb{X})$. The interpretation of $\mathcal{F}$ should to be clear from the context.

The transformation $\mathcal{F} : \mathbb{P}(\mathbb{X}) \to \mathbb{P}(\mathbb{X})$ is a strict contraction, with contractivity factor

$$\bar{l} = p_1 l_1 + p_2 l_2 + \cdots + p_N l_N$$

with respect to the metric $d_{\mathbb{P}}$ on $\mathbb{P}(\mathbb{X})$. It is also strictly contractive with contractivity factor l with respect to the metric $\widehat{d}_{\mathbb{P}}$.

DEFINITION 4.3.1   Let $\mathbb{X}$ be a compact metric space and let

$$\mathcal{F} = \{\mathbb{X}; f_1, f_2, \ldots, f_N; p_1, p_2, \ldots, p_N\}$$

be a hyperbolic IFS with probabilities. Then the unique fixed point $\mu \in \mathbb{P}(\mathbb{X})$ of $\mathcal{F} : \mathbb{P}(\mathbb{X}) \to \mathbb{P}(\mathbb{X})$ is called the **measure attractor** of the IFS.

The measure attractor μ of a hyperbolic IFS with probabilities obeys the self-referential equation

$$\mu = \sum_{n=1}^{N} p_n f_n(\mu).$$

This says that the measure is a weighted sum of the transformations of the IFS applied to it.

An example of a measure attractor, represented as a picture, is given on the right in Figure 4.3. It can be seen that this picture is a superposition of two transformed copies of itself, weighted by the probabilities in Table 4.1. In Section 4.5 we explain how, with the aid of the chaos game, this picture was computed.

Table 4.2   *Affine IFS code for the IFS $\mathcal{F}_1$. This is an example of a just-touching IFS*

n	a	b	c	d	e	f	p
1	$\frac{1}{2}$	0	0	$\frac{1}{2}$	0	0	$\frac{1}{3}$
2	$\frac{1}{2}$	0	0	$\frac{1}{2}$	$\frac{1}{2}$	0	$\frac{1}{3}$
3	$\frac{1}{2}$	0	0	$\frac{1}{2}$	0	$\frac{1}{2}$	$\frac{1}{3}$

Let $\mathcal{F}$ denote a hyperbolic IFS with probabilities, as discussed above. Let A denote the set attractor of $\mathcal{F}$. Let

$$O_{\mathcal{F}} := \{ f_i(A) \cap f_j(A) : i, j \in \{1, 2, \ldots, N\}, i \neq j \}.$$

Then we may refer to $O_{\mathcal{F}}$ as 'the set of overlapping points in the attractor of the IFS'. We say that $\mathcal{F}$ is **totally disconnected** iff $O_{\mathcal{F}} = \varnothing$. We say that $\mathcal{F}$ is **overlapping** iff $O_{\mathcal{F}}$ contains a nonempty set that is open in the relative topology on A. We say that $\mathcal{F}$ is **just-touching** iff it is not totally disconnected and it is not overlapping.

EXERCISE 4.3.2   *Let $\mathbb{X}$ be a compact metric space and let $\mathcal{F}$ denote a hyperbolic IFS on $\mathbb{X}$. Let $A \in \mathbb{H}(\mathbb{X})$ denote the set attractor and $\mu \in \mathbb{P}(\mathbb{X})$ denote the measure attractor of $\mathcal{F}$. Show that the support of μ is strictly contained in A and that it equals A when the probabilities are all strictly positive.*

EXERCISE 4.3.3   *Show that the IFSs represented in Tables 4.2 and 4.3 are just-touching. Show that the IFS represented in Table 4.4 is overlapping.*

EXERCISE 4.3.4   *Show that a hyperbolic IFS is totally disconnected iff its attractor is totally disconnected. Give an example of a totally disconnected IFS.*

4.4   IFS codes

Here we digress to give examples of the notation used to represent IFSs of projective, Möbius and other transformations. This is mainly for reference.

Table 4.3 *A projective IFS code. This is used in Figure 4.4*

n	a_n	b_n	c_n	d_n	e_n	f_n	g_n	h_n	j_n	p_n
1	6	0	0	0	6.5	0	−3	−2	15	$\frac{1}{4}$
2	1	−2	6	−3	1.5	6.5	−3	−2	10	$\frac{1}{4}$
3	7	2	4	0	6.5	0	3	2	8	$\frac{1}{4}$
4	6	0	0	3	5.5	3.5	3	2	7	$\frac{1}{4}$

Table 4.4 *Affine IFS code for a filled square. The IFS is strictly contractive, but not with respect to the usual euclidean metric. This is an example of an overlapping IFS*

n	a	b	c	d	e	f	p
1	$\frac{3}{4}$	0	0	0	$\frac{1}{2}$	0	$\frac{3}{8}$
2	$\frac{3}{4}$	0	0	0	$\frac{1}{2}$	$\frac{1}{2}$	$\frac{3}{8}$
3	0	$\frac{1}{3}$	$\frac{2}{3}$	1	0	0	$\frac{1}{4}$

An example of an affine hyperbolic IFS is

$$\mathcal{F}_1 = \left\{ \mathbb{R}^2; \left(\tfrac{1}{2}x, \tfrac{1}{2}y\right), \left(\tfrac{1}{2}(x+1), \tfrac{1}{2}y\right), \left(\tfrac{1}{2}x, \tfrac{1}{2}(y+1)\right) \right\}.$$

Here the transformations are defined by their actions on the point $(x, y) \in \mathbb{R}^2$. We identify them by the labels 1, 2 and 3, as in f_1, f_2 and f_3 respectively.

Notice that although the space $(\mathbb{R}^2, d_{euclidean})$ is not compact, closed bounded subsets of it are. It is straightforward to show that there exists a compact subset $\mathbb{X}$ of $\mathbb{R}^2$ such that $f_n : \mathbb{X} \to \mathbb{X}$ for $n = 1, 2, \ldots, N$. See Exercise 4.4.1.

The IFS $\mathcal{F}_1$ may be specified succinctly by means of the array in Table 4.2, which we refer to as an affine **IFS code**. The affine transformations are denoted

Table 4.5 *Projective IFS code for Figures 4.9 and 4.10*

n	a_n	b_n	c_n	d_n	e_n	f_n	g_n	h_n	j_n	p_n
1	9.17	−1.39	−6.92	−4.33	−1.97	2.59	12.17	−1.83	−10.61	$\frac{1}{2}$
2	5.66	−2.22	−1.15	−0.88	4.84	−1.45	3.23	−1.71	4.14	$\frac{1}{2}$

by their coefficients according to

$$f(x, y) = (ax + by + c, \; dx + ey + f).$$

We may include probabilities in an IFS code even when the IFS has not been specified to be an IFS with probabilities. The default values could be $p_n = 1/N$ for $n = 1, 2, \ldots, N$.

An example of a hyperbolic IFS with probabilities is

$$\mathcal{F}_3 = \left\{ \mathbb{R}^2; \left(\tfrac{3}{4}x, \tfrac{1}{2}y \right), \left(\tfrac{3}{4}x, \tfrac{1}{2}(y + 1) \right), \left(\tfrac{1}{3}(y + 2), x \right); \tfrac{3}{8}, \tfrac{3}{8}, \tfrac{1}{4} \right\}.$$

It is defined by the affine IFS code in Table 4.4. This IFS consists of three affine transformations. The third transformation does not contract all distances with respect to the euclidean metric. But all three transformations are strictly contractive with respect to the metric defined in Exercise 4.4.2.

We say that IFS codes such as those for $\mathcal{F}_1$ and $\mathcal{F}_3$ are 'simple' because they involve 'small' amounts of information: in each case there are three transformations, each of which can be represented by a small set of numbers, which themselves can be written down briefly.

An example of a projective hyperbolic IFS is

$$\mathcal{F}_4 = \left\{ \square \subset \mathbb{R}^2; f_n(x, y) = \left(\frac{a_n x + b_n y + c_n}{g_n x + h_n y + j_n}, \; \frac{d_n x + e_n y + f_n}{g_n x + h_n y + j_n} \right), \right.$$

$$\left. n = 1, 2, 3, 4; \tfrac{1}{4}, \tfrac{1}{4}, \tfrac{1}{4}, \tfrac{1}{4} \right\},$$

where the transformations are given by the projective IFS code in Table 4.3. Here the underlying space, on which the transformations are strictly contractive, is taken to be a specified subset $\square \subset \mathbb{R}^2$.

The IFS defined in Table 4.3 is used in Figures 4.4 and 4.20 to illustrate IFS colouring and colour-stealing respectively. Another example of a projective hyperbolic IFS is represented by the IFS code in Table 4.5.

Table 4.6   *Example of a Möbius IFS code*

n	a_R	a_I	b_R	b_I	c_R	c_I	d_R	d_I	p
1	1	1	0	0	0	0	2	0	$\frac{1}{2}$
2	-3	5	8	0	2	0	8	0	$\frac{1}{2}$

Figure 4.4 This illustates IFS colouring applied to two different IFSs with the same attractor; see the main text. It is suggestive of a homeomorphism between the two pictures. But the fractal dimensions of the level sets may not be the same.

Notice that an affine IFS can be represented by a projective IFS code in which $g = h = 0$ and $j = 1$.

An example of a Möbius hyperbolic IFS is

$$\mathcal{F}_5 = \left\{ \bigcirc \subset \mathbb{C}; \frac{(1+i)z}{2}, \frac{(-3+5i)z+8}{2z+8}; \frac{1}{2}, \frac{1}{2} \right\},$$

where $\bigcirc = \{z \in \mathbb{C} : |z| \leq 1\}$. This IFS may be represented by the Möbius IFS code in Table 4.6, where the coefficients reference transformations written in the form

$$f(z) = \frac{(a_R + ia_I)z + (b_R + ib_I)}{(c_R + ic_I)z + (d_R + id_I)}.$$

Note that when the transformations of an IFS belong to a particular geometry then so do their attractors, in the sense described in Section 3.7.

EXERCISE 4.4.1   *Let $\mathcal{F}$ be a hyperbolic IFS. Show that there exists a closed ball of finite radius in $\mathbb{X}$ that is mapped into itself by the transformations of the IFS. Show therefore that if $\mathbb{X}$ is a locally compact metric space then there exists $\widetilde{\mathbb{X}}$ such that $\{\widetilde{\mathbb{X}}; f_1, f_2, \ldots, f_N\}$ is a hyperbolic IFS, where $\widetilde{\mathbb{X}}$ is compact.*

EXERCISE 4.4.2   *Prove that all three transformations referenced in Table 4.4 are strictly contractive with respect to the metric*

$$d((x_1, y_1), (x_2, y_2)) = \sqrt{(x_1 - x_2)^2 + \tfrac{1}{4}(y_1 - y_2)^2}.$$

Find two points that are not moved closer together, in the euclidean metric, by the third transformation.

4.5   The chaos game

The 'chaos game' is our name for a well-known type of algorithm, namely the Markov Chain Monte Carlo (MCMC) or **random iteration** algorithm. The scholarly history of the chaos game is discussed in [91] and in [55] and it appears that it began in 1935 with the work of Onicescu and Mihok, [76]. Its usage in computing approximations to the invariant probability measure and the set attractor of a hyperbolic IFS is justified by the following theorem, which can be proved with the aid of Birkhoff's ergodic theorem; see for example [39]. See also [22], and [28]. It was introduced to fractal geometry in [64] and [4]; see also [7].

THEOREM 4.5.1   *Let $(\mathbb{X}, d)$ be a compact metric space. Let $\{\mathbb{X}; f_1, f_2, \ldots, f_N; p_1, p_2, \ldots, p_N\}$ be a hyperbolic IFS with probabilities, and let $\mu \in \mathbb{P}(\mathbb{X})$ denote its measure attractor. Specify a starting point $x_1 \in \mathbb{X}$. Define a random orbit of the IFS to be $\{x_l\}_{l=1}^{\infty}$ where $x_{l+1} = f_m(x_l)$ with probability p_m. Then for almost all random orbits $\{x_l\}_{l=1}^{\infty}$ we have*

$$\mu(B) = \lim_{l \to \infty} \frac{|B \cap \{x_1, x_2, \ldots, x_l\}|}{l}, \qquad (4.5.1)$$

for all $B \in \mathbb{B}(\mathbb{X})$ such that $\mu(\partial B) = 0$, where ∂B denotes the boundary of B.

This is equivalent, by standard arguments, to the following: for any $x_1 \in \mathbb{X}$ and almost all random orbits the sequence of point measures $l^{-1}(\delta_{x_1} + \delta_{x_2} + \cdots + \delta_{x_l})$ converges in the weak sense to μ; see for example [24], pp. 11–12. The weak convergence of probability measures is the same as convergence in the Monge–Kantorovitch metric; see [29], pp. 310–11.

The conclusion of Theorem 4.5.1 applies under the more general condition that the underlying space is locally compact and the transformations are contractive on the average, that is $0 \leq \bar{l} < 1$; see [33]. Similar results hold in much more general circumstances; see for example the review article [91].

Theorem 4.5.1 says that, almost always, if we follow the orbit of an IFS, where the underlying space is two dimensional, and we keep track of the fraction of the total number of iterations for which the current point is contained within the domain of a particular pixel, we obtain in the limit the value of the invariant probability measure for that pixel domain. (It is as though the chaos game distributes the magic dust of Figure 2.10.) But we have to be careful when the invariant measure of the boundary of the pixel domain is nonzero: to see this, consider the case where the attractor of the IFS is a line segment that coincides with the boundaries of the domains of some pixels. In this regard, we note that, when rendered measure-theoretically using the chaos game, lines and curves that are attractors of IFSs tend to be anti-aliased [47].

Algorithms based on the chaos game have the benefits, when compared with deterministic iteration, of low memory requirement and high accuracy; the iterated point can be kept at a precision much higher than the resolution of the attractor. Also they allow the efficient computation of zooms into small parts of pictures of attractors. However, as in the case of deterministic algorithms, the images produced depend on the computational details of image resolution, the precision to which the points $\{x_1, x_2, \ldots\}$ are computed, the contractivity of the transformations, the choice of colours, the way in which Equation (4.5.1) is evaluated etc. Different implementations can produce different results; see for example [79]. Very often, over years of studying IFSs, I have used one form or another of this robust algorithm both to guide intuition and to compute pictures.

As an example of practical implementation, the right-hand side of Figure 4.3 shows two pictures of the invariant measure of the Möbius IFS in Table 4.1 computed using a discrete version of the chaos game. The measure is depicted in shades of green, from 0 (black) to 255 (bright green). These pictures were computed according to the following scheme.

Pixels corresponding to a discrete model for $\square \subset \mathbb{R}^2$ are assigned the colour white. Successive floating-point coordinates of points in $\square$ are computed by random iteration and the first (say) one hundred points are discarded. Thereafter, as each new point is calculated the pixel to which it belongs is assigned the component values $R = G = B = 0$, i.e. black. This phase of the computation continues until the pixels cease to change, and it produces a black image of the support of the measure, the set attractor of the IFS, against a white background. Then the random iteration process is continued and, as each new point is computed, the green component of the pixel to which the latest point belongs is brightened by a fixed amount. Once a pixel is at brightest green, its value is not changed when later points are added to it. The computation is continued until a balance is obtained between that part of the image which is brightest green and that is least green, i.e. darkest, and it is then stopped.

Table 4.7   *Example of an IFS color code*

n	α_n	β_n	γ_n	a_n	b_n	c_n
1	100	0	75	0.5	0.52	0.51
2	100	0	0	0.5	0.1	0.1
3	200	0	10	0.6	0.2	0.6
4	200	150	50	0.7	0.5	0.35

EXERCISE 4.5.2   *Write and execute a computer program that uses the chaos game to make a digital picture of the invariant measure for one of the IFS codes in Section 4.4.*

4.6   IFS colouring of set attractors

A simple way to assign colour to the attractor A of a hyperbolic IFS $\mathcal{F}$ in the case of two-dimensional transformations is to modify $\mathcal{F}$ so that it acts in five dimensions, as follows. What we describe here is *not* colour-stealing.

Table 4.7 gives an example of an IFS colour code. Each row of this table describes a colour transformation, namely a mapping from the colour space $\mathfrak{C} = \mathbb{R}^3$ into itself. These transformations are written in the form

$$C_n(R, G, B) = (\alpha_n + a_n R, \ \beta_n + b_n G, \ \gamma_n + c_n B),$$

for $n = 1, 2, \ldots, N$. The coefficients are chosen so that $\{\mathfrak{C} : C_1, C_2, C_3, C_4\}$ is a hyperbolic IFS. This ensures that the IFS

$$\widehat{\mathcal{F}} = \{\mathbb{X} \times \mathfrak{C} : (f_n, C_n), n = 1, 2, \ldots, N\}$$

is also hyperbolic and possesses a unique attractor, G. In general G is the graph of a multivalued function from A into $\mathbb{R}^3$. But when A is totally disconnected this graph is single-valued and assigns a unique colour to each point in A. We discretize the colour values so that they are triples of integers in $[0, 255]^3$.

In practice we do not worry about whether $\mathcal{F}$ is overlapping. We simply use the chaos game applied to the IFS $\widehat{\mathcal{F}}$, at each step plotting the projection of the latest point on A in the colour defined by the discretized values of the remaining three coordinates. That is, let $F_n = (f_n, C_n)$ and start at a point $X_0 = (x_0, y_0, R_0, G_0, B_0) \in \mathbb{X} \times \mathfrak{C}$. Compute a random orbit $\{X_0, X_1, \ldots\}$ by following the chaos game; for k sufficiently large, start by plotting the points

$$X_{k+1} = (x_{k+1}, y_{k+1}, R_{k+1}, G_{k+1}, B_{k+1})$$
$$= F_{\sigma_{k+1}}(x_k, y_k, R_k, G_k, B_k) \quad \text{for } k = 0, 1, 2, \ldots$$

That is, the point (x_k, y_k) is plotted with colour values (R_k, G_k, B_k).

Figure 4.5 Attractors of IFSs belonging to similitude geometry, Möbius geometry and projective geometry are shown on the left, in the middle and on the right respectively. The pictures are coloured by IFS colouring.

Figure 4.4 shows IFS colouring of the attractors of two different just-touching IFSs. In both cases the attractor is a square and the IFS colour code is defined in Table 4.7. The IFS for the right-hand image is given by Table 4.3 while that for the left-hand image consists of four transformations, each of which maps the square into one of its quadrants.

Notice how the right-hand picture looks as though it is a continuous transformation of the left-hand picture. This illustrates what we call a 'fractal transformation'; see Section 4.15. In this case the implied mathematical transformation is a homeomorphism. But it is not differentiable and does not provide a metric equivalence between the two pictures.

Figure 4.5 illustrates various attractors, belonging to different geometries, coloured using IFS colouring (in contrast with colour-stealing, which we come to in Section 4.10.) Points that lie in the set of overlapping points of the attractor will tend to be grainy and to change continually as the chaos game progresses.

EXERCISE 4.6.1   *Use the IFS colouring algorithm to render your example in Exercise 4.5.2.*

4.7   The collage theorem

How does one go about finding an IFS, in a two-dimensional setting, whose attractor is equal to, or 'looks like', a given compact target set $T \subset \mathbb{R}^2$? Sometimes we can simply spot a set of contractive transformations $f_1, f_2, \ldots, f_N$ taking $\mathbb{R}^2$ into itself, such that

$$T = f_1(T) \cup f_1(T) \cup \cdots \cup f_N(T). \qquad (4.7.1)$$

If so then the unique solution T to Equation (4.7.1) is the attractor of the IFS $\{\mathbb{R}^2; f_1, f_2, \ldots, f_N\}$.

If T has properties which belong to a certain geometry then it makes sense to seek the required IFS among transformations that belong to that geometry. For example, if T is a polygon in $\mathbb{R}^2$ then it is a good idea to restrict attention to projective IFSs.

But in computer-graphical modelling, image approximation and biological modelling applications, it is often not possible to find an IFS, with a restricted number of transformations belonging to a given geometry, such that Equation (4.7.1) holds. Nonetheless, we may seek an IFS that makes Equation (4.7.1) approximately true. That is, we may *try* to make T out of transformations of itself. The following theorem gives an upper bound to the distance between the attractor of the resulting IFS and T. The upper bound depends only on the distance from T to $\mathcal{F}(T)$.

THEOREM 4.7.1 *(The collage theorem [5])   (i) Let $(\mathbb{X}, d)$ be a complete metric space. Let $T \in \mathbb{H}(\mathbb{X})$ be given and let $\epsilon \geq 0$ be given. Suppose that a hyperbolic IFS $\mathcal{F} = \{\mathbb{X}; f_1, f_2, \ldots, f_N\}$ of contractivity factor $0 \leq l < 1$ can be found such that*

$$d_{\mathbb{H}}(T, \mathcal{F}(T)) \leq \epsilon,$$

where $d_{\mathbb{H}}$ denotes the Hausdorff metric. Then

$$d_{\mathbb{H}}(T, A) \leq \frac{\epsilon}{1 - l},$$

where A is the set attractor of the IFS.

(ii)   Similarly,   let $(\mathbb{X}, d)$ be   a   compact   metric   space.   Let $\upsilon \in \mathbb{P}(\mathbb{X})$ be   given.   Suppose   that   a   hyperbolic   IFS   with   probabilities $\mathcal{F} = \{\mathbb{X}; f_1,$

Figure 4.6 Illustrations of the collage theorem. The target set T is the leaf at top left. Next to it are shown three different collages of T, superimposed on T. Below each collage is shown the attractor of the corresponding projective IFS, also superimposed on T. Each attractor is rendered in four colours, showing how it may be seen as a union of transformations of itself.

$f_2, \ldots, f_N; p_1, p_2, \ldots, p_N\}$ *with average contractivity* $0 \le l < 1$ *can be found such that*

$$d_{\mathbb{P}}(\upsilon, \mathcal{F}(\upsilon)) \le \epsilon,$$

where $d_{\mathbb{P}}$ *denotes the Monge–Kantorovitch metric on* $\mathbb{P}(\mathbb{X})$. *Then*

$$d_{\mathbb{P}}(\upsilon, \mu) \le \frac{\epsilon}{1 - \bar{l}},$$

where μ *is the measure attractor of the IFS.*

PROOF   Hint: Sum the series $1 + l + l^2 + \cdots$. Otherwise see [9]. $\square$

The collage theorem is an expression of the general principle that the attractor of a hyperbolic IFS depends continuously on its defining parameters, such as the coefficients in an IFS code. In practice, once we have an IFS whose attractor A resembles a given target T we can adjust the parameters in the IFS code to move A closer to T. Here you might find it useful to recall our discussion, in Section 1.12, of paths of steepest descent for the Hausdorff distance between two sets.

The process of seeking a small collection of 'simple' contractive transformations with which to form a collage of a given target set can be very surprising and rewarding. When preparing to write this section, I chose the leaf shown in Figure 3.1 to define the target set T. The digital picture was photographed several years ago beside Lake Padden in Washington State. So, restricting myself to four projective transformations, I made the illustrative collages shown in Figure 4.6. Then I realized that it was more efficient to use only three leaves, in a different type of configuration; see Figure 4.7.

Figure 4.7   From left to right this illustrates a target set, a collage and the corresponding attractor. In contrast with Figure 4.6, here only three transformations are used. You can probably find an even better approximation.

Next, I was further rewarded by finding a nearby IFS of three similitudes of great simplicity, namely

$$\left\{ \mathbb{C} : \left(\frac{x}{2}, \frac{y+1}{2} \right), \left(\frac{x+y}{2}, \frac{-x+y}{2} \right), \left(\frac{x-y}{2}, \frac{x+y}{2} \right) \right\}. \quad (4.7.2)$$

This IFS is also overlapping, and it is at first sight hard to see that its attractor, shown in Figure 4.8, is the union of three similitudes applied to itself. But it is. This attractor, while being quite close to T in the Hausdorff metric, possesses a completely different topology from T. I think that it is very beautiful because of the diversity of shapes that it contains. Does it suggest a model for the way in which the veins in the leaf grew?

The same idea of making collages applies equally well to the approximation of a given target set T using an orbital set or to the approximation of a given probability measure, or a picture of one, either by using the measure attractor of an IFS with probabilities or by using an orbital measure. Variants of the collage theorem have been applied to fractal image compression, fractal interpolation and vector IFS image modelling. See for example [9], [38], [97] and references therein.

We do not know of a good analogue to the collage theorem for orbital pictures or for fractal tops. This is the case despite the fact that, in practice, perhaps somewhat intuitively and perhaps only in the case of suitable target pictures, it seems possible to manipulate orbital pictures and rendered fractal tops so that they 'look like' the target.

Notice that approaches based on the collage theorem do not, *per se*, provide a control of fractal dimension or a topology of the approximate attractor. The topology and the fractal dimension of features of attractors may vary wildly with tiny changes in parameters, unless constraints are imposed. In fractal interpolation, the approximating IFS is constrained so that its attractor is the graph of a continuous

Figure 4.8   From left to right this illustrates the set attractor, the measure attractor and the fractal top for the very simple IFS in Equation (4.7.2). You can use the collage theorem to decode this fractal. It was inspired by the Lake Padden leaf in Figure 4.7.

function; it may be further constrained so that the fractal dimension of the attractor is equal to a given value, or even so that it is piecewise differentiable.

In IFS-based approximation methods it is clearly important to constrain the approximating IFS in such a way that its attractor has the same topology as the target. For example, in an imaging application we may want to constrain the attractor to be a surface, and in a biological modelling application we may want to constrain the attractor to be tree-like or leaf-like. The fractal homeomorphism theorem, to be discussed in Section 4.14, tells exactly how this can be achieved in some cases and provides an approach in others.

In making a collage analysis of an image, look for transformations that map the image into itself without worrying about contractivity. Then try the chaos game to see whether the resulting IFS provides a model for the image. If so, it is likely that there is a metric with respect to which the IFS is contractive. In [60] it is shown how the collage theorem may be used to approximate solutions of some differential equations.

EXERCISE 4.7.2   *Find an IFS of three similitudes in $\mathbb{R}^2$ whose attractor is represented in Figure 4.8.*

EXERCISE 4.7.3   *Show that any triangle is a union of three orthogonal projections applied to itself. Explain why, even though these transformations are not strictly contractive, the chaos game will, almost always, produce a picture of the triangle.*

4.8   Deterministic calculation of attractors

Deterministic algorithms for the computation of set attractors and measure attractors are based on the strict contractivity of $\mathcal{F}$. The speed of convergence of these algorithms is determined by the contractivity factor of the IFS, as described in the following theorem.

THEOREM 4.8.1 *(i) Let $(\mathbb{X}, d)$ be a complete metric space and let $\mathcal{F}$ be a hyperbolic IFS with probabilities, contractivity factor l and average contractivity factor $\bar{l}$. Let A denote the set attractor of $\mathcal{F}$. Let $A_0 \in \mathbb{H}(\mathbb{X})$ and define recursively*

$$A_k = \mathcal{F}(A_{k-1}),$$

for $k = 1, 2, \ldots$ respectively; then

$$\lim_{k \to \infty} A_k = A. \tag{4.8.1}$$

The rate of convergence is geometrical, according to

$$d(A_k, A) \le l^k \cdot d(A_0, A) \quad \text{for all } k \in \mathbb{N}.$$

(ii) Let $(\mathbb{X}, d)$ be a complete metric space and let $\mathcal{F}$ be a hyperbolic IFS with contractivity factor l. Let μ denote the measure attractor of $\mathcal{F}$. Let $\mu_0 \in \mathbb{P}(\mathbb{X})$ and define recursively

$$\mu_k = \mathcal{F}(\mu_{k-1}),$$

for $k = 1, 2, \ldots$ respectively; then

$$\lim_{k \to \infty} \mu_k = \mu.$$

The rate of convergence is geometrical with respect to both the uniform Prokhorov metric $\widehat{d_{\mathbb{P}}}$ and the Monge–Kantorovitch metric $d_{\mathbb{P}}$ on $\mathbb{P}(\mathbb{X})$. Specifically,

$$d_{\mathbb{P}}(A_k, A) \le \bar{l}^k \cdot d_{\mathbb{P}}(A_0, A) \quad \text{and} \quad \widehat{d_{\mathbb{P}}}(A_k, A) \le l^k \cdot \widehat{d_{\mathbb{P}}}(A_0, A) \quad \text{for } k = 1, 2, \ldots$$

PROOF   This is a straightforward consequence of the way in which contraction mappings work in complete metric spaces. See for example [9], [44] or [48].
□

Deterministic calculation of set attractors

In practical applications of Theorem 4.8.1 to two-dimensional computer graphics, the transformations and the spaces upon which they act must be discretized. The precise behaviour of computed sequences of approximations A_{k+1}, starting with $A_0 \in \mathbb{H}(\mathbb{X})$, is defined by

$$A_{k+1} = \mathcal{F}(A_k) = f_1(A_k) \cup f_2(A_k) \cup \cdots \cup f_1(A_k) \quad \text{for } k = 0, 1, 2, \ldots$$

It depends on the details of the implementation and is generally quite complicated; for example, the discrete IFS may have multiple attractors; see [79], Chapter 4.

Figure 4.9   Illustration of the action of a deterministic algorithm for calculating a sequence of approximants for the attractor of an IFS. Shown, from left to right and top to bottom, are some of the sets $\mathcal{F}^{\circ n}(leaf)$ for $n = 0, 1, 2, \ldots$ The last panel shows the attractor of the IFS, rendered using IFS colouring. The IFS is the one given in Table 4.5.

An example of a deterministic sequence of approximations is shown in Figure 2.21. Another example is shown in Figure 4.9.

Now notice this. Define

$$\mathcal{H}_k(A_0) := \bigcup_{r=k}^{\infty} \mathcal{F}^{\circ k}(A_0).$$

Then $\{\overline{\mathcal{H}_k(B)}\}_{k=0}^{\infty}$ is a decreasing sequence of compact sets and so converges to a compact set. This limit is invariant under $\mathcal{F}$ and so must be the attractor A. This provides the basis for other deterministic algorithms for computing approximations

Figure 4.10   Read this picture from left to right and from top to bottom. The first panel shows the set, a leaf, and the second panel shows the union of the elements of its orbit under an IFS semigroup. The $(k+1)$th panel shows the set $\bigcup_{n\geq k}\mathcal{F}^{\circ n}$ (*leaf*), that is, the union of the elements in the 'tail' of the set orbit. See also Figure 4.9, which uses the same IFS.

to the attractor, as illustrated for example in Figure 4.10. Here it is seen that the approximants $\mathcal{H}_k(A_0)$, $k = 1, 2, \ldots$, may themselves provide interesting models for biological objects and may have their own applications in computer graphics.

Deterministic calculation of measure attractors

In practice, in two-dimensional situations it is quite difficult to compute deterministic sequences of approximations to the measure attractor μ of a hyperbolic IFS $\mathcal{F}$. The sequence of approximations, starting with μ_0, is defined by

$$\mu_{k+1} = \mathcal{F}(\mu_k) = p_1 f_1(\mu_k) + p_2 f_2(\mu_k) + \cdots + p_N f_N(\mu_k) \quad \text{for } k = 0, 1, 2, \ldots$$

The computed approximation to μ_k will necessarily be discretized and will not provide the value of μ_k on sets other than those defined by discretization; this leads to difficulties in the iterative step.

For example, suppose we know that the support of μ is contained in $\square \subset \mathbb{R}^2$ and we want to construct approximations $\{\widetilde{\mu}_k\}$ to $\{\mu_k\}$, where $\widetilde{\mu}_k$ is defined by an array of pixels $\{\mathfrak{P}^{(k)}_{w,h} : w \in \{1, 2, \ldots, W\}, h \in \{1, 2, \ldots, H\}\}$, of fixed resolution $W \times H$. The domains of the pixels $\{\square_{w,h} : w \in \{1, 2, \ldots, W\}, h \in \{1, 2, \ldots, H\}\}$, form a partition of $\square$. Then in order to compute $\mathfrak{P}^{(k)}_{w,h}$ we need to know the value of $\mu_k(f_n^{-1}(\square_{w,h}))$, which is not available except in the special case where each f_n^{-1} maps the set of boundaries of the pixel domains into itself. So, in practice, where we are constrained to a fixed resolution it is generally necessary to make further approximations over and above the discretization step.

Nonetheless, sensible sequences of approximations can be obtained in this way, as illustrated in Figure 2.23. As in the case of the set attractor, when dealing with two-dimensional hyperbolic IFSs a much easier approach is to use algorithms based on the chaos game, as was explained and justified in Section 4.5.

Deterministic calculation of pictures?

Let $\mathcal{F}$ denote a hyperbolic IFS of one-to-one transformations. Then we define $\mathcal{F} : \Pi_{\mathfrak{C}}(\mathbb{X}) \to \Pi_{\mathfrak{C}}(\mathbb{X})$ by

$$\mathcal{F}(\mathfrak{P}) = f_1(\mathfrak{P}) \uplus f_2(\mathfrak{P}) \uplus \cdots \uplus f_N(\mathfrak{P}) \quad \text{for all } \mathfrak{P} \in \Pi_{\mathfrak{C}}(\mathbb{X}). \tag{4.8.2}$$

Can we find pictures that are invariant under the IFS $\mathcal{F}$ by recursive application? In particular, does the sequence of pictures, starting from a given $\mathfrak{P}_0 \in \Pi_{\mathfrak{C}}(\mathbb{X})$, specified by

$$\mathfrak{P}_{k+1} = \mathcal{F}(\mathfrak{P}_k) = f_1(\mathfrak{P}_k) \uplus f_2(\mathfrak{P}_k) \uplus \cdots \uplus f_N(\mathfrak{P}_k) \quad \text{for } k = 0, 1, 2, \ldots,$$

converge to a single limiting picture?

The answer is: 'Generally, it does not!' But the way in which the sequence $\{\mathfrak{P}_k\}_{k=0}^{\infty}$ may fail to converge is very interesting, because it illustrates the ergodic theorem and because it inspired the discovery of fractal tops and colour-stealing.

Let us look at a two-dimensional pictorial example, illustrated in Figure 4.11. $\mathcal{F}$ is defined by the IFS code in Table 4.8. It consists of four transformations. The initial picture $\mathfrak{P}_0$ is illustrated at the top left of Figure 4.11. From left to right and from top to bottom the figure shows approximate pictures of some of the sequence $\mathfrak{P}_0, \mathfrak{P}_1, \mathfrak{P}_2, \ldots$ Let the domain of $\mathfrak{P}_k$ be denoted by A_k, and suppose that A_0 is compact. Then by Equation (4.8.1) in Theorem 4.8.1 the sequence of sets $A_{k+1} = \mathcal{F}(A_k)$ converges to the set attractor A of $\mathcal{F}$. It is apparent in Figure 4.11 that the domains of the pictures converge towards some limiting leaf-shaped set. Indeed, to the printed resolution, the domain of the last panel is probably an accurate representation of the set attractor. But what of the colours?

Table 4.8   *Projective IFS code used in Figures 4.11–4.13, as illustrated below*

n	a_n	b_n	c_n	d_n	e_n	f_n	g_n	h_n	j_n	p_n
1	0.8	0	0.1	0	0.8	0.04	0	0	1	$\frac{1}{4}$
2	0.5	0	0.25	0	0.5	0.4	0	0	1	$\frac{1}{4}$
3	3.55	−3.55	2.66	3.55	3.55	0.78	0	0	10	$\frac{1}{4}$
4	3.55	3.55	3.78	−3.55	3.5	4.34	0	0	10	$\frac{1}{4}$

Note that the initial image $\mathfrak{P}_0$ is partitioned into three subsets corresponding to the colours green, yellow and blue. Each successive computed image is made of pixels belonging to a discrete model for $\square$ and consists of green pixels, yellow pixels and blue pixels. Each pixel corresponds to a set of points in $\mathbb{R}^2$. But, for the purposes of computation, only one point corresponding to each pixel is used. When points with different colours, belonging to say $\mathfrak{P}_k$, are mapped under one of the transformations f_n of $\mathcal{F}$ to points in the same pixel in $\mathfrak{P}_{k+1}$, a choice has to be made about which colour, green, yellow or blue, to assign to the new pixel of $\mathfrak{P}_{k+1}$. In the computation of Figure 4.11 we chose to make the new pixel of $\mathfrak{P}_{k+1}$ the same colour as that of the pixel containing the last point in $\mathfrak{P}_k$, encountered in the course of running the computer program, to be mapped to the new pixel. The result is that although the sequence of pictures converges to the set attractor of the IFS the colours themselves do not settle down. See Figures 4.12 and 4.13: the successive colours of the same pixel change, iteration after iteration, in a seemingly random manner, even though the set attractor, the support of the sequence of pictures, has stabililized. This behaviour, which we call the **texture effect**, occurs in many examples at which we have looked; it occurs also in underneath pictures, as noted in Figure 3.26.

A pleasing explanation for the texture effect is provided by the following theorem, which expresses the ergodicity of the hyperbolic IFS $\mathcal{F}$.

THEOREM 4.8.2   *Suppose that μ is the unique measure attractor for the IFS $\mathcal{F}$. Suppose that $B \in \mathbb{B}(\mathbb{X})$ is such that $f_m(B) \subset B$ for all $m \in \{1, 2, \ldots, M\}$. Then $\mu(B) = 0$ or 1.*

PROOF   See [16]. The proof depends centrally on the uniqueness of the measure attractor. A variant of this theorem, weaker in the constraints on the IFS but stronger in the conditions on the set B and stated in the language of stochastic processes, is proved in [33]. We prefer the present version for its simple statement and direct measure-theoretic proof. $\square$

Figure 4.11 Some elements of a sequence of pictures $\mathcal{F}^{\circ n}$ acting on a picture $\mathfrak{P}_0$. Whereas corresponding sequences of sets and measures both converge, this sequence of pictures never settles down. See Figure 4.12. The reader will recognize a sequence of underneath pictures . . .

Theorem 4.8.2 provides a simple model explanation for the texture effect, as follows. Assume that the yellow pixels and the blue pixels both correspond to sets of points of positive measure, both invariant under $\mathcal{F}$. Then we have a contradiction to Theorem 4.8.2. So neither the yellow set nor the blue set can be invariant under $\mathcal{F}$. Hence, either one of the sets disappears – which occurs in some other examples – or the pixels must jump around. A similar argument applied to powers of $\mathcal{F}$ shows that the way in which the pixels jump around cannot be periodic and hence must be 'random'. The same argument applies whenever there is more than one colour present in the successive iterates. A more careful explanation involves numerical and statistical analysis of the specific computation.

4.9   Fractal tops

Theorem 4.8.2 tells us that it is tricky to define a picture in such a way that it is invariant under the transformation $\mathcal{F} : \Pi_{\mathfrak{C}}(\mathbb{X}) \to \Pi_{\mathfrak{C}}(\mathbb{X})$. But we would like to find a unique canonical picture that is somehow invariant under $\mathcal{F}$. It turns out that we can achieve this by taking the colour space $\mathfrak{C}$ to be code space Ω and extending the definition of the hyperbolic IFS $\mathcal{F}$ to a new operator $\mathcal{F}_{TOP}$ that acts on the colour component as well as the spatial part of a picture. The resulting unique invariant picture is called a fractal top, first mentioned in the Introduction.

In Theorem 3.3.12 we learnt that there exists a *continuous* transformation

$$\phi : \Omega_{\{1,2,...,N\}} \to A$$

Figure 4.12   The restless sequence of textures is revealed in this continuation of Figure 4.11.

Figure 4.13   Close-up on the restless textures.

from the code space $\Omega_{\{1,2,\ldots,N\}}$ *onto* the set attractor A of the hyperbolic IFS $\mathcal{F} = \{\mathbb{X}; f_1, f_2, \ldots, f_N\}$. This transformation is defined by

$$\phi(\sigma) = \lim_{n \to \infty} f_{\sigma_1 \sigma_2 \cdots \sigma_n}(x) \quad \text{for } \sigma = \sigma_1 \sigma_2 \cdots \in \Omega_{\{1,2,\ldots,N\}},$$

for any $x \in \mathbb{X}$, where we recall that

$$f_{\sigma_1 \sigma_2 \cdots \sigma_n}(x) := f_{\sigma_1} \circ f_{\sigma_2} \circ \cdots \circ f_{\sigma_n}(x).$$

Note that $\phi : \Omega_{\{1,2,\ldots,N\}} \to A$ interacts with the shift transformation $S : \Omega_{\{1,2,\ldots,N\}} \to \Omega_{\{1,2,\ldots,N\}}$ according to the expression

$$\phi(\sigma) = f_{\sigma_1}(\phi(S\sigma)) \quad \text{for all } \sigma \in \Omega_{\{1,2,\ldots,N\}}.$$

The code space $\Omega = \Omega_{\{1,2,\ldots,N\}}$ is an ordered space when equipped with the relation $<$ defined by

$$\sigma < \omega \quad \text{iff} \quad \sigma_k > \omega_k,$$

where k is the least index for which $\sigma_k \neq \omega_k$. Notice that all elements of Ω are less than or equal to $\overline{1} = 11111\cdots$ and greater than or equal to $\overline{N} = NNNNN\cdots$. Also, any pair of distinct elements of Ω is such that one member of the pair is strictly greater than the other.

Now notice that the set of addresses of a point $x \in A$, defined to be $\phi^{-1}(x)$, is closed. The reason is that A is closed, so $A\backslash\{x\}$ is open and $\phi^{-1}(A\backslash\{x\}) = \phi^{-1}(A)\backslash\phi^{-1}(\{x\}) = \Omega\backslash\phi^{-1}(x)$ is also open. It follows that $\phi^{-1}(x)$ is both bounded above and closed, so it must possess a unique largest element. We denote this element by $\tau(x)$.

DEFINITION 4.9.1　Let $\mathcal{F}$ be a hyperbolic IFS with set attractor A and code space function $\phi : \Omega \to A$. Then the **tops function** of $\mathcal{F}$ is $\tau : A \to \Omega$, defined by

$$\tau(x) = \max\{\sigma \in \Omega : \phi(\sigma) = x\}.$$

The set of points $G_\tau := \{(x, \tau(x)) : x \in A\}$ is called the **graph of the top of the IFS** or simply the (fractal) **top** of the IFS.

The top of an IFS may be described as follows: consider the **lifted** IFS, which we define to be

$$\widehat{\mathcal{F}} = \{\mathbb{X} \times \Omega : \widehat{f}_1, \widehat{f}_2, \ldots, \widehat{f}_N\},$$

where

$$\widehat{f}_n(x, \sigma) = (f_n(x), s_n(\sigma))$$

and $s_n(\sigma) := n\sigma := \omega$, with $\omega_1 = n$ and $\omega_{n+1} = \sigma_n$ for $n = 1, 2, \ldots$ Then $\widehat{\mathcal{F}}$ is a hyperbolic IFS with respect to the metric

$$d_{\mathbb{X}\times\Omega}((x, \sigma), (y, \theta)) := d_{\mathbb{X}}(x, y) + d_\Omega(\sigma, \theta)$$

for all pairs of points $(x, \sigma), (y, \theta) \in \mathbb{X} \times \Omega$, where d_Ω is the code space metric defined in Equation (1.6.1). Notice that $d_\Omega(s_n(\sigma), s_n(\theta)) \leq \frac{1}{2}d_\Omega(\sigma, \theta)$ for each $n \in \{1, 2, \ldots, N\}$, and so it follows that

$$d_{\mathbb{X}\times\Omega}(\widehat{f}_n(x, \sigma), \widehat{f}_n(y, \theta)) \leq \max\left\{\tfrac{1}{2}, l\right\} d_{\mathbb{X}\times\Omega}((x, \sigma), (y, \theta)),$$

where l is a contractivity factor for $\mathcal{F}$.

Let $\widehat{A}$ denote the set attractor of $\widehat{\mathcal{F}}$. Then the projections of $\widehat{A}$ onto $\mathbb{X}$ and Ω are A and Ω respectively. The top of $\mathcal{F}$ is related to $\widehat{A}$ according to

$$G_\tau = \{(x, \sigma) \in \widehat{A} : (x, \omega) \in \widehat{A} \implies \omega \leq \sigma\}.$$

This formulation is useful because it makes it obvious how we can use the chaos game, applied to $\widehat{\mathcal{F}}$, to approximate G_τ and hence τ. We simply keep track of the 'highest' values encountered along random orbits. In practice, in the two-dimensional case, we assign a set of pixels to a discretized approximation to the attractor A of $\mathcal{F}$. The value of each pixel is initialized to $\overline{N}$, the lowest possible value. Then, starting at a point of the form $X_0 = (x_0, \overline{N}) \in A \times \Omega$, we simply follow a random orbit $X_1, X_2, \ldots$, namely

$$X_{k+1} = (x_{k+1}, \sigma_{k+1}\sigma_k \cdots \sigma_1 \overline{N}) = \widehat{f}_{\sigma_{k+1}}(X_k) \quad \text{for } k = 0, 1, 2, \ldots$$

generated by the chaos game. At the $(k+1)$th step, the value of the pixel in which the point x_{k+1} lies is updated to become $\sigma_{k+1}\sigma_k \cdots \sigma_1 \overline{N}$ if the latter is greater than the current value of the pixel. The values of the pixels are kept truncated to a fixed length.

Many variations on this algorithm are feasible. Another method for computing approximations to τ follows the deterministic orbits of an associated dynamical system, called the tops dynamical system, as described in Corollary 4.11.4.

As an example of a fractal top and its associated structures, consider the IFS

$$\mathcal{F}_1 = \{[0, 1] \subset \mathbb{R}; f_1(x) = \alpha x + (1 - \alpha), f_2(x) = \alpha x\}, \tag{4.9.1}$$

where

$$\tfrac{1}{2} < \alpha < 1.$$

The attractor of $\mathcal{F}_1$ is the closed real interval $[0, 1]$. It follows that $\mathcal{F}_1$ is overlapping because $f_1([0, 1]) \cap f_2([0, 1])$ contains a nonempty open set. The lifted IFS associated with $\mathcal{F}_1$ is

$$\widehat{\mathcal{F}}_1 = \big\{[0, 1] \times \Omega_{\{1,2\}} \subset \mathbb{R}; \widehat{f}_1(x, \sigma) = (\alpha x + (1 - \alpha), 1\sigma), \widehat{f}_2(x, \sigma) = (\alpha x, 2\sigma)\big\}. \tag{4.9.2}$$

In order to compute a picture of the attractor $\widehat{A}_1$ of $\widehat{\mathcal{F}}_1$ we embed the code space $\Omega_{\{1,2\}}$ in $[0, 1]$ by identifying it with the Cantor set generated by the IFS

$$\{[0, 1]; w_1(y) = 0.49999y + 0.50001, \quad w_2(y) = 0.49999y\}.$$

Accordingly, in Figure 4.14 we illustrate the attractor $\widetilde{A}_1$ of the IFS

$$\widetilde{\mathcal{F}}_1 = \{[0, 1] \times [0, 1] \subset \mathbb{R}^2; (\alpha x, w_1(y)), (\alpha x + (1 - \alpha), w_2(y))\},$$

for the case $\alpha = \tfrac{2}{3}$. This was computed using the chaos game. The top of $\mathcal{F}_1$ corresponds to the literal top, shown in red, of the set $\widetilde{A}_1$ plotted in Figure 4.14.

EXERCISE 4.9.2 *It is sometimes convenient to think of the real interval $[0, 1]$, written in binary notation, as a representation of code space. This does not lead to*

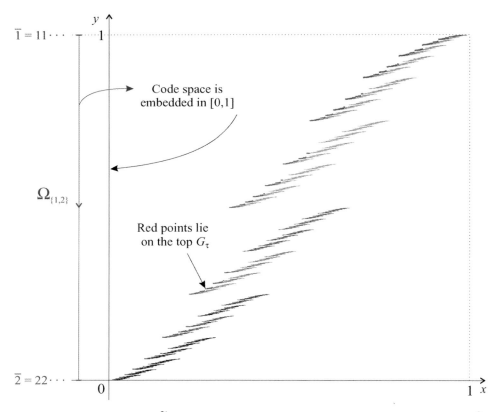

Figure 4.14 The attractor $\widetilde{A}_1$ of the IFS in Equation (4.9.2). This attractor is related to the attractor $\widehat{A}_1$ of the lifted IFS corresponding to Equation (4.9.1) by the embedding of the code space $\Omega_{\{1,2\}}$ in the interval [0, 1] on the y-axis, as described in the text. The top of the IFS is indicated in red.

problems except possibly at those points on [0, 1] *that have two representations. Explain how we have avoided such problems in the example based on Equation (4.9.1).*

Now we mention the following result, because it is in a similar vein to the one above. It spells out how each measure attractor of $\mathcal{F}$ is the transformation under $\phi : \Omega \to A$ of a canonical measure on code space.

THEOREM 4.9.3   *Let $\mu \in \mathbb{P}(\mathbb{X})$ denote the measure attractor of the hyperbolic IFS $\mathcal{F} = \{\mathbb{X}; f_1, f_2, \ldots, f_N; p_1, p_2, \ldots, p_N\}$ and let $\mu_\Omega \in \mathbb{P}(\Omega)$ denote the measure attractor of the IFS $\mathcal{S} = \{\Omega; s_1, s_2, \ldots, s_N; p_1, p_2, \ldots, p_N\}$, where $s_n : \Omega \to \Omega$ is the branch transformation defined by $s_n(\sigma) = n\sigma$ for all $\sigma \in \Omega$, and let $\widehat{\mu} \in \mathbb{P}(\mathbb{X} \times \Omega)$ denote the measure attractor of the hyperbolic IFS $\widehat{\mathcal{F}} = \{\mathbb{X} \times \Omega : \widehat{f}_1, \widehat{f}_2 \ldots, \widehat{f}_N; p_1, p_2, \ldots, p_N\}$, where $\widehat{f}_n = (f_n, s_n)$, for $n = 1, 2, \ldots, N$. Then* *the projections of $\widehat{\mu} \in \mathbb{P}(\mathbb{X} \times \Omega)$ onto $\mathbb{P}(\mathbb{X})$ and $\mathbb{P}(\Omega)$ are μ and μ_Ω respectively. Moreover,*

$$\mu = \phi(\mu_\Omega).$$

PROOF   This is straightforward and is based on the uniqueness of the three measure attractors and the continuity of ϕ. See for example [48], Theorem 4.4(3), (4).   □

4.10   Pictures of tops: colour-stealing

In later sections we will see that the mathematics of fractal tops is very deep. But we have, in some ways, hurried to get to this section, where we shall explain how to make beautiful pictures of fractal tops using colour-stealing. The colour-stealing method has potential applications in computer graphics, image processing, image watermarking and image compression.

Here we are concerned with picture functions of the form $\mathfrak{P} : D_{\mathfrak{P}} \subset \mathbb{R}^2 \to \mathfrak{C}$, where $\mathfrak{C}$ is a colour space; for example, $\mathfrak{C} = [0, 255]^3 \subset \mathbb{R}^3$. For colour-stealing applications we may choose

$$D_{\mathfrak{P}} = \square := \{(x, y) \in \mathbb{R}^2 : 0 \le x, y \le 1\}.$$

Let two hyperbolic IFSs

$$\mathcal{F}_D := \{\square; f_1, f_2, \dots, f_N\} \quad \text{and} \quad \mathcal{F}_C := \{\square; \widetilde{f}_1, \widetilde{f}_2, \dots, \widetilde{f}_N\}$$

and a picture function

$$\mathfrak{P}_C : \square \to \mathfrak{C}$$

be given. The index 'D' stands for 'drawing' and the index 'C' stands for 'colouring'. Let A_D denote the attractor of $\mathcal{F}_D$ and let A_C denote the attractor of $\mathcal{F}_C$. Let

$$\tau_D : A_D \to \Omega$$

denote the tops function for $\mathcal{F}_D$ and

$$\phi_C : \Omega \to A_C \subset \square$$

denote the addressing function for $\mathcal{F}_C$. Then we define a new picture

$$\mathfrak{P}_D : A_D \to \mathfrak{C}$$

by

$$\mathfrak{P}_D = \mathfrak{P}_C \circ \phi_C \circ \tau_D.$$

The picture $\mathfrak{P}_D$ is uniquely defined by $\mathcal{F}_D$, $\mathcal{F}_C$ and the picture $\mathfrak{P}_C$. We say that $\mathfrak{P}_D$ has been produced by **tops plus colour-stealing** and we sometimes call it the

stolen picture. We think in this way: first we 'steal' colours from the picture $\mathfrak{P}_C$, to 'paint' code space, that is, we make a code space picture, $\mathfrak{P}_C \circ \phi_C : \Omega \to \mathfrak{C}$; then we combine the latter with the top of $\mathcal{F}_D$ to 'paint' the attractor A_D.

The stolen picture $\mathfrak{P}_C \circ \phi_C \circ \tau_D$ may be computed by algorithms based on a variant of the chaos game, in which the lifted IFS associated with $\mathcal{F}_D$ is coupled to the IFS $\mathcal{F}_C$. This is the method used to compute many of the rendered pictures of fractal tops in this book and is described in more detail in [15] and [18]. Briefly, we play two chaos games simultaneously, one using $\mathcal{F}_C$ in which the current point x_k^C dances around on the picture $\mathfrak{P}_C$ and the other, using $\mathcal{F}_D$, in which the current point x_k^D dances around on □, so to speak 'drawing' on a blank sheet of paper while keeping track of the code space index, say $\sigma_k \sigma_{k-1} \cdots \sigma_1 \overline{N}$. At the $(k+1)$th roll of the die, the same selected map index σ_{k+1} is used to fix the transformation to apply in each game; a point x_{k+1}^D which either has not been coloured or has already been coloured, and for which the new code space value $\sigma_{k+1} \sigma_k \sigma_{k-1} \cdots \sigma_1 \overline{N}$ is greater than all previous code space values associated with that location, is assigned the colour $\mathfrak{P}_C(x_k^C)$.

For example, for Figure 4.15, $\mathcal{F}_D$ was the Möbius IFS, consisting of four transformations obtained by composing the Möbius IFS represented in Table 4.1 with itself. The colouring IFS was

$$\mathcal{F}_C = \left\{ \square; \left(\tfrac{1}{2}x, \tfrac{1}{2}y\right), \left(\tfrac{1}{2}(x+1), \tfrac{1}{2}y\right), \left(\tfrac{1}{2}x, \tfrac{1}{2}(y+1)\right), \left(\tfrac{1}{2}(x+1), \tfrac{1}{2}(y+1)\right) \right\}.$$

$$(4.10.1)$$

In each case the picture $\mathfrak{P}_C$ was resized to make its domain equal to □. Compare the two rendered fractal tops here with the set attractor and the measure attractor in Figure 4.3. Only the upper portion of the picture $\mathfrak{P}_C$, used to make the right-hand image in Figure 4.15, is shown. But by using a digital camera and by reversing the roles of $\mathcal{F}_C$ and $\mathcal{F}_D$ you can approximate the missing portion.

Notice the following facts. (i) Picture functions have properties that are determined by their sources: digital pictures of natural scenes such as clouds and sky, fields of flowers and grasses, seascapes, thick foliage etc. have their own distinctive palettes, relationships between colour and position, apparent continuities and discontinuities and so on. (ii) Addressing functions are continuous. (iii) Tops functions have their own special properties, as we will discuss more fully in the following sections. For example, a tops function is 'nearly' continuous, with discontinuities that recapitulate hidden geometrical structures in the underlying set attractor. Thus the pictures produced by tops plus colour-stealing may carry a natural palette and so possess continuity and discontinuity displayed in harmonious and beautiful ways.

These facts are demonstrated in Figure 4.16. Here $\mathcal{F}_D$ is the projective IFS given in Table 0.1 in the Introduction, and the colouring IFS $\mathcal{F}_C$ corresponds to a partition

Figure 4.15   Two pictures of fractal tops produced by colour-stealing. The colours were 'stolen' from the pictures shown inset. Although these colour transformations are not continuous, they 'nearly' are, which accounts for the beautiful groupings of colours.

of $\square$ into four rectangles whose areas are proportional to the probabilities in Table 0.1. The picture from which the colours were stolen is shown on the right. The whole rendered top, which looks like a leafy tree, as well as a zoom into the region of a tiny hole in the picture, is shown. Notice how the colours tend to be grouped in much the same way as they are in the picture $\mathfrak{P}_C$, yet none of the structure of $\mathfrak{P}_C$ is preserved. Notice the intricate geometry of the fractal top, revealed in the colouring as well as in the boundaries of its domain. In Figure 4.17 we show, in the centre and right-hand panels, two other colourings for part of the region shown, zoomed, at the left of Figure 4.16. Again, see how different aspects of the top are bought into view by different choices for $\mathfrak{P}_C$. The panel on the left in Figure 4.17 shows the fractal top partially rendered; this was obtained by stopping the coupled chaos game algorithm before it had run sufficiently long for the picture to stop changing. Another rendering of the same picture is shown in Figure 4.18, which includes an additional zoom and reveals more, potentially endless, diverse beautiful detail. In each case the picture $\mathfrak{P}_C$ was a digital photograph of a real-world scene.

Figure 4.16 A top rendered by colour-stealing (middle image). A zoom is shown on the left. The picture from which the colours were stolen is on the right. A tiny hole in the full image is revealed to have fascinating detail, biological, mysterious.

Figure 4.17 Three renderings of the zoom on the fractal top in Figure 4.16 (also shown in Figure 0.2), computed using a version of the chaos game and colour-stealing. On the left, the computation has been stopped early to produce the effect of mist. Different aspects of the fractal top are revealed by different choices of the picture from which the colours are stolen.

In Figure 4.19 various renderings of the attractor A of a linked projective IFS are shown. A **linked IFS** is one in which there exist various relations of the form $f_i(P_i) = f_j(P_j)$ between pairs of points $P_i, P_j \in A$, with $i, j \in \{1, 2, \ldots, N\}$, $i \neq j$. Examples of linked IFSs occur often in connection with biological modelling applications; for example, the fractal fern is produced by a linked IFS. They also occur in fractal interpolation and in the construction of space-filling curves (a space-filling curve is an infinitely long curve that winds here and there in a space, eventually visiting every point in the space.). The geometrical properties of projective transformations make them particularly suitable for designing interesting linked IFSs.

Figure 4.20 illustrates a 'nearly' continuous invertible transformation between two pictures, at upper left and upper right, achieved by colour-stealing. Clearly, though, this transformation is *not* a homeomorphism, in contrast with the situation

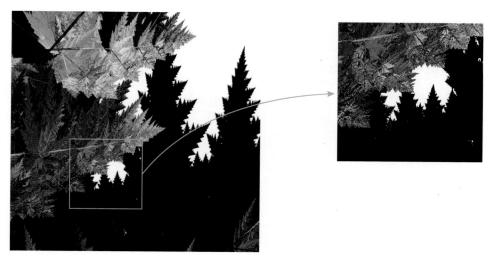

Figure 4.18 Another colouring photograph $\mathfrak{P}_C$ has been chosen to render this version of the same fractal top as that shown in Figures 4.16 and 4.17. This picture includes an additional zoom, revealing more of the 'nearly' continuous transformation of $\mathfrak{P}_C$ and showing something of the intricacy of this single fractal top. The fractal boundaries of the holes may be seen.

discussed in Section 4.15. Here $\mathcal{F}_D$ is the projective IFS defined in Table 4.3 and $\mathcal{F}_C$ is given in Equation (4.10.1). The two zooms at the bottom of Figure 4.20 illustrate how the natural clutter of foliage seems more unchanged by the transformation than does the man-made car and the face of the little girl.

In Figure 4.21 two different renderings of the same portion of the same fractal top are shown. The only difference between the two pictures is the picture $\mathfrak{P}_C$ from which the colours were stolen. If you compare the two pictures carefully you will see that there is an underlying geometrical similarity between them. Perhaps they were painted by the same artist? A different example of the same kind is shown in Figure 4.22. You may be struck by the elegant styles of these pictures and their balance and harmony. These effects are typical of the many instances we have observed. Indeed, this 'beauty effect' is very pronounced when you put $\mathfrak{P}_C$ as a window into a larger picture, which is shifted continuously in time. Then the stolen picture changes continuously and beautifully, shifting and shimmering, shapes appearing and going away, like patterns in smoke, maintaining the underlying geometry in a very subtle way. The results are breathtaking. Try it!

Colour-stealing has recently been applied in computer graphics; see [71].

EXERCISE 4.10.1   *What happens when you carry out colour-stealing and two of the transformations in the IFS $\mathcal{F}_D$ are the same? What happens when two of the transformations in $\mathcal{F}_C$ are the same?*

Figure 4.19 Four different ways of rendering the attractor of a projective IFS are shown here for comparison. The figure shows the set attractor (top left), the measure attractor (bottom left), the set attractor rendered by IFS colouring (bottom right) and the set attractor rendered by colour-stealing (top right). Colours for the latter were stolen from the top left picture in Figure 4.20. But you can treat all these pictures as having been produced by tops plus colour stealing!

EXERCISE 4.10.2   *Look again at Figure 4.4. These two pictures were actually produced by IFS colouring, as described in Section 4.6. But either could equally well have been obtained from the other via fractal tops plus colour stealing. Which two IFSs would you use to achieve this?*

4.11   The tops dynamical system

Here we assume that the transformations of $\mathcal{F}$ are invertible. Then we show how the tops function provides a natural dynamical system $T : A \rightarrow A$. It is related to another dynamical system $\widehat{T} : G_\tau \rightarrow G_\tau$ and to a shift-invariant subspace Ω_τ of code space. The notation is the same as above.

Figure 4.20   The relationship between the top two pictures here is the same as that between the two pictures in Figure 4.4, obtained by IFS colouring. Why is this? The zooms at the bottom illustrate differing visual effects on the background clutter, the man-made objects and the human face.

LEMMA 4.11.1   *Let $(x, \sigma) \in G_\tau$. Then $(f_{\sigma_1}^{-1}(x), S\sigma) \in G_\tau$.*

PROOF   Notice that $(f_{\sigma_1}^{-1}(x), S\sigma)$ is uniquely defined and belongs to $\mathbb{X} \times \Omega$ because $f_n : \mathbb{X} \to \mathbb{X}$ is one-to-one and $S : \Omega \to \Omega$. Suppose that $(f_{\sigma_1}^{-1}(x), S\sigma) \notin G_\tau$. Then there exists a point $(f_{\sigma_1}^{-1}(x), \omega) \in \widehat{A}$ where $\omega > S\sigma$. But since $\widehat{A} = \bigcup_n \widehat{f}_n(\widehat{A})$ it follows that $\widehat{f}_{\sigma_1}(f_{\sigma_1}^{-1}(x), \omega) \in \widehat{A}$. However, $\widehat{f}_{\sigma_1}(f_{\sigma_1}^{-1}(x), \omega) = (x, \sigma_1\omega)$ and $\sigma_1\omega > \sigma$, which is a contradiction.   □

LEMMA 4.11.2   *Let $(x, \sigma) \in G_\tau$. Then there exists $(y, \omega) \in G_\tau$ such that $(f_{\sigma_1}^{-1}(y), S\omega) = (x, \sigma) \in G_\tau$.*

PROOF   Let $(y, \omega) = (f_1(x), 1\sigma)$. Clearly $(f_{\sigma_1}^{-1}(y), S(1\sigma)) = (x, \sigma)$ and $(f_1(x), 1\sigma) \in \widehat{A}$. We just have to show that $(f_1(x), 1\sigma) \in G_\tau$. Suppose not. Then

Figure 4.21 This shows two different renderings of the same portion of the same fractal top. The only difference between the two pictures is the picture $\mathfrak{P}_C$ from which the colours were stolen. There is an underlying geometrical similarity between them. See also Figure 4.22.

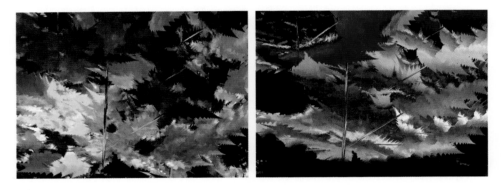

Figure 4.22 See also Figure 4.21. Here too a portion of a fractal top is rendered using the same colouring IFS, $\mathcal{F}_C$, but different pictures $\mathfrak{P}_C$. Breathtaking animated textures can be produced by smoothly varying $\mathfrak{P}_C$, as in a continuous video sequence.

there exists a point $(f_1(x), \upsilon) \in \widehat{A}$ with $\upsilon > 1\sigma$. It follows that $\upsilon_1 = 1$ and thus that $S\upsilon > \sigma$. Also, since $(f_1(x), \upsilon) \in \widehat{A}$ we have $\widehat{f}_1^{-1}(f_1(x), \upsilon) \in \widehat{A}$ and hence $(x, S\upsilon) \in \widehat{A}$. It follows that $(x, \sigma) \notin G_\tau$, which is a contradiction. $\square$

We may conclude from these two lemmas that the mapping

$$\widehat{T} : G_\tau \to G_\tau \quad \text{defined by} \quad \widehat{T}(x, \sigma) = \left(w_{\sigma_1}^{-1}(x), S\sigma \right)$$

is well defined and onto. It can be treated as a dynamical system, which we refer to as $\{G_\tau, \widehat{T}\}$. As such we may explore its invariant sets, invariant measures, other types of invariants such as entropies and information dimensions and its ergodic properties, using 'standard' terminology and machinery, as mentioned in the discussion of the symbolic invariants of orbital pictures in Section 3.5.

We can project $\{G_\tau, \widehat{T}\}$ onto the Ω-direction, as follows. Let

$$\Omega_\tau := \{\sigma \in \Omega : (x, \sigma) \in G_\tau \text{ for some } x \in \mathbb{X}\}.$$

Then Ω_τ is a shift-invariant subspace of Ω, that is, $S : \Omega_\tau \to \Omega_\tau$ with

$$S(\Omega_\tau) = \Omega_\tau,$$

and we see that $\{\Omega_\tau, S\}$ is a **symbolic dynamical system**; see for example [77]. We call Ω_τ the **tops code space**. (We will use the notation $\Omega_{\mathcal{F}}$ to denote the tops code space of the IFS $\mathcal{F}$.) Indeed, $\{\Omega_\tau, S\}$ is the symbolic dynamical system corresponding to a partition of the domain of yet a third dynamical system $\{A, T\}$ corresponding to a mapping $T : A \to A$ that is obtained by projecting $\{G_\tau, \widehat{T}\}$ onto A. This system is defined by

$$T(x) = \begin{cases} f_1^{-1}(x) & \text{if } x \in D_1 := f_1(A), \\ f_2^{-1}(x) & \text{if } x \in D_2 := f_2(A) \backslash f_1(A), \\ \quad \vdots & \quad \vdots \qquad \qquad \vdots \\ f_N^{-1}(x) & \text{if } x \in D_N := f_N(A) \backslash \bigcup_{n=1}^{N-1} f_n(A), \end{cases} \qquad (4.11.1)$$

for all $x \in A$, and we have

$$T(A) = A.$$

We call $\{A, T\}$ the **tops dynamical system** associated with the IFS. $\{\Omega_\tau, S\}$ is the symbolic dynamical system obtained by starting from the tops dynamical system $\{A, T\}$ and partitioning A into the disjoint sets $D_0, D_1, \ldots, D_{N-1}$ defined in Equation (4.11.1), where

$$A = \bigcup_{n=0}^{N-1} D_n \quad \text{and} \quad D_i \cap D_j = \varnothing \text{ for } i \neq j.$$

An example of such a partition is illustrated in Figure 4.23. This illustrates the domains D_1, D_2, D_3, D_4 for the tops dynamical system associated with the IFS given in the Introduction in Table 0.1. This was the IFS used in Figures 4.16–4.18.

If the domains $\{D_n : n = 1, 2, \ldots, N\}$ are known then it is easy to compute the tops function. Just follow the orbit of x under the tops dynamical system and keep track of the sequence of indices $\sigma_1 \sigma_2 \sigma_3 \cdots$ visited by the orbit! This is stated formally in Corollary 4.11.4 below.

THEOREM 4.11.3   *The tops dynamical system $\{A, T\}$ and the symbolic dynamical system $\{\Omega_\tau, S\}$ are conjugate. The identification between them is provided by the tops function $\tau : A \to \Omega_y$. That is,*

$$T(x) = \phi \circ S \circ \tau(x)$$

Figure 4.23 Illustrates the domains D_1, D_2, D_3, D_4 for the tops dynamical system associated with the IFS given used to make Figure 4.16. Once this 'picture' has been computed it is easy then to compute the tops function. Just follow the orbits of the tops dynamical system!

for all $x \in A$. Moreover, $\xi \in \mathbb{P}(\mathbb{X})$ is an invariant probability measure for $\{A, T\}$ iff $\tau(\xi) \in \mathbb{P}(\Omega)$ is an invariant probability measure for $\{\Omega_\tau, S\}$.

PROOF   This follows directly from everything we have said above. But do not confuse an invariant measure $\xi \in \mathbb{P}(\mathbb{X})$ of T, which obeys $\xi = T(\xi)$, with the unique measure attractor $\mu \in \mathbb{P}(\mathbb{X})$ of $\mathcal{F}$, which obeys $\mu = \mathcal{F}(\mu)$. Compare with Theorem 4.9.3. $\square$

In the special case where the IFS is totally disconnected and the f_n are one-to-one then $T : A \to A$ is defined by $T(x) = f_n^{-1}(x)$, where n is the unique index such that $x \in f_n(A)$. This dynamical system has been considered elsewhere, for example in [14], [4] and [58]. It is interesting because in this case $\phi : \Omega \to A$ is a homeomorphism and T is conjugate to the shift transformation on code space, according to

$$T = \phi \circ S \circ \phi^{-1}.$$

We see that in this case $\tau = \phi^{-1}$. In this special case the topological entropy of T is $\log_2 N$.

COROLLARY 4.11.4   *The value of $\tau(x)$ may be computed by following the orbit of $x \in A$ as follows. Let $x_1 = x$ and $x_{k+1} = T(x_k)$ for $k = 1, 2, \ldots$, so that the orbit of x is $\{x_k\}_{k=1}^{\infty}$. Then $\tau(x) = \sigma$ where $\sigma_k \in \{0, 1, \ldots, N-1\}$ is the unique index such that $x_k \in D_k$ for all $k = 1, 2, \ldots$*

PROOF   This follows directly from Theorem 4.11.3. At the risk of being repetitive, notice the following points. (i) The IFS $\widehat{\mathcal{F}}$ has totally disconnected attractor $\widehat{A}$. Hence there is a well-defined shift dynamical system $\widehat{T} : \widehat{A} \to \widehat{A}$, which is equivalent to $S : \Omega \to \Omega$. $\widehat{A}$ is the disjoint union of the sets $\widehat{f}_n(\widehat{A})$, and the code space address of each point $x \in \widehat{A}$ can be obtained by following its orbit under $\widehat{T}$ and concatenating the indices of the successive regions $\widehat{f}_n(A)$ in which it lands. The corollary follows because $\widehat{\widehat{T}} = \widehat{T}|_{G\tau}$. (ii) We may follow directly the orbit of x under T, keeping track of the indices as described and then use the strict contractivity of the maps to show that $\lim\limits_{n \to \infty} f_{\sigma_1} \circ f_{\sigma_2} \circ \cdots \circ f_{\sigma_n}(y) = x$ for all $y \in A$. The key observation is that the orbit of a point on the top stays on the top, and so the code given by its orbit is the value of the tops function applied to the point.   □

The following theorem explains precisely what we mean when we say that tops functions are 'nearly' continuous.

THEOREM 4.11.5   *Let $\mathcal{F} = \{\mathbb{X}; f_1, f_2, \ldots, f_N\}$ be a hyperbolic IFS, where $f_n : A \to D_n$ is a homeomorphism, for $n = 1, 2, \ldots, N$. Let*

$$A^{inside} = A \setminus \bigcup_{k=0}^{\infty} T^{\circ(-k)} \left(\bigcup_{n=1}^{N} \partial D_n \right),$$

where ∂D_n denotes the boundary of D_n, as defined in Equation (4.11.1). Then the restricted tops dynamical system $T|_{A^{inside}} : A^{inside} \to A^{inside}$ is continuous. Also, the restricted tops function $\tau|_{A^{inside}} : A^{inside} \to \Omega$ is continuous.

Here and elsewhere we use the notation $T^{\circ(-k)}(S)$ (i.e. T^{-1} composed with itself k times) to mean the set $\{x \in A : T^{\circ k}(x) \in S\}$.

PROOF   Since the function $f_n^{-1}(x)$ is continuous on the interior of D_n, T must be continuous on $A \setminus \bigcup_{n=1}^{N} \partial D_n$. It follows *a fortiori* that T is continuous on A^{inside}. It is readily verified that $T(A^{inside}) = A^{inside}$, which proves the first assertion.

From the continuity of T on $A \setminus \bigcup_{n=1}^{N} \partial D_n$, $T^{\circ k}$ must be continuous on $A \setminus T^{\circ(-k)}(\bigcup_{n=1}^{N} \partial D_n)$ for $k = 0, 1, 2, \ldots$ It follows that $T^{\circ l}$ is continuous on A^{inside} for $l = 0, 1, 2, \ldots$

Let $x \in A^{inside}$ and let $\varepsilon > 0$. Choose M so that $2^{-M} < \varepsilon$. Let $\tau(x) = \sigma$. Then, since each of the functions $T^{\circ k}(x)$ is continuous on A^{inside}, we can find $\delta > 0$ such that $T^{\circ k}(y)$ belongs to the interior of D_{σ_k} for $k = 0, 1, \ldots, M$ whenever $d(x, y) < \delta$. It follows from Corollary 4.11.4 that $\tau(y)$ agrees with $\tau(x)$

through the first M symbols, that is, $(\tau(y))_k = (\tau(x))_k = \sigma_k$ for $k = 0, 1, \ldots, M$ whenever $d(x, y) < \delta$. It follows that $d_\Omega(\tau(x), \tau(y)) < 2^{-M} < \varepsilon$ whenever $d(x, y) < \delta$. $\qquad\square$

EXERCISE 4.11.6 *Show that the tops dynamical system for the IFS $\mathcal{F}_1$ in Equation (4.9.1) is correctly defined by Equation (4.14.2) and correctly illustrated in Figure 4.27. (This dynamical system is closely related to the one studied by A. Renyi [82] in connection with information theory. The IFS $\mathcal{F}_1$ also appears in the context of Bernoulli convolutions; quite recently, a longstanding problem regarding the absolute continuity or otherwise of the measure attractor of $\mathcal{F}_1$ was solved by Boris Solomyak; see [89].)*

EXERCISE 4.11.7 *Look up 'Markov partitions' on the internet. Important research publications in this general area have been written by David Ruelle, Y. G. Sinai, Rufus Bowen, Caroline Series, Brian Marcus and many others. What are the objectives of research in symbolic dynamics?*

4.12 The fractal top is the fixed point of $\mathcal{F}_{TOP}$

Next we show that $\tau(x)$ is the unique fixed point of a contractive transformation, closely related to $\mathcal{F} : \Pi_\Omega(\mathbb{X}) \to \Pi_\Omega(\mathbb{X})$. Here we take $\Omega = \Omega_{\{1,2,\ldots,N\}} \cup \Omega'_{\{1,2,\ldots,N\}}$. We recall, from Section 1.6, that (Ω, d_Ω) is a compact metric space.

We define $\mathcal{F}_{TOP} : \Pi_\Omega(\mathbb{X}) \to \Pi_\Omega(\mathbb{X})$ as follows. Let $\mathfrak{P} \in \Pi_\Omega(\mathbb{X})$, and let $D_\mathfrak{P}$ denote the domain of $\mathfrak{P}$. Then we define the set of disjoint sets by

$$D_1 = f_1(D_\mathfrak{P}),$$
$$D_2 = f_2(D_\mathfrak{P})\backslash D_1,$$
$$D_3 = f_3(D_\mathfrak{P})\backslash(D_1 \cup D_2),$$
$$\vdots$$
$$D_N = f_N(D_\mathfrak{P})\backslash(D_1 \cup D_2 \cup \cdots \cup D_{N-1}).$$

Let the domain of $\mathcal{F}_{TOP}(\mathfrak{P})$ be $D_1 \cup D_2 \cup \cdots \cup D_N$, which is the same as the domain of $\mathcal{F}(\mathfrak{P})$. Then we define

$$(\mathcal{F}_{TOP}(\mathfrak{P}))(x) = n\mathfrak{P}\big(f_n^{-1}(x)\big) \quad \text{when } x \in D_n. \tag{4.12.1}$$

Let $\Gamma_\Omega(A) \subset \Pi_\Omega(A) \subset \Pi_\Omega(\mathbb{X})$ denote the set of functions with domain A and range in Ω. Then it is easy to see that $\mathcal{F}_{TOP}(\Gamma_\Omega(A)) \subset \Gamma_\Omega(A)$. Also, you can readily check that $(\Gamma_\Omega(A), d_{sup})$ is a complete metric space with respect to the distance function

$$d_{sup}(\mathfrak{P}_1, \mathfrak{P}_2) = \sup_{x \in A} d_\Omega(\mathfrak{P}_1(x), \mathfrak{P}_2(x)).$$

THEOREM 4.12.1   *Let $\mathcal{F}$ be a hyperbolic IFS of invertible transformations, and let $\mathcal{F}_{TOP} : \Gamma_\Omega(A) \to \Gamma_\Omega(A)$ denote the restriction of $\mathcal{F}_{TOP}$, as defined above, to $\Gamma_\Omega(A)$. Then $\mathcal{F}_{TOP}$ is a contraction mapping, with contractivity factor 0.5, on the complete metric space $(\Gamma_\Omega(A), d_{sup})$. Its unique fixed point is τ, the tops function of $\mathcal{F}$.*

PROOF   Let $\mathfrak{P}_1$ and $\mathfrak{P}_2$ belong to $\Gamma_\Omega(A)$. In this case the domain D_n defined above is the same for both $\mathfrak{P}_1$ and $\mathfrak{P}_2$ and is given by

$$D_n = f_n(A) \backslash \big(f_1(A) \cup \cdots \cup f_{n-1}(A) \big) \quad \text{for } n = 1, 2, \ldots, N.$$

Then

$$
\begin{aligned}
d_{sup}\big(\mathcal{F}_{TOP}(\mathfrak{P}_1), \mathcal{F}_{TOP}(\mathfrak{P}_2)\big) &= \sup_{x \in A} d_\Omega\big(\mathcal{F}_{TOP}(\mathfrak{P}_1(x)), \mathcal{F}_{TOP}(\mathfrak{P}_2(x))\big). \\
&\leq \max_{n=1,2,\ldots,N} \sup_{x \in D_n} d_\Omega(n\mathfrak{P}_1(x), n\mathfrak{P}_2(x)) \\
&\leq \max_{n=1,2,\ldots,N} (0.5) \sup_{x \in A} d_\Omega(\mathfrak{P}_1(x), \mathfrak{P}_2(x)) \\
&= (0.5) d_{sup}(\mathfrak{P}_1, \mathfrak{P}_2).
\end{aligned}
$$

It remains to verify that $\mathcal{F}_{TOP}(\tau) = \tau$. Suppose that $x \in D_n = f_n(A) \backslash f_1(A) \cup \cdots \cup f_{n-1}(A)$. Then, by definition,

$$(\mathcal{F}_{TOP}(\tau))(x) = n\tau\big(f_n^{-1}(x)\big).$$

From Lemmas 4.11.1 and 4.11.2, $\tau(x) = n\tau\big(f_n^{-1}(x)\big)$ for all $x \in D_n$.   □

Consequently we can obtain a deterministic sequence $\{\tau_k(x)\}_{k=0}^\infty$ of approximations to the tops function $\tau(x)$ by starting from any function $\tau_0 : A \to \Omega$ and iteratively defining

$$\tau_{k+1}(x) = (\mathcal{F}_{TOP}(\tau_k))(x) \quad \text{for } k = 0, 1, 2, \ldots,$$

for all $x \in A$. An example is illustrated in Figure 4.24. This is based on the same IFS as used in Figure 4.11, where we encountered the texture effect. In this example we took $\tau_0 \in \Gamma_\Omega(A)$ to be a constant,

$$\tau_0(x) = \sigma^{(0)} \in \Omega_{\{1,2,3,4\}} \quad \text{for all } x \in A$$

where $\sigma^{(0)} = \overline{1}$. It is easy to see that in this case we have $\tau_k : A \to \Omega_{\{1,2,3,4\}}$ rather than $\tau_k : A \to \Omega_{\{1,2,3,4\}} \cup \Omega'_{\{1,2,3,4\}}$. This allows us to represent the approximant τ_k by colouring its graph using colour-stealing, that is, by rendering the point $x \in A$ in the colour $\mathfrak{P}_C(\phi_D(\tau_k(x)))$. The picture $\mathfrak{P}_C$ is a digital photograph of some grasses, shown at the bottom of Figure 4.24. In contrast with the situation in Figure 4.11, this sequence of pictures converges to a stable (no visible texture effect) picture, despite discretization errors, after about sixteen iterations. That is, rendered to viewing resolution, $\mathfrak{P}_C(\phi_D(\tau_{16})) = \mathfrak{P}_C(\phi_D(\tau_{17})) = \cdots$ To check our

Figure 4.24 This illustrates a sequence of six pictures converging to the unique fixed point of the fractal tops operator $\mathcal{F}_{TOP}$. From left to right, the first, second, third, fourth, eighth and thirteenth iterations are shown. At the printed resolution, the image ceases to change with further iterations. The picture from which the colours were stolen is at the bottom. See the main text. No texture effect here!

results, we used a coupled chaos game to compute $\mathfrak{P}_C(\phi_D(\tau))$ and found the same picture.

So now it is possible to see how the texture effect, and the general lack of 'closure' that we first noticed in connection with orbital pictures and underneath pictures, led to the discovery of fractal tops!

4.13   Relationship between fractal tops and some orbital pictures

To get a feel for what will be discussed in this section, look back at Figures 3.32, 3.49 and 3.50. The idea, formalized below, is that in some cases the set of accumulation points of the addresses of the panels in an orbital picture, $\Omega_{\mathfrak{P}_0}$, is equal to the closure of Ω_τ. Consider the family of IFSs

$$\mathcal{F} = \left\{ \square; \left(\lambda x, \tfrac{1}{2}y\right), \left(\lambda x + 1 - \lambda, \tfrac{1}{2}y\right) \right\} \tag{4.13.1}$$

parameterized by $\lambda \in (0, 1)$. This is similar in spirit to the family of IFSs in Equation (3.5.17), which was used to generate Figure 3.32. Let $\mathfrak{P}_0$ be a picture whose domain is the line segment that joins the two points $(0, 1)$ and $(1, 1)$ in $\square$,

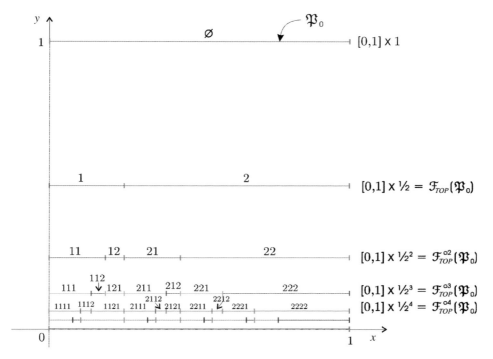

Figure 4.25   Here some panels of an orbital picture $\mathfrak{P}(\mathfrak{P}_0)$ are labelled with their addresses. Each labelled line segment $\mathcal{F}^{\circ k}(\mathfrak{P}_0)$ defines an approximation $\tau_k(x)$ to the top $\tau(x)$ of the IFS $\mathcal{F}$ in Equation (4.13.1) with $\lambda = 0.25$. The attractor of the IFS is the line segment [0, 1] contained in the x-axis. In such examples $\Omega_{\mathfrak{P}_0} = \overline{\Omega_\tau}$.

namely the upper pair of corners of $\square$. The first few generations of panels and their addresses are illustrated in Figure 4.25. Let $[0, 1] \times 2^{-k}$ denote the line segment that joins the pair of points $(0, 2^{-k}) \in \mathbb{R}^2$ and $(1, 2^{-k}) \in \mathbb{R}^2$, for $k = 1, 2, \ldots$ Then the domain of the orbital picture $\mathfrak{P}(\mathfrak{P}_0)$ is the union of these line segments, and we can define a piecewise constant function $\tau_k(x)$ whose value is the address of the panel to which $(x, 2^{-k})$ belongs. In this case we recognize that $\tau_k(x) = \mathcal{F}_{TOP}^{\circ k}(\tau_0(x))$ where $\tau_0(x) = \varnothing$ for all $x \in [0, 1]$. But $[0, 1]$ is the set attractor of $\mathcal{F}$ and so, by Theorem 4.12.1, $\{\tau_k(x)\}_{k=0}^\infty$ converges to $\tau(x)$. In particular, as proved in Theorem 4.13.1 below, the set of accumulation points of the set of addresses of all the panels here, that is, the set $\Omega_{\mathfrak{P}_0}$, is the same as the closure of Ω_τ.

Next we generalize this example. This provides a link between some orbital pictures and tops functions. Many other specific results connecting $\Omega_{\mathfrak{P}_0}$ and Ω_τ can be obtained in a similar vein, but we do not have a more general simple theorem.

Let $\mathcal{F} = \{\mathbb{X}; f_1, f_2, \ldots, f_N\}$ be a hyperbolic IFS on a compact metric space $(\mathbb{X}, d)$. Let A denote the set attractor of $\mathcal{F}$ and let Ω_τ denote the associated tops code space. Let $\mathbb{X}^* = \mathbb{X} \times [0, 1]$ and define a metric d^* on $\mathbb{X}^*$ by $d^*((x_1, x_2), (y_1, y_2)) = d(x_1, x_2) + |y_1 - y_2|$, so that $(\mathbb{X}^*, d^*)$ is a compact metric space. Let $\mathcal{F}^*$ denote the hyperbolic IFS $\{\mathbb{X}^*; f_1^*, f_2^*, \ldots, f_N^*\}$, where $f_n^*(x, y) = (f_n(x), \frac{1}{2}y)$ for $n = 1, 2, \ldots, N$. Then we look at the orbital picture $\mathfrak{P}^*(\mathfrak{P}_0^*)$ produced when the IFS semigroup generated by $\mathcal{F}^*$ acts on a condensation picture $\mathfrak{P}_0^* \in \Pi_{\mathfrak{C}}(\mathfrak{P}^*)$ with domain

$$D_{\mathfrak{P}_0^*} = A \times \{1\}.$$

Then the domains of $(\mathcal{F}^*)^{\circ k}(\mathfrak{P}_0^*)$, namely

$$\mathcal{F}^{\circ k}(A) \times \{2^{-k}\} = A \times \{2^{-k}\}$$

for $k = 0, 1, 2, \ldots$, are disjoint, and we have

$$D_{\mathfrak{P}^*(\mathfrak{P}_0^*)} = \bigcup_{k=0}^{\infty} \left(A \times \{2^{-k}\}\right).$$

In particular, the union of the domains of the set of panels of $\mathfrak{P}^*(\mathfrak{P}_0^*)$ having code length k is precisely the set $A \times \{2^{-k}\}$. Accordingly we can define a function $\mathfrak{P}_k : A \to \Omega'$, where $\Omega' = \Omega'_{\{1,2,\ldots,N\}}$, by

$$\mathfrak{P}_k(x) = \text{address of the unique panel of } \mathfrak{P}^*(\mathfrak{P}_0^*) \text{ whose domain contains } (x, 2^{-k}),$$

for $k = 0, 1, 2, \ldots$, for all $x \in A$. But it will be recognized by the alert reader that

$$\mathfrak{P}_k = \mathcal{F}_{TOP}^{\circ k}(\mathfrak{P}_0) \tag{4.13.2}$$

for $k = 0, 1, 2, \ldots$, where we note that $\mathfrak{P}_0 \in \Pi_{\Omega}(A)$ is given by

$$\mathfrak{P}_0(x) = \varnothing \quad \text{for all } x \in A.$$

Now let $\Omega_{\mathfrak{P}_0^*}$ denote the code space of the orbital picture $\mathfrak{P}^*(\mathfrak{P}_0^*)$.

THEOREM 4.13.1 *Let Ω_τ denote the range of the tops function $\tau : A \to \Omega$ for a hyperbolic IFS $\mathcal{F}$. Let $\Omega_{\mathfrak{P}_0^*}$ be the set of addresses of the orbital picture $\mathfrak{P}^*(\mathfrak{P}_0^*)$, constructed above. Then*

$$\Omega_{\mathfrak{P}_0^*} = \overline{\Omega_\tau}.$$

PROOF Suppose that $\sigma \in \Omega_{\mathfrak{P}_0}$. Then there is an infinite strictly increasing sequence of positive integers $\{k_l\}_{l=1}^{\infty}$ such that

$$\sigma_1 \sigma_2 \cdots \sigma_{k_l} \in \Omega'_{\mathfrak{P}_0}$$

for $l = 1, 2, \ldots$ But Equations (4.12.1) and (4.13.2) together imply that

$$\mathfrak{P}_k(x) = (\tau(x))_1 (\tau(x))_2 \cdots (\tau(x))_k \quad \text{for all } x \in A \tag{4.13.3}$$

and for $k = 0, 1, 2, \ldots$ It follows that there exists $x_{k_l} \in A$ such that the first k_l symbols in the address $\tau(x_{k_l})$ agree with the first k_l symbols in σ. It follows that $\Omega_\tau = \tau(A)$ contains points arbitrarily close to σ and that $\sigma \in \overline{\Omega_\tau}$.

Conversely, suppose that $\sigma \in \overline{\Omega_\tau}$. Then there is a infinite sequence of points $\{\omega^{(k_l)}\}_{l=0}^\infty$ in Ω_τ such that $\sigma_1\sigma_2\cdots\sigma_{k_l}\omega^{(k_l)} \in \Omega_\tau$ for $l = 1, 2, \ldots$ Let $x_{k_l} = \tau^{-1}(\sigma_1\sigma_2\cdots\sigma_{k_l}\omega^{(k_l)})$. Then, by Equation (4.13.3), $\mathfrak{P}_{k_l}(x_{k_l}) = \sigma_1\sigma_2\cdots\sigma_{k_l}$. Hence $\sigma \in \Omega_{\mathfrak{P}_0}$. $\qquad\qquad\square$

In some other cases, such those suggested by Figure 3.43, we have found that $\Omega_{\mathfrak{P}_0} \subset \overline{\Omega_\tau}$. This is interesting because, as described in Theorem 3.5.13, $S(\Omega_{\mathfrak{P}_0}) \subset \Omega_{\mathfrak{P}_0}$. It suggests that in some cases we can use orbital pictures to probe the structure of fractal tops and find interesting shift-invariant subsets of $\overline{\Omega_\tau}$. We are interested in such shift-invariant subsets because of their association with homeomorphisms between fractals, which we discuss in the remainder of this chapter.

In yet other situations we find that $\Omega_{\mathfrak{P}_0} \supset \overline{\Omega_\tau}$, as illustrated in the following exercise.

EXERCISE 4.13.2 *Let* $\mathcal{F} = \{\square; (0.75x, 0.5y), (0.75x + 0.25, 0.5y)\}$ *and let* $\mathfrak{P}_0 \in \Pi_\Omega(\square)$ *denote the picture whose domain is the point* $(0.12179\cdots, 1)$ *and whose value is* $111111\cdots$ *Consider the associated orbital picture* $\mathfrak{P}(\mathfrak{P}_0)$. *Show that if the number* $0.12179\cdots$ *is irrational then* $\Omega_{\mathfrak{P}_0} = \Omega_{\{1,2\}}$, $\overline{\Omega_\tau} \neq \Omega_{\{1,2\}}$ *and, in particular,* $\Omega_{\mathfrak{P}_0} \supset \overline{\Omega_\tau}$.

4.14 The fractal homeomorphism theorem

Sometimes, when looking at different stolen pictures rendered using the same code space picture $\mathfrak{P}_C \circ \phi_C$, you will gain the impression that some pairs of them are homeomorphic while others are not. Such is the case with the pictures in Figure 4.26. Look also at Figure 4.2. This shows two colourful fractal ferns, produced by tops plus colour-stealing, corresponding to different IFSs, say $\mathcal{F}_{D1}$ and $\mathcal{F}_{D2}$, but using the same picture $\mathfrak{P}_C$ and the same colouring IFS $\mathcal{F}_C$. Here it seems that the underlying attractors are not homeomorphic because, intuitively, any continuous mapping H from one to the other would be forced to identify the set of points where the two lowest fronds join the main stem; thus H would not be invertible. In this way we are led to ask: when are IFS set attractors homeomorphic? Also, when does there exist a one-to-one continuous invertible transformation between one fractal top and another? When are two tops dynamical systems topologically conjugate?

These questions turn out to have a wonderful answer, which is stated formally in Theorems 4.14.5 and 4.14.9 and Corollary 4.14.8 below: roughly, two set attractors are homeomorphic if and only if they can be expressed as the set attractors of two IFSs whose tops code spaces have the same structure up to permutation, regardless of the underlying spaces on which the IFSs act. Also, in this case, their tops functions and tops dynamical systems are topologically conjugate. Of course, we need to explain carefully what we mean by the structure of a tops code space.

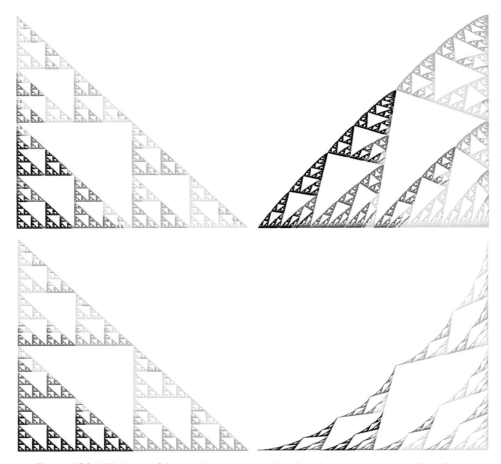

Figure 4.26 Which two of these stolen pictures may be related by a continuous invertible deformation? All four pictures of IFS set attractors were rendered using the same mapping, $\mathfrak{P}_C \circ \phi_C : \Omega \to A_C$. By Theorem 4.14.5, IFS set attractors are homeomorphic iff they possess the same code space structure.

This was a new realization to me: it says that one can handle the topology of IFS set attractors in a manner that has parallels to the way in which coordinate geometry provides an algebraic approach to classical geometry. The fractal homeomorphism theorem, Theorem 4.14.9, has a specific application to special effects in digital imaging produced by continuous deformations of video and still pictures by colour-stealing, as discussed further in Section 4.15. In these applications, in effect, we adjust the topology of set attractors by controlling their code structures! That is, we use mainly symbolic means to control topology. This result is, moreover, a special case of a much more general theorem, related to directed IFSs, discussed in Section 4.16: code structure is a total topological invariant!

So what is the code structure of a hyperbolic IFS? Let $\mathcal{F} = \{\mathbb{X}; f_1, f_2, \ldots, f_N\}$ denote a hyperbolic IFS on a compact metric space. Let $A_{\mathcal{F}}$ denote the attractor of

$\mathcal{F}$ and let $\phi_{\mathcal{F}} : \Omega_{\{1,2,...,N\}} \rightarrow A_{\mathcal{F}}$. Let $\Omega_{\mathcal{F}} \subset \Omega_{\{1,2,...,N\}}$ denote the tops code space for $\mathcal{F}$. That is,

$$\Omega_{\mathcal{F}} = \tau_{\mathcal{F}}(A_{\mathcal{F}})$$

where $\tau_{\mathcal{F}} : A_{\mathcal{F}} \rightarrow \Omega_{\mathcal{F}}$ is the tops function defined by

$$\tau_{\mathcal{F}}(x) = \max \left\{ \phi_{\mathcal{F}}^{-1}(x) \right\} \quad \text{for each } x \in A_{\mathcal{F}}.$$

DEFINITION 4.14.1   Let $\mathcal{F}$ denote a hyperbolic IFS on a compact metric space and let $\Omega_{\mathcal{F}}$ denote its tops code space. Let $\overline{\Omega}_{\mathcal{F}}$ denote the closure of $\Omega_{\mathcal{F}}$ in the natural (product) topology on code space. Let $\phi_{\mathcal{F}} : \overline{\Omega}_{\mathcal{F}} \rightarrow A_{\mathcal{F}}$ denote the restriction of the code space mapping to $\overline{\Omega}_{\mathcal{F}}$. Then the **code structure** of $\mathcal{F}$ is defined to be the set of sets $\{\phi_{\mathcal{F}}^{-1}(x) \cap \overline{\Omega}_{\mathcal{F}} : x \in A_{\mathcal{F}}\}$.

We will say such things as 'the set A possesses a code structure Q' when we mean that there exists an IFS $\mathcal{F}$ with set attractor A and code structure Q. It is important to realize that a set attractor of an IFS may have many different code structures, because it may be the set attractor of many different IFSs.

For example, the closed line segment $[0, 1]$ possesses many different code structures. Firstly, each member of the family of just-touching hyperbolic IFSs

$$\mathcal{F} = \{[0, 1]; f_1(x) = \lambda x, \ f_2(x) = (1 - \lambda)x + \lambda\} \quad \text{for } 0 < \lambda < 1$$

has tops code space $\Omega_{\mathcal{F}}$ such that $\overline{\Omega}_{\mathcal{F}} = \Omega_{\{1,2\}}$ and each nonempty set in its code structure either contains exactly one point or is of the form $\{\sigma 1\overline{0}, \ \sigma 0\overline{1}\}$ for some $\sigma \in \Omega'_{\{1,2\}}$. Secondly, it also possesses a diverse collection of code structures provided by the family of overlapping IFSs

$$\mathcal{F}_1 = \{[0, 1]; f_1(x) = \alpha x + 1 - \alpha, \ f_2(x) = \alpha x\} \quad \text{for } \tfrac{1}{2} < \alpha < 1. \quad (4.14.1)$$

These can be calculated, as in Corollary 4.11.4, by following the orbits of the tops dynamical system $T : [0, 1] \rightarrow [0, 1]$ defined by

$$T(x) = \begin{cases} \dfrac{x}{\alpha} & \text{for } x \in D_2 = [0, 1 - \alpha), \\[2mm] \dfrac{x + \alpha - 1}{\alpha} & \text{for } x \in D_1 = [1 - \alpha, 1]. \end{cases} \quad (4.14.2)$$

See Figure 4.27. Recall that the tops code $\tau(x)$ is defined by $(\tau(x))_k =$ index of the interval, D_1 or D_2, to which $T^{\circ k}(x)$ belongs, for each $k = 0, 1, 2, \ldots$ For instance, if α is the golden mean $\frac{1}{2}(\sqrt{5} - 1)$ then it is found that Ω_G consists of $\overline{1} = 11 \cdots, \overline{2} = 22 \cdots$ and all strings $\sigma \in \Omega_{\{1,2\}}$ that can be written in the form

$$11 \cdots 122 \cdots 2122 \cdots 2122 \cdots 2122 \cdots 21 \cdots,$$

where $11 \cdots 1$ is either empty or a finite-length sequence of ones and where each $22 \cdots 2$ is a finite-length, nonempty, sequence of twos. It follows that $\overline{\Omega}_{\mathcal{F}}$ consists

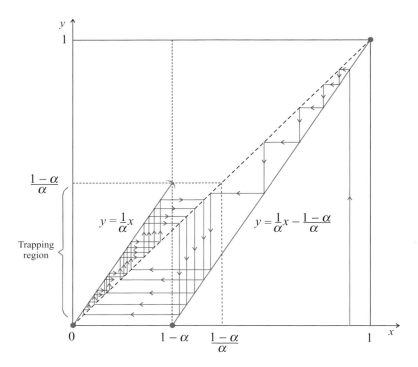

Figure 4.27 The tops dynamical system associated with the IFS $\mathcal{F}_1$ in Equations (4.9.1) and (4.14.2). The tops code space and its closure can be found, in quite a straightforward way, by thinking about the possible structures of orbits such as the one illustrated in the figure.

of $\Omega_{\mathcal{F}}$ together with all strings that can be written in the form

$$11 \cdots 122 \cdots 2122 \cdots 2122 \cdots 21\overline{2}.$$

Code structures associated with dynamical systems of the form seen in Figure 4.27 have been extensively studied in the context of information theory and symbolic dynamics; see for example papers in the literature that cite [82]. The complexity and richness of this one very simple example suggests that endless wonders await those who explore simple nontrivial two-dimensional situations.

EXERCISE 4.14.2 *Find the code space structure for the IFS, given in Table 4.2, which makes a Sierpinski triangle.*

EXERCISE 4.14.3 *Describe, roughly, the code space structure for the overlapping IFS in Table 4.4.*

DEFINITION 4.14.4 Let $\mathcal{F}$ and $\mathcal{G}$ be two hyperbolic IFSs. We say that the **code structures of $\mathcal{F}$ and $\mathcal{G}$ are homeomorphic** if there is a homeomorphism $\chi : \overline{\Omega}_{\mathcal{F}} \to \overline{\Omega}_{\mathcal{G}}$ which respects the code structures, that is, which maps each set in the code structure of $\mathcal{F}$ onto a set in the code structure of $\mathcal{G}$.

If $\mathcal{F}$ is a totally disconnected IFS and $\mathcal{G}$ is obtained from $\mathcal{F}$ by permuting the order of the functions in $\mathcal{F}$ then clearly the code space structures of $\mathcal{F}$ and $\mathcal{G}$ are homeomorphic. Other examples of homeomorphic code structures, involving overlapping IFSs, can also be exhibited. But for the most part it is convenient to think of χ as the identity map.

THEOREM 4.14.5   *Let the code structures of two hyperbolic IFSs $\mathcal{F}$ and $\mathcal{G}$ be homeomorphic. Then their set attractors, $A_\mathcal{F}$ and $A_\mathcal{G}$, are homeomorphic. That is, there exists a continuous one-to-one invertible transformation $H : A_\mathcal{F} \to A_\mathcal{G}$ such that*

$$H(A_\mathcal{F}) = A_\mathcal{G}.$$

We remark that if the tops code spaces of two hyperbolic IFSs $\mathcal{F}$ and $\mathcal{G}$ are the same then their set attractors are related by the one-to-one invertible transformation $\widetilde{H} : A_\mathcal{F} \to A_\mathcal{G}$ defined by

$$\widetilde{H} = \tau_\mathcal{G}^{-1} \circ \tau_\mathcal{F}.$$

But, in general, this transformation is not continuous because of the possible discontinuity of $\tau_\mathcal{F}$.

PROOF   Let us denote the two IFSs by $\mathcal{F} = \{\mathbb{X}_\mathcal{F}; f_1, f_2, \ldots, f_{N_\mathcal{F}}\}$ and $\mathcal{G} = \{\mathbb{X}_\mathcal{G}; g_1, g_2, \ldots, g_{N_\mathcal{G}}\}$, where $\mathbb{X}_\mathcal{F}$ and $\mathbb{X}_\mathcal{G}$ are complete metric spaces. Let the attractors be denoted by $A_\mathcal{F} \subset \mathbb{X}_\mathcal{F}$ and $A_\mathcal{G} \subset \mathbb{X}_\mathcal{G}$ respectively. Let the code space functions be denoted by $\phi_\mathcal{F} : \Omega_{\{1,2,\ldots,N_\mathcal{F}\}} \to A_\mathcal{F}$ and $\phi_\mathcal{G} : \Omega_{\{1,2,\ldots,N_\mathcal{G}\}} \to A_\mathcal{G}$ respectively and let the tops code spaces be denoted by $\Omega_\mathcal{F}$ and $\Omega_\mathcal{G}$ respectively. Let the tops functions be denoted by $\tau_\mathcal{F} : A_\mathcal{F} \to \Omega_\mathcal{F}$ and $\tau_\mathcal{G} : A_\mathcal{G} \to \Omega_\mathcal{G}$ respectively. Now let the homeomorphism between code structures, asserted in the theorem, be denoted by

$$\chi : \overline{\Omega}_\mathcal{F} \to \overline{\Omega}_\mathcal{G}.$$

Then we claim that the transformation $H : A_\mathcal{F} \to A_\mathcal{G}$ defined by

$$H = \phi_\mathcal{G} \circ \chi \circ \tau_\mathcal{F}$$

provides the desired homeomorphism between the attractors.

It is useful to notice at this stage that H is a one-to-one invertible transformation between the two attractors. You can readily verify that its inverse is given by

$$H^{-1} = \phi_\mathcal{F} \circ \chi^{-1} \circ \tau_\mathcal{G} = \tau_\mathcal{F}^{-1} \circ \chi^{-1} \circ \tau_\mathcal{G}.$$

We need to show that it is continuous. This follows from a subtle observation and a theorem in elementary topology.

The subtle observation, by Mendelson [70], p. 194, is this. *If $F : X \to Y$ is a continuous mapping of a compact space X onto a Hausdorff space Y then the topology of Y is the same as the identification topology induced by F.*

This tells us that the identification topology on $A_{\mathcal{F}}$, which is induced by the code space mapping $\phi_{\mathcal{F}} : \Omega_{\{1,2,...,N_{\mathcal{F}}\}} \to A_{\mathcal{F}}$ restricted to $\overline{\Omega}_{\mathcal{F}}$, is the same as the natural topology on $A_{\mathcal{F}}$, as a subset of the metric space $\mathbb{X}_{\mathcal{F}}$.

The theorem we apply is Proposition 7.4 of [70], p. 195. *Let X, Y and Z be topological spaces. Let $F : X \to Y$ be onto and let Y have the identification topology induced by F. Then the function $H : Y \to Z$ is continuous iff the composition $G := H \circ F : X \to Z$ is continuous.*

We choose $F : X \to Y$ to be $\phi_{\mathcal{F}} : \overline{\Omega}_{\mathcal{F}} \to A_{\mathcal{F}}$ and $H : Y \to Z$ to be

$$H = \phi_{\mathcal{G}} \circ \chi \circ \tau_{\mathcal{F}} : A_{\mathcal{F}} \to A_{\mathcal{G}}.$$

Then $F : X \to Y$ is onto and, as we pointed out above, the topology of $Y = A_{\mathcal{F}}$ is the same as the identification topology and also the natural topology. Now look at the function

$$G := H \circ F = \phi_{\mathcal{G}} \circ \chi \circ \tau_{\mathcal{F}} \circ \phi_{\mathcal{F}} : \overline{\Omega}_{\mathcal{F}} \to A_{\mathcal{G}}.$$

If $\sigma \in \overline{\Omega}_{\mathcal{F}}$ then $(\tau_{\mathcal{F}} \circ \phi_{\mathcal{F}})(\sigma)$ belongs to the same set in the code space structure of $\mathcal{F}$ as σ. Since χ respects the code space structure, both $\chi(\sigma)$ and $(\chi \circ \tau_{\mathcal{F}} \circ \phi_{\mathcal{F}})(\sigma)$ belong to the same set in the code space structure of $\mathcal{G}$. It follows that

$$(\phi_{\mathcal{G}} \circ \chi \circ \tau_{\mathcal{F}} \circ \phi_{\mathcal{F}})(\sigma) = (\phi_{\mathcal{G}} \circ \chi)(\sigma) \quad \text{for all } \sigma \in \overline{\Omega}_{\mathcal{F}}.$$

But $\phi_{\mathcal{G}} \circ \chi : \overline{\Omega}_{\mathcal{F}} \to A_{\mathcal{G}}$ is the composition of two continuous maps, and so it is continuous. It follows that G is continuous and therefore that H is continuous. Similarly, its inverse is continuous. Hence H is a homeomorphism. $\qquad\square$

EXERCISE 4.14.6 *Prove Mendelson's subtle observation, above. The key is to note that if $f : \mathbb{X} \to \mathbb{Y}$ is continuous, where $\mathbb{X}$ and $\mathbb{Y}$ are topological spaces, and if C is a compact subset of $\mathbb{X}$ then $f(C)$ is a compact subset of $\mathbb{Y}$.*

EXERCISE 4.14.7 *Prove Proposition 7.4 of [70], quoted above.*

The idea that the tops code space is roughly, but not precisely, the same as the code structure of an IFS is a convenient *aide memoire*, but remember to take the closure.

COROLLARY 4.14.8 *Let $\mathcal{A}$ denote the set of all set attractors of all hyperbolic IFSs. Then*

$$A, B \in \mathcal{A} \text{ are homeomorphic} \iff A \text{ and } B \text{ possess the same code}$$
$$\text{space structure.}$$

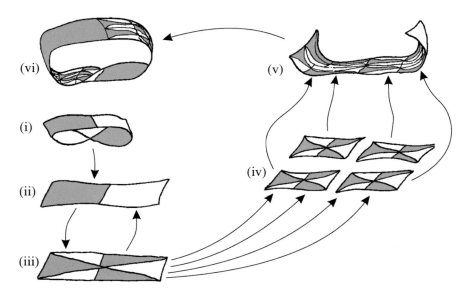

Figure 4.28   To find an IFS whose attractor is a Möbius strip, first colour the Möbius strip pink and white, as in (i), to help illustrate the transformations. Next, flatten the strip to make a rectangle whose top looks like (ii) and whose bottom looks like (iii). Then make four shrunken copies (iv) of the rectangle and join them to form a new rectangle (v). Then twist and join the composite figure (v) to form a second Möbius strip (vi). Note that the code structure of the resulting IFS *cannot* be the same as that of any IFS whose attractor lies in $\mathbb{R}^2$.

PROOF   Theorem 4.14.5 yields $\Leftarrow$. To prove $\Rightarrow$, suppose that $A, B \in \mathcal{A}$ are homeomorphic. Then there exists a hyperbolic IFS $\mathcal{F} = \{A; f_1, f_2, \ldots, f_N\}$, with set attractor $A_{\mathcal{F}} = A$, and a metric d such that (A, d) is a compact metric space. There also exists a homeomorphism $H : A \to B$ with $H(A) = B$. Then it is readily verified that the IFS

$$\mathcal{G} = \{B; H \circ f_1 \circ H^{-1}, H \circ f_2 \circ H^{-1}, \ldots, H \circ f_N \circ H^{-1}\}$$

is strictly contractive with respect to the metric d_B defined by $d_B(x, y) = d(H^{-1}(x), H^{-1}(y))$. It is also readily verified that the set attractor of $\mathcal{G}$ is B and that its code structure is the same as that of $\mathcal{F}$.   □

One consequence of the corollary is that there can exist no code structure, for a set such as a circle, a figure eight or a Sierpinski triangle, that is also a code structure for a set that is contained in $\mathbb{R}$; for otherwise the circle, figure-eight or Sierpinski triangle would be homeomorphic to a subset of $\mathbb{R}$, which clearly is not true. In Figure 4.28 we suggest how you might find a code structure for a Möbius strip. The resulting code structure could not also belong to an IFS on $\mathbb{R}^2$.

THEOREM 4.14.9 (*Fractal homeomorphism theorem*)   *With the same set-up and notation as above, let the homeomorphism* $\chi : \overline{\Omega}_{\mathcal{F}} \to \overline{\Omega}_{\mathcal{G}}$ *have the property* $\chi(\Omega_{\mathcal{F}}) = \Omega_{\mathcal{G}}$. *Then*

$$\tau_{\mathcal{G}} = \chi \circ \tau_{\mathcal{F}} \circ H^{-1},$$

where $H : A_{\mathcal{F}} \to A_{\mathcal{G}}$ *is the continuous one-to-one invertible transformation from* $A_{\mathcal{F}}$ *onto* $A_{\mathcal{G}}$ *given in Theorem 4.14.5. Moreover the tops dynamical systems* $T_{\mathcal{F}} : A_{\mathcal{F}} \to A_{\mathcal{F}}$ *and* $T_{\mathcal{G}} : A_{\mathcal{G}} \to A_{\mathcal{G}}$ *are topologically conjugate according to*

$$T_{\mathcal{G}} = H \circ T_{\mathcal{F}} \circ H^{-1}.$$

PROOF   This follows immediately from the proof of Theorem 4.14.5. The stated identities turn on the facts that $(\tau_{\mathcal{F}} \circ \phi_{\mathcal{F}})(\sigma) = \sigma$ for all $\sigma \in \Omega_{\mathcal{F}}$ and that we have restricted the homeomorphism $\chi : \overline{\Omega}_{\mathcal{F}} \to \overline{\Omega}_{\mathcal{G}}$ so that it maps from $\Omega_{\mathcal{F}}$ onto $\Omega_{\mathcal{G}}$.   □

When dynamical systems are topologically conjugate they may share many properties, as discussed in Section 3.5.

When the drawing IFS $\mathcal{F}_D$ and the colouring IFS $\mathcal{F}_C$ have the same code structure, the relationship between the stolen picture $\mathfrak{P}_D$ and $\mathfrak{P}_C$ is precisely

$$H(\mathfrak{P}_D) = \mathfrak{P}_C|_{A_C}.$$

That is, $\mathfrak{P}_D$ is a continuous invertible transformation of $\mathfrak{P}_C$ restricted to the domain of the colouring IFS. We will discuss this relationship further in Section 4.15. A consequence, which has never failed to surprise colleagues to whom I have demonstrated it, in video applications, is that if $\mathcal{F}_D = \mathcal{F}_C$ then $\mathfrak{P}_D = \mathfrak{P}_C$. This is illustrated in Figure 4.29.

Here, briefly, let me suppose that a hyperbolic IFS $\mathcal{F}$ in $\mathbb{R}^3$ is a model of botanical meristem: in the development of the set attractor of $\mathcal{F}$, which I think of as a geometrical model for an entity in plant physiology, such as phloem or the veins within a leaf, I see the expression of the code structure of $\mathcal{F}$ in the topology of the intricate tangle of proteins that comprises the botanical entity. The locations in space of the amino acids in the proteins, both in sequences along strands and adjacently at different points along the same or different strands, may then be considered as a physical manifestation of the identification topology induced by $\phi_{\mathcal{F}} : \overline{\Omega}_{\mathcal{F}} \to A_{\mathcal{F}}$. Is it possible, in a realistic practical way, to make a mathematical model for the veins of a leaf using a simple IFS and so provide a reasonable connection between the physical development, via DNA and meristem, and the decoding of the IFS?

Figure 4.29 In this example of colour-stealing, the same IFS has been used both to draw and to colour-steal. The attractor of the IFS is the domain of this picture.

4.15 **Fractal transformations**

In this section we study more closely the relationship, implicit in colour-stealing, between some pairs of set attractors. In so doing we alert you, the sharp-eyed reader, to some areas of application of fractal tops and colour-stealing. For simplicity of presentation we restrict the discussion to the case where there is a single underlying compact metric space $\mathbb{X}$. Let $\mathcal{F} = \{\mathbb{X}; f_1, f_2, \ldots, f_N\}$ and $\mathcal{G} = \{\mathbb{X}; g_1, g_2, \ldots, g_N\}$ denote hyperbolic IFSs on $\mathbb{X}$. Let $A_{\mathcal{F}}$ and $A_{\mathcal{G}}$ denote their respective set attractors, $\Omega_{\mathcal{F}} \subset \Omega$ and $\Omega_{\mathcal{G}} \subset \Omega$ denote their tops code spaces, $\tau_{\mathcal{F}} : A_{\mathcal{F}} \to \Omega_{\mathcal{F}}$ and $\tau_{\mathcal{G}} : A_{\mathcal{G}} \to \Omega_{\mathcal{G}}$ denote their tops functions and $\phi_{\mathcal{F}} : \Omega \to A_{\mathcal{F}}$ and $\phi_{\mathcal{G}} : \Omega \to A_{\mathcal{G}}$ denote their respective code space functions.

DEFINITION 4.15.1   The transformation $F_{\mathcal{F}\mathcal{G}} : A_{\mathcal{F}} \to A_{\mathcal{G}}$ defined by

$$F_{\mathcal{F}\mathcal{G}} = \phi_{\mathcal{G}} \circ \tau_{\mathcal{F}}$$

is called a **fractal transformation** from $A_{\mathcal{F}}$ into $A_{\mathcal{G}}$.

Fractal transformations underlie colour-stealing. To see this, suppose that we choose $\mathcal{F}_D = \mathcal{F}$ and $\mathcal{F}_C = \mathcal{G}$. Also, suppose without loss of generality that $D_{\mathfrak{P}_C} = A_{\mathcal{G}}$. Then the relationship between the stolen picture $\mathfrak{P}_{\mathcal{F}} := \mathfrak{P}_D$ and the picture $\mathfrak{P}_{\mathcal{G}} := \mathfrak{P}_C$ from which the colours were stolen is

$$\mathfrak{P}_{\mathcal{F}} = \mathfrak{P}_{\mathcal{G}} \circ F_{\mathcal{F}\mathcal{G}}. \tag{4.15.1}$$

There are three quite distinct situations which can occur. Each corresponds to a different kind of relationship between $\mathfrak{P}_{\mathcal{F}}$ and $\mathfrak{P}_{\mathcal{G}}$. We treat these situations separately.

(i) $\Omega_{\mathcal{F}} \neq \Omega_{\mathcal{G}}$. In this case, in general the tops transformation $F_{\mathcal{F}\mathcal{G}}$ is neither one-to-one nor onto. But from Theorem 4.11.5 we know that it is 'nearly' continuous, at least when f_n is a homeomorphism between $\mathbb{X}$ and $f_n(\mathbb{X})$ for each $n \in \{1, 2, \ldots, N\}$. Thus, in this case $F_{\mathcal{F}\mathcal{G}}$ is continuous on $A_{\mathcal{F}}^{inside}$; its discontinuities are restricted to certain pre-images, possibly a countable dense set of lines, of the boundary of $A_{\mathcal{F}}$. This partially explains the look-and-feel of some stolen pictures, as follows. A set of fractal boundaries, transformed copies of the boundary of $A_{\mathcal{F}}$, provide what we refer to as 'line-art' on $A_{\mathcal{F}}$, and $F_{\mathcal{F}\mathcal{G}}$ is continuous at all other points of $A_{\mathcal{F}}$; hence, as x varies over $A_{\mathcal{F}}^{inside}$, $F_{\mathcal{F}\mathcal{G}}(x)$ varies continuously over the domain of the picture $\mathfrak{P}_{\mathcal{G}}$ while the colour at $F_{\mathcal{F}\mathcal{G}}(x)$ is given to the point x. Insofar as the colours $\mathfrak{P}_{\mathcal{G}}(y)$ depend continuously on y, the colours painted on $A_{\mathcal{F}}$ will vary continuously away from the line-art.

It is important, though, to notice that in general, in this case, we cannot think of $\mathfrak{P}_{\mathcal{F}}$ as being the result of applying a transformation to $\mathfrak{P}_{\mathcal{G}}$ as described in Chapter 2, because $F_{\mathcal{F}\mathcal{G}}$ is not invertible.

A simple example of colour-stealing when the tops transformation is not one-to-one is illustrated by the top right image in Figure 4.30. Here the line-art consists of portions of boundaries of many rectangles. You will see that bits of the rocks are missing. It is as though parts of the stolen picture have been put underneath other parts. The picture $\mathfrak{P}_C$ from which colours were stolen is shown at the top left. The IFSs $\mathcal{F} = \mathcal{F}_1$ and $\mathcal{G}$ are illustrated in Figure 4.31. The code structure of $\mathcal{G}$ is essentially that of the unit square written in quadtree representation, but the code structure of $\mathcal{F}_1$ omits many addresses in the code structure of $\mathcal{G}$ because its tops dynamical system behaves in much the same way as that of the IFS in Equation (4.14.1) above, in both the x- and y- directions. Many potential addresses are simply wiped out by the overlapping region. In Figure 4.31, at upper right, we

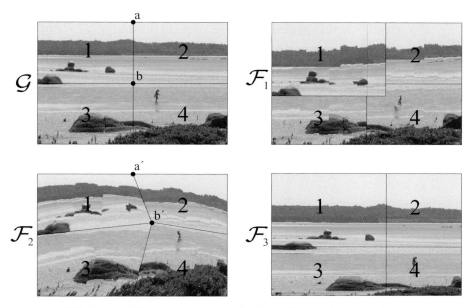

Figure 4.30 The top left image shows the original image. The top right image illustrates the result of applying a fractal transformation that is not one-to-one. The bottom left image shows the result of applying a fractal transformation that is one-to-one invertible but not continuous. The bottom right image shows the result of applying a fractal homeomorphism. See text.

Figure 4.31 The result of 'inverting' the fractal transformations used in Figure 4.30.

show the result of reversing the colour-stealing process, with $\mathfrak{P}_C = \mathfrak{P}_\mathcal{F}, \mathcal{F}_C = \mathcal{F}_1$ and $\mathcal{F}_D = \mathcal{G}$; clearly we do not get back to the original picture!

(ii) $\Omega_\mathcal{F} = \Omega_\mathcal{G}$, *but the code structures of $\mathcal{F}$ and $\mathcal{G}$ are different.* In this case the fractal transformation $F_{\mathcal{F}\mathcal{G}}$ maps $A_\mathcal{F}$ one-to-one onto $A_\mathcal{G}$, but in general it is not continuous. Its inverse is $F_{\mathcal{F}\mathcal{G}}^{-1} = F_{\mathcal{G}\mathcal{F}} = \phi_\mathcal{F} \circ \tau_\mathcal{G}$. In particular, we have

$$\mathfrak{P}_\mathcal{G} = F_{\mathcal{G}\mathcal{F}}(\mathfrak{P}_\mathcal{F}).$$

In this case, we do get back to the original picture from the stolen one when we reverse the colour-stealing process; specifically,

$$\mathfrak{P}_\mathcal{F} = F_{\mathcal{F}\mathcal{G}}(\mathfrak{P}_\mathcal{G}).$$

In practice we achieve this inversion by swapping the colouring and drawing IFSs and using the previously stolen picture as the one from which to steal the colours. This technique has immediate application to image encryption and to copyright protection. It is also provides a novel method for changing the distribution of information in a picture, by first transforming to a stolen picture, then filtering and then transforming back. This is relevant to image-compression applications.

An example of case (ii) is shown in the bottom left panel of Figure 4.30. Here you can see that $F_{\mathcal{F}\mathcal{G}}$ is not continuous; it has discontinuities associated with the line-art, which consists of segments of conic sections. The picture $\mathfrak{P}_C$ is shown at the top left, and the IFSs $\mathcal{F} = \mathcal{F}_2$ and $\mathcal{G}$ are illustrated in Figure 4.31. The latter figure also illustrates the result (see the bottom left image) of applying the inverse of the transformation to the transformed image; although not exactly the same as the original, it is accurate to within computational errors. These errors are due mainly to discretization.

How do we know that $\Omega_{\mathcal{F}_2} = \Omega_\mathcal{G}$? By considering the tops dynamical systems of $\mathcal{F}_2$ and $\mathcal{G}$. It is straightforward to prove that σ is the code of an orbit of $T_{\mathcal{F}_2}$ iff it is an orbit of $T_\mathcal{G}$. How do we know that the code structures of $\Omega_{\mathcal{F}_2}$ and $\Omega_\mathcal{G}$ are different, in this example? One can prove this by comparing the sets of addresses of points on the line ab in Figure 4.31, provided by the IFS $\mathcal{G}$, with those of points on the line $a'b'$, provided by the IFS $\mathcal{F}_2$. These structures are quite different, because the former are provided by superimposing two standard binary codings of a line segment, whereas the latter correspond to the superposition of two mismatched 'projective' binary codings of a line segment.

(iii) *The code structures of $\mathcal{F}$ and $\mathcal{G}$ are the same.* In this case $F_{\mathcal{F}\mathcal{G}}$ is precisely the homeomorphism $H : A_\mathcal{F} \to A_\mathcal{G}$ provided in Theorems 4.14.5 and 4.14.9 when $\chi = 1$. An example is shown in the bottom right panel of Figure 4.30. Here you can see that $F_{\mathcal{F}\mathcal{G}}$ is continuous. The IFSs $\mathcal{F} = \mathcal{F}_3$ and $\mathcal{G}$ are illustrated in

Figure 4.32 Example of a fractal transformation. These two pictures are homeomorphic, by the fractal homeomorphism theorem.

Figure 4.33 Before and after a fractal transformation. Which is before?

Figure 4.31, which also shows the result (see the bottom right image) of applying the inverse of the homeomorphism to the transformed image.

The code structures of $\mathcal{F}_3$ and $\mathcal{G}$ are the same because both collages are made of affine transformations. The problem of mismatching binary codings, mentioned at the end of point (ii), is resolved; in this case they match.

Two other examples of homeomorphic fractal transformations are illustrated in Figures 4.32 and 4.33. These use similar IFSs to the pair used in (iii) above.

4.16 Directed IFSs and general deterministic fractals

Here we introduce a natural generalization of the IFS concept called a **directed IFS**. The basic idea is very simple and has the virtue of being easy to understand and apply.

This is not generalization for its own sake! Firstly we will show that there exists a unique attractor that obeys a self-referential equation; thus we obtain a wider class of deterministic fractals. Secondly, the concept of a directed IFS handles multiple fractals simultaneously and is such that the theory of fractal tops and fractal homeomorphisms, adjusted to the more general setting, continues to hold. This means that we can deal topologically with multiple set attractors within a single coordinate system. Thirdly, the directed IFS framework subsumes the other main generalizations of the IFS framework, including deterministic graph-directed IFSs [68], recurrent IFSs [10] and local IFS theory [12]. Fourthly, this generalization arose in an entirely natural manner upon consideration of fractal tops and of the boundaries of set attractors of IFSs: how might one describe the essential features of a fractal top without having to go through all the usual IFS machinery?

In this brief introduction to directed IFSs we concentrate on aspects related to set attractors and fractal tops. We leave it to the reader to apply the same underlying idea to generalize orbital pictures, orbital sets, orbital measures and so on, as needed.

Let $\mathcal{F}$ denote a hyperbolic IFS, with attractor A, code space Ω and code space mapping $\phi_{\mathcal{F}} : \Omega \to A$. Let $\Omega_0 \subset \Omega$ denote any shift-invariant subspace of code space, and let Ω_0 be closed. A shift-invariant subspace is one that is unchanged by the application of the shift transformation $S : \Omega \to \Omega$. So we have

$$S : \Omega \to \Omega, \quad \Omega_0 \subset \Omega, \quad S|_{\Omega_0} : \Omega_0 \to \Omega_0 \quad \text{and} \quad S|_{\Omega_0}(\Omega_0) = \Omega_0.$$

DEFINITION 4.16.1   A **directed IFS** is an IFS $\mathcal{F} = \{\mathbb{X}; f_1, f_2, \ldots, f_N\}$ together with a closed nonempty shift-invariant subspace $\Omega_0 \subset \Omega = \Omega_{\{1,2,\ldots,N\}}$. It is denoted by $(\mathcal{F}, \Omega_0)$ or by

$$(\{\mathbb{X}; f_1, f_2, \ldots, f_N\}, \Omega_0).$$

For simplicity we assume throughout that $\mathbb{X}$ is a compact metric space and that $\mathcal{F}$ is hyperbolic.

DEFINITION 4.16.2   The **set attractor** of the directed IFS $(\mathcal{F}, \Omega_0)$ is the set

$$A_0 = \phi_{\mathcal{F}}(\Omega_0).$$

We think of the order in which the functions in the IFS may be applied as being 'directed' by the shift-invariant subspace Ω_0. This is distinct from a graph-directed IFS, where the order in which the functions may be applied is restricted to orderings specified by a (combinatorial) graph. It follows easily from the above definitions that the set attractor of a directed IFS $(\mathcal{F}, \Omega_0)$ is a compact subset of the set attractor of $\mathcal{F}$. Think no less of it for that. After all, consider how many interesting compact sets there are in the filled unit square!

The following theorem tells us that a compact set obeys a self-referential equation of a very general kind, Equation (4.5.11), iff it is the set attractor of a directed IFS.

THEOREM 4.16.3   *Let* $\mathcal{F} = \{\mathbb{X}; f_1, f_2, \ldots, f_N\}$ *be a hyperbolic IFS, and let* $\phi_{\mathcal{F}} : \Omega_{\{1,2,\ldots,N\}} \to A_{\mathcal{F}}$ *denote the corresponding code space mapping. Then there exist compact sets* $A_n \subset \mathbb{X}$ *for* $n = 1, 2, \ldots, N$, *at least one of which is nonempty, such that*

$$A_1 \cup A_2 \cup \cdots \cup A_N = f_1(A_1) \cup f_2(A_2) \cup \cdots \cup f_N(A_N) \qquad (4.16.1)$$

iff there is a closed set $\Omega_0 \subset \Omega_{\{1,2,\ldots,N\}}$ *such that* $(\mathcal{F}, \Omega_0)$ *is a directed IFS whose set attractor is* $A_0 = A_1 \cup A_2 \cup \cdots \cup A_N$.

PROOF   Suppose that $(\{\mathbb{X}; f_1, f_2, \ldots, f_N\}, \Omega_0)$ is a directed IFS whose attractor is $A_0 \subset \mathbb{X}$. Then

$$A_0 = \phi_{\mathcal{F}}(\Omega_0) = \phi_{\mathcal{F}}(\Omega_1) \cup \phi_{\mathcal{F}}(\Omega_2) \cup \cdots \cup \phi_{\mathcal{F}}(\Omega_N),$$

where $\Omega_n = \{\sigma \in \Omega_0 : \sigma_1 = n\}$ for $n = 1, 2, \ldots, N$. But this is the same as

$$A_0 = f_1\big(\phi_{\mathcal{F}}(S(\Omega_1))\big) \cup f_2\big(\phi_{\mathcal{F}}(S(\Omega_2))\big) \cup \cdots \cup f_N\big(\phi_{\mathcal{F}}(S(\Omega_N))\big).$$

So, we define $A_n = \phi_{\mathcal{F}}(S(\Omega_n))$. Then we have

$$A_0 = f_1(A_1) \cup f_2(A_2) \cup \cdots \cup f_N(A_N).$$

But the shift invariance of Ω_0 implies that

$$\Omega_0 = S(\Omega_0) = S(\Omega_1 \cup \Omega_2 \cup \cdots \cup \Omega_N) = S(\Omega_1) \cup S(\Omega_2) \cup \cdots \cup S(\Omega_N),$$

which provides us with

$$A_0 = \phi_{\mathcal{F}}(\Omega_0) = \phi_{\mathcal{F}}(S(\Omega_1)) \cup \phi_{\mathcal{F}}(S(\Omega_2)) \cup \cdots \cup \phi_{\mathcal{F}}(S(\Omega_N))$$
$$= A_1 \cup A_2 \cup \cdots \cup A_N.$$

Since $\phi_{\mathcal{F}}$ and S are continuous and Ω_0 is compact and nonempty, it follows that the sets $A_1, A_2, \ldots, A_N$ are compact and at least one of them is nonempty.

Conversely, suppose that there exist compact sets $A_n \subset \mathbb{X}$ for $n = 1, 2, \ldots, N$ such that

$$A_1 \cup A_2 \cup \cdots \cup A_N = f_1(A_1) \cup f_2(A_2) \cup \cdots \cup f_N(A_N).$$

Let $x_1 \in f_1(A_1) \cup f_2(A_2) \cup \cdots \cup f_N(A_N)$. Then there exist $\sigma_1 \in \{1, 2, \ldots, N\}$ and $x_2 \in A_{\sigma_1}$ such that $f_{\sigma_1}(x_2) = x_1$. But then $x_2 \in f_1(A_1) \cup f_2(A_2) \cup \cdots \cup f_N(A_N)$, so there exist $\sigma_2 \in \{1, 2, \ldots, N\}$ and $x_3 \in A_{\sigma_1}$ such that $f_{\sigma_2}(x_3) = x_2$. Continuing in this manner we find a sequence of σ_ks and x_ks such that

$$x_1 = f_{\sigma_1} \circ f_{\sigma_2} \circ \cdots \circ f_{\sigma_k}(x_{k+1}).$$

But the sequence on the right converges to $\phi_{\mathcal{F}}(\sigma)$, where $\sigma = \sigma_1 \sigma_2 \cdots$. It follows that we can find a subset Ω_0 of code space such that $\phi_{\mathcal{F}}(\Omega_0)$. This subset can be chosen to be closed and shift invariant. The latter is immediate and the former is achieved by taking the closure of Ω_0 in case it is not already closed. Finally, Ω_0 is nonempty because at least one A_n is nonempty. $\square$

In view of Theorem 4.16.3 and the content of Theorems 4.16.6 and 4.16.7 below, and of Section 4.17, we make the following definition. The concept of a directed IFS is a convenient theoretical structure for handling the diverse properties of all sets that, it appears to me, it is natural to call 'deterministic fractals'.

DEFINITION 4.16.4   A compact nonempty subset A_0 of a complete metric space $\mathbb{X}$ is called a **deterministic fractal** (set) iff there are nonempty compact sets $A_n \subset A$ and strictly contractive functions $f_n : A_n \to A$ such that Equation (4.16.1) holds, with $A_0 = \bigcup_{n=1}^{N} A_n$.

Why do we call these objects 'fractals'? The set attractor of an overlapping IFS has the same dimension as the space in which it lies, say $\mathbb{R}^2$. We call the set attractors of directed IFSs 'fractal' because they possess distinctive subsets, such as their boundaries, that do indeed appear typically to possess non-integer Hausdorff dimension.

As a simple example of a directed IFS let $\mathcal{F}_1 = \{\mathbb{R}^2; f_1, f_2\}$ have attractor A_1. Let

$$\mathcal{F}_2 = \{\mathbb{R}^2; f_3, f_4, f_5\}$$

have attractor A_2 that intersects A_1. For example see Figure 4.34, where A_1 is a Sierpinski triangle and A_2 is a line segment. Let

$$\Omega_0 = \Omega_{\{1,2\}} \cup \Omega_{\{3,4,5\}} \subset \Omega_{\{1,2,3,4,5\}}.$$

Then it is readily verified that Ω_0 is shift invariant and that

$$A_0 = A_1 \cup A_2.$$

So here the attractor of a directed IFS is the union of the attractors of two IFSs. This might all seem rather trivial. But it is not. As we will show in Section 4.17, the theory of fractal tops and related homeomorphisms can be adapted to directed IFSs: two deterministic fractals have the same code structure iff they are homeomorphic. This means that we can handle multiple fractals and their intersections, using

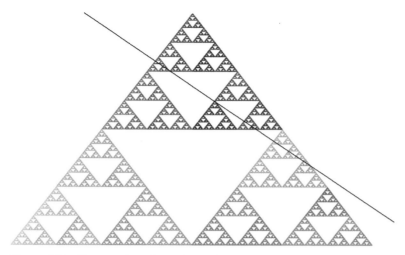

Figure 4.34 The natural topology of $A_1 \cup A_2$ is the same as the identification topology induced by $\phi_0 : \Omega_0 \to A_0$. Here A_1 is a Sierpinski triangle and A_2 is a line segment.

code space manipulation, in ways that are somewhat analogous to the handling of intersections of geometrical objects such as parabolas and straight lines using coordinate geometry. But our conclusions will of course be topological rather than geometrical!

The next simple examples show that the attractors of directed IFSs may be much more complicated. Let

$$\Omega_0 = \big\{\sigma \in \Omega_{\{1,2,\ldots,5\}} : \sigma_k \in \{1, 2, 4\} \Rightarrow \sigma_{k+1} \in \{1, 2, 3\},$$
$$\text{otherwise } \sigma_{k+1} \in \{4, 5\}, \text{ for all } k\big\}.$$

Then it is easy to see that, by the very form of its definition, $\Omega_0 = S(\Omega_0)$. Let

$$\mathcal{F}_1 = \big\{\mathbb{R}^2; \big(\tfrac{1}{2}(x + 2), \tfrac{1}{2}(y + 1)\big), \big(\tfrac{1}{2}(x + 3), \tfrac{1}{2}y\big),$$
$$\big(\tfrac{1}{2}(x + 4), \tfrac{1}{2}y\big), (x - 2, y), (-x + 1, -y + 1)\big\}.$$

Then the set attractor of the directed IFS $(\mathcal{F}_1, \Omega_0)$ is precisely the set pictured in Figure 0.6 in the Introduction, consisting of a triangle and a square. Look back and see how the chaos game was used to compute this set attractor. Let

$$\Omega_0' = \big\{\sigma_1\sigma_2\cdots\sigma_k : \sigma \in \Omega_0, k \in \{0, 1, \ldots\}\big\}.$$

Then you will see that at the kth step of the chaos game we generate a string $\sigma \in \Omega_0'$ of length k and compute the point $f_\sigma(x_0)$. Clearly these approximations approach elements of A_0 geometrically fast, and distribute themselves ergodically upon it, in the usual manner. In Figure 4.35 we show a decomposition of A_0 into sets A_1, A_2, A_3, A_4 and A_5, corresponding to Equation (4.16.1).

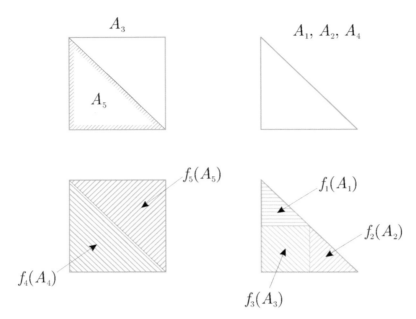

Figure 4.35 The upper two images show the set attractor A_0 of the directed IFS $(\mathcal{F}_1, \Omega_0)$ and the sets A_1, A_2, A_3, A_4 and A_5 in the self-referential Equation (4.16.1). The lower two images show how A_0 is the union of $f_1(A_1)$, $f_2(A_2)$, $f_3(A_3)$, $f_4(A_4)$ and $f_5(A_5)$.

Two related examples are illustrated in Figures 4.36 and 4.37. In Tables 4.9 and 4.10 we give the IFS codes for the set of projective transformations described in Figure 4.38. Note that neither IFS on its own makes a fern! The IFS in Table 4.9 represents a 'fern' in which the fronds do not meet the main stem, while the IFS in Table 4.10 represents a 'fern' in which the stalks of the fronds stick out of the opposite side of the main stem. Here we choose the IFS to be $\{\mathbb{R}^2; f_1, f_2, \ldots, f_8\}$. But we choose the shift-invariant space to be

$$\Omega_0 = \big\{\sigma \in \Omega_{\{1,2,\ldots,8\}} : \sigma_k \in \{1, 2, 3, 4\} \Rightarrow \sigma_{k+1} \in \{1, 2, 7, 8\},$$
$$\text{otherwise } \sigma_{k+1} \in \{3, 4, 5, 6\}, \text{ for all } k\big\}.$$

It is readily verified that $S(\Omega_0) = \Omega_0$. In fact, if we write $\Omega_0 = \Omega_1 + \Omega_2$, where

$$\Omega_1 = \{\sigma \in \Omega_0 : \sigma_1 \in \{1, 2, 7, 8\}\}$$

and

$$\Omega_2 = \{\sigma \in \Omega_0 : \sigma_1 \in \{3, 4, 5, 6\}\},$$

then the two individual ferns, which appear superimposed in Figure 4.37, are given by $\phi_{\mathcal{F}}(\Omega_1)$ and $\phi_{\mathcal{F}}(\Omega_2)$, and we note that $S(\Omega_1) = S(\Omega_2) = \Omega_0$.

How do we know that the above description is correct? Really, this follows from noting that these examples can be re-expressed in the language of recurrent

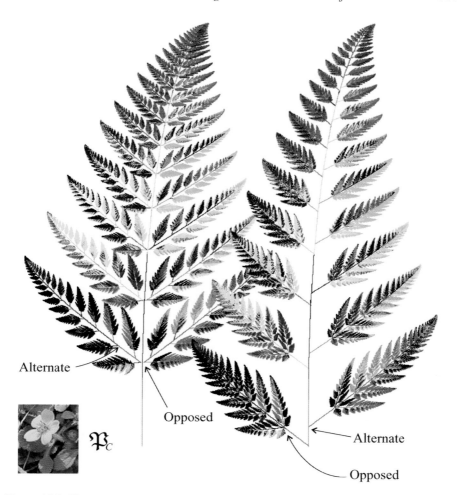

Figure 4.36   The two main images constitute a single set attractor of a directed IFS, but neither fern on its own is an attractor. It has been coloured by colour-stealing from the picture $\mathfrak{P}_C$ shown inset.

IFS theory; see [10] or [13]. But many shift-invariant subspaces do not correspond to finite-order Markov processes, as will be explained in Section 4.18.

We will use the following definition several times in the next few pages.

DEFINITION 4.16.5   Let $\mathbb{X}$ and $\mathbb{Y}$ be topological spaces. A transformation $f : \mathbb{X} \to \mathbb{Y}$ is called **open** iff $f(\mathcal{O})$ is open whenever $\mathcal{O} \subset \mathbb{X}$ is open.

The following theorem says that the boundary of the set attractor of an IFS may contain an interesting deterministic fractal, provided that all the transformations involved are open.

THEOREM 4.16.6   *Let $\mathcal{F} = \{\mathbb{X}; f_1, F_2, \dots, f_N\}$ be an IFS such that each transformation f_n is open. Let A denote the set attractor of $\mathcal{F}$ and let ∂A denote*

Interwoven fronds
illustrate that this is a
single fractal object

$\mathfrak{P}_C$

Figure 4.37 A recurrent IFS, corresponding to the eight transformations in Tables 4.9 and 4.10, generated this single attractor (main figure). This is a picture of the fractal top of this directed IFS, computed using the chaos game. The panel at top left shows the intricate interplay of the object *with itself*. The picture $\mathfrak{P}_C$ from which colours were stolen is shown inset.

its boundary. Then there exist closed subsets $\partial A_n \subset \partial A$ such that

$$\partial A = f_1(\partial A_1) \cup f_2(\partial A_2) \cup \cdots \cup f_N(\partial A_N).$$

Let $\phi_{\mathcal{F}} : \Omega \to A$ denote the code space function associated with $\mathcal{F}$. If ∂A is nonempty and $\Omega_0 = \phi_{\mathcal{F}}^{-1}(\partial A)$ then $S(\Omega_0) \subset \Omega_0$ and consequently Ω_0 contains a closed nonempty shift-invariant subset $\widetilde{\Omega}_0 = \lim_{k \to \infty} S^{\circ k}(\Omega_0)$. The attractor of the directed IFS $(\mathcal{F}, \widetilde{\Omega}_0)$ is contained in ∂A.

PROOF   Let us write $\mathrm{Int}(V)$ to denote the interior of a set $V \subset \mathbb{X}$. The key observation is that $A = \bigcup_{n=1}^{N} f_n(A)$ implies

$$\mathrm{Int}(A) = \mathrm{Int}\left(\bigcup_{n=1}^{N} f_n(A)\right) \supset \bigcup_{n=1}^{N} \mathrm{Int}(f_n(A)) \supset \bigcup_{n=1}^{N} f_n(\mathrm{Int}(A)).$$

Table 4.9   *A projective IFS code. This, together with the code in
Table 4.10, is used in Figure 4.37. Part of the attractor is shown
below. A similar but different pair of IFS codes was used to produce
Figure 4.36*

n	a_n	b_n	c_n	d_n	e_n	f_n	g_n	h_n	j_n	p_n
1	85	4	0	−4	85	160	0	0	100	$\frac{7}{10}$
2	1	0	0	0	16	0	0	0	100	$\frac{1}{10}$
3	20	−26	−40	23	22	80	0	0	100	$\frac{1}{10}$
4	−15	28	30	26	24	40	0	0	100	$\frac{1}{10}$

Table 4.10   *A projective IFS code. Part of the attractor is pictured
below. This code, together with the values in Table 4.9, is used in
Figure 4.37*

n	a_n	b_n	c_n	d_n	e_n	f_n	g_n	h_n	j_n	p_n
5	85	4	30	−4	85	160	0	0	100	$\frac{7}{10}$
6	2	0	200	0	16	0	0	0	100	$\frac{1}{10}$
7	20	−26	200	23	22	80	0	0	100	$\frac{1}{10}$
8	−15	28	200	26	24	40	0	0	100	$\frac{1}{10}$

In the last step here we have used the fact that the transformations f_n are open.
It now follows that no point in the boundary of A is the image under any f_n of a
point in the interior of A. Hence each point on the boundary must be the image
of a point on the boundary under one of the f_n. This proves the first assertion.
It also follows that $S(\Omega_0) \subset \Omega_0$. Consequently $\{S^{\circ k}(\Omega_0)\}_{k=1}^{\infty}$ is a nested sequence
of nonempty compact sets which converges to a nonempty compact set $\widetilde{\Omega}_0$ such
that $S(\widetilde{\Omega}_0) = \widetilde{\Omega}_0$. The attractor of the directed IFS $(\mathcal{F}, \widetilde{\Omega}_0)$, namely $\phi_{\mathcal{F}}(\widetilde{\Omega}_0)$, is
clearly contained in ∂A. I have drawn attention to this special subset of the
boundary of the attractor of $\mathcal{F}$ because I think it is the key to understanding
the boundary as a whole; it may be treated as a condensation set from which

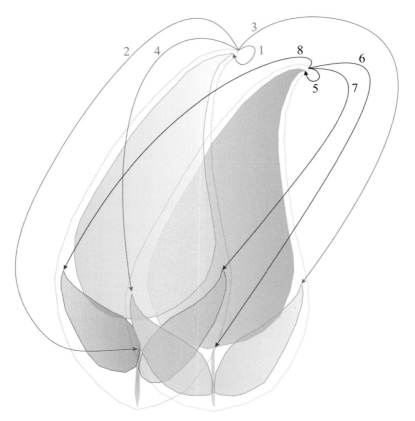

Figure 4.38 The actions of the eight transformations, given in Tables 4.9 and 4.10, of the directed IFS used in Figure 4.37 are illustrated here by showing how they act on leaf-shaped regions corresponding to the whole or part of the ferns in Figure 4.37.

the whole boundary may be constructed, quite naturally, by iterated application of $\mathcal{F}$. □

A **local IFS** is an IFS that may be denoted by

$$\mathcal{F} = \{\mathbb{X}; f_n : \mathbb{X}_n \to \mathbb{X}, n = 1, 2, \ldots, N\},$$

where $\bigcup_{n=1}^{N} \mathbb{X}_n = \mathbb{X}$, $\mathbb{X}$ is a compact metric space, each $\mathbb{X}_n$ is a member of $\mathbb{H}(\mathbb{X})$, the set of nonempty compact subsets of $\mathbb{X}$, and each f_n is a strict contraction. A set $A \in \mathbb{H}(\mathbb{X})$ is called a set attractor of the local IFS $\mathcal{F}$ iff it obeys

$$A = f_1(\mathbb{X}_1 \cap A) \cup f_2(\mathbb{X}_2 \cap A) \cup \cdots \cup f_N(\mathbb{X}_N \cap A). \tag{4.16.2}$$

Local IFSs are used extensively in fractal image compression, see [12], and there are efficient algorithms for computing their attractors, including pixel-chaining; see for example [62], pp. 207–10. The latter can be readily adapted to the computation of the 'top' of the attractor of a local IFS. We mention local IFSs here because they provide a rich source of deterministic fractals.

THEOREM 4.16.7   *Let $\mathcal{F} = \{\mathbb{X}; f_n : \mathbb{X}_n \to \mathbb{X}, n = 1, 2, \ldots, N\}$ be a local IFS. Then it possesses at least one attractor A. Each attractor of a local IFS is a deterministic fractal.*

PROOF   Define $K_0 = \mathbb{X}$ and, recursively,

$$K_k = f_1(\mathbb{X}_1 \cap K_{k-1}) \cup f_2(\mathbb{X}_2 \cap K_{k-1}) \cup \cdots \cup f_N(\mathbb{X}_N \cap K_{k-1})$$

for $k = 1, 2, \ldots$ Then the sequence of sets $\{K_k \in \mathbb{H}(\mathbb{X})\}_{k=0}^{\infty}$ is a decreasing sequence of compact sets and so possesses a unique limit A. This limit obeys Equation (4.16.2). Let $A_n = \mathbb{X}_n \cap A$. Then

$$\bigcup_{n=1}^{N} f_n(A_n) = \bigcup_{n=1}^{N} A_n = \bigcup_{n=1}^{N} (\mathbb{X}_n \cap A) = \mathbb{X} \cap A = A.$$

$\square$

Note that the union of two attractors of a local IFS is also an attractor. In particular, a local IFS possesses a unique 'largest' attractor, that mentioned in the proof of Theorem 4.16.7 above.

A rich class of directed IFSs is provided by fractal tops. If $\Omega_{\mathcal{F}}$ is the tops code space of a hyperbolic IFS $\mathcal{F}$ then $\overline{\Omega_{\mathcal{F}}}$ is shift invariant and $(\mathcal{F}, \overline{\Omega_{\mathcal{F}}})$ is a directed IFS. Its set attractor is the same as the set attractor of $\mathcal{F}$. Typically, when the IFS $\mathcal{F}$ is overlapping, the symbolic dynamical system formed by the shift transformation acting on the closure of the tops code space, $S : \overline{\Omega_{\mathcal{F}}} \to \overline{\Omega_{\mathcal{F}}}$, does not correspond to any finite-order Markov chain, because the mapping $S : \overline{\Omega_{\mathcal{F}}} \to \overline{\Omega_{\mathcal{F}}}$ is not open. This means that techniques used to explore invariant measures and attractors associated with recurrent IFSs [10] and graph-directed IFSs [68] may not be generally applicable to directed IFSs.

A specific example is provided by

$$\mathcal{F} = \{[0, 1] \subset \mathbb{R}; f_1(x) = \beta x + (1 - \beta), f_2(x) = \alpha x\},$$

where $0.5 < \alpha < 1$, $0.5 < \beta < 1$, and is illustrated in Figure 4.39. We find that $A_{\mathcal{F}} = [0, 1]$ and that

$$\Omega_{\mathcal{F}} = \left\{ \sigma \in \Omega_{\{1,2\}} : S^{\circ k+1}(\sigma) \geq \tau_{\mathcal{F}}\left(\frac{1 - \beta}{\alpha}\right) \quad \Rightarrow \quad \sigma_k = 2, \text{ for } k = 1, 2, \ldots \right\}.$$

It is possible to establish many cases for which $\tau_{\mathcal{F}}((1 - \beta)/\alpha)$ does not end in $\overline{1}$, for example if we choose $\alpha = \beta = 2/3$. Then $S : \overline{\Omega_{\mathcal{F}}} \to \overline{\Omega_{\mathcal{F}}}$ is not open. See [77].

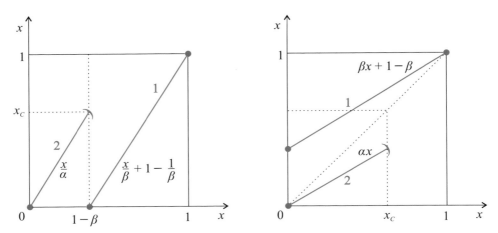

Figure 4.39   The tops dynamical system and the branches of its inverse, which define a restricted IFS.

The idea of a directed IFS is new to fractal geometry. But a more general structure, of iterated set maps, has been investigated in [69] and is quite closely related.

4.17   The top of a directed IFS

The following will tell us that most of the theory of transformations between fractal tops goes through for directed IFSs. Hence lies our success in colour-stealing, using directed IFSs in place of hyperbolic IFS, as illustrated in Figure 4.37. The main difference, in general, is that there is no tops dynamical system.

We present the main ideas somewhat concisely, and in such a way that they can be used in Chapter 5 in connection with superfractals. These key ideas are the same as in Section 4.14, and the proofs, which we omit, are entirely analogous. I want to stress here that although these ideas are very simple, they may look complicated because they require quite a few symbols for their expression.

Let $\mathcal{F} = \{\mathbb{X}; f_1, f_2, \ldots, f_N\}$ be a hyperbolic IFS. Let $A_{\mathcal{F}}$ denote its attractor and let $\phi_{\mathcal{F}} : \Omega_{\{1,2,\ldots,N\}} \to A_{\mathcal{F}}$ denote the associated addressing function. Let $\Sigma \subset \Omega_{\{1,2,\ldots,N\}}$ be closed and define

$$A_{(\mathcal{F},\Sigma)} = \phi_{\mathcal{F}}(\Sigma).$$

When Σ is shift invariant, $(\mathcal{F}, \Sigma)$ is of course a directed IFS and $A_{(\mathcal{F},\Sigma)}$ is a deterministic fractal, but we want to discuss sets of the form of $A_{(\mathcal{F},\Sigma)}$ quite generally. Indeed, in Chapter 5 we will represent certain V-variable and random fractal sets in just this way. Now let $\tau_{(\mathcal{F},\Sigma)} : A_{(\mathcal{F},\Sigma)} \to \Sigma$ be defined by

$$\tau_{(\mathcal{F},\Sigma)}(x) = \max\{\sigma \in \Sigma : \phi_{\mathcal{F}}(\sigma) = x\} \quad \text{for all } x \in A_{(\mathcal{F},\Sigma)}.$$

Notice that

$$(\phi_{\mathcal{F}} \circ \tau_{(\mathcal{F},\Sigma)})(x) = x \quad \text{for all } x \in A_{(\mathcal{F},\Sigma)}.$$

We define the **restricted tops code space** $\Omega_{(\mathcal{F},\Sigma)}$ to be the set

$$\Omega_{(\mathcal{F},\Sigma)} := \tau_{(\mathcal{F},\Sigma)}\big(A_{(\mathcal{F},\Sigma)}\big) = \tau_{(\mathcal{F},\Sigma)}(\phi_{\mathcal{F}}(\Sigma))$$

and we define

$$\phi_{(\mathcal{F},\Sigma)} : \overline{\Omega}_{(\mathcal{F},\Sigma)} \to A_{(\mathcal{F},\Sigma)}$$

to be the restriction of $\phi_{\mathcal{F}}$ to the closure of $\Omega_{(\mathcal{F},\Sigma)}$. The latter maps the closure of $\Omega_{(\mathcal{F},\Sigma)}$ continuously onto $A_{(\mathcal{F},\Sigma)}$.

DEFINITION 4.17.1   The set of sets $Q_{(\mathcal{F},\Sigma)} := \big\{\phi_{(\mathcal{F},\Sigma)}^{-1}(x) : x \in A_{(\mathcal{F},\Sigma)}\big\}$ is called a **restricted code structure** of the set $A_{(\mathcal{F},\Sigma)}$.

When $(\mathcal{F}, \Sigma)$ is a directed IFS we call $Q_{(\mathcal{F},\Sigma)}$ *the* restricted code structure of the directed IFS $(\mathcal{F}, \Sigma)$. We remark that clearly on the one hand not all sets possess restricted code structures and on the other hand a set may possess many restricted code structures. Moreover, we may consider 'projective restricted code structures' and 'Möbius restricted code structures', with obvious meanings; then we discover that projective restricted code structure is a property of projective geometry, and so on, along the lines discussed at the end of Chapter 3.

DEFINITION 4.17.2   We say that **two restricted code structures** $Q_{(\mathcal{F},\Sigma)}$ **and** $Q_{(\mathcal{G},\Xi)}$ **are homeomorphic** iff there is a homeomorphism $\chi : \overline{\Omega}_{(\mathcal{F},\Sigma)} \to \overline{\Omega}_{(\mathcal{G},\Xi)}$ that respects the code structures, that is, such that $q \in Q_{(\mathcal{F},\Sigma)} \iff \chi(q) \in Q_{(\mathcal{G},\Xi)}$.

THEOREM 4.17.3   *Let the two restricted code structures* $Q_{(\mathcal{F},\Sigma)}$ *and* $Q_{(\mathcal{G},\Xi)}$ *be homeomorphic. Then* $A_{(\mathcal{F},\Sigma)}$ *and* $A_{(\mathcal{G},\Xi)}$ *are homeomorphic. That is, there exists a homeomorphism* $H : A_{(\mathcal{F},\Sigma)} \to B_{(\mathcal{G},\Xi)}$ *such that*

$$H\big(A_{(\mathcal{F},\Sigma)}\big) = A_{(\mathcal{G},\Xi)}.$$

If the homeomorphism $\chi : \overline{\Omega}_{(\mathcal{F},\Sigma)} \to \overline{\Omega}_{(\mathcal{G},\Xi)}$ *has the property that* $\chi(\Omega_{(\mathcal{F},\Sigma)}) = \Omega_{(\mathcal{G},\Xi)}$ *then*

$$\tau_{(\mathcal{G},\Xi)} \circ H = \chi \circ \tau_{(\mathcal{F},\Sigma)}.$$

PROOF   This follows similar lines to the discussion in Section 4.14 and is therefore omitted.   □

Analogous results to those concerning fractal transformations apply to directed IFSs and, as we will see in Chapter 5, in connection with superfractals. This extends our ability to construct homeomorphisms between pictures and between objects. One immediate application is to the construction of synthetic imagery via colour-stealing plus tops, where now 'tops' has a more general meaning. (For example, the continuous deformations between the pictures in Figure 5.13 rely on Theorem 4.17.3.)

4.18   A very special case: $S : \Omega \to \Omega$ is open

Here we quote the brilliant work of William Parry, showing that a directed IFS is essentially a graph-directed IFS if and only if $S : \Omega \to \Omega$ is open.

We consider the symbolic dynamical system $S : \Omega \to \Omega$ where $\Omega \subset \Omega_{\{1,2,\dots,N\}}$ is a closed set with $S(\Omega) = \Omega$. The topology on Ω is the restriction of the natural topology on $\Omega_{\{1,2,\dots,N\}}$ to Ω. Equivalently, the topology on Ω is the natural topology of the compact metric space (Ω, d_Ω). See Theorem 1.9.6.

Recall that a cylinder set of $\Omega_{\{1,2,\dots,N\}}$ is a subset of $\Omega_{\{1,2,\dots,N\}}$ that can be written in the form

$$\mathcal{C}(\sigma) := \big\{\omega \in \Omega_{\{1,2,\dots,N\}} : \omega_n = \sigma_n \text{ for all } n = 1, 2, \dots, |\sigma|\big\},$$

for some $\sigma \in \Omega'_{\{1,2,\dots,N\}}$. A set $\mathcal{C} \subset \Omega$ is called a cylinder set of Ω when it is the same as the intersection of a cylinder set of $\Omega_{\{1,2,\dots,N\}}$ with Ω; that is, when there exists a cylinder set $\widetilde{\mathcal{C}} \subset \Omega_{\{1,2,\dots,N\}}$ such that $\mathcal{C} = \widetilde{\mathcal{C}} \cap \Omega$. Let us introduce the notation

$$\Omega' := \big\{\sigma \in \Omega'_{\{1,2,\dots,N\}} : \mathcal{C}(\sigma) \cap \Omega \neq \varnothing\big\} \cup \{\varnothing\}.$$

That is, Ω' consists of the set of all finite-length 'beginnings' of strings which belong to Ω, together with the empty string. An important property of Ω is that its cylinders are both open and closed. The transformation $S : \Omega \to \Omega$ is continuous, but it is not necessarily open, as illustrated at the end of Section 4.16.

Parry [77] defines the dynamical system $S : \Omega \to \Omega$ to be an **intrinsic Markov chain of order r** when the following condition is satisfied: whenever k is a positive integer with $k \geq r$, then

$$\sigma_1 \sigma_2 \cdots \sigma_k \in \Omega' \quad \text{and} \quad \sigma_{k-r+1}\sigma_{k-r+2} \cdots \sigma_k \sigma_{k+1} \in \Omega'$$
$$\text{imply that } \sigma_1 \sigma_2 \cdots \sigma_{k-r+1} \cdots \sigma_k \sigma_{k+1} \in \Omega'. \tag{4.18.1}$$

THEOREM 4.18.1   *Let $\Omega \subset \Omega_{\{1,2,\dots,N\}}$ be closed and shift invariant. Then the dynamical system $S : \Omega \to \Omega$ is an intrinsic Markov chain iff S is open.*

PROOF   See [77], p. 370.   □

In the course of his proof, Parry shows that the pair of equivalent assertions in the statement of the theorem are also equivalent to the following: there exists a finite set of cylinders $\{\mathcal{C}(\sigma^{(m)}) \subset \Omega : m = 1, 2, \dots, M\}$, where $\sigma^{(m)} \in \Omega'$ and

$|\sigma^{(m)}| = |\sigma^{(1)}| \geq 1$ for $m = 1, 2, \ldots, M$, such that

$$\Omega = \bigcup_{m=1}^{M} \mathcal{C}(\sigma^{(m)}) \quad \text{and} \quad S(\mathcal{C}(\sigma^{(m)})) = \mathcal{C}(S(\sigma^{(m)})) \quad \text{for each } m = 1, 2, \ldots, M.$$

This is equivalent, back on the attractor A of the directed IFS, to the statement that there exists a finite set $\{B_1, B_2, \ldots, B_L\}$ of compact subsets of A such that $A = \bigcup_{m=1}^{M} B_m$, where each B_m can be expressed as $B_m = \bigcup_{(n,l) \in I_m} f_n(B_l)$ and where, for each m, $I_m \subset \{1, 2, \ldots, N\} \times \{1, 2, \ldots, L\}$. This structure is of the kind that, in essence, underlies recurrent IFSs, see [10], and graph-directed IFSs, as described in [68] and [98]. This shows that the concept of directed IFSs subsumes these other well-known generalizations of IFS theory.

4.19   Invariant measures for tops dynamical systems

Here, most briefly, we alert the reader to the rich literature that exists regarding invariant measures associated with symbolic dynamical systems. In the case where the system derives from a fractal top, the tops dynamical system and the corresponding symbolic dynamical system are often equivalent, and invariant measures that assign zero to each single point can be moved painlessly back and forth between the two systems. These measures are relevant to the design of efficient algorithms for computing tops functions; the question when such measures have maximum entropy, or are nice and smoothly distributed on the fractal top, is very interesting. In this regard we note the work of Lasota and Yorke [61] and more recent studies that cite it.

We are interested in normalized Borel measures defined on the Borel field generated by the cylinders of Ω. Quite generally, given any closed shift-invariant code space $\Omega \subset \Omega_{\{1,2,\ldots,N\}}$ there exists at least one measure $\mu \in \mathbb{P}(\Omega)$ such that $S(\mu) = \mu$; see [56], p. 139.

The symbolic dynamical system $S : \Omega \to \Omega$ is called **regionally transitive** iff, whenever σ and ω belong to Ω', there exists an integer n such that

$$S^{\circ n}(\mathcal{C}(\sigma)) \cap \mathcal{C}(\omega) \neq \varnothing.$$

We quote the following from [77], p. 371.

THEOREM 4.19.1   *If $S : \Omega \to \Omega$ is a regionally transitive symbolic dynamical system then there is a normalized Borel invariant measure μ, with respect to which S is ergodic, such that*

$$h_\mu(S) = e(S),$$

where

$$h_\mu(S) = \lim_{n \to \infty} -\frac{1}{n} \sum_{\{\sigma \in \Omega' : |\sigma| = n\}} \mu(C(\sigma)) \log \mu(C(\sigma))$$

and

$$e(S) = \lim_{n \to \infty} \left(-\frac{1}{n} \log \theta(n) \right),$$

where $\theta(n) = \left| \{\sigma \in \Omega' : |\sigma| = n\} \right|$. *For all normalized invariant measures,* $h_\mu(S_0) \leq e(S)$. *When S is open,* μ *is unique.*

By means of the tops transformation, we can use such measures to define corresponding invariant measures for corresponding tops dynamical systems. These are relevant to the efficient computation of fractal tops with algorithms that are of the chaos game type. We note the recent review by Zbigniew Nitecki [75] on the topological entropy and pre-image structure of symbolic dynamical systems.

CHAPTER 5

Superfractals

5.1 Introduction

In this chaper we introduce the theory and some applications of superfractals. Superfractals are families of sometimes beautiful fractal objects which can be explored by means of the chaos game (see Figure 5.1) and which span the gap between fully 'random' fractal objects and deterministic fractal objects. Our presentation is via elementary examples and theory together with brief descriptions of natural feasible extensions. This chapter depends heavily on the earlier material. You may grasp intuitively the key ideas of superfractals and '2-variability' by studying the experiment described in Section 5.2. But be careful not to miss subtleties such as those that enable the construction of superfractals whose elements are vast collections of homeomorphic pictures, as for example those illustrated in Figures 5.13 and 5.17.

A superfractal (see Figure 5.2) is associated with a single underlying hyperbolic IFS. It has its own underlying logical structure, called the 'V-variability' of the superfractal, for some $V \in \{1, 2, \ldots\}$, which enables us to sample the superfractal by means of the chaos game and produce generalized fractal objects such as fractal sets, pictures, measures and so on, one after another. The property of V-variability enables us to 'dance on the superfractal', sometimes producing wondrous objects in splendid succession.

Deterministic fractal sets are, in some practical ways, not rich enough to describe real objects in the real world. No two clouds are ever exactly the same, are they? You might say 'These leaves are from a beech tree, and these are oak leaves', because you see underlying common patterns as well as randomness. We desire models that display both random and deterministic aspects.

Superfractals were discovered during an exciting and intense collaboration between John Hutchinson, Örjan Stenflo and myself at the Australian National University in Canberra, Australia during the last four months of 2002. We had never all worked together before but were united in our interest in random fractals. See for example [49]–[52]. In our experience they were hard to compute. By a combination

Figure 5.1 A superfractal of seascapes is represented here by diverse triangular pictures surrounding the square ocean picture. The triangular pictures correspond to samples of a particular superfractal, produced by the chaos game. See the text for more details.

of ideas from all of us we soon had superfactals up and running on a computer, and we realized that they were significant and, in the case $V \geq 2$, new. The important case of '1-variable' sets is a special case of a type of random fractal investigated by Kifer and others; see [43], [3], [59] and [90]. See also [2]. The computational experiment described below in Section 5.2 was in essence the first experiment we ran. The basic theory of V-variable sets and measures and some applications first appeared in [16], which is the main source for the presentation here. Subsequent developments are reported in [19], [20] and [21]; see the end of Section 5.19.

5.2 Computational experiment: glimpse of a superfractal

We begin by describing a computational experiment that gives you the basic idea of how to compute V-variable fractals. This experiment was first described in published form in [16]. We start with two projective IFSs, $\mathcal{F}_1 = \{\Box; f_1^1, f_2^1\}$ and $\mathcal{F}_2 = \{\Box; f_1^2, f_2^2\}$, where $\Box := [0, 1] \times [0, 1] \subset \mathbb{R}^2$. The IFS codes are given in Tables 5.1 and 5.2.

In Figure 5.3 we illustrate the action of $\mathcal{F}_1$ and $\mathcal{F}_2$ on a triangle ABC. Both $\mathcal{F}_1$ and $\mathcal{F}_2$ are linked IFSs, and their attractors, pictures of which are included

Table 5.1   *Projective IFS code for one of the two IFSs used in the computational experiment. The attractor is pictured below, as in the tables that follow. See Figure 5.3*

n	a_n	b_n	c_n	d_n	e_n	f_n	g_n	h_n	j_n	p_n
1	8	-6	5	8	6	3	0	0	16	$\frac{1}{2}$
2	8	6	3	-8	6	11	0	0	16	$\frac{1}{2}$

h is an example of a superfractal

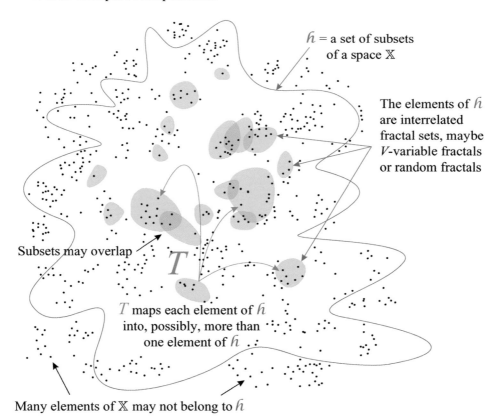

h = a set of subsets of a space $\mathbb{X}$

The elements of h are interrelated fractal sets, maybe V-variable fractals or random fractals

Subsets may overlap

T

T maps each element of h into, possibly, more than one element of h

Many elements of $\mathbb{X}$ may not belong to h

Figure 5.2   Illustation of some aspects of a superfractal. See also Figure 5.22. The transformation T is explained in Section 5.18.

Table 5.2   *The other IFS code used in the computational experiment*

n	a_n	b_n	c_n	d_n	e_n	f_n	g_n	h_n	j_n	p_n
1	8	-6	5	-8	-6	13	0	0	16	$\frac{1}{2}$
2	8	6	3	8	-6	5	0	0	16	$\frac{1}{2}$

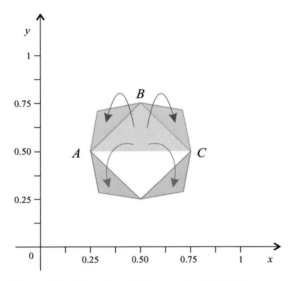

Figure 5.3   The actions of the two affine IFSs $\mathcal{F}_1$ (upper arrows) and $\mathcal{F}_2$ (lower arrows) used in the computational experiment.

in Figure 5.6, may be represented as the graphs of continuous mappings g_m : $[0, 1] \to \mathbb{R}^2$, $m = 1, 2$, such that $g_1(0) = g_2(0)$ and $g_1(1) = g_2(1)$. We say that these continuous paths are **tethered** at A and C.

To set up the experiment we need two pairs of digital buffers, i.e. memory displays, ($\square_1$, $\square_2$) and ($\square_{1'}$, $\square_{2'}$). Each buffer is the same size and may be used to represent binary, black-and-white, images. The buffers are discretized copies of $\square \subset \mathbb{R}^2$. The value 0 is assigned to white pixels and the value 1 is assigned to black pixels. We refer to these pairs of buffers as the input screens and output screens respectively. We also need a digital processor that can read from and write to each pair of buffers.

We initialize the experiment by setting some of the pixels to 0 and some to 1 on each input screen, as illustrated by the two fish in the top pair of images on the left in Figure 5.4. We also clear both output screens by setting all their pixels to zero.

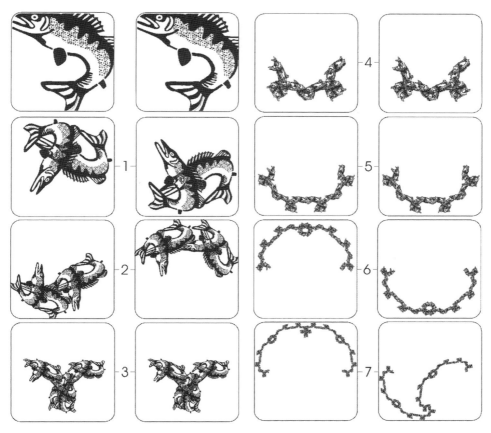

Figure 5.4 Shows successive pairs of images produced in the course of the computational experiment described in Section 5.2.

We now construct a sequence of pairs of images, which appear successively in pairs on the output screens, as follows.

(i) Pick randomly one of the IFSs $\mathcal{F}_1$ and $\mathcal{F}_2$, say $\mathcal{F}_{n_1}$. Apply $f_1^{n_1}$ to the image on either $\square_1$ or $\square_2$, selected randomly, to make an image on $\square_{1'}$. Then apply $f_2^{n_1}$ to the image on either $\square_1$ or $\square_2$, also selected randomly, and overlay the resulting image I on the image now already on $\square_{1'}$. That is, form the union of the black region of I with the black region on $\square_{1'}$ and put the result back onto $\square_{1'}$.

(ii) Again pick randomly one of the IFSs $\mathcal{F}_1$ and $\mathcal{F}_2$, say $\mathcal{F}_{n_2}$. Apply $f_1^{n_2}$ to the image on either $\square_1$ or $\square_2$, selected randomly, to make an image on $\square_{2'}$. Also apply $f_2^{n_2}$ to the image on $\square_1$ or $\square_2$, also selected randomly, and overlay the resulting image on the image now already on $\square_{2'}$.

(iii) Switch input and output, clear the new output screens and repeat steps (i) and (ii).

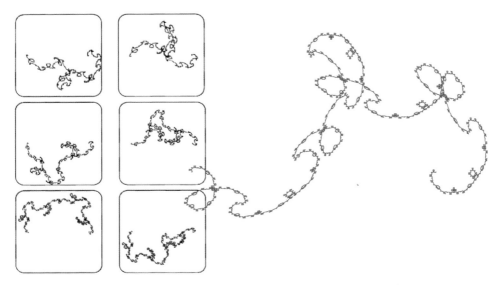

Figure 5.5 This illustrates successive pairs of images on the two output screens after a certain number $L > 20$ of iterations. The red image is a close-up of another image in the stationary state distribution. Such pictures are typical of the 'stationary state' at the printed resolution.

(iv) Repeat step (iii) many times, to allow the system to settle into its 'stationary state'.

In Figure 5.4 we observe the start of the sequence of pairs of images obtained in a particular trial, for the first seven iterations. Notice that some pairs are the same! Then in Figure 5.5 we show three successive pairs of computed screens, obtained after more than twenty iterations. These latter images are typical of those obtained after twenty or more iterations. Notice how the two images in the second pair of panels in Figure 5.5 consist of the union of smaller affine copies of the images in the top pair of panels. We observe that these images are very diverse but that they always appear to represent continuous 'random' paths in $\mathbb{R}^2$ tethered at A and C; they correspond to the stationary state, at the resolution of the images. More precisely, with probability 1, the empirically obtained distribution of such images over a long experimental run corresponds to the stationary state distribution.

In order to illustrate the intricate structure of the observed curves, which in general possess both disordered and ordered aspects, Figure 5.5 includes a close-up in red of one such curve. We also observe that the images produced in the stationary state are independent of the starting images. For example, if the initial images in the example had been dots or lines instead of fish, and the same sequence of random choices had been made, then the images corresponding to those in Figure 5.5 would have been the same at the printed resolution.

Figure 5.6 illustrates the attractors of the IFSs $\mathcal{F}_1$ and $\mathcal{F}_2$.

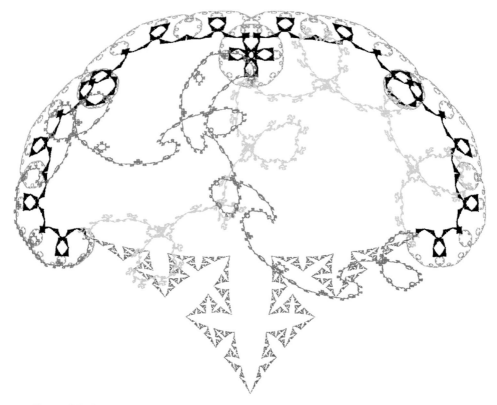

Figure 5.6 The attractors of the two IFSs used in the computational experiment are illustrated in red and green. The black curve is a 1-variable fractal set while the colours lavender and yellow indicate two 2-variable fractal sets associated with the same superIFS.

5.3   SuperIFSs and superfractals

Our computational experiment suggests that we begin our mathematical treatment of superfractals by defining a compact metric space $\mathbb{X}$ together with a collection of hyperbolic IFSs $\{\mathcal{F}_m : m = 1, 2, \ldots, M\}$ with probabilities, where

$$\mathcal{F}_m = \left\{ \mathbb{X}; f_1^m, f_2^m, \ldots, f_{L_m}^m; p_1^m, p_2^m, \ldots, p_{L_m}^m \right\}$$

and $M \geq 1$ is an integer, to be a **superIFS**. We denote it by

$$\{\mathbb{X}; \mathcal{F}_1, \mathcal{F}_2, \ldots, \mathcal{F}_M\} \quad \text{or} \quad \{\mathbb{X}; \mathcal{F}_1, \mathcal{F}_2, \ldots, \mathcal{F}_M; P_1, P_2, \ldots, P_M\}, \quad (5.3.1)$$

where the P_m are probabilities, with

$$\sum_{m=1}^{M} P_m = 1, \, P_m \geq 0 \quad \text{for all } m \in \{1, 2, \ldots, M\}.$$

Notice that a superIFS is not an IFS: each IFS $\mathcal{F}_m$ defines a transformation on $\mathbb{H}(\mathbb{X})$, $\Pi_{\mathscr{C}}(\mathbb{X})$ and $\mathbb{P}(\mathbb{X})$, see Equations (4.3.1), (4.8.2) and (4.3.2), but we do not specify how it might map $\mathbb{X}$ into itself.

We will use a given superIFS to define IFSs acting on higher-order spaces such as $\mathbb{H}(\mathbb{X})$, $\mathbb{P}(\mathbb{X})$ and

$$\mathbb{H}(\mathbb{X})^V = \underbrace{\mathbb{H}(\mathbb{X}) \times \mathbb{H}(\mathbb{X}) \times \cdots \mathbb{H}(\mathbb{X})}_{V \text{ times}}, \quad \text{where } V \in \{1, 2, \dots\}.$$

The metric space $(\mathbb{H}(\mathbb{X}), d_{\mathbb{H}(\mathbb{X})})$ consists of the nonempty compact subsets of $\mathbb{X}$, with the Hausdorff metric. The metric space $(\mathbb{P}(\mathbb{X}), d_{\mathbb{P}(\mathbb{X})})$ consists of the normalized Borel measures on $\mathbb{X}$, with the Monge–Kantorovitch metric. The metric of each higher-order space is deduced from the metric of the space from which it is built, in an obvious way, as discussed in Chapter 1.

The IFSs on these new higher-order spaces arise in a very natural manner. The attractors of *these* IFSs provide the sets that we call **superfractals**. A superfractal is thus a collection of fractal objects, such as fractal sets, generalized orbital pictures, relative tops or measures. The structure of the higher-order space provides the variability V of the superfractal, ties its elements together and constrains the underlying code space, as we will explain. Superfractals have useful properties, not the least of which is that they may be sampled by means of the chaos game in various different settings. We describe the objects by adjectives such as V-variable, 2-variable and 1-variable. A certain superfractal may consist of 2-variable projective fractal sets while another might consist of 1-variable orbital pictures.

5.4   1-variable IFSs

The superIFS in Equation (5.3.1) may be used to define the hyperbolic IFS

$$\mathcal{F}^{(1)} = \{\mathbb{H}(\mathbb{X}); \mathcal{F}_1, \mathcal{F}_2, \dots, \mathcal{F}_M; P_1, P_2, \dots, P_M\}. \tag{5.4.1}$$

How do we know that $\mathcal{F}^{(1)}$ is indeed a hyperbolic IFS? Firstly, we know that $(\mathbb{H}(\mathbb{X}), d_{\mathbb{H}})$ is a compact metric space, by Exercise 1.13.3. Secondly, here each of the IFSs $\mathcal{F}_m$ acts as a transformation

$$\mathcal{F}_m : \mathbb{H}(\mathbb{X}) \to \mathbb{H}(\mathbb{X})$$

defined by

$$\mathcal{F}_m(B) = \bigcup_{l=1}^{L_m} f_l^m(B) \quad \text{for } m = 1, 2, \dots, M.$$

Each transformations is strictly contractive on $\mathbb{H}(\mathbb{X})$ with respect to the Hausdorff metric, by Theorem 2.4.8. That is,

$$d_{\mathbb{H}}(\mathcal{F}_m(X), \mathcal{F}_m(Y)) \le \lambda_m d_{\mathbb{H}}(X, Y) \quad \text{for all } X, Y \in \mathbb{H}(\mathbb{X}),$$

for some $\lambda_m \in [0, 1)$ and all $m = 1, 2, \ldots, M$. For reasons that will become increasingly clear, we call $\mathcal{F}^{(1)}$ a **1-variable IFS**.

5.5   The set attractor $A^{(1)}$ of the 1-variable IFS $\mathcal{F}^{(1)}$

The set attractor $A^{(1)}$ of the IFS $\mathcal{F}^{(1)}$ must be an element of $\mathbb{H}(\mathbb{H}(\mathbb{X}))$. That is, it must be a nonempty set, compact with respect to the metric $d_{\mathbb{H}(\mathbb{X})}$, whose elements are themselves compact nonempty subsets of $\mathbb{X}$. Furthermore, this set attractor must be given by

$$A^{(1)} = \phi_{\mathcal{F}^{(1)}}\big(\Omega_{\{1,2,\ldots,M\}}\big),$$

where

$$\phi_{\mathcal{F}^{(1)}} : \Omega_{\{1,2,\ldots,M\}} \to \mathbb{H}(\mathbb{X})$$

is the code space mapping associated with $\mathcal{F}^{(1)}$. Following the definition of ϕ in Theorem 3.3.12, we see that the elements of $A^{(1)}$ must be precisely the points $A_\sigma \in \mathbb{H}(\mathbb{X})$ that can be written in the form

$$A_\sigma = \phi_{\mathcal{F}^{(1)}}(\sigma) = \lim_{k \to \infty} \mathcal{F}_{\sigma_1} \circ \mathcal{F}_{\sigma_2} \circ \cdots \circ \mathcal{F}_{\sigma_k}(\mathbb{X})$$

for $\sigma \in \Omega_{\{1,2,\ldots,M\}}$. We call A_σ a **1-variable fractal set**. The set

$$A^{(1)} = \big\{A_\sigma : \sigma \in \Omega_{\{1,2,\ldots,M\}}\big\}$$

is an example of a **1-variable superfractal**.

Next we look at two examples of 1-variable fractal sets. Setting $\sigma = \overline{1}$ we find

$$\phi_{\mathcal{F}^{(1)}}(\overline{1}) = \lim_{k \to \infty} \mathcal{F}_1^{\circ k}(\mathbb{X}) = A_{\overline{1}},$$

the attractor of the IFS $\mathcal{F}_1$. Supposing that $N \geq 3$ and setting $\sigma = 13\overline{12}$ we find that

$$\phi_{\mathcal{F}^{(1)}}(13\overline{12}) = \mathcal{F}_1 \circ \mathcal{F}_3\Big(\lim_{k \to \infty} (\mathcal{F}_1 \circ \mathcal{F}_2)^{\circ k}(\mathbb{X})\Big) = \mathcal{F}_1 \circ \mathcal{F}_3(A_{\overline{12}}),$$

where $A_{\overline{12}}$ is the attractor of the IFS

$$\mathcal{F}_1 \circ \mathcal{F}_2 := \big\{\mathbb{X}; f_{l_1}^1 \circ f_{l_2}^2, l_1 = 1, 2, \ldots, L_1, l_2 = 1, 2, \ldots, L_2\big\}.$$

Here and elsewhere we use the continuity of the transformations $\mathcal{F}_m : \mathbb{H}(\mathbb{X}) \to \mathbb{H}(\mathbb{X})$.

By considering such examples you will see that $A^{(1)}$ contains the attractors of all the IFSs that can be constructed by composing finite sequences of the $\mathcal{F}_m$. It also contains the images of these attractors under such finite compositions of the $\mathcal{F}_m$. To be precise, let $\mathcal{F}_\sigma$ denote the hyperbolic IFS defined, in an obvious manner, by

$$\mathcal{F}_\sigma = \mathcal{F}_{\sigma_1} \circ \mathcal{F}_{\sigma_2} \circ \cdots \circ \mathcal{F}_{\sigma_{|\sigma|}} \quad \text{for all } \sigma \in \Omega'_{\{1,2,\ldots,M\}}\backslash\{`\varnothing'\}.$$

Let $A_\sigma \in \mathbb{H}(\mathbb{X})$ denote the set attractor of $\mathcal{F}_\sigma$ for all $\sigma \in \Omega'_{\{1,2,\ldots,M\}} \backslash \{`\varnothing'\}$. Then

$$A^{(1)} \supset A^{(1)}_{rational} := \left\{ \mathcal{F}_\sigma(A_\omega) : \sigma, \omega \in \Omega'_{\{1,2,\ldots,M\}}, \omega \neq `\varnothing' \right\}.$$

Here $\mathcal{F}_\varnothing : \mathbb{H}(\mathbb{X}) \to \mathbb{H}(\mathbb{X})$ is taken to be the identity transformation. It is straight-forward to prove that $A^{(1)}$ is the closure of $A^{(1)}_{rational}$ in the metric $d_{\mathbb{H}(\mathbb{X})}$. You should not find the latter too hard to envisage, after your experiences with $\mathbb{H}(\mathbb{X})$ at the end of Chapter 1.

In order to see what 1-variable fractal sets look like, and to obtain a better understanding of the corresponding superfractal, we can use the chaos game.

5.6 Chaos game reveals 1-variable fractal sets

The 1-variable IFS $\mathcal{F}^{(1)}$ is a hyperbolic IFS with probabilities. Hence it possesses a unique invariant probability measure $\mu^{(1)} \in \mathbb{P}(\mathbb{H}(\mathbb{X}))$. Here $\mathbb{P}(\mathbb{H}(\mathbb{X}))$ is the space of normalized Borel measures defined on $\mathbb{H}(\mathbb{X})$. The chaos game, adapted to the present setting, yields sequences of 'points' $\{X_k\}_{k=1}^\infty$, $X_k \in \mathbb{H}(\mathbb{X})$, that almost always converge to the measure attractor $\mu^{(1)}$ of $\mathcal{F}^{(1)}$. The manner in which this convergence occurs is governed by Theorem 4.5.1. Here it is, transcribed to the present setting.

THEOREM 5.6.1   *Let* $\mathcal{F}^{(1)} = \{\mathbb{H}(\mathbb{X}); \mathcal{F}_1, \mathcal{F}_2, \ldots, \mathcal{F}_M; P_1, P_2, \ldots, P_M\}$ *be a 1-variable IFS and let* $\mu^{(1)} \in \mathbb{P}(\mathbb{H}(\mathbb{X}))$ *denote its measure attractor. Specify a starting set* $X_1 \in \mathbb{H}(\mathbb{X})$. *Define a random orbit of the IFS to be* $\{X_k\}_{k=1}^\infty$, *where* $X_{k+1} = \mathcal{F}_m(X_k)$ *with probability* P_m *independently of all other choices. Then for almost all random orbits* $\{X_k\}_{k=1}^\infty$ *we have*

$$\mu^{(1)}(B) = \lim_{k\to\infty} \frac{|B \cap \{X_1, X_2, \ldots, X_k\}|}{k}. \tag{5.6.1}$$

for all $B \in \mathbb{B}(\mathbb{H}(\mathbb{X}))$ *such that* $\mu(\partial B) = 0$, *where* ∂B *denotes the boundary of* B.

This tells us that, almost always, the random orbit $\{X_k\}_{k=1}^\infty$ is asymptotically distributed according to $\mu^{(1)}$. In practice, working to some level of approximation, say to within the accuracy specified by a parameter $\epsilon > 0$, the sequence of sets $\{X_k\}_{k=1}^\infty$ will be such that, after a readily estimated number of steps K, each of the sets $X_{K+1}, X_{K+2}, \ldots$ will lie to within the Hausdorff distance ϵ of an element of $A^{(1)}$.

Let us think more precisely how this occurs. Since $A^{(1)}$ is a point in the metric space $(\mathbb{H}(\mathbb{H}(\mathbb{X})), d_{\mathbb{H}(\mathbb{H}(\mathbb{X}))})$ we can find a finite set of points $A_1, A_2, \ldots, A_{\mathcal{N}(\epsilon)}$ in $\mathbb{H}(\mathbb{X})$ such that every point of $A^{(1)}$ is contained in one of the balls $B(A_n, \epsilon)$ of radius ϵ, each of which is centred on one of the points A_n. What does such a ball $B(A_n, \epsilon)$ look like? A_n is a fractal set belonging to the set attractor $A^{(1)}$. The ball $B(A_n, \epsilon)$ is the set of all nonempty compact subsets of $\mathbb{X}$ that, when dilated by ϵ, contain A_n and moreover are such that each is contained in A_n dilated by ϵ.

Now suppose that the probabilities $\{P_m : m = 1, 2, \ldots, M\}$ are strictly positive. Then the support of $\mu^{(1)}$ is $A^{(1)}$. Theorem 5.6.1 then, generally speaking, tells us that our random orbit $\{X_k\}_{k=1}^{\infty}$ will visit $B(A_n, \epsilon)$ at a proportion of its random steps that is equal asymptotically to $\mu^{(1)}(B(A_n, \epsilon)) > 0$. Exactly how this occurs, of course, is stated exactly in Theorem 5.6.1.

So, in practical two-dimensional pictorial examples, we may choose $\epsilon > 0$ to be sufficiently small that compact sets, when represented as images on the screen or on paper in our chosen experimental set-up, are visually indistinguishable when they are at a distance less than ϵ apart in the metric $d_{\mathbb{H}(\mathbb{R}^2)}$. Then we may estimate a number $K > 0$ such that the elements of $\{X_k\}_{k=K+1}^{\infty}$ are all indistinguishable, at viewing resolution, from elements of $A^{(1)}$.

If we discard sufficiently many initial iterates of the random orbit $\{X_k\}_{k=1}^{\infty}$ then we should find that our chaos-game orbit produces, one after another, fractal sets that, to within viewing resolution, are all indistinguishable from elements of $A^{(1)}$, and our orbit will dance wildly about yielding at each step one of a finite but probably huge number of representatives of the actual elements of $A^{(1)}$ over and over again, with relative frequencies controlled by $\mu^{(1)}$. And this is exactly what does happen! It has the same flavour as our computational experiment.

In Figure 5.7 we show some pictures of approximations to some of the 1-variable fractal sets produced by following just such a chaos-game orbit in a case for which $M = 2$. The attractor $A_{\bar{1}}$ of the IFS $\mathcal{F}_1$ is illustrated at the top right in Figure 5.23, from which you can deduce the form of the three affine transformations that comprise $\mathcal{F}_1$. The IFS $\mathcal{F}_2$ consists of four affine transformations on $\mathbb{R}^2$ and its attractor $A_{\bar{2}}$ is illustrated at the top left of Figure 5.23. You can see that the individual set attractors in Figure 5.7 look like semi-random mixtures of $A_{\bar{1}}$ and $A_{\bar{2}}$. The images in this figure were rendered by colour-stealing, with relative tops functions produced, quite remarkably, using the chaos game, as we will explain in Section 5.9.

In Figure 5.8 we show examples of 1-variable fractal interpolation functions, again produced using the chaos game. Here $M = 2$ and we have used two IFSs, $\mathcal{F}_1$ and $\mathcal{F}_2$. The attractor of each is the graph of a standard fractal interpolation function; see Section 5.14. Both graphs have the same endpoints. It is easy to show that the elements of the corresponding 1-variable superfractal are the graphs of continuous functions that connect the same pair of endpoints.

Another example is illustrated in Figure 5.9. Here $M = 2$ and $\mathcal{F}_m = \alpha_m \circ \mathcal{F} \circ \alpha_m^{-1}$, where α_m, for $m = 1$ and 2, is an affine rescaling in the x-direction and $\mathcal{F}$ is the IFS given in Table 0.1 in the Introduction. Here too the successive 2-variable fractal sets have been rendered by colour-stealing, to demonstrate how successive random iterates of approximate tops functions converge towards accurate tops functions; see Section 5.9. What you should notice, at this juncture, is how the silhouettes of the pictures, which represent successive 1-variable fractal

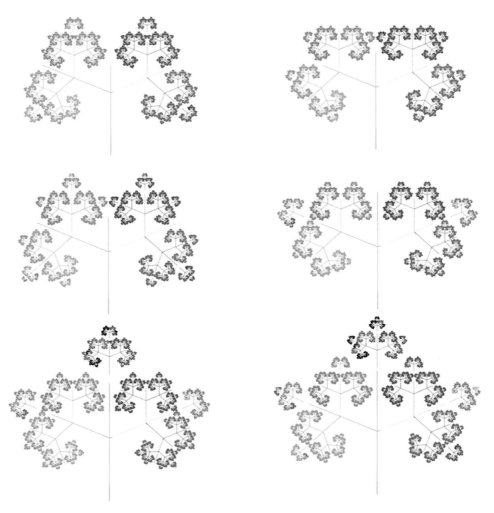

Figure 5.7 Some 1-variable fractals belonging to the same superfractal, computed by the chaos game. The sets are rendered by a variant of IFS colouring. Look closely, and do not hurry by. Note the differences.

sets produced by the chaos game, change from one step to the next. In this manner, over time we can obtain a visual sample of 1-variable fractal sets belonging to the corresponding superfractal, distributed according to the probability measure $\mu^{(1)}$.

5.7   Hausdorff dimension of some 1-variable fractal sets

DEFINITION 5.7.1   The superIFS $\{\mathbb{X}; \mathcal{F}_1, \mathcal{F}_2, \ldots, \mathcal{F}_M; P_1, P_2, \ldots, P_M\}$ is said to obey the **uniform open set condition** if there exists a nonempty open set $\mathcal{O} \subset \mathbb{X}$ such that

$$\mathcal{F}_m(\mathcal{O}) \subset \mathcal{O}$$

and

$$f_k^m(\mathcal{O}) \cap f_l^m(\mathcal{O}) = \varnothing \quad \text{if } k \neq l, \quad \text{for all } k, l \in \{1, 2, \ldots, L_m\}$$

and for all $m \in \{1, 2, \ldots, M\}$.

The following result due to Hambly [43] is noted in [16]. Here we refer to orthonormal transformations. An orthonormal transformation is a linear transformation that preserves euclidean distances and orthogonality.

THEOREM 5.7.2   *Let $\mathcal{N} > 1$ be a positive integer. Let the superIFS*

$$\{\mathbb{R}^{\mathcal{N}}; \mathcal{F}_1, \mathcal{F}_2, \ldots, \mathcal{F}_M; P_1, P_2, \ldots, P_M\}$$

obey the uniform open set condition. Let the functions that comprise the IFS $\mathcal{F}_m$ be similitudes of the form

$$f_l^m(x) = s_l^m O_l^m x + t_l^m,$$

where O_l^m is an orthonormal transformation, $s_l^m \in (0, 1)$ and $t_l^m \in \mathbb{R}^{\mathcal{N}}$, for all $l \in \{1, 2, \ldots, L_m\}$ and $m \in \{1, 2, \ldots, M\}$. Then, for almost all $A_\sigma \in A^{(1)}$,

$$\dim_H A_\sigma = D,$$

where D is the unique solution of

$$\sum_{m=1}^{M} P_m \ln \sum_{l=1}^{L_m} \left(s_l^m\right)^D = 0.$$

The case $M = 1$ of this theorem was first proved by Moran in 1946 [72]; see [48] and also [35], p. 118.

As you will have noticed, this book does not emphasize the subject of fractal dimensions. There are three principal reasons for this. Firstly, the subject is well covered in many other books on fractal geometry; see for example [31] and [35]. Secondly, in this book we focus on overlapping IFSs, where little is generally known about the fractal dimensions of, for example, the boundaries of attractors. Thirdly, in my own experience with the use of IFS in digital imaging, despite substantial efforts in this direction by my coworkers of which I have been aware, for example in connection with fractal image compression, I have not found notable applications of specific fractal dimension formulas. Lastly, my passion is in the direction of homeomorphisms between pictures and more protoplasmic ideas such as identification topologies on code space. So, while fractal dimensions are certainly of great general importance, I have not found them to be so in the kinds of problems of image representation on which I have chosen to work.

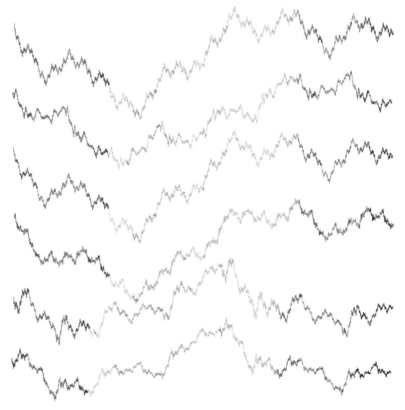

Figure 5.8 A few of the uncountably many 1-variable fractal interpolation functions belonging to a single superfractal. The graphs at the top and bottom are the attractors of the two IFSs that comprise the superIFS. Such graphs are produced in rapid succession by the chaos game. Can V-variable fractal interpolants be used to improve simulations of stockmarket portfolio performance, as proposed by Mandelbrot [65]?

5.8   The underlying IFS of a superIFS

In order to discuss sets of 1-variable tops and other objects associated with $\mathcal{F}^{(1)}$, as well as to prepare the way for V-variable fractals, we introduce some notation related to the IFS

$$\mathcal{F}_{underlying} = \left\{ \mathbb{X}; f_1^1, f_2^1, \ldots, f_{L_1}^1, f_1^2, f_2^2, \ldots, f_{L_2}^2, \ldots, f_1^M, f_2^M, \ldots, f_{L_M}^M \right\},$$

which we call the **underlying IFS**. We will let $A_{underlying}$ denote its attractor. We also write

$$\mathcal{F}_{underlying} = \left\{ \mathbb{X}; f_1, f_2, \ldots, f_{N_1}, f_{N_1+1}, f_{N_1+2}, \ldots, \right.$$
$$\left. f_{N_2}, \ldots, f_{N_{M-1}+1}, f_{N_{M-1}+2}, \ldots, f_N \right\},$$

where $N = L_1 + L_2 + \cdots + L_M$,

$$N_0 = 0, \quad N_1 = L_1, \quad N_2 = L_1 + L_2, \quad \ldots, \quad N_M = L_1 + L_2 + \cdots + L_M$$

Figure 5.9   A few elements of a random sequence of 1-variable tops, $\{\tau_k\}_{k=1}^{\infty}$, rendered by colour-stealing. Observe the differences in the silhouettes.

and

$$f_{N_{m-1}+l} = f_l^m \quad \text{for } l = 1, 2, \ldots, L_m \text{ and } m = 1, 2, \ldots, M.$$

We also write

$$I_m = \{N_{m-1} + 1, N_{m-1} + 2, \ldots, N_m\} \quad \text{for } m = 1, 2, \ldots, M, \quad (5.8.1)$$

to denote the members of the obvious partition of the set $\{1, 2, \ldots, N\}$.

All the V-variable fractal objects discussed in this chapter are built on or in some way connected to subsets of the attractor $A_{underlying}$ of the underlying IFS. You will know that you are dealing with the underlying IFS by the occurrence of the symbol N, which we reserve in this chapter for the number of functions f_n in the underlying IFS, whereas the symbol M is used to define the number of IFSs $\mathcal{F}_m$ in the superIFS. In particular, we will define relative tops functions of 1-variable fractal sets and establish homeomorphisms between them with the aid of the underlying IFS. We do this next.

5.9   Tops of 1-variable fractal sets

It follows from the discussion in Section 5.5 that the set attractor $A^{(1)}$ of the 1-variable IFS $\mathcal{F}^{(1)}$ can be written as

$$A^{(1)} = \left\{ A_\sigma : \sigma \in \Omega_{\{1,2,\ldots,M\}} \right\},$$

where

$$A_\sigma = \phi_{\mathcal{F}^{(1)}}(\sigma)$$

and

$$\phi_{\mathcal{F}^{(1)}}(\sigma) = \lim_{k \to \infty} \mathcal{F}_{\sigma_1} \circ \mathcal{F}_{\sigma_2} \circ \cdots \circ \mathcal{F}_{\sigma_k}(\mathbb{X})$$

for all $\sigma \in \Omega_{\{1,2,\ldots,M\}}$. Then the following key expression can be readily verified:

$$A_\sigma = \phi_{\mathcal{F}_{underlying}}(\Omega_\sigma),$$

where

$$\Omega_\sigma = I_{\sigma_1} \times I_{\sigma_2} \times I_{\sigma_3} \times \cdots \subset \Omega_{\{1,2,\ldots,N\}}$$

for all $\sigma = \sigma_1 \sigma_2 \sigma_3 \cdots \in \Omega_{\{1,2,\ldots,M\}}$ and the sets I_{σ_k} are defined in Equation (5.8.1). Be careful to distinguish between N and M here: $\Omega_{\{1,2,\ldots,N\}}$ is the code space associated with the underlying IFS $\mathcal{F}_{underlying}$ while $\Omega_{\{1,2,\ldots,M\}}$ is the code space associated with the 1-variable IFS $\mathcal{F}^{(1)}$.

Notice that Ω_σ is a closed subset of $\Omega_{\{1,2,\ldots,N\}}$. Hence A_σ is precisely of the form of $A_{(\mathcal{F},\Sigma)}$ in Section 4.17; we simply replace $\mathcal{F}$ by $\mathcal{F}_{underlying}$ and Σ by Ω_σ. That is, we have

$$A_\sigma = A_{(\mathcal{F}_{underlying},\Omega_\sigma)} \quad \text{for all } \sigma \in \Omega_{\{1,2,\ldots,M\}}. \tag{5.9.1}$$

So we can follow the development given in Section 4.17, but we must adapt the notation to fit the present circumstances. The restricted tops function $\tau_{(\mathcal{F},\Sigma)} : A_{(\mathcal{F},\Sigma)} \to \Sigma$ becomes here

$$\tau_\sigma : A_\sigma \to \Omega_\sigma,$$

where

$$\tau_\sigma(x) = \max \left\{ \omega \in \Omega_\sigma : \phi_{\mathcal{F}_{underlying}}(\omega) = x \right\} \quad \text{for all } x \in A_\sigma.$$

We call $\{\tau_\sigma : \sigma \in \Omega_{\{1,2,\ldots,M\}}\}$ the **set of 1-variable tops** associated with the IFS $\mathcal{F}^{(1)}$. This may be thought of as a set of pictures in the space $\Pi_{\Omega_{\{1,2,\ldots,M\}}}(\mathbb{X})$. That is, $\tau_\sigma \in \Pi_{\Omega_{\{1,2,\ldots,N\}}}(\mathbb{X})$ for each $\sigma \in \Omega_{\{1,2,\ldots,M\}}$ is treated as a picture whose domain is A_σ and whose colour space is $\Omega_{\{1,2,\ldots,N\}}$. As in Section 4.15, these tops functions may be used to render the 1-variable fractal sets A_σ via tops plus colour-stealing. Specifically, if $\mathfrak{P}_C$ is a picture from which colours are stolen and $\mathcal{F}_C$ is an IFS of N functions, which need not be distinct, with attractor $A_C \subset D_{\mathfrak{P}_C}$ and code space function $\phi_{\mathcal{F}_C} : \Omega_{\{1,2,\ldots,N\}} \to A_C$, then we assign the colour $\mathfrak{P}_C \circ \phi_{\mathcal{F}_C}(\tau_\sigma(x))$ to the point $x \in A_\sigma$.

The following wonderful result tells us that we can compute random orbits of 1-variable tops with the aid of the chaos game. Let $\mathcal{F}_m$ be a member of the superIFS in Equation (5.3.1). Then we define, consistently with Equation (4.12.1),

$$\mathcal{F}_{m,TOP} : \Pi_{\Omega_{\{1,2,\dots,N\}}}(\mathbb{X}) \to \Pi_{\Omega_{\{1,2,\dots,N\}}}(\mathbb{X}) \qquad (5.9.2)$$

as follows. Let $\tau \in \Pi_{\Omega_{\{1,2,\dots,N\}}}(\mathbb{X})$ and let D_τ denote the domain of τ. Then, for $m = 1, 2, \dots, M$, we define a set of disjoint sets by

$$D_{m,1} = f_{N_{m-1}+1}(D_\tau),$$
$$D_{m,2} = f_{N_{m-1}+2}(D_\tau) \backslash D_{m,1},$$
$$D_{m,3} = f_{N_{m-1}+3}(D_\tau) \backslash (D_{m,1} \cup D_{m,2}),$$
$$\vdots$$
$$D_{m,L_m} = f_{N_m}(D_\tau) \backslash \big(D_{m,1} \cup D_{m,2} \cup \cdots \cup D_{m,L_m-1}\big).$$

Let the domain of $\mathcal{F}_{m,TOP}(\tau)$ be $D_{m,1} \cup D_{m,2} \cup \cdots \cup D_{m,L_m}$, which is the same as the domain of $\mathcal{F}_m(\tau)$. Then we define

$$\mathcal{F}_{m,TOP}(\tau)(x) = (N_{m-1} + l)\tau\big(f_{(N_{m-1}+l)}^{-1}(x)\big) \quad \text{when } x \in D_{m,l},$$

for $l = 1, 2, \dots, L_m$ and $m = 1, 2, \dots, M$. Here $(N_{m-1} + l)\tau(f_{(N_{m-1}+l)}^{-1}(x))$ means the string obtained by putting the symbol corresponding to the value of $N_{m-1} + l$ in front of the infinite string $\tau(f_{(N_{m-1}+l)}^{-1}(x)) \in \Omega_{\{1,2,\dots,N\}}$.

T H E O R E M 5.9.1 *Let $\{\tau_\sigma : \sigma \in \Omega_{\{1,2,\dots,M\}}\}$ be the set of 1-variable tops associated with the IFS $\mathcal{F}^{(1)} = \{\mathbb{H}(\mathbb{X}); \mathcal{F}_1, \mathcal{F}_2, \dots, \mathcal{F}_M\}$. Let $\mathcal{F}_{m,TOP} : \Pi_{\Omega_{\{1,2,\dots,N\}}}(\mathbb{X}) \to \Pi_{\Omega_{\{1,2,\dots,N\}}}(\mathbb{X})$ be defined as in Equation (5.9.2). Then*

$$\mathcal{F}_{m,TOP}(\tau_\sigma) = \tau_{m\sigma} \quad \text{*for all* } \sigma \in \Omega_{\{1,2,\dots,M\}} \text{ *and* } m = 1, 2, \dots, M.$$

Furthermore, if $\tau_1, \tau_2 \in \Pi_{\Omega_{\{1,2,\dots,N\}}}(\mathbb{X})$ have the same domain, $D \subset \mathbb{X}$, then

$$\sup_{x \in \mathcal{F}_m(D)} |\mathcal{F}_{m,TOP}(\tau_1)(x) - \mathcal{F}_{m,TOP}(\tau_2)(x)| \le \tfrac{1}{2} \sup_{x \in D} |\tau_1(x) - \tau_2(x)|.$$

P R O O F The proof is completely mechanical and follows the same lines as the proof of Theorem 4.12.1. □

Theorem 5.9.1 tells us two things. (i) By means of the chaos game we can produce random orbits $\{\tau_k\}_{k=1}^\infty$ of 1-variable tops functions that are almost always asymptotically distributed according to the measure attractor $\mu^{(1)}$ of $\mathcal{F}^{(1)}$. We refer to the resulting probability measure on the set of 1-variable tops as the **push-forward** of $\mu^{(1)}$ on the set of 1-variable tops. Specifically, we start from any tops function $\tau_1 \in \{\tau_\sigma : \sigma \in \Omega_{\{1,2,\dots,M\}}\}$ and define

$$\tau_{k+1} = \mathcal{F}_{\sigma_k,TOP}(\tau_k) \quad \text{for } k = 1, 2, \dots, \qquad (5.9.3)$$

where, at the kth step, σ_k is chosen equal to m with probability P_m independently of all the other choices. (ii) The second part of Theorem 5.9.1 tells us that if

Figure 5.10 The image at top left represents a function $\tau_1 \in \Pi_{\Omega_{\{1,2,\dots,M\}}}(\mathbb{R}^2)$ and the images down the diagonal illustrate elements of a chaos game orbit $\{\tau_k\}_{k=1}^{\infty}$ produced by randomly applying transformations from the set $\{\mathcal{F}_{m,TOP} : m = 1, 2\}$. The images are rendered using colour-stealing. This illustrates Theorem 5.9.1.

we start the random iteration Equation (5.9.3) from any $\tau_1 \in \Pi_{\Omega_{\{1,2,\dots,N\}}}(\mathbb{X})$ whose domain belongs to $A^{(1)}$, the sequence $\{\tau_k\}_{k=1}^{\infty}$ will almost always approach, in the sense of Equation (5.6.1), the push-forward of $\mu^{(1)}$ on the set of 1-variable tops of $\mathcal{F}^{(1)}$. It also tells us that if we start from any $\tau_1 \in \Pi_{\Omega_{\{1,2,\dots,N\}}}(\mathbb{X})$ then the sequence $\{\tau_k\}_{k=1}^{\infty}$ will almost always approach, in the sense of Equation (5.6.1), a probability distribution on $\Pi_{\Omega_{\{1,2,\dots,N\}}}(\mathbb{X})$ that depends only on the domain of τ_1.

We illustrate these conclusions in Figures 5.9–5.11. Figure 5.9 corresponds to the 1-variable IFS $\mathcal{F}^{(1)}$ described at the end of Section 5.6. The top row illustrates τ_3, τ_5 and τ_7 from a random orbit $\{\tau_k\}_{k=1}^{\infty}$ rendered using colour-stealing, where the domain of τ_1 is approximately the attractor $A_{\bar{1}} \in A^{(1)}$ and the value of τ_1 is constant, say $\tau_1(x) = \bar{1}$ for all $x \in A_{\bar{1}}$. The bottom row illustrates τ_k for three values of k between 10 and 20. We emphasize how simple it is to produce such sequences of images consecutively while apparently maintaining accuracy to within viewing resolution.

Figures 5.10 and 5.11 also correspond to the 1-variable IFS $\mathcal{F}^{(1)}$ described at the end of Section 5.6; they were produced in a similar manner to the previous example, but the colours were stolen from a different picture. In this case the

Figure 5.11   Samples from the random orbit started in Figure 5.10 are illustrated here. There are lots of little differences between these rendered 1-variable 'tops' functions. Can you find some of them?

domain of $\tau_1 \in \Pi_{\Omega_{\{1,2,\dots,N\}}}(\mathbb{R}^2)$ is not an element of $A^{(1)}$. Successive images down the diagonal in Figure 5.10 illustrate τ_1, τ_{k_2}, τ_{k_3}, τ_{k_4} and τ_{k_5}, where $1 < k_2 < k_3 < k_4 < k_5 < 15$. Figure 5.11 shows several members of the same random sequence, for much larger values of k. Notice the differences between the images. These later images are approximations to pictures of 1-variable tops, despite the fact that the domain of τ_1 does not belong to $A^{(1)}$.

A fascinating illustration of the feasibility of computing random orbits of 1-variable tops, and of rendering them by colour-stealing 'on-the-fly', is the whole-sale production of sequences of varied pictures that are all homeomorphic, similar to the homeomorphic pictures described in Section 4.15. But now, instead of one picture we have a potentially endless sequence of them, and so we have a balance between orderliness and randomness.

5.10   Homeomorphisms between 1-variable fractal sets and between their tops

We continue the application of Section 4.17 to the present context. We previously mentioned this development just before Equation (5.9.1) above.

The **restricted tops code space** $\Omega_{(\mathcal{F},\Sigma)}$ is here the set

$$\Sigma_\sigma := \tau_\sigma(A_\sigma).$$

We define

$$\phi_\sigma : \overline{\Sigma}_\sigma \to A_\sigma$$

to be the restriction of $\phi_{\mathcal{F}_{underlying}} : \Omega_{\{1,2,\dots,N\}} \to A_{underlying}$ to the closure of Σ_σ. As in Definition 4.17.1, here we may refer to the set of sets

$$Q_\sigma := \left\{ \phi_\sigma^{-1}(x) : x \in A_\sigma \right\}$$

as a **restricted code structure** of the 1-variable fractal set A_σ, for each $\sigma \in \Omega_{\{1,2,\dots,M\}}$. Also, we say that **two restricted code structures Q_σ and Q_ω**

Figure 5.12 Which house do Peter and Elizabeth live in? The others are 1-variable fractal homeomorphisms of it, produced by following a random orbit on a superfractal.

are homeomorphic iff there is a homeomorphism $\chi : \overline{\Sigma}_\sigma \to \overline{\Sigma}_\omega$ such that $q \in Q_\sigma \iff \chi(q) \in Q_\omega$. Then Theorem 4.17.3, transcribed to the present setting, may be expressed as follows.

THEOREM 5.10.1   *Let* $\mathcal{F}^{(1)} = \{\mathbb{H}(\mathbb{X}); \mathcal{F}_1, \mathcal{F}_2, \dots, \mathcal{F}_M\}$ *be a 1-variable IFS, let* $A^{(1)} = \{A_\sigma : \sigma \in \Omega_{\{1,2,\dots,M\}}\}$ *be its set attractor and let* $\{\tau_\sigma : \sigma \in \Omega_{\{1,2,\dots,M\}}\}$ *be the associated set of 1-variable tops. Let two restricted code structures* Q_σ *and* Q_ω *be homeomorphic for some pair of addresses* $\sigma, \omega \in \Omega_{\{1,2,\dots,M\}}$. *Then* A_σ *and* A_ω *are homeomorphic. That is, there exists a homeomorphism* $H : A_\sigma \to A_\omega$ *such that*

$$H(A_\sigma) = A_\omega.$$

If the homeomorphism $\chi : \overline{\Sigma}_\sigma \to \overline{\Sigma}_\omega$ *has the property* $\chi(\Sigma_\sigma) = \Sigma_\omega$ *then*

$$\tau_\omega \circ H = \chi \circ \tau_\sigma.$$

Figures 5.12, 5.13, 5.15, 5.17 and 5.18 provide illustrations of the power of Theorem 5.10.1. In the first three of these figures $\mathcal{F}^{(1)} = \{\mathbb{H}(\mathbb{R}^2); \mathcal{F}_1, \mathcal{F}_2\}$ where the two IFSs $\mathcal{F}_1$ and $\mathcal{F}_2$ are of the form in Table 5.3, for say $\lambda = 0.4$ and $\lambda = 0.6$. The attractors of two such IFSs are illustrated in Figure 5.14, where they are rendered by IFS colouring and overlay the picture from which the colours in Figure 5.15 were stolen. It is reasonably straightforward to prove that all the IFSs represented by Table 5.3 have the same code structures, which I call 'tops quadtree'. Note, however, that a decomposition of a square into four 'just-touching' rectangles does not yield typically an IFS whose code structure is homeomorphic to those of the IFSs represented by Table 5.3.

Table 5.3 *A parameterized family of affine IFS codes, all with the same code structure. This family is used to produce the 1-variable continuous fractal transformations illustrated in Figures 5.12, 5.13 and 5.15*

n	a_n	b_n	c_n	d_n	e_n	f_n	p_n
1	λ	0	0	λ	0	0	$\frac{1}{4}$
2	$1-\lambda$	0	0	λ	λ	0	$\frac{1}{4}$
3	$1-\lambda$	0	0	$1-\lambda$	λ	λ	$\frac{1}{4}$
4	λ	0	0	$1-\lambda$	0	λ	$\frac{1}{4}$

Figure 5.13 Four different 1-variable fractal homeomorphisms applied to a picture of New York at dusk. This illustrates Theorem 5.10.1. See the main text. The original was obtained from BigStockPhoto.com. The copyright to the original is owned by Brian Kelly.

Figure 5.14 Attractors of the two IFSs $\mathcal{F}_1$ and $\mathcal{F}_2$ used in the 1-variable superIFS $\mathcal{F}^{(1)}$ to construct the 'stolen' pictures in Figure 5.15.

Figure 5.15 Examples of 1-variable superfractal transformations of surf at Waratah Bay in January 2005. The original photograph from which the colours were stolen sits in the background in Figure 5.14, which also illustrates the IFSs, whose code structures are homeomorphic, used to generate these and many other seascapes.

Figure 5.16   The attractors of two IFSs are illustrated here. They are both rendered using IFS colouring and are superimposed on a picture of the sea. Two such IFSs, $\mathcal{F}_1$ and $\mathcal{F}_2$, comprise the superIFS used to make Figures 5.17 and 5.18.

In Figure 5.12 you can see six pictures. Five were selected from one random orbit of 1-variable tops, rendered by colour-stealing. The stealing IFS also corresponds to Table 5.3 with $\lambda = 0.5$, and the picture $\mathfrak{P}_C$ from which colours were stolen is illustrated at lower left. As predicted by Theorem 5.10.1 and our earlier discussion of homeomorphisms, all these pictures are homeomorphic to one another and to the picture from which the colours were stolen. It is astonishing to see these images appear on the screen of a computer, one after another, each one momentarily a new delight. Figures 5.13 and 5.15 were produced similarly, but in neither of these cases is the picture $\mathfrak{P}_C$ included. You'll have to guess what it looks like. Can you?

Figures 5.16 and 5.18 illustrate a different 1-variable IFS, where each of two IFSs involved is constructed using four affine transformations that form a collage of a triangle. Figure 5.16 shows pictures of the attractors, both rendered using the same IFS colouring, of two such IFSs. A similar IFS was used for colour-stealing. The IFSs were chosen to have the same code space structure, ensuring that all the resulting 1-variable pictures are homeomorphic. Figure 5.17 involves more extreme distortions but each mathematical picture is homeomorphic to the original.

Imagine that the superfractal to which the pictures in Figure 5.18 belong is an art gallery full of triangular seascapes. By means of the chaos game you may rush about in the gallery, from picture to picture, sampling its contents according to the push-forwards of an invariant measure of the IFS $\mathcal{F}^{(1)}$.

5.11   Other sets of 1-variable fractal objects

In addition to providing the superfractal of 1-variable fractal sets $A^{(1)} = \{A_\sigma : \sigma \in \Omega_{\{1,2,...,M\}}\}$ and the set of 1-variable tops, $\{\tau_\sigma : \sigma \in \Omega_{\{1,2,...,M\}}\}$, the superIFS $\{\mathbb{X}; \mathcal{F}_1, \mathcal{F}_2, \ldots, \mathcal{F}_M\}$ may be used to describe many other collections of 1-variable fractal objects and, more generally, V-variable fractal objects. All these collections arise quite naturally and are interesting, at the very least, because of their potential for applications in computer graphics. Some of these collections are easiest to describe in the 1-variable case, so we do this next. Then their generalizations may be inferred after we formalize the concept of V-variablility for $V = 1, 2, 3, \ldots$

Figure 5.17 Pictures belonging to a 1-variable superfractal. Three IFSs were used, one for colour-stealing and two to construct $\mathcal{F}^{(1)}$. Each IFS is of the form illustrated in Figure 5.16, and each has the same code structure, which ensures that the mathematical pictures are homeomorphic.

The superfractal of 1-variable fractal measures

The superIFS in Equation (5.3.1) may be used to define the hyperbolic IFS

$$\mathcal{F}^{P(1)} = \{\mathbb{P}(\mathbb{X}); \mathcal{F}_1, \mathcal{F}_2, \ldots, \mathcal{F}_M; P_1, P_2, \ldots, P_M\}.$$

We may call this a **1-variable measure IFS**. How do we know that $\mathcal{F}^{P(1)}$ is indeed a hyperbolic IFS? Firstly, we know that $(\mathbb{P}(\mathbb{X}), d_{\mathbb{P}(\mathbb{X})})$ is a compact metric space, by Theorem 2.4.15. Secondly, here each of the IFSs $\mathcal{F}_m$ acts as a transformation

$$\mathcal{F}_m : \mathbb{P}(\mathbb{X}) \to \mathbb{P}(\mathbb{X})$$

defined by

$$\mathcal{F}_m(\mu) = \bigcup_{l=1}^{L_m} p_l^m f_l^m(\mu) \quad \text{for } m = 1, 2, \ldots, M.$$

By Theorem 2.4.21, each of these transformations is strictly contractive on $\mathbb{P}(\mathbb{X})$ with respect to the Hausdorff metric. That is,

$$d_{\mathbb{P}}(\mathcal{F}_m(X), \mathcal{F}_m(Y)) \leq \lambda_m d_{\mathbb{P}}(X, Y) \quad \text{for all } X, Y \in \mathbb{H}(\mathbb{X}),$$

for some $\lambda_m \in [0, 1)$ and $m = 1, 2, \ldots, M$.

The set attractor $A^{P(1)}$ of the IFS $\mathcal{F}^{P(1)}$ must be an element of $\mathbb{H}(\mathbb{P}(\mathbb{X}))$. That is, it must be a nonempty set, compact with respect to the metric $d_{\mathbb{P}(\mathbb{X})}$, whose elements are themselves normalized Borel measures on $\mathbb{X}$. Furthermore, this set attractor must be given by

$$A^{P(1)} = \phi_{\mathcal{F}^{P(1)}}\left(\Omega_{\{1,2,\ldots,M\}}\right),$$

where

$$\phi_{\mathcal{F}^{P(1)}} : \Omega_{\{1,2,\ldots,M\}} \to \mathbb{P}(\mathbb{X})$$

is the code space mapping associated with $\mathcal{F}^{P(1)}$. The elements of $A^{P(1)}$ are the points in $\mathbb{P}(\mathbb{X})$ that can be written in the form

$$\mu_\sigma = \phi_{\mathcal{F}^{P(1)}}(\sigma) = \lim_{k \to \infty} \mathcal{F}_{\sigma_1} \circ \mathcal{F}_{\sigma_2} \circ \cdots \circ \mathcal{F}_{\sigma_k}(\mu_{\mathbb{X}}),$$

where $\mu_{\mathbb{X}}$ is an element of $\mathbb{P}(\mathbb{X})$. The particular choice of $\mu_{\mathbb{X}}$ makes no difference since the convergence on the right-hand side, with respect to the metric $d_{\mathbb{P}(\mathbb{X})}$, is uniform with respect to both $\mu_{\mathbb{X}}$ and σ. We will refer to the measure μ_σ as a **1-variable fractal measure**. We have

$$A^{P(1)} = \left\{\mu_\sigma : \sigma \in \Omega_{\{1,2,\ldots,M\}}\right\},$$

which we may refer to as a **superfractal of 1-variable fractal measures** associated with the superIFS $\{\mathbb{X}; \mathcal{F}_1, \mathcal{F}_2, \ldots, \mathcal{F}_M\}$.

Examples of elements of $A^{P(1)}$ are $\mu_{\overline{1}} \in \mathbb{P}(\mathbb{X})$, which is the measure attractor of the hyperbolic IFS $\mathcal{F}_1 = \{\mathbb{X}; f_1^1, f_2^1, \ldots, f_{L_1}^1; p_1^1, p_2^1, \ldots, p_{L_1}^1\}$, and $\mu_{\overline{1312}} \in$

$\mathbb{P}(\mathbb{X})$, which is the same as $\mathcal{F}_1 \circ \mathcal{F}_3(\mu_{\overline{12}})$, where $\mu_{\overline{12}}$ is the measure attractor of the hyperbolic IFS with probabilities

$$\mathcal{F}_1 \circ \mathcal{F}_1 = \Big\{ \mathbb{X}; f_{l_1}^1 \circ f_{l_2}^2 \text{ with probability } p_{l_1}^1 p_{l_2}^2$$
$$\text{for } l_1 = 1, 2, \dots, L_1 \text{ and } l_2 = 1, 2, \dots, L_2 \Big\}.$$

Just as in the case of the 1-variable IFS $\mathcal{F}^{(1)}$, the chaos game may be applied to $\mathcal{F}^{P(1)}$ and provides a means for sampling $A^{P(1)}$. The difference is that now we obtain random orbits of measures in place of random orbits of sets. There is nothing difficult in this, but it is necessary to note carefully the natures of the various objects involved.

To demonstrate how the chaos game applies to $\mathcal{F}^{P(1)}$ we here transcribe the discussion at the start of Section 5.6 to the present setting. $\mathcal{F}^{P(1)}$ is a hyperbolic IFS with probabilities. Hence it possesses a unique invariant probability measure $\mu^{P(1)} \in \mathbb{P}(\mathbb{P}(\mathbb{X}))$; here $\mathbb{P}(\mathbb{P}(\mathbb{X}))$ is the space of normalized Borel measures defined on $\mathbb{P}(\mathbb{X})$. Each element of $\mathbb{P}(\mathbb{P}(\mathbb{X}))$ is a probability measure defined on a set of normalized measures. The chaos game, adapted to the present setting, yields sequences of points $\{\mu_k\}_{k=1}^\infty$, $\mu_k \in \mathbb{P}(\mathbb{X})$, which almost always converge to the measure attractor $\mu^{P(1)}$. The manner in which this convergence occurs is governed by Theorem 4.5.1, which, in the present setting, reads as follows.

THEOREM 5.11.1 *Let* $\{\mathbb{P}(\mathbb{X}); \mathcal{F}_1, \mathcal{F}_2, \dots, \mathcal{F}_N; P_1, P_2, \dots, P_N\}$ *be a 1-variable measure IFS and let* $\mu^{P(1)} \in \mathbb{P}(\mathbb{P}(\mathbb{X}))$ *denote its measure attractor. Specify a starting measure* $\mu_1 \in \mathbb{P}(\mathbb{X})$. *Define a random orbit of the IFS to be* $\{\mu_k\}_{k=1}^\infty$ *where* $\mu_{k+1} = \mathcal{F}_m(\mu_k)$ *with probability* P_m, *independently of all other choices. Then for almost all random orbits* $\{\mu_k\}_{k=1}^\infty$ *we have*

$$\mu^{P(1)}(B) = \lim_{k \to \infty} \frac{|B \cap \{\mu_1, \mu_2, \dots, \mu_k\}|}{k}$$

for all $B \in \mathbb{B}(\mathbb{P}(\mathbb{X}))$ *such that* $\mu(\partial B) = 0$, *where* ∂B *denotes the boundary of* B.

Equivalently, the sequence of measures

$$\Big\{ K^{-1}\big(\delta_{\mu_1} + \delta_{\mu_2} + \cdots + \delta_{\mu_K}\big) \in \mathbb{P}(\mathbb{P}(\mathbb{X})) \Big\}_{K=1}^\infty$$

converges weakly to $\mu^{P(1)}$.

As an example see Figure 5.35, which actually illustrates three 2-variable fractal measures, belonging to a random orbit, rather than 1-variable measures. But I am sure you get the idea.

EXERCISE 5.11.2 *Transcribe to the context of* $\mu^{P(1)}$ *the six paragraphs of discussion immediately following Theorem 5.6.1.*

Figure 5.18 Similar to Figure 5.17, but the picture from which the colours were stolen is different. Check whether you agree that these pictures could indeed be homeomorphic.

Collections of 1-variable orbital pictures

The superIFS $\{\mathbb{X}; \mathcal{F}_1, \mathcal{F}_2, \ldots, \mathcal{F}_M\}$ may be used to define sets of what we call **1-variable orbital pictures**, in the following manner. Let $\mathfrak{P}_0 \in \Pi_{\mathfrak{C}}(\mathbb{X})$ and let

$$\mathfrak{P}_\sigma(\mathfrak{P}_0) = \mathfrak{P}_0 \uplus \mathcal{F}_{\sigma_1}(\mathfrak{P}_0) \uplus \mathcal{F}_{\sigma_1 \sigma_2}(\mathfrak{P}_0) \uplus \cdots \uplus \mathcal{F}_{\sigma_1 \sigma_2 \cdots \sigma_{|\sigma|}}(\mathfrak{P}_0)$$

for all $\sigma \in \Omega'_{\{1,2,\ldots,M\}}$, where $\mathcal{F}_\omega(\mathfrak{P}_0) := \mathcal{F}_{\omega_1} \circ \mathcal{F}_{\omega_2} \circ \cdots \circ \mathcal{F}_{\omega_{|\omega|}}$ for all $\omega \in \Omega'_{\{1,2,\ldots,M\}}$. Then it is straightforward to show that, for given $\sigma_1 \sigma_2 \cdots \in \Omega_{\{1,2,\ldots,M\}}$,

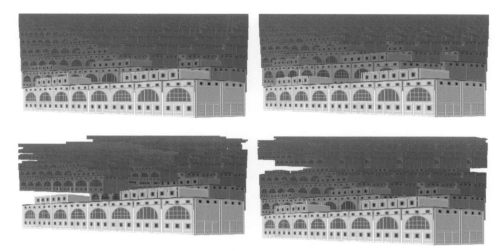

Figure 5.19 The top two images are examples of 1-variable orbital pictures, rendered using a version of IFS colouring. The condensation picture is in the foreground of each image. The bottom two images represent 2-variable orbital pictures. All four pictures are associated with the same superIFS, which consists of two affine IFSs on $\mathbb{R}^2$.

the sequence $\{\mathfrak{P}_{\sigma_1\sigma_2\cdots\sigma_k}(\mathfrak{P}_0)\}_{k=1}^{\infty}$ provides a unique picture, much as in Theorem 3.5.3. We denote this picture by

$$\mathfrak{P}_{\sigma}(\mathfrak{P}_0) = \lim_{k\to\infty}\mathfrak{P}_{\sigma_1\sigma_2\cdots\sigma_k}(\mathfrak{P}_0).$$

We refer to $\mathfrak{P}_{\sigma}(\mathfrak{P}_0)$ as the **1-variable orbital picture** associated with $\mathcal{F}^{(1)}$, the condensation picture $\mathfrak{P}_0$ and the string $\sigma \in \Omega_{\{1,2,\ldots,M\}}$.

Here is the crucial bit of algebra: for all $\mathfrak{P}_0 \in \Pi_{\mathfrak{C}}(\mathbb{X})$, $m \in \{1, 2, \ldots, M\}$ and $\sigma \in \Omega'_{\{1,2,\ldots,M\}} \cup \Omega_{\{1,2,\ldots,M\}}$ we have

$$\mathfrak{P}_{m\sigma}(\mathfrak{P}_0) = \mathfrak{P}_0 \uplus \mathcal{F}_m(\mathfrak{P}_{\sigma}(\mathfrak{P}_0)).$$

This tells us that we may use the chaos game to approximate and sample the **set of 1-variable orbital pictures** $\{\mathfrak{P}_{\sigma}(\mathfrak{P}_0) : \sigma \in \Omega_{\{1,2,\ldots,M\}}\}$. We define a random orbit $\mathfrak{P}_k(\mathfrak{P}_0)$ according to

$$\mathfrak{P}_{k+1}(\mathfrak{P}_0) = \mathfrak{P}_0 \uplus \mathcal{F}_{\sigma_k}(\mathfrak{P}_k(\mathfrak{P}_0)) \quad \text{with } \mathfrak{P}_1(\mathfrak{P}_0) = \mathfrak{P}_0$$

for $k = 1, 2, \ldots$, where σ_k is chosen equal to m with probability P_m, independently of all other choices.

The upper two images in Figure 5.19 provide two examples of 1-variable orbital pictures. The condensation picture $\mathfrak{P}_0$ is represented by the pale-blue building in the foreground. I modified the colours of successive panels by IFS colouring to yield a stable picture on the 'horizon'. To see more correctly the mathematical orbital pictures you should imagine that each building is the same colour as the

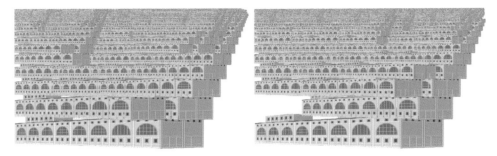

Figure 5.20   Two different 1-variable underneath pictures, generated using a superIFS and a condensation picture; the latter is represented by the building in the foreground. Many different sequences of underneath pictures may be obtained by random iteration. Such sequences of pictures may in practice display a 'texture effect' as illustrated here, near the horizon.

one in the foreground. The 1-variable IFS $\mathcal{F}^{(1)}$ used here consists of two affine IFSs, which in turn consist of two transformations each. You should be able to deduce these transformations approximately by studying the two images. I was for a while quite mesmerized by the sequence of fantasy cities which appeared on my computer screen.

By now, you should be able to make a good guess at the definition of a 2-variable orbital picture. Examples of 2-variable orbital pictures are given in the bottom two panels of Figure 5.19 and also in Figure 5.26.

Collections of 1-variable underneath pictures

The superIFS in Equation (5.4.1) may be used to define a set of transformations $\overleftarrow{\mathcal{F}}_m : \Pi_{\mathscr{C}}(\mathbb{X}) \to \Pi_{\mathscr{C}}(\mathbb{X})$ by

$$\overleftarrow{\mathcal{F}}_m(\mathfrak{P}_0) = f_{L_m}^m(\mathfrak{P}_0) \uplus f_{L_m-1}^m(\mathfrak{P}_0) \uplus \cdots \uplus f_1^m(\mathfrak{P}_0)$$

for all $\mathfrak{P}_0 \in \Pi_{\mathscr{C}}(\mathbb{X})$ and for $m = 1, 2, \ldots, M$. Then we define

$$\overleftarrow{\mathcal{F}}_\sigma := \overleftarrow{\mathcal{F}}_{\sigma_1} \circ \overleftarrow{\mathcal{F}}_{\sigma_2} \circ \cdots \circ \overleftarrow{\mathcal{F}}_{\sigma_{|\sigma|}}$$

for all $\sigma \in \Omega'_{\{1,2,\ldots,M\}}$. We may generate corresponding random orbits of **1-variable underneath pictures** $\{\mathfrak{P}''_k(\mathfrak{P}_0)\}_{k=1}^\infty$ with the aid of the chaos game, according to

$$\mathfrak{P}''_{k+1}(\mathfrak{P}_0) = \overleftarrow{\mathcal{F}}_{\sigma_k}(\mathfrak{P}''_k(\mathfrak{P}_0)) \quad \text{with } \mathfrak{P}''_1(\mathfrak{P}_0) = \mathfrak{P}_0$$

for $k = 1, 2, \ldots$, where, as elsewhere, σ_k is chosen as equal to m with probability P_m, independently of all other choices.

In computational examples these random orbits behave in a very similar manner to the sequences of underneath pictures discussed in Section 3.5. For example, Figure 5.20 illustrates the 'texture effect' in several 1-variable underneath pictures corresponding to the same superIFS.

Random orbits of 1-variable pictures

The superIFS

$$\{\mathbb{X}; \mathcal{F}_1, \mathcal{F}_2, \ldots, \mathcal{F}_M; P_1, P_2, \ldots, P_M\}$$

may be used to define transformations $\mathcal{F}_m : \Pi_{\mathfrak{C}}(\mathbb{X}) \to \Pi_{\mathfrak{C}}(\mathbb{X})$ by

$$\mathcal{F}_m(\mathfrak{P}_0) = f_1^m(\mathfrak{P}_0) \uplus f_2^m(\mathfrak{P}_0) \uplus \cdots \uplus f_{L_m}^m(\mathfrak{P}_0)$$

for all $\mathfrak{P}_0 \in \Pi_{\mathfrak{C}}(\mathbb{X})$ and for $m = 1, 2, \ldots, M$. Then we define

$$\mathcal{F}_\sigma(\mathfrak{P}_0) := \mathcal{F}_{\sigma_1} \circ \mathcal{F}_{\sigma_2} \circ \cdots \circ \mathcal{F}_{\sigma_{|\sigma|}}(\mathfrak{P}_0)$$

for $\sigma \in \Omega'_{\{1,2,\ldots,M\}}$ to be a **1-variable picture**. We may generate corresponding random orbits of 1-variable pictures $\{\mathfrak{P}_k(\mathfrak{P}_0)\}_{k=1}^\infty$ with the aid of the chaos game, according to

$$\mathfrak{P}_{k+1}(\mathfrak{P}_0) = \mathcal{F}_{\sigma_k}(\mathfrak{P}_k(\mathfrak{P}_0)) \quad \text{with } \mathfrak{P}_1(\mathfrak{P}_0) = \mathfrak{P}_0,$$

for $k = 1, 2, \ldots$, where, as elsewhere, σ_k is chosen equal to m with probability P_m, independently of all other choices. Similarly defined random orbits of measures, starting from a measure $\mu_0 \in \mathbb{P}(\mathbb{X})$, approach elements of $A^{P(1)}$ and similarly defined random orbits of sets, starting from a set $C_0 \in \mathbb{H}(\mathbb{X})$, approach elements of $A^{(1)}$, in each case in the manner already described. The random orbit of pictures $\{\mathfrak{P}_k(\mathfrak{P}_0)\}_{k=1}^\infty$, however, typically does not converge to a limiting superfractal of pictures, in the sense of asymptotically dancing around on some limiting object, for essentially the same reason that the deterministic sequence of pictures $\{\mathcal{F}^{\circ k}(\mathfrak{P}_0)\}_{k=1}^\infty$ does not in general converge to a well-defined picture, as discussed in Section 4.8. In computational examples we find that random orbits of 1-variable pictures often display texture effects, as do the analogous random orbits of V-variable pictures. The restless pattern of purple dots on green 2-variable 'ti-trees' in Figure 5.24 illustrates this phenomenon.

Directed sets of 1-variable fractal sets

We can use the superIFS $\{\mathbb{X}; \mathcal{F}_1, \mathcal{F}_2, \ldots, \mathcal{F}_M\}$ together with a shift-invariant closed subset Ω of $\Omega_{\{1,2,\ldots,M\}}$ to define the **directed 1-variable IFS** $(\mathcal{F}^{(1)}, \Omega)$. The theory from Section 4.16 and onwards in Chapter 4 comes into play and may be interpreted in the present setting. The deterministic fractal set associated with $(\mathcal{F}^{(1)}, \Omega)$ consists of a set of compact subsets of $\mathbb{X}$.

To show that it is useful to think about directed 1-variable IFSs, we note the following. By choosing the structure of Ω appropriately, say to have a recurrent or graph-directed form, it is possible to design algorithms, along the lines of the chaos game, which yield for example families of '1-variable' overlapping ferns that generalize those illustrated in Figure 4.37. Also, it appears straightforward to provide conditions on the superIFS under which the boundaries of the elements of the attractor $A^{(1)}$ of the IFS $\mathcal{F}^{(1)}$ are related to the deterministic fractal set associated

Figure 5.21 This illustrates symbolically the 2-variable superfractal associated with the computational experiment in Section 5.2. What you actually see is the superposition of many of the 2-variable sets, tethered curves, treated as measures. The bright points correspond to a high density of curves. The red and green curves represent the attractors of the two IFSs that generate the superfractal.

with a directed 1-variable IFS of the form $(\mathcal{F}^{(1)}, \Omega)$, Ω being defined appropriately. Such a result may be proved similarly to the proof of Theorem 4.16.6.

5.12   *V*-variable IFSs

We begin by describing the transformations that are implicit in the computational experiment in Section 5.2, generalized to the case where we have V buffers in place of two buffers. So if you have trouble initially in seeing what the following formulas mean, take $V = 2$ and think about how you would write down one of the transformations between the 'input screens' and 'output screens' in Section 5.2. See Figure 5.21.

Let $V \geq 1$ be an integer and let the superIFS

$$\{\mathbb{X}; \mathcal{F}_1, \mathcal{F}_2, \ldots, \mathcal{F}_M; P_1, P_2, \ldots, P_M\}$$

be defined as above. Let $\mathcal{A}$ denote the set of indices

$$\mathcal{A} := \Big\{ a = (\mathbf{m}, \mathbf{v}) : \mathbf{m} = (m_1, m_2, \ldots, m_V), \mathbf{v} = \big(v_{1,1}, v_{1,2}, \ldots, v_{1,L_{m_1}},$$

$$v_{2,1}, v_{2,2}, \ldots, v_{2,L_{m_2}}, \ldots, v_{V,1}, v_{V,2}, \ldots, v_{V,L_{m_V}}\big),$$

$$v_{v,l} \in \{1, 2, \ldots, V\} \text{ for } l = 1, 2, \ldots, L_{m_v},$$

$$m_v = 1, 2, \ldots, M \text{ and } v = 1, 2, \ldots, V \Big\}. \tag{5.12.1}$$

For each $a \in \mathcal{A}$ we define

$$f_a : \mathbb{H}(\mathbb{X})^V \to \mathbb{H}(\mathbb{X})^V$$

by

$$f^a(B_1, B_2, \ldots, B_V)$$
$$= \Big(f_1^{m_1}\big(B_{v_{1,1}}\big) \cup f_2^{m_1}\big(B_{v_{1,2}}\big) \cup \cdots \cup f_{L_{m_1}}^{m_1}\big(B_{v_{1,L_{m_1}}}\big),$$
$$f_1^{m_2}\big(B_{v_{2,1}}\big) \cup f_2^{m_2}\big(B_{v_{2,2}}\big) \cup \cdots \cup f_{L_{m_2}}^{m_2}\big(B_{v_{2,L_{m_2}}}\big),$$
$$\ldots,$$
$$f_1^{m_V}\big(B_{v_{V,1}}\big) \cup f_2^{m_V}\big(B_{v_{V,2}}\big) \cup \cdots \cup f_{L_{m_V}}^{m_V}\big(B_{v_{V,L_{m_V}}}\big)\Big)$$
$$= \Big(\bigcup_{l=1}^{L_{m_1}} f_l^{m_1}\big(B_{v_{1,l}}\big), \bigcup_{l=1}^{L_{m_2}} f_l^{m_2}\big(B_{v_{2,l}}\big), \ldots, \bigcup_{l=1}^{L_{m_V}} f_l^{m_V}\big(B_{v_{V,l}}\big)\Big)$$

for each $B_1, B_2, \ldots, B_V \in \mathbb{H}(\mathbb{X})^V$.

We use these transformations to define the *V*-**variable IFS**

$$\mathcal{F}^{(V)} = \big\{\mathbb{H}(\mathbb{X})^V; f_a, p_a, a \in \mathcal{A}\big\}.$$

The probabilities p_a are given by

$$p_a = p_{(\mathbf{m}, \mathbf{v})} = \frac{P_{m_1} P_{m_2} \cdots P_{m_V}}{V^{L_{m_1} + L_{m_2} + \cdots + L_{m_V}}}. \tag{5.12.2}$$

We assume that the transformations that comprise this IFS are actually presented in a fixed order. The metric in $\mathbb{H}(\mathbb{X})^V$ is denoted $d_{\mathbb{H}^V}$ and is given by

$$d_{\mathbb{H}^V}(B, C) = \max_{v \in \{1, 2, \ldots, V\}} d_{\mathbb{H}(\mathbb{X})}(B_v, C_v)$$

for all $B = (B_1, B_2, \ldots, B_V)$ and $C = (C_1, C_2, \ldots, C_V)$ in $\mathbb{H}^V = \mathbb{H}(\mathbb{X})^V$. Since $(\mathbb{X}, d)$ is a compact metric space it follows that $(\mathbb{H}(\mathbb{X})^V, d_{\mathbb{H}^V})$ is a compact metric space.

THEOREM 5.12.1   *$\mathcal{F}^{(V)}$ is a hyperbolic IFS, for all $V = 1, 2, 3, \ldots$*

PROOF   We suppose that the IFS $\mathcal{F}_m$ has contractivity factor $\lambda \in [0, 1)$ for all $m = 1, 2, \ldots, M$. Then we prove that the mapping $f^a : \mathbb{H}^V \to \mathbb{H}^V$ is contractive, with contractivity factor λ, for all $a \in \mathcal{A}$. Note that, for $v = 1, 2, \ldots, V$ and for all $B = (B_1, B_2, \ldots, B_{L_{m_v}})$ and $C = (C_1, C_2, \ldots, C_{L_{m_v}}) \in \mathbb{H}(\mathbb{X})^{L_{m_v}}$, we have

$$
\begin{aligned}
d_\mathbb{H}\Big(&f_1^{m_v}(B_1) \cup f_2^{m_v}(B_2) \cup \cdots \cup f_{L_{m_v}}^{m_v}\big(B_{L_{m_v}}\big), \\
&f_1^{m_v}(C_1) \cup f_2^{m_v}(C_2) \cup \cdots \cup f_{L_{m_v}}^{m_v}\big(C_{L_{m_v}}\big)\Big) \\
&\leq \max_{l \in \{1,2,\ldots,L_{m_v}\}} \big\{ d_\mathbb{H}\big(f_l^{m_v}(B_l), f_l^{m_v}(C_l)\big)\big\} \\
&\leq \max_{l \in \{1,2,\ldots,L_{m_v}\}} \big\{ \lambda \cdot d_\mathbb{H}(B_l, C_l)\big\} = \lambda \cdot d_{\mathbb{H}^{L_{m_v}}}(B, C).
\end{aligned}
$$

Here we have used Theorem 1.12.15. Hence, for all $B_1, B_2, \ldots, B_V$ and $C_1, C_2, \ldots, C_V \in \mathbb{H}^V$,

$$
\begin{aligned}
&d_{\mathbb{H}^V}\big(f_a(B_1, B_2, \ldots, B_V), f_a(C_1, C_2, \ldots, C_V)\big) \\
&= \max_{v \in \{1,2,\ldots,V\}} \left\{ d_\mathbb{H}\left(\bigcup\nolimits_{l=1}^{L_{m_v}} f_l^{m_v}(B_{v,l}), \bigcup\nolimits_{l=1}^{L_{m_v}} f_l^{m_v}(C_{v,l})\right)\right\} \\
&\leq \max_{v \in \{1,2,\ldots,V\}} \left\{ \lambda \cdot d_{\mathbb{H}^{L_{m_v}}}\big((B_{v,1}, B_{v,2}, \ldots, B_{v,L_{m_v}}), (C_{v,1}, C_{v,2}, \ldots, C_{v,L_{m_v}})\big)\right\} \\
&\leq \lambda \cdot d_{\mathbb{H}^V}\big((B_1, B_2, \ldots, B_V), (C_1, C_2, \ldots, C_V)\big).
\end{aligned}
$$

$\square$

It follows that $\mathcal{F}^{(V)}$ possesses a unique set attractor

$$
\mathbf{A}^{(V)} \in \mathbb{H}\big(\mathbb{H}(\mathbb{X})^V\big).
$$

It must also possess a unique measure attractor

$$
\boldsymbol{\mu}^{(V)} \in \mathbb{P}\big(\mathbb{H}(\mathbb{X})^V\big).
$$

We refer to the set

$$
A^{(V)} = \big\{ B \in \mathbb{H}(\mathbb{X}) : B \text{ is a component of a point in } \mathbf{A}^{(V)}\big\}
$$

as the *V*-variable superfractal associated with $\mathcal{F}^{(V)}$ and we refer to its elements as *V*-variable fractal sets. Notice that $\mathbf{A}^{(1)} = A^{(1)}$ and that

$$
A^{(1)} \subset A^{(2)} \subset A^{(3)} \subset \cdots \subset \mathbb{H}(\mathbb{H}(\mathbb{X})).
$$

For $V \geq 2$ the elements of $A^{(V)}$ are not in general directly related to set attractors of compositions of the $\mathcal{F}_m$. But, for example, if $M \geq 4$ then $A^{(2)}$ contains such vectors of sets as $(A_{\overline{12}}, A_{34\overline{12}})$ and $(A_{\overline{1}}, A_\sigma)$, where $A_\sigma \in A^{(1)}$; $A^{(V)}$ contains many other types of set, however.

The chaos game corresponding to the IFS $\mathcal{F}^{(V)}$ may be used to produce sequences of points, V-tuples of compact sets, in $\mathbf{A}^{(V)}$, asymptotically distributed according to the probability measure $\boldsymbol{\mu}^{(V)}$.

THEOREM 5.12.2 *For each* $v \in \{1, 2, \ldots, V\}$ *we have*

$$A^{(V)} = A_v^{(V)},$$

where $A_v^{(V)}$ *denotes the set comprising the vth components of the elements of* $\mathbf{A}^{(V)}$. *If the probabilities in* $\mathcal{F}^{(V)}$ *are given by Equation (5.12.2) then, starting from any initial V-tuple of nonempty compact subsets of* $\mathbb{X}$, *the random distribution of the sets comprising the vth components of the first K vectors produced by the chaos game converges weakly to the marginal probability measure*

$$\mu^{(V)}(\mathcal{B}) := \mu^{(V)}(\mathcal{B}, \mathbb{H}, \mathbb{H}, \ldots, \mathbb{H}) \quad \text{for all } \mathcal{B} \in \mathbb{B}(\mathbb{H}), \qquad (5.12.3)$$

independently of v, almost always, as $K \to \infty$.

PROOF   See [16].   □

Theorem 5.12.2 tells us that we can sample the superfractal $A^{(V)}$ by means of the chaos game applied to the IFS $\mathcal{F}^{(V)}$. To do this we define sequences of vectors of nonempty compact subsets of $\mathbb{X}$ by

$$\left(B_1^{k+1}, B_2^{k+1}, \ldots, B_V^{k+1}\right) = f_{a_k}\left(B_1^k, B_2^k, \ldots, B_V^k\right) \quad \text{for } k = 1, 2, \ldots,$$

starting from any point $(B_1^1, B_2^1, \ldots, B_V^1) \in \mathbb{H}(\mathbb{X})^V$. Here

$$a_k = a \in \mathcal{A} \quad \text{with probability } p_a$$

independently of all other choices, a is given by Equation (5.12.1) and p_a is given by Equation (5.12.2). The sequence of measures

$$\mu_K^{(V)} = K^{-1}\left(\delta_{B_1^1} + \delta_{B_1^2} + \cdots + \delta_{B_1^K}\right)$$

belonging to the space $\mathbb{P}(\mathbb{H}(\mathbb{X}))$ converges weakly as $K \to \infty$, almost always, to the same probability measure in $\mathbb{P}(\mathbb{H}(\mathbb{X}))$, namely the measure $\mu^{(V)}$ defined by Equation (5.12.3).

An example in which we use the chaos game to compute samples of a 2-variable superfractal is provided by the computational experiment in Section 5.2. In this case the superfractal is equivalent to a family of continuous paths tethered at two points, and the observed stationary state corresponds to the measure $\mu^{(2)}$.

In Figure 5.22 we illustrate some 2-variable fractal sets associated with the superIFS used to produce Figure 5.7, discussed in Section 5.6. See Figure 5.23 for a close-up of one of these sets. The 2-variable fractal sets here have been rendered using a version of colour-stealing. When $V \geq 2$ colour-stealing works somewhat differently from the case $V = 1$, because the chaos game does not preserve even a 'local' tops structure, as mentioned in Section 5.13 below.

The arrows in Figure 5.22 help to explain the 2-variability of these images. The two arrows attached to a set point to the parents of the set. Each set is a union of three or four transformations applied to one or other of its parents. Furthermore,

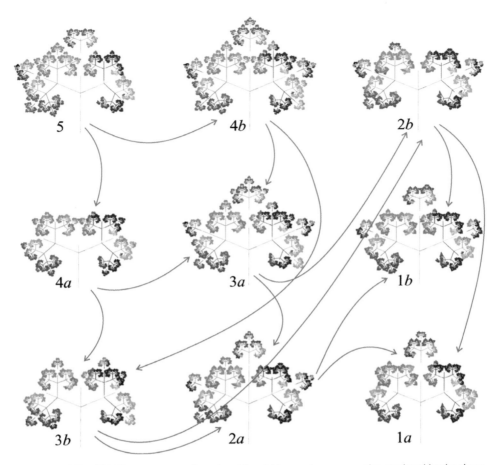

Figure 5.22 This illustrates successive pairs of 2-variable sets, in reverse order, produced by the chaos game. Each arrow points from a 'child' to one of its 'parents'; the pair of arrows emanating from each set shows the result of applying the transformation *T* to the set, as explained in Section 5.18.

the parents are siblings, both having the same parents. And so on back through history. Thus, if you look closely at any set in the figure you can see either the parent or, when there are two parents, both parents. You can also see, as though homunculi within homunculi, all the grandparents and all the great-grandparents. There are at most two forebears in each generation. The arrows also illustrate the structure of a dynamical system belonging to the superfractal, which we discuss in more detail in Section 5.18.

During the chaos game, of course, we travel in the opposite direction, from parents to children. The images 1*a* and 1*b* were used to compute 2*a* and 2*b*, which were used to compute 3*a* and 3*b*, which were used to compute 4*a* and 4*b*, which were used to compute 5.

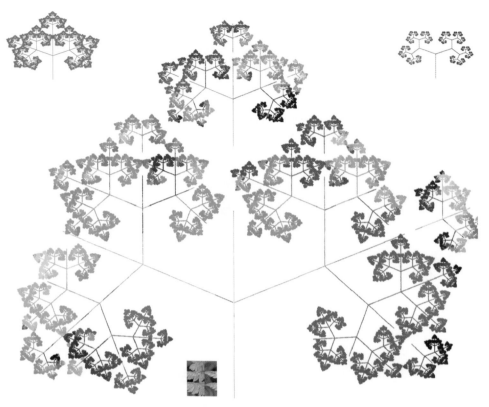

Figure 5.23 Example of a 2-variable fractal top rendered using colour-stealing. The picture from which the colours were stolen is shown inset. The attractors of the two IFSs that comprise the superIFS are illustrated at top left and top right.

Another example of fractal sets belonging to a 2-variable superfractal is illustrated in Figure 5.24. In this case $M = 2$ and the projective IFSs used are those in Tables 5.4 and 5.5. One of the goals of this example is to illustrate how closely similar images can be produced, with 'random' variations, so the two IFSs were chosen to be quite similar. Let us refer to images such as those in the bottom row of Figure 5.24 as 'ti-trees'. Then each transformation maps approximately the unit square $\square := \{(x, y) \mid 0 \leq x \leq 1, \, 0 \leq y \leq 1\}$, in which each ti-tree lies, into itself. Both $f_2^1(x, y)$ and $f_2^2(x, y)$ map ti-trees to lower right branches of ti-trees. Both $f_1^1(x, y)$ and $f_1^2(x, y)$ map a ti-tree to a ti-tree minus the lower right branch. In this case, in place of computing successive pairs of sets we actually computed successive pairs of pictures $(\mathfrak{P}_{1,k}, \mathfrak{P}_{2,k})$, for $k = 1, 2, \ldots$, according to

$$(\mathfrak{P}_{1,k+1}, \mathfrak{P}_{2,k+1}) = f^{a_k}(\mathfrak{P}_{1,k}, \mathfrak{P}_{2,k}),$$

where a_k was chosen equal to $a \in \mathcal{A}$ with probability p_a independently of all other choices, in the usual manner. Here $\mathfrak{P}_{1,1}$ represents a green filled rectangle with rounded corners and $\mathfrak{P}_{2,1}$ is similar to $\mathfrak{P}_{1,1}$ but purple. This illustrates how we may generalize the concept of random orbits of 1-variable pictures to V-variable pictures for $V \geq 1$. It also illustrates the 'texture effect', discussed in Section 5.11; see for example Figure 5.20.

5.13   *V*-variable pictures with stolen colours, and *V*-variable orbital pictures

We assume here that the functions that comprise the IFSs in the superIFS are all one-to-one and thus invertible on their ranges. This assumption is always needed when we apply transformations to pictures.

The transformations $f^a : \mathbb{H}(\mathbb{X})^V \to \mathbb{H}(\mathbb{X})^V$ for $a \in \mathcal{A}$ may be used to define, consistently with Equation (4.12.1),

$$f^a_{TOP} : \Pi_{\Omega_{\{1,2,\dots,N\}}}(\mathbb{X})^V \to \Pi_{\Omega_{\{1,2,\dots,N\}}}(\mathbb{X})^V.$$

Let $\mathfrak{P} = (\mathfrak{P}_1, \mathfrak{P}_2, \dots, \mathfrak{P}_V) \in \Pi_{\Omega_{\{1,2,\dots,N\}}}(\mathbb{X})^V$ and let D_v denote the domain of $\mathfrak{P}_v : D_v \to \Omega_{\{1,2,\dots,N\}}$ for $v = 1, 2, \dots, V$. $\mathfrak{P}$ is a vector of picture functions whose colour values are points in code space. Then we define

$$
\begin{aligned}
&f^a(\mathfrak{P}_1, \mathfrak{P}_2, \dots, \mathfrak{P}_V) \\
&= \Big(\big(((N_{m_1-1}+1)f_1^{m_1}(\mathfrak{P}_{v_{1,1}})) \uplus ((N_{m_1-1}+2)f_2^{m_1}(\mathfrak{P}_{v_{1,2}})) \\
&\quad \uplus \cdots \uplus \big(N_{m_1} f_{L_{m_1}}^{m_1}(\mathfrak{P}_{v_{1,L_{m_1}}})\big), \\
&\quad ((N_{m_2-1}+1)f_1^{m_2}(\mathfrak{P}_{v_{2,1}})) \uplus ((N_{m_2-1}+2)f_2^{m_2}(\mathfrak{P}_{v_{2,2}})) \\
&\quad \uplus \cdots \uplus \big(N_{m_2} f_{L_{m_2}}^{m_2}(\mathfrak{P}_{v_{2,L_{m_2}}})\big), \\
&\quad \cdots, \\
&\quad ((N_{m_V-1}+1)f_1^{m_V}(\mathfrak{P}_{v_{V,1}})) \uplus ((N_{m_V-1}+2)f_2^{m_V}(\mathfrak{P}_{v_{V,2}})) \\
&\quad \uplus \cdots \uplus \big(N_{m_V} f_{L_{m_V}}^{m_V}(\mathfrak{P}_{v_{V,L_{m_V}}})\big) \Big),
\end{aligned}
\tag{5.13.1}
$$

for each $(\mathfrak{P}_1, \mathfrak{P}_2, \dots, \mathfrak{P}_V) \in \Pi_{\Omega_{\{1,2,\dots,N\}}}(\mathbb{X})^V$. To show what this notation means let us consider the picture function $((N_{m_2-1}+1)f_1^{m_2}(\mathfrak{P}_{v_{2,1}}))$. We start at the right of the expression: $f_1^{m_2}(\mathfrak{P}_{v_{2,1}})$ is the function whose domain is $f_1^{m_2}(D_{v_{2,1}})$ and whose value at $x \in f_1^{m_2}(D_{v_{2,1}})$ is $\mathfrak{P}_{v_{2,1}}((f_1^{m_2})^{-1}(x))$. This value is an element of the code space $\Omega_{\{1,2,\dots,N\}}$, namely an infinite string of symbols from the alphabet $\{1, 2, \dots, N\}$. Then the value of the picture function $((N_{m_2-1}+1)f_1^{m_2}(\mathfrak{P}_{v_{2,1}}))$ is obtained by putting the symbol for the number $N_{m_2-1}+1$ at the beginning of this string. See also the discussion at the start of Section 5.9.

Notice that the domain of the function $f^a(\mathfrak{P})$ is $f^a(D)$, where $D = (D_1, D_2, \dots, D_V)$.

Table 5.4   *A projective IFS code. This is used in Figure 5.24*

n	a_n	b_n	c_n	d_n	e_n	f_n	g_n	h_n	j_n	p_n
1	1.629	0.135	−1.99	−0.505	−1.935	0.216	−0.780	0.864	−2.569	$\frac{1}{2}$
2	1.616	−2.758	3.678	2.151	0.567	2.020	1.664	−0.944	3.883	$\frac{1}{2}$

Table 5.5   *A projective IFS code. This is used in Figure 5.24*

n	a_n	b_n	c_n	d_n	e_n	f_n	g_n	h_n	j_n	p_n
1	1.667	0.098	−2.005	−0.563	−2.064	0.278	−0.773	0.790	−2.575	$\frac{1}{2}$
2	1.470	−2.193	3.035	1.212	0.686	2.059	2.432	−0.581	2.872	$\frac{1}{2}$

THEOREM 5.13.1   *Let the V-variable IFS $\mathcal{F}^{(V)} = \{\mathbb{H}(\mathbb{X})^V; f^a, P_a, a \in \mathcal{A}\}$ be given, as above. Let $\mathcal{F}^a_{TOP} : \Pi_{\Omega_{\{1,2,\ldots,N\}}}(\mathbb{X})^V \to \Pi_{\Omega_{\{1,2,\ldots,N\}}}(\mathbb{X})^V$ be defined as in Equation (5.13.1). If $\mathfrak{P}_1, \mathfrak{P}_2 \in \Pi_{\Omega_{\{1,2,\ldots,N\}}}(\mathbb{X})^V$ have the same domain, $D \subset \mathbb{X}^V$, then*

$$\sup_{x \in f^a(D)} |f^a(\mathfrak{P}_1)(x) - f^a(\mathfrak{P}_2)(x)| \le \tfrac{1}{2} \sup_{x \in D} |\mathfrak{P}_1(x) - \mathfrak{P}_2(x)|.$$

PROOF   The proof is completely mechanical and follows the same lines as the proof of Theorem 4.12.1.   □

Theorem 5.13.1 tells us much less about the case $V > 1$ than Theorem 5.9.1 tells us about the case $V = 1$. But it does tell us just enough to be useful: that the 'colours' of any sequence of pictures which we obtain via the chaos game converge and that the sequence of pictures which is obtained depends asymptotically only on the domains of the initial pictures and not on their 'colours', namely their code space values. Of course, we also know that the domains of the pictures converge to V-variable fractal sets belonging to the appropriate superfractal.

We are certainly able to compute sequences of pictures of V-variable fractal sets, rendered using colour-stealing, that have a reasonable level of stability. So they clearly have applications to the creation of synthetic content and textures.

Figure 5.24 Some elements of a sequence of images that are converging towards 2-variable fractals. Convergence of the silhouttes, to within the numerical resolution, has occurred in the lower left and centre images. Note the subtle but real differences between the silhouettes of these two sets. A variant of the texture effect can also be seen: the purple points appear to dance forever on the green ti-trees, while the ti-trees dance forever on the superfractal.

New techniques in computer graphics are playing an increasingly important role in the digital-content creation industry, as evidenced by the succession of successes of computer-generated films. Part of the appeal of such films is the artistic quality of the graphics. It appears that V-variable fractals can provide, efficiently, new types of rendered digital imagery, significantly extending standard IFS graphics, as discussed for example in [27] and [8]. Figures 5.25–5.27 illustrate a few rendered 2-variable fractal sets to hint at the diversity of possibilities.

V-variable fractal sets may have applications to biological modelling. This theme is illustrated in Figure 5.28, which is a smaller version of Figure 0.5 in the Introduction. The top twelve pictures illustrate elements of a random sequence of 2-variable fractal sets belonging to a superfractal of fern-like sets associated with the two IFSs given in Tables 5.6 and 5.7. After approximately 130 iterations one of the IFSs was changed in a subtle way, see Figure 5.29, and the picture from which colours were stolen was switched.

We can imagine that two types of fern were growing close together a long time ago. They are distinguished by the amount of tilt in the main frond after the initial

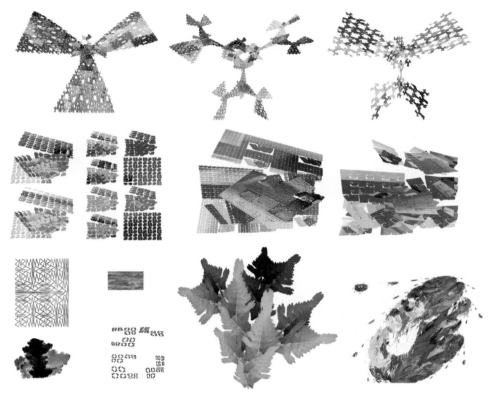

Figure 5.25 Various 2-variable fractal objects. This hints at the diversity of textures that can be obtained using the chaos game on a superfractal together with colour-stealing.

Figure 5.26 Illustration of a 2-variable orbital picture, with modified colours, belonging to a superfractal. In contrast with the orbital pictures generated by IFS semigroups, here we have a curious overlap. Why? See Section 5.16.

Table 5.6   *A projective IFS code. This is used in Figure 5.28*

n	a_n	b_n	c_n	d_n	e_n	f_n	g_n	h_n	j_n	p_n
1	85	−2	0	2	85	160	0	0	100	$\frac{7}{10}$
2	2	0	0	0	16	0	0	0	100	$\frac{1}{10}$
3	20	−26	0	23	22	80	0	0	100	$\frac{1}{10}$
4	−15	28	0	26	24	40	0	0	100	$\frac{1}{10}$

Figure 5.27   Examples of 1- and 2-variable fractal sets rendered by IFS colouring. The two images on the left correspond to the two separate IFSs used to define the superIFS. The image at lower right represents a 1-variable fractal set. The images at upper middle, upper right and lower middle represent fractal sets that are 2-variable but not 1-variable.

Table 5.7   *Another projective IFS code. This too is used in Figure 5.28*

n	a_n	b_n	c_n	d_n	e_n	f_n	g_n	h_n	j_n	p_n
1	85	6	0	−30	85	160	0	0	100	$\frac{7}{10}$
2	2	0	0	0	16	0	0	0	100	$\frac{1}{10}$
3	20	−26	0	23	22	40	0	0	100	$\frac{1}{10}$
4	−15	28	0	26	24	40	0	0	100	$\frac{1}{10}$

pair of fronds has been produced. This tilting is a function of the meristem and is supposed, in this story, to be activated by a gene which is switched on and off at each successive generation, up the fern and along the fronds, by factors that relate to the fern type. We can further imagine that the two types of fern interact by sharing their DNA in a 2-variable manner, resulting in numerous attempted new types of fern, each of which switches on and off the tilting mechanism at each level in the successive meristems according to its own 2-variable pattern. Further sharing of DNA maintains 2-variability. The resulting types strive for longer-term existence by being bountiful with their own spores. Sadly, none of them survives to tell the tale. But a random mutation of the gene led to a new sequence of trials. Perhaps a better informed botanist can tell a real story like this one?

5.14   *V*-variable fractal interpolation

The technique of fractal interpolation, see for example [6], [11], [66] and [74], has many applications, including the modelling of speech signals, altitude maps in geophysics and stock-market indices. A simple version of this technique adapted to the *V*-variable setting is as follows. Let a set of real interpolation points

$$-\infty < x_0 < x_1 < \cdots < x_L < \infty$$

and a set of data

$$\left(x_l, y_l^m\right) \in \mathbb{R}^2 \quad \text{for } l = 0, 1, 2, \ldots, L \quad \text{and} \quad m = 1, 2, \ldots, M$$

be given, such that

$$y_0^m = y_0 \quad \text{and} \quad y_L^m = y_L$$

Figure 5.28   The top twelve pictures belong to a random sequence of 2-variable sets, starting from the set at the top left. After about 130 iterations, one of the IFSs in the superIFS was altered in a subtle manner and the picture from which colours were stolen was changed. The bottom four pictures show elements of the continuing random orbit, now moving onto a new superfractal. Perhaps this new species will be more successful?

are independent of $m \in \{1, 2, \ldots, M\}$. We may or may not also require that y_l^m is independent of m, and say equal to y_l, for all $l \in \{1, 2, \ldots, L - 1\}$ and $m \in \{1, 2, \ldots, M\}$. We will suppose that $L \geq 2$ and that $M \geq 1$.

It is desired to find a superfractal of continuous functions $f_\sigma : [x_0, x_L] \to \mathbb{R}$ such that $f_\sigma(x_0) = y_0$ and $f_\sigma(x_L) = y_L$ for all $\sigma \in \Omega$, where Ω is an appropriate code space. It may also be desired, when the values y_l^m are independent of m for all $l \in \{1, 2, \ldots, L\}$ and all $m \in \{1, 2, \ldots, M\}$, that $f_\sigma(x_l) = y_l$ for all

Figure 5.29　Pictures of the attractors of the three IFSs which were used in connection with Figure 5.28. We may think of these attactors as representing three fern phenotypes.

$l \in \{1, 2, \ldots, L - 1\}$ and all $\sigma \in \Omega$. Furthermore, we require that $G_\sigma = \{(x, y) \in \mathbb{R}^2 : y = f_\sigma(x)\}$ is a V-variable fractal set for all $\sigma \in \Omega$, possibly with a specified Hausdorff dimension.

To achieve this goal we introduce the superIFS $\{\mathbb{R}^2; \mathcal{F}_1, \mathcal{F}_2, \ldots, \mathcal{F}_M\}$, where $L_m = L$ for $m = 1, 2, \ldots, M$ and where each of the functions f_l^m that comprise the IFS $\mathcal{F}_m$ is an affine transformation of the special form

$$f_l^m(x, y) = \left(a_l^m x + e_l^m, \ c_l^m x + d_l^m y + g_l^m\right),$$

the real coefficients $a_l^m, e_l^m, c_l^m, d_l^m$ and g_l^m being chosen so that

$$f_l^m(x_0, y_0) = y_{l-1}^m, \quad f_l^m(x_L, y_L) = y_l^m$$

and $d_l^m \in (-1, 1)$, for $m \in \{1, 2, \ldots, M\}$ and $l \in \{1, 2, \ldots, L\}$. Then the superfractal $A^{(V)}$ associated with the corresponding IFS $\mathcal{F}^{(V)}$ is a set of graphs of functions with the desired properties and may be explored by means of the chaos game in the usual manner. The superfractal and the corresponding set of functions $\{f_\sigma : \sigma \in \Omega\}$ depend on the free parameters $\{d_l^m : m = 1, 2, \ldots, M, l = 1, 2, \ldots, L\}$, which may be used to control the distribution of fractal dimensions of the graphs or the Hölder exponents of the functions. Some 1-variable fractal interpolation functions are illustrated in Figure 5.8.

In the case $M = 1$ this situation reduces to standard affine fractal interpolation, as described in [9]. The Hausdorff dimension $D = \dim_H G$ of the only graph G belonging to the superfractal is either equal to 1 or, in non-degenerate cases, to the

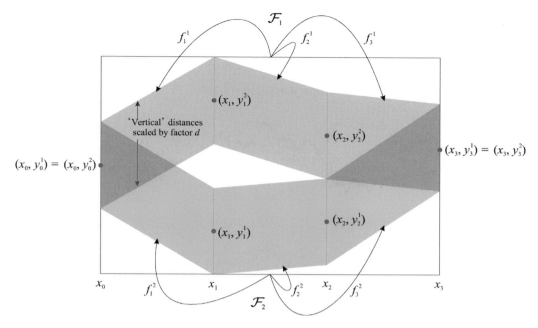

Figure 5.30 A superIFS may be constructed using IFSs $\mathcal{F}_1 = \{\mathbb{R}^2; f_1^1, f_2^1, f_3^1\}$ and $\mathcal{F}_2 = \{\mathbb{R}^2; f_1^2, f_2^2, f_3^2\}$, where the transformations act on the outlined box as illustrated. These IFSs are of the type used to provide fractal interpolation functions. Such a superIFS may be used to define superfractals of V-variable interpolation functions whose graphs have specified Hausdorff dimensions. In the present case, all the graphs belonging to the superfractal will have the same fractal dimension; the latter depends on the vertical scaling factor d. See the main text.

positive real solution of the equation

$$\sum_{l=1}^{L} \left(\frac{x_l - x_{l-1}}{x_L - x_0} \right)^{(D-1)} \left| d_l^1 \right| = 1. \tag{5.14.1}$$

If $\left| d_l^1 \right| = d \geq 1/N > 0$ for all $l = 1, 2, \ldots, L$ and the x_l are equally spaced then, in non-degenerate cases such as occur when none of the interpolated data lie on a straight line,

$$D = \frac{\log(N^2 d)}{\log N}.$$

Now suppose that $M > 1$, that the interpolation points are equally spaced and that $\left| d_l^m \right| = d \geq 1/N > 0$ for all $l = 1, 2, \ldots, L$ and for all $m = 1, 2, \ldots, M$. Then we can take $\Omega = \Omega_{\{1,2,\ldots,M\}}$, and it is intuitively immediate and also readily proved, following the line of the proof of Equation (5.14.1) described in [9], that

$$\dim_H G_\sigma = \frac{\log(N^2 d)}{\log N} \quad \text{for almost all } \sigma \in \Omega_{\{1,2,\ldots,M\}}.$$

In Figure 5.30 we illustrate the transformations used to construct such a super-fractal and in Figure 5.31 we illustrate two sets of 2-variable fractal interpolation

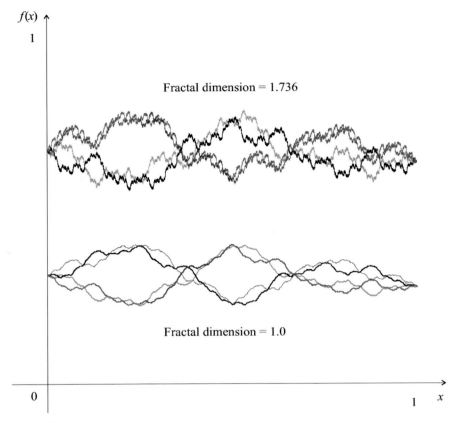

Figure 5.31 Two sets of superfractal interpolation functions, each with equally spaced interpolation points. The vertical scaling factor in the top set is 0.45 while in the lower set it is 0.25. The colours red and green indicate the attractors of the two IFSs; black indicates a 1-variable IFS and blue indicates a 2-variable IFS.

functions corresponding to two different values of the Hausdorff, or fractal, dimension. Each set includes the attractors (red and green) of the two IFSs used to create the corresponding superfractal, together with a 1-variable graph (black) and also 2-variable graph (blue). In each set the members have the same vertical scaling factor d_l^m. For some recent developments in V-variable fractal interpolation, see [85].

5.15   *V*-variable space-filling curves

Space-filling curves may be constructed with the aid of IFS theory; see for example [84], Chapter 9. These curves have many applications, including adaptive multigrid methods for the numerical computation of solutions of PDEs and the hierarchical watermarking of digital images. Here we note that interesting V-variable space-filling curves, and finite-resolution approximants to them, can be produced.

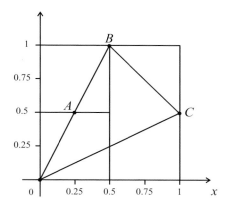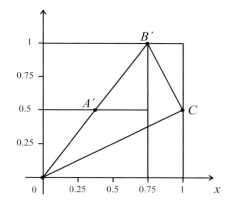

Figure 5.32   The two diagrams shown are used to define IFSs $\mathcal{F}_1 = \{\Box; f_1^1, f_2^1, f_3^1\}$ and $\mathcal{F}_2 = \{\Box : f_1^2, f_2^2, f_3^2\}$ for a superIFS $\{\Box; \mathcal{F}_1, \mathcal{F}_2\}$ that provides V-variable space-filling curves. The affine transformations f_m^n are such that $f_1^n(\Box)$, $f_2^n(\Box)$, and $f_3^n(\Box)$, for $n = 1$ (on the left) and $n = 2$ (on the right) provide rectangular tilings. Moreover, $\mathcal{F}_1(\overrightarrow{OC}) = \{\overrightarrow{OA}, \overrightarrow{BA}, \overrightarrow{BC}\}$ and $\mathcal{F}_2(\overrightarrow{OC}) = \{\overrightarrow{OA'}, \overrightarrow{B'A'}, \overrightarrow{B'C'}\}$.

As an example we choose $V = 2$, $M = 2$ and $\mathcal{F}_m = \{\Box; f_1^m, f_2^m, f_3^m\}$, where

$$f_1^1(x, y) = \left(\tfrac{1}{2}y, \tfrac{1}{2}x\right), \quad f_2^1(x, y) = \left(-\tfrac{1}{2}y + \tfrac{1}{2}, -\tfrac{1}{2}x + 1\right),$$
$$f_3^1(x, y) = \left(\tfrac{1}{2}x + \tfrac{1}{2}, -y + 1\right)$$

and

$$f_1^2(x, y) = \left(\tfrac{2}{3}y, \tfrac{1}{2}x\right), \quad f_2^2(x, y) = \left(-\tfrac{2}{3}y + \tfrac{2}{3}, -\tfrac{1}{2}x + 1\right),$$
$$f_3^2(x, y) = \left(\tfrac{1}{3}x + \tfrac{2}{3}, -y + 1\right).$$

See Figure 5.32. Neither $\mathcal{F}_1$ nor $\mathcal{F}_2$ is strictly contractive but each is contractive on average, for any assignment of positive probabilities to the constituent functions.

An initial image consisting of the line segment $\overrightarrow{OC}$ is chosen on both screens, and the random iteration algorithm is applied; typical images produced after five iterations are illustrated in Figure 5.33; an image produced after seven iterations is shown in Figure 5.34.

5.16   Fractal transformations between the elements of V-variable superfractals of 'maybe-not-tops'

It is not true, in general, that the chaos game preserves local tops in the sense of Theorem 5.13.1. To convince yourself of this, randomly iterate a few steps of Equation (5.13.1) in a decently overlapping case with say $M = 2$, $L_1 = L_2 = 2$ and $V = 2$. You can readily construct examples in which 'higher' code values become buried under 'lower' ones.

This has the consequence that the analogues of the sequences of 1-variable orbital pictures described in Section 5.11 do not have a tidy structure, as illustrated

Figure 5.33 Low-order approximants to two 2-variable space-filling curves generated by the IFS illustrated in Figure 5.32. Both images represent elements of the same superfractal.

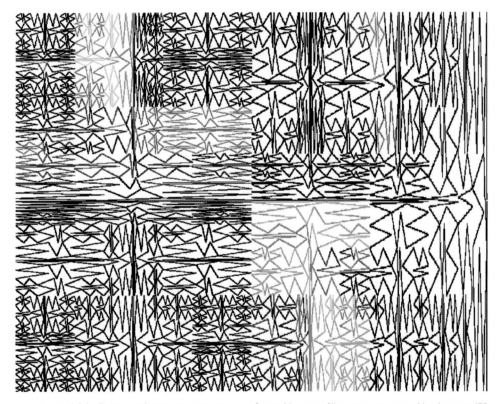

Figure 5.34 Finite-resolution approximation to a 2-variable space-filling curve generated by the superIFS of Figure 5.32.

by the two 2-variable orbital pictures in the second line of Figure 5.19 and by the now not-so-curious overlap in Figure 5.26.

Despite this, Theorem 5.13.1 assures us that we can use the chaos game to define sequences of transformations $\{\widetilde{\tau}_k : A_k \to \Omega_{\{1,2,\ldots,N\}}\}_{k=1}^{\infty}$ where the sequence $\{A_k\}_{k=1}^{\infty}$ is a chaos-game orbit of V-variable fractal sets belonging to $A^{(V)}$. We simply need to start the random iteration at a vector of tops functions belonging to an attractor of an IFS in the superIFS. If the code structures associated with the closures of the ranges of these transformations are homeomorphic then the tops functions $\widetilde{\tau}_k$ are also homeomorphic to one another, for $k = 1, 2, \ldots$, and we can still obtain sequences of pictures that are homeomorphic. But it is a little more difficult.

5.17 The superfractal of V-variable fractal measures

Just as in Section 5.11 we defined the 1-variable measure IFS $\mathcal{F}^{P(1)}$ similarly to the way in which we defined the 1-variable IFS $\mathcal{F}^{(1)}$, so here we define the **V-variable measure IFS** $\mathcal{F}^{P(V)}$ similarly to the way in which we defined $\mathcal{F}^{(V)}$. The key difference is that now we work in the space $\mathbb{P}^V = \mathbb{P}(\mathbb{X})^V$ instead of the space $\mathbb{H}(\mathbb{X})^V$. As elsewhere, we tie all metrics back to the metric spaces $(\mathbb{X}, d_{\mathbb{X}})$, $(\mathbb{H}(\mathbb{X}), d_{\mathbb{H}(\mathbb{X})})$ and $(\mathbb{P}(\mathbb{X}), d_{\mathbb{P}(\mathbb{X})})$, in the manner described in Section 1.13. Here we merely point out the form of the **V-variable measure IFS**.

Let $V \in \mathbb{N}$, let $\mathcal{A}$ be the index set introduced in Equation (5.12.1), let the superIFS

$$\{\mathbb{X}; \mathcal{F}_1, \mathcal{F}_2, \ldots, \mathcal{F}_M; P_1, P_2, \ldots, P_M\}$$

be as above and let probabilities $\{\mathcal{P}^a | a \in \mathcal{A}\}$ be given as in Equation (5.12.2); here we use $\mathcal{P}^a$ in place of f^a. Then we define, in the manner which you might already have guessed,

$$f^a : \mathbb{P}(\mathbb{X})^V \to \mathbb{P}(\mathbb{X})^V$$

by

$$f^a(\mu) = \left(\sum_{l=1}^{L_{m_1}} p_l^{m_1} f_l^{m_1}\left(\mu_{v_{1,l}}\right), \ \sum_{l=1}^{L_{m_2}} p_l^{m_2} f_l^{m_2}\left(\mu_{v_{2,l}}\right), \ldots, \ \sum_{l=1}^{L_{m_V}} p_l^{m_V} f_l^{m_V}\left(\mu_{v_{V,l}}\right) \right)$$

(5.17.1)

for all $\mu = (\mu_1, \mu_2, \ldots, \mu_V) \in \mathbb{P}(\mathbb{X})^V$. We define the V-variable measure IFS $\mathcal{F}^{P(V)}$ to be

$$\mathcal{F}^{P(V)} := \left\{ \mathbb{P}(\mathbb{X})^V; f^a, \mathcal{P}^a, a \in \mathcal{A} \right\}.$$

(5.17.2)

THEOREM 5.17.1 *For $V = 1, 2, \ldots$, $\mathcal{F}^{P(V)}$ is a hyperbolic IFS.*

Figure 5.35 Three successive fractal measures belonging to a 2-variable superfractal. The pixels in the support of each 2-variable measure are coloured either black or a shade of green. The intensity of the green of a pixel is a monotonic increasing function of the measure of the pixel.

PROOF   See [16].   □

Everything works analogously to the case of $\mathcal{F}^{(V)}$ but now the underlying space consists of measures instead of sets. The set attractor $\mathbf{A}^{P(V)}$ of $\mathcal{F}^{P(V)}$ is a set of measures in $\mathbb{P}(\mathbb{X})^V$. The set of components of the elements of $\mathbf{A}^{P(V)}$ is a super-fractal, which we may denote by $A^{P(V)}$. It consists of V-variable fractal measures. The elements of $\mathbf{A}^{P(V)}$ are distributed on $\mathbb{P}(\mathbb{X})^V$ according to the probability measure $\boldsymbol{\mu}^{P(V)} \in \mathbb{P}(P(X)^V)$, which is the measure attractor of $\mathcal{F}^{P(V)}$. The measure $\boldsymbol{\mu}^{P(V)}$ defines a marginal probability distribution $\mu^{P(V)}$, obtained by projecting it onto a single component, and this measure describes the asymptotic distribution of measures obtained, almost always, by following chaos-game orbits for $\mathcal{F}^{P(V)}$ and keeping only the first components.

In Figure 5.35 we show some examples of 2-variable fractal measures, rendered in shades of green according to pixel mass. This example corresponds to the same superIFS as that used in Figure 5.24. The probabilities of the functions in the IFSs are $p_1^1 = p_1^2 = 0.74$ and $p_2^1 = p_2^2 = 0.26$. The IFSs are assigned probabilities $P_1 = P_2 = 0.5$.

I think that by now you will have got the idea. There are many fascinating kinds of V-variable objects that may be defined and explored both mathematically and experimentally. If you do this in the context of either scientific or engineering applications, rich rewards may be obtained. This is new territory!

5.18   Code trees and (general) V-variability

Let $\Omega \in \mathbb{H}(\Omega_{\{1,2,...,N\}})$. That is, Ω is a nonempty compact subset of $\Omega_{\{1,2,...,N\}}$. Then we define the **code tree** of Ω to be the set $\Omega' \subset \Omega'_{\{1,2,...,N\}}$ given by

$$\Omega' = \left\{ \sigma_1 \sigma_2 \cdots \sigma_k \in \Omega'_{\{1,2,...,N\}} : \sigma_1 \sigma_2 \cdots \in \Omega \text{ and } k \in \{0, 1, 2, \ldots\} \right\}.$$

We use $k = 0$ here to say that the empty string is an element of Ω'. If we know Ω' then we know Ω because the latter is the set of accumulation points of Ω', that is,

$$\Omega = \overline{\Omega'} \cap \Omega_{\{1,2,...,N\}}.$$

Thus, each element of $\mathbb{H}(\Omega_{\{1,2,...,N\}})$ can be represented by its code tree.

A code tree can be described precisely in botanical terms. To do this we treat the tree as being embedded in $\mathbb{R}^2$, much as we did in Section 2.8. See Figure 5.36. The bottom of the tree consists of a single point, which we call the level-1 node. This node is connected, by a finite set of upward reaching straight-line segments, to a set of level-2 nodes, each of which in turn is similarly connected to a finite set of level-3 nodes, and so on without end, as illustrated. We call the straight-line segments **limbs**. Optionally we include an additional limb, which we call the **trunk**, below the level-1 node. The figure makes clear what we mean when we refer to 'level-1 limbs', or to 'the set of level-k limbs attached to a particular level-k node'. The set corresponding to the code tree seen in Figure 5.36 contains only strings that commence with the numbers 1, 2 or 5. At the next level of precision, it contains only strings that commence with 12, 14, 24, 25, 51, 52 or 53.

Each limb except the trunk is labelled by one of the indices $\{1, 2, \ldots, N\}$. The trunk is labelled by the symbol for the empty set. The labels of the level-k limbs attached to a level-k node are all distinct. To provide a unique representation we write the labels of the level-k limbs from a particular level-k node in increasing order, from left to right. The tree spreads upwards without limit. We call the part of the tree which is connected to a node and which lies above the node a **level-k branch** of the tree. Notice that each branch defines a code tree. The elements of the code space $\Omega \subset \Omega_{\{1,2,...,N\}}$ corresponding to the tree Ω' are provided by all sequences of labels that may be obtained by starting at the base of Ω' and steadily climbing up the tree from one level to the next, from one node to another that is connected to it by a limb, and reading off the string of codes that is encountered.

Now we note the following. Let $S : \Omega'_{\{1,2,...,N\}} \to \Omega'_{\{1,2,...,N\}}$ be the shift transformation. Then there exists a finite set of code trees, $\Omega'_{1,\lambda_1}, \Omega'_{1,\lambda_2}, \ldots, \Omega'_{1,\lambda_K}$, derived from the level-2 branches of Ω', such that

$$S(\Omega') = \Omega'_{1,\lambda_1} \cup \Omega'_{1,\lambda_2} \cup \cdots \cup \Omega'_{1,\lambda_K}. \tag{5.18.1}$$

The shift transformation maps any code tree into a finite union of distinct code trees. It follows that there exists a finite set of distinct code trees $\Omega'_{k,\lambda_1}, \Omega'_{k,\lambda_2}, \ldots, \Omega'_{k,\lambda_{K_k}}$, derived from the level-$(k+1)$ branches of Ω, such that

$$S^{\circ k}(\Omega) = \Omega'_{k,\lambda_1} \cup \Omega'_{k,\lambda_2} \cup \cdots \cup \Omega'_{k,\lambda_{K_k}}$$

for $k = 1, 2, \ldots$ where $1 \leq K_1 \leq K_2 \leq \cdots$.

DEFINITION 5.18.1   Let $\Omega \in \mathbb{H}(\Omega_{\{1,2,...,N\}})$, and let K_k denote the number of distinct level-k branches of the code tree Ω', for $k = 1, 2, \ldots$ If there exists a

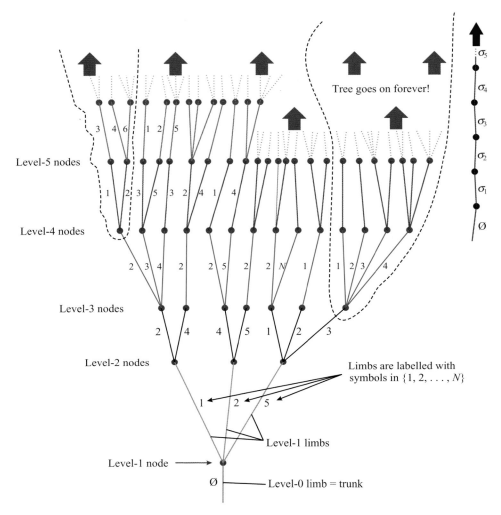

Figure 5.36 Any element Ω of $\mathbb{H}(\Omega_{\{1,2,\dots,N\}})$ may be represented by a code tree Ω' such as this. The strings constituting points of Ω may be discovered by climbing up the tree, along all possible connected paths. The sets of dotted lines on the left and on the right each enclose a branch of the code tree. It should be noted that any branch of a code tree is itself a code tree. The dark blue vertical object at top right represents the code tree of the point $\sigma \in \Omega_{\{1,2,\dots,N\}}$.

positive integer V such that $K_k \leq V$ for all k then Ω is called a **general V-variable subset of** $\Omega_{\{1,2,\dots,N\}}$ and Ω' is called a **general V-variable code tree**.

Figure 5.37 illustrates a general 1-variable code tree.

For $V = 1, 2, \dots$ let $\mathbb{H}^{(V)}(\Omega_{\{1,2,\dots,N\}})$ denote the set of all general V-variable subsets of $\Omega_{\{1,2,\dots,N\}}$. Then $\mathbb{H}^{(V)}(\Omega_{\{1,2,\dots,N\}}) \in \mathbb{H}(\mathbb{H}(\Omega_{\{1,2,\dots,N\}}))$ and

$$\mathbb{H}^{(1)}\big(\Omega_{\{1,2,\dots,N\}}\big) \subset \mathbb{H}^{(2)}\big(\Omega_{\{1,2,\dots,N\}}\big) \subset \cdots \subset \mathbb{H}\big(\Omega_{\{1,2,\dots,N\}}\big).$$

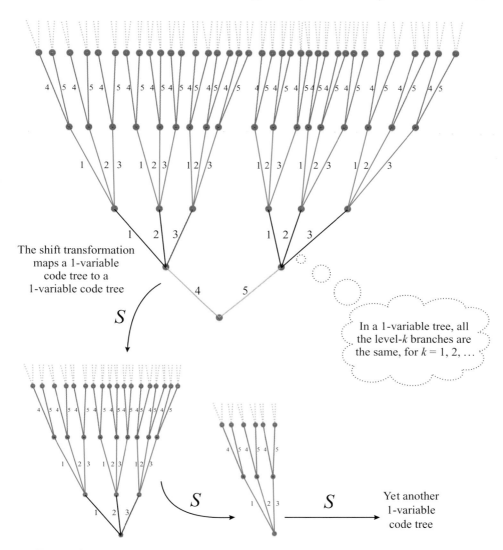

The shift transformation maps a 1-variable code tree to a 1-variable code tree

In a 1-variable tree, all the level-k branches are the same, for $k = 1, 2, \ldots$

Yet another 1-variable code tree

Figure 5.37 Illustration of a general 1-variable code tree. Also shown is the action of the shift transformation S on the tree: it maps the tree into one of its level-1 branches. The transformation is well defined because all level-1 branches are the same and each is itself a 1-variable code tree.

DEFINITION 5.18.2   Let $(\mathbb{X}, d)$ be a compact metric space, let $B \in \mathbb{H}(\mathbb{X})$ and let $\mathcal{F} = \{\mathbb{X}; f_1, f_2, \ldots, f_N\}$ be a hyperbolic IFS. Then B is called a **general V-variable subset of** $\mathbb{X}$, associated with $\mathcal{F}$, iff there exists $\Sigma \in \mathbb{H}^{(V)}(\Omega_{\{1,2,\ldots,N\}})$ such that $B = \phi_{\mathcal{F}}(\Sigma)$.

Notice that $B = A_{(\mathcal{F}, \Sigma)}$ in the nomenclature of Section 4.17 and that consequently Theorem 4.17.3 applies to general V-variable sets.

Next we describe V-variability without the adjective 'general', that is, general V-variability *restricted by the form of the superIFS* $\{\mathbb{X}; \mathcal{F}_1, \mathcal{F}_2, \ldots, \mathcal{F}_M\}$. We will

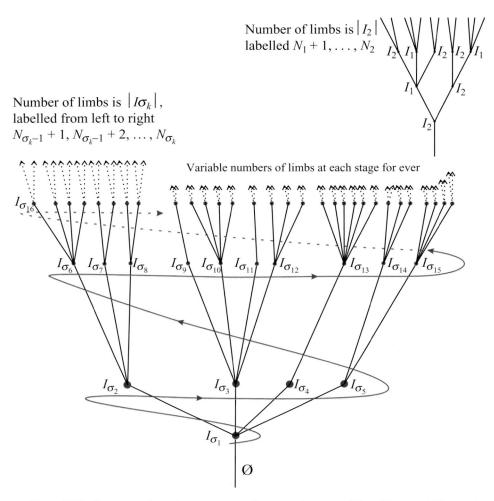

Number of limbs is $|I_2|$
labelled $N_1 + 1, \ldots, N_2$

Number of limbs is $|I\sigma_k|$,
labelled from left to right
$N_{\sigma_{k-1}} + 1, N_{\sigma_{k-1}} + 2, \ldots, N_{\sigma_k}$

Variable numbers of limbs at each stage for ever

Figure 5.38 Illustration of a code tree corresponding to an element σ of $\mathbb{H}_{super}(\Omega_{\{1,2,\ldots,N\}})$. The purple arrow shows the ordering of the nodes of the tree. The example at top right corresponds to the case $N = 5$ and $M = 2$, so that $M = \sigma_k = 1$ or 2 and therefore there are only two possible labels, I_1 and I_2; here we have chosen $I_1 = 3$ and $I_2 = 2$.

hold this superIFS fixed for the rest of this section. We continue to use the notation introduced in Section 5.8.

We define the set $\mathbb{H}_{super}(\Omega_{\{1,2,\ldots,N\}})$ to be the set of $\Omega \in \mathbb{H}(\Omega_{\{1,2,\ldots,N\}})$ such that, given any level-k node of the code tree Ω', there is an $m \in \{1, 2, \ldots, M\}$ such that the level-k limbs, which are connected to the node, are labelled from left to right $N_{m-1} + 1, N_{m-1} + 2, \ldots, N_m$. Figure 5.38 illustrates the code tree of an element of $\mathbb{H}_{super}(\Omega_{\{1,2,\ldots,N\}})$. In this figure each σ_k belongs to $\{1, 2, \ldots, M\}$ and thus plays the role of m above.

Notice that if $\Omega \in \mathbb{H}_{super}(\Omega_{\{1,2,...,N\}})$ then $S(\Omega)$ is a finite union of elements of $\mathbb{H}_{super}(\Omega_{\{1,2,...,N\}})$.

DEFINITION 5.18.3   Let $\{\mathbb{X}; \mathcal{F}_1, \mathcal{F}_2, \ldots, \mathcal{F}_M\}$ be a superIFS. Then, for $V = 1, 2, \ldots$, the space

$$\widetilde{\Omega^{(V)}} = \mathbb{H}_{super}(\Omega_{\{1,2,...,N\}}) \cap \mathbb{H}^{(V)}(\Omega_{\{1,2,...,N\}})$$

is called the set of V-**variable subsets** of $\Omega_{\{1,2,...,N\}}$. A point $A \in \mathbb{H}(\mathbb{X})$ is said to be V-**variable** iff it can be written in the form

$$A = \phi_{\mathcal{F}_{underlying}}(\Sigma)$$

for some $\Sigma \in \widetilde{\Omega^{(V)}}$. We denote the set of all V-variable points in $\mathbb{H}(\mathbb{X})$ by $\widetilde{A^{(V)}}$.

What do V-variable points in $\mathbb{H}(\mathbb{X})$ look like? How can we compute them? Actually you already know the answer. $\widetilde{A^{(V)}}$ is precisely $A^{(V)}$. But by now we have some understanding of the nature of the associated code space. The transformation T illustrated in Figures 5.2 and 5.22 is just the manifestation of the shift transformation in Equation (5.18.1), acting on the underlying code trees.

THEOREM 5.18.4   *For all $V \in \{1, 2, \ldots\}$ we have $\widetilde{A^{(V)}} = A^{(V)}$. That is, the set of all V-variable points in $\mathbb{H}(\mathbb{X})$ is the same as the set of all first components of the points belonging to the attractor of the hyperbolic IFS $\mathcal{F}^{(V)}$.*

PROOF   This follows by applying $\phi_{\mathcal{F}_{underlying}}$ to Theorem 5.18.5 below.   □

We now introduce the superIFS

$$\{\Omega_{\{1,2,...,N\}}; \mathcal{S}_1, \mathcal{S}_2, \ldots, \mathcal{S}_M; P_1, P_2, \ldots, P_M\},$$

where $\mathcal{S}_m$ is the hyperbolic IFS

$$\mathcal{S}_m = \{\Omega_{\{1,2,...,N\}}; s_1^m, s_2^m, \ldots, s_{L_m}^m\}$$

and $s_l^m : \Omega_{\{1,2,...,N\}} \to \Omega_{\{1,2,...,N\}}$ is the contraction mapping

$$s_l^m(\sigma) = (N_{m-1} + l)\sigma$$

for all $l = 1, 2, \ldots, L_m$ and all $m = 1, 2, \ldots, M$. Let the associated V-variable IFS be denoted by

$$\mathcal{S}^{(V)} = \{\mathbb{H}(\Omega_{\{1,2,...,N\}})^V; s^a, \mathcal{P}^a, a \in \mathcal{A}\},$$

where $\mathcal{A}$ is the index set defined in Equation (5.12.1). Let $\Omega^{(V)}$ denote the corresponding superfractal of V-variable fractal sets and $\rho^{(V)}$ denote the corresponding invariant measure, namely the projection of the measure attractor $\rho^{(V)} \in \mathbb{P}(\mathbb{H}(\Omega_{\{1,2,...,N\}}))$ onto one component.

THEOREM 5.18.5 *For all $V \in \{1, 2, \dots\}$ we have $\widetilde{\Omega^{(V)}} = \Omega^{(V)}$. That is, the set of all V-variable points in $\mathbb{H}(\Omega_{\{1,2,\dots,N\}})$ is the same as the set of all first components of the points belonging to the attractor of the hyperbolic IFS $\mathcal{S}^{(V)}$.*

PROOF $\Leftarrow$ is easy. The other direction is more subtle but should not cause you much difficulty. Otherwise consult [16]. $\qquad\square$

5.19 *V*-variability and what happens as $V \to \infty$

Here we complete the explanation of how V-variable fractals provide a bridge between deterministic fractal sets and fully random fractal sets.

We can address the elements of $\mathbb{H}_{super}(\Omega_{\{1,2,\dots,N\}})$ by means of the continuous onto-mapping

$$\xi_{super} : \Omega_{\{1,2,\dots,M\}} \to \mathbb{H}_{super}\big(\Omega_{\{1,2,\dots,N\}}\big)$$

defined by

$$\xi_{super}(\sigma_1 \sigma_2 \cdots) = \Omega_{I_{\sigma_1} I_{\sigma_2} \dots} \quad \text{for all } \sigma \in \Omega_{\{1,2,\dots,M\}},$$

where the code tree of $\Omega_{I_{\sigma_1} I_{\sigma_2} \dots} \in \mathbb{H}_{super}(\Omega_{\{1,2,\dots,N\}})$ is the unique one whose kth node, reading up from the bottom of the tree in the ordering shown in Figure 5.38, is associated with the set of integers I_{σ_k} for $k = 1, 2, \dots$

We use ξ_{super} to define a probability measure ρ_{super} on $\mathbb{H}_{super}(\Omega_{\{1,2,\dots,N\}})$, according to

$$\rho_{super} = \xi_{super}(\rho).$$

Here $\rho \in \mathbb{P}(\Omega_{\{1,2,\dots,M\}})$ is uniquely defined by its values on the cylinder sets $C_\omega \subset \Omega_{\{1,2,\dots,M\}}$:

$$\rho(C_\omega) = P_{\omega_1} P_{\omega_2} \cdots P_{\omega_{|\omega|}}$$

for all $\omega \in \Omega'_{\{1,2,\dots,M\}}$.

THEOREM 5.19.1 *Let $\Omega^{(V)}$ denote the set of V-variable subsets of $\Omega_{\{1,2,\dots,N\}}$, and let $\rho^{(V)}$ denote the associated probability distribution for the IFS $\mathcal{S}^{(V)}$ defined at the end of Section 5.18. Then*

$$\lim_{V \to \infty} \Omega^{(V)} = \mathbb{H}_{super}\big(\Omega_{\{1,2,\dots,N\}}\big)$$

with respect to the metric $d_{\mathbb{H}(\mathbb{H}(\Omega_{\{1,2,\dots,N\}}))}$ and

$$\lim_{V \to \infty} \rho^{(V)} = \rho_{super}$$

with respect to the metric $d_{\mathbb{P}(\mathbb{H}(\Omega_{\{1,2,\dots,N\}}))}$.

PROOF   See Theorem 12 of [16], which provides the full proof in the case where $L_m = L$ for all $m = 1, 2, \ldots, M$.   □

Accordingly, we introduce the notation

$$\Omega^{(\infty)} = \mathbb{H}_{super}\big(\Omega_{\{1,2,\ldots,N\}}\big), \quad \rho^{(\infty)} = \rho_{super},$$

$$A^{(\infty)} = \phi_{\mathcal{F}underlying}\big(\Omega^{(\infty)}\big), \quad \mu^{(\infty)} = \phi_{\mathcal{F}underlying}\big(\rho^{(\infty)}\big).$$

Then in the spirit of Falconer [34], [35], Graf [41] and Mauldin and Williams [67] we make the following definition.

DEFINITION 5.19.2   We refer to the set of fractal sets $A^{(\infty)}$ distributed according to the probability distribution $\mu^{(\infty)}$ as the **random fractals** associated with the superIFS $\{\mathbb{X}; \mathcal{F}_1, \mathcal{F}_2, \ldots, \mathcal{F}_M; P_1, P_2, \ldots, P_M\}$.

Finally, we state the main result.

THEOREM 5.19.3   *Let the superIFS* $\{\mathbb{X}; \mathcal{F}_1, \mathcal{F}_2, \ldots, \mathcal{F}_M; P_1, P_2, \ldots, P_M\}$ *be given. Let* $A^{(V)}$ *denote the corresponding superfractal of V-variable sets and* $\mu^{(V)}$ *denote the corresponding probability distribution. Then*

$$\lim_{V \to \infty} A^{(V)} = A^{(\infty)}$$

with respect to the metric $d_{\mathbb{H}(\mathbb{H}(\mathbb{X}))}$, *and*

$$\lim_{V \to \infty} \rho^{(V)} = \rho^{(\infty)}$$

with respect to the metric $d_{\mathbb{P}(\mathbb{H}(\mathbb{X}))}$.

PROOF   This is just $\phi_{\mathcal{F}underlying}$ applied to Theorem 5.19.1.   □

What's the point? Simply this. We can compute approximations to, and study, random fractals by working with V-variable fractals. The latter can be explored by means of the chaos game on superfractals and lead to a wealth of insights into random fractals. In particular, we see how random fractals may be thought of as V-variable fractals, but of infinite variability.

Similar results also relate V-variable fractal measures to the random fractal measures introduced by Arbeiter [1]; see [16]. In [19] the theory of V-variable fractal sets and measures, as presented in this chapter, is strengthened to admit the uniform Prokhorov metric in place of the Monge–Kantorovitch metric, in order to allow a separable complete metric space $\mathbb{X}$ to replace the compact metric space used here, and to admit IFSs that are 'on average' contractive. The Hausdorff dimensions of some V-variable fractals and other recent developments are discussed in [20] and [21].

5.20 Final section

So, dear Diana and Rose and gentle reader, there you have it! When I started, three years ago, I hoped to weave more closely a relationship between art, biology and mathematics, to exhibit a new geometry of colour and space and to make the vision so compelling that it would almost leave the abstract world where it lives and become instead part of yours. You must be the judge of how far this book fulfils my aim; and I will keep trying to develop these ideas further at www.superfractals.com.

REFERENCES

[1] Arbeiter, Matthias. Random recursive construction of self-similar fractal measures. The noncompact case. *Probab. Theory Related Fields* **88** (1991), no. 4, 497–520.

[2] Asai, T. Fractal image generation with iterated function set. Ricoh Technical Report No. 24, November 1998, pp. 6–11.

[3] Barlow, M. T.; Hambly, B. M. Transition density estimates for Brownian motion on scale irregular Sierpinski gaskets. *Ann. Inst. H. Poincaré Probab. Statist.* **33** (1997), no. 5, 531–557.

[4] Barnsley, M. F.; Demko, S. Iterated function systems and the global construction of fractals. *Proc. Roy. Soc. London Ser. A* **399** (1985), no. 1817, 243–275.

[5] Barnsley, M. F.; Ervin, V.; Hardin, D.; Lancaster, J. Solution of an inverse problem for fractals and other sets. *Proc. Nat. Acad. Sci. USA* **83** (1986), no. 7, 1975–1977.

[6] Barnsley, Michael F. Fractal functions and interpolation. *Constr. Approx.* **2** (1986), no. 4, 303–329.

[7] Barnsley, Michael F.; Elton, J. H. A new class of Markov processes for image encoding. *Adv. Appl. Probab.* **20** (1988), no. 1, 14–32.

[8] Barnsley, M. F.; Reuter, L.; Jacquin, A.; Malassenet, F.; Sloan, A. Harnessing chaos for image synthesis. *Computer Graphics* **22** (1988), 131–140.

[9] Barnsley, Michael. *Fractals Everywhere*. Boston MA, Academic Press, 1988.

[10] Barnsley, Michael F.; Elton, John H.; Hardin, Douglas P. Recurrent iterated function systems. Fractal approximation. *Constr. Approx.* **5** (1989), no. 1, 3–31.

[11] Barnsley, Michael F.; Harrington, Andrew N. The calculus of fractal interpolation functions. *J. Approx. Theory* **57** (1989), no. 1, 14–34.

[12] Barnsley, Michael F.; Hurd, Lyman P. *Fractal Image Compression*. Illustrations by Louisa F. Anson. Wellesley MA, A. K. Peters, 1993.

[13] Barnsley, Michael F. *Fractals Everywhere*, second edition. Revised with the assistance of and a foreword by Hawley Rising, III. Boston MA, Academic Press Professional, 1993.

[14] Barnsley, Michael F. Iterated function systems for lossless data compression. In *Fractals in Multimedia* (Minneapolis MN, 2001), pp. 33–63. IMA Vol. Math. Appl., vol. 132. New York, Springer, 2002.

[15] Barnsley, Michael F.; Barnsley, Louisa F. Fractal transformations. In *The Colours of Infinity: The Beauty and Power of Fractals*, pp. 66–81. London, Clear Books, 2004.

[16] Barnsley, Michael; Hutchinson, John; Stenflo, Örjan. A fractal valued random iteration algorithm and fractal hierarchy. *Fractals* **13** (2005), no. 2, 111–146.

[17] Barnsley, M. F. Theory and application of fractal tops. Preprint, Australian National University, 2005.

[18] Barnsley, M. F. Theory and application of fractal tops. In *Fractals in Engineering: New Trends in Theory and Applications*, J. Lévy-Véhel; E. Lutton (eds.), pp. 3–20. London, Springer-Verlag, 2005.

[19] Barnsley, M. F.; Hutchinson, J. E.; Stenflo, Ö. *V*-variable fractals. In preparation.

[20] Barnsley, M. F.; Hutchinson, J. E.; Stenflo, Ö. *V*-variable fractals and dimensions. In preparation.

[21] Barnsley, M. F.; Hutchinson, J. E.; Stenflo, Ö. *V*-variable fractals and correlated random fractals. In preparation.

[22] Berger, Marc A. *An Introduction to Probability and Stochastic Processes.* Springer Texts in Statistics. New York, Springer-Verlag, 1993.

[23] Berger, Marcel. *Geometry*, vols. I and II. Translated from the French by M. Cole and S. Levy. Universitext. Berlin, Springer-Verlag, 1987.

[24] Billingsley, Patrick. *Ergodic Theory and Information.* New York, London, Sydney, John Wiley & Sons, 1965.

[25] Brannan, David A.; Esplen, Matthew F.; Gray, Jeremy J. *Geometry.* Cambridge, Cambridge University Press, 1999.

[26] Coxeter, H. S. M. *The Real Projective Plane.* New York, McGraw-Hill, 1949.

[27] Demko, S.; Hodges, L.; Naylor, B. Constructing fractal objects with iterated function systems. *Computer Graphics* **19** (1985), 271–278.

[28] Diaconis, Persi; Freedman, David. Iterated random functions. *SIAM Rev.* **41** (1999), no. 1, 45–76.

[29] Dudley, Richard M. *Real Analysis and Probability.* Pacific Grove CA, Wadsworth & Brooks/Cole Advanced Books & Software, 1989.

[30] Dunford, N.; Schwartz, J. T. *Linear Operators. Part I: General Theory*, third edition. New York, John Wiley & Sons, 1966.

[31] Edgar, Gerald A. *Integral, Probability, and Fractal Measures.* New York, Springer, 1998.

[32] Eisen, Martin. *Introduction to Mathematical Probability Theory.* Englewood Cliffs NJ, Prentice-Hall, 1969.

[33] Elton, John H. An ergodic theorem for iterated maps. *Ergodic Theory Dynam. Systems* **7** (1987), no. 4, 481–488.

[34] Falconer, Kenneth. Random fractals. *Math. Proc. Cambridge Philos. Soc.* **100** (1986), no. 3, 559–582.

[35] Falconer, Kenneth. *Fractal Geometry. Mathematical Foundations and Applications*. Chichester, John Wiley & Sons, 1990.

[36] Fathauer, R. *Dr. Fathauer's Encyclopedia of Fractal Tilings*. Version 1.0, 2000. http://members.cox.net/fractalenc.

[37] Feller, William. *An Introduction to Probability Theory and its Applications*, vol. I, third edition. New York, London, Sydney, John Wiley & Sons, 1968.

[38] Fisher, Yuval (ed.) *Fractal Image Compression. Theory and Application*. New York, Springer-Verlag, 1995.

[39] Forte, B.; Mendivil, F. A classical ergodic property for IFS: a simple proof. *Ergodic Theory Dynam. Systems* **18** (1998), no. 3, 609–611.

[40] Gardner, Martin. Rep-tiles. *The Colossal Book of Mathematics: Classic Puzzles, Paradoxes and Problems*. pp. 46–58. New York, London, W. W. Norton & Co., 2001.

[41] Graf, Siegfried. Statistically self-similar fractals. *Probab. Theory Related Fields* **74** (1987), no. 3, 357–392.

[42] Grünbaum, Branko; Shephard, G. C. *Tilings and Patterns*. New York, W. H. Freeman and Co., 1987.

[43] Hambly, B. M. Brownian motion on a random recursive Sierpinski gasket. *Ann. Probab.* **25** (1997), no. 3, 1059–1102.

[44] Hata, Masayoshi, On the structure of self-similar sets. *Japan J. Appl. Math.* **2** (1985), no. 2, 381–414.

[45] Hénon, M. A two-dimensional mapping with a strange attractor. *Comm. Math. Phys.* **50** (1976), no. 1, 69–77.

[46] Hoggar, S. G. *Mathematics for Computer Graphics*. Cambridge, Cambridge University Press, 1992.

[47] Horn, Alistair N. IFSs and the interactive design of tiling structures. In *Fractals and Chaos*, pp. 119–144. New York, Springer, 1991.

[48] Hutchinson, John E. Fractals and self-similarity. *Indiana Univ. Math. J.* **30** (1981), no. 5, 713–747.

[49] Hutchinson, John E.; Rüschendorf, Ludger. Random fractal measures via the contraction method. *Indiana Univ. Math. J.* **47** (1998), no. 2, 471–487.

[50] Hutchinson, John E. Deterministic and random fractals. *In Complex Systems*, pp. 127–166, Cambridge, Cambridge University Press, 2000.

[51] Hutchinson, John E.; Rüschendorf, Ludger. Selfsimilar fractals and selfsimilar random fractals. In *Fractal Geometry and Stochastics, II* (Greifswald/Koserow, 1998), pp. 109–123. Progress in Probability, vol. 46. Basel, Birkhäuser, 2000.

[52] Hutchinson, John E.; Rüschendorf, Ludger. Random fractals and probability metrics. *Adv. in Appl. Probab.* **32** (2000), 925–947.

[53] Jacquin, Arnaud. Image coding based on a fractal theory of iterated contractive image transformations. *IEEE Trans. Image Proc.* **1** (1992), 18–30.

[54] Kaandorp, Jaap A. *Fractal modelling. Growth and Form in Biology.* With a forward by P. Prusinkiewicz. Berlin, Springer-Verlag, 1994.

[55] Kaijser, Thomas. On a new contraction condition for random systems with complete connections. *Rev. Roumaine Math. Pures Appl.* **26** (1981), no. 8, 1075–1117.

[56] Katok, Anatole; Hasselblatt, Boris. Introduction to the modern theory of dynamical systems. With a supplementary chapter by Anatole Katok and Leonardo Mendoza. In *Encyclopedia of Mathematics and Its Applications*, vol. 54. Cambridge, Cambridge University Press, 1995.

[57] Keeton, W. T.; Gould, J. L.; Gould, C. G. *Biological Science,* fifth edition. New York, London, W. W. Norton & Co., 1993.

[58] Kieninger, B. *Iterated Function Systems on Compact Hausdorff Spaces.* Aachen, Shaker-Verlag, 2002.

[59] Kifer, Yuri. Fractals via random iterated function systems and random geometric constructions. In *Fractal Geometry and Stochastics* (Finsbergen, 1994), pp. 145–164. Progress in Probability, vol. 37. Basel, Birkhäuser, 1995.

[60] Kunze, H.; Vrscay, E. Inverse problems for ODEs using the Picard contraction mapping. *Inverse Problems* **15** (1999).

[61] Lasota, A.; Yorke, James A. On the existence of invariant measures for piecewise monotonic transformations. *Trans. Amer. Math. Soc.* **186** (1973), 481–488.

[62] Lu, N. *Fractal Imaging.* San Diego, Academic Press, 1997.

[63] Mandelbrot, Benoit B. *Fractals: Form, Chance, and Dimension.* Translated from the French. Revised edition. San Francisco, W. H. Freeman and Co., 1977.

[64] Mandelbrot, Benoit B. *The Fractal Geometry of Nature.* San Francisco, W. H. Freeman, 1983.

[65] Mandelbrot, Benoit B. A multifractal walk down Wall Street. *Scientific American*, February 1999, 70–73.

[66] Massopust, Peter R. *Fractal Functions, Fractal Surfaces, and Wavelets.* San Diego CA, Academic Press, 1994.

[67] Mauldin, R. Daniel; Williams, S. C. Random recursive constructions: asymptotic geometrical and topological properties. *Trans. Amer. Math. Soc.* **295** (1986), no. 1, 325–346.

[68] Mauldin, R. Daniel; Williams, S. C. Hausdorff dimension in graph directed constructions. *Trans. Amer. Math. Soc.* **309** (1988), no. 2, 811–829.

[69] McGhehee, Richard. Attractors for closed relations on compact Hausdorff spaces. *Indiana Univ. Math. J.* **41** (1992), no. 4, 1165–1209.

[70] Mendelson, Bert. *Introduction to Topology,* British edition. London, Glasgow, Blackie & Son, 1963.

[71] Mochizuki S.; Horie, D.; Cai, D. *Stealing Autumn Color.* ACM SIGGRAPH poster, 2005. See http://mochi.jpn.org/temp/mochipdf.zip.

[72] Moran, P. A. P. Additive functions of intervals and Hausdorff measure. *Proc. Cambridge Philos. Soc.* **42** (1946), 15–23.

[73] Mumford, David; Series, Caroline; Wright, David. *Indra's Pearls. The Vision of Felix Klein.* New York, Cambridge University Press, 2002.

[74] Navascués, M. A. Fractal polynomial interpolation. *Z. Anal. Anwendungen* **24** (2005), no. 2, 401–418.

[75] Nitecki, Zbigniew H. Topological entropy and the preimage structure of maps. *Real Anal. Exchange* **29** (2003/4), no. 1, 9–41.

[76] Onicescu, O.; Mihok, G. Sur les chaînes de variables statistiques. *Bull. Sci. Math. de France* **59** (1935), 174–192.

[77] Parry, William. Symbolic dynamics and transformations of the unit interval. *Trans. Amer. Math. Soc.* **122** (1966), 368–378.

[78] Peitgen, H.-O.; Richter, P. H. *The Beauty of Fractals. Images of Complex Dynamical Systems.* Berlin, Springer-Verlag, 1986.

[79] Peruggia, Mario. *Discrete Iterated Function Systems.* Wellesley MA, A. K. Peters, 1993.

[80] Prusinkiewicz, Przemyslaw; Lindenmayer, Aristid. *The Algorithmic Beauty of Plants.* With the collaboration of James S. Hanan, F. David Fracchia, Deborah R. Fowler, Martin J. M. de Boer and Lynn Mercer. The Virtual Laboratory. New York, Springer-Verlag, 1990.

[81] Rachev, Svetlozar T. *Probability Metrics and the Stability of Stochastic Models.* Wiley Series in Probability and Statistics. Chichester, John Wiley & Sons, 1991.

[82] Rényi, A. Representations for real numbers and their ergodic properties. *Acta Math. Acad. Sci. Hungar.* **8** (1957), 477–493.

[83] Ruelle, David. Zeta-functions for expanding maps and Anosov flows. *Invent. Math.* **34** (1976), no. 3, 231–242.

[84] Sagan, Hans. *Space-Filling Curves.* Universitext. New York, Springer-Verlag, 1994.

[85] Scealy, R. *V*-Variable Fractal Interpolation. In preparation.

[86] Schattschneider, D. *M. C. Escher: Visions of Symmetry*, second edition. New York, Harry N. Abrams, 2004.

[87] *Scientific Workplace 4.0*. MacKichan Software, 2002.

[88] Shields, Paul C. *The Ergodic Theory of Discrete Sample Paths*. Graduate Studies in Mathematics, vol. 13. Providence RI, American Mathematical Society, 1996.

[89] Solomyak, Boris. Dynamics of self-similar tilings. *Ergodic Theory and Dynam. Systems* **17** (1997), no. 3, 695–738. Corrections in *Ergodic Theory and Dynam. Systems* **19** (1999), no. 6, 1685.

[90] Stenflo, Örjan. Markov chains in random environments and random iterated function systems. *Trans. Amer. Math. Soc.* **353** (2001), no. 9, 3547–3562.

[91] Stenflo, Örjan. Uniqueness of invariant measures for place-dependent random iterations of functions. In *Fractals in Multimedia* (Minneapolis MN, 2001), pp. 13–32. IMA Vol. Math. Appl. New York, Springer-Verlag, 2002.

[92] Stewart, Ian; Clarke, Arthur C.; Mandelbrot, Benoît; *et al.* In *The Colours of Infinity: The Beauty and Power of Fractals*. London, Clear Books, 2004.

[93] Szoplik, T. (ed.) *Selected Papers on Morphological Image Processing: Principles and Optoelectronic Implementations*. Vol. MS 127, SPIE. Optical Engineering Press, 1996.

[94] The history of mathematics. In *The New Encyclopaedia Britannica*, fifteenth edition, vol. II, pp. 656–657. Chicago, London, 1979.

[95] Tosan, Eric; Excoffier, Thierry; Rondet-Mignotte, Martine. Création de formes et de couleurs avec les IFS. Preprint 2005. Université Claude Bernard, France.

[96] Tricot, Claude. *Curves and Fractal Dimension*. With a foreword by Michel Mendès France. Translated from the 1993 French original. New York, Springer-Verlag, 1995.

[97] Vrscay, Edward R. From fractal image compression to fractal-based methods in mathematics. In *Fractals in Multimedia* (Minneapolis MN, 2001), pp. 65–106. IMA Vol. Math. Appl., vol. 132. New York, Springer, 2002.

[98] Werner, Ivan. Ergodic theorem for contractive Markov systems. *Nonlinearity* **17** (2004), no. 6, 2303–2313.

INDEX